CAMBRIDGE LIBRARY COLLECTION

Books of enduring scholarly value

Physical Sciences

From ancient times, humans have tried to understand the workings of the world around them. The roots of modern physical science go back to the very earliest mechanical devices such as levers and rollers, the mixing of paints and dyes, and the importance of the heavenly bodies in early religious observance and navigation. The physical sciences as we know them today began to emerge as independent academic subjects during the early modern period, in the work of Newton and other 'natural philosophers', and numerous sub-disciplines developed during the centuries that followed. This part of the Cambridge Library Collection is devoted to landmark publications in this area which will be of interest to historians of science concerned with individual scientists, particular discoveries, and advances in scientific method, or with the establishment and development of scientific institutions around the world.

Notes on Recent Researches in Electricity and Magnetism

This 1893 publication is a central text in the work of the Nobel prize winning physicist Sir Joseph John Thomson (1858–1940). Intended as an extension of James Clerk Maxwell's *Treatise on Electricity and Magnetism*, it documents the important shift in Thomson's thinking towards the model of the atomic electric field, a theory that would eventually lead to his discovery of the electron. In Chapter 1, Thomson documents his experiments with Faraday tubes, using them to physically demonstrate a 'molecular theory of electricity'. Chapter 2 considers the discharge of electricity through gases, Chapter 3 theories of electrostatics, and Chapters 4–6 are primarily concerned with alternating currents. In addition to providing crucial insight into Thomson's evolving theory of the atom, *Recent Researches* underscores his commitment to experimental physics, which offers 'all the advantages in vividness which arise from concrete qualities rather than abstract symbols'.

Notes on Recent Researches in Electricity and Magnetism

Intended as a Sequel to Professor Clerk-Maxwell's Treatise on Electricity and Magnetism

Joseph John Thomson

CAMBRIDGE
UNIVERSITY PRESS

CAMBRIDGE UNIVERSITY PRESS

Cambridge, New York, Melbourne, Madrid, Cape Town, Singapore,
São Paolo, Delhi, Dubai, Tokyo

Published in the United States of America by Cambridge University Press, New York

www.cambridge.org
Information on this title: www.cambridge.org/9781108015202

This edition first published 1893
This digitally printed version 2010

ISBN 978-1-108-01520-2 Paperback

Clarendon Press Series

RECENT RESEARCHES

IN

ELECTRICITY AND MAGNETISM

J. J. THOMSON

London

HENRY FROWDE

OXFORD UNIVERSITY PRESS WAREHOUSE

AMEN CORNER, E.C.

New York

MACMILLAN & CO., 112 FOURTH AVENUE

NOTES

ON

RECENT RESEARCHES IN

ELECTRICITY AND MAGNETISM

INTENDED AS A SEQUEL TO

PROFESSOR CLERK-MAXWELL'S TREATISE
ON ELECTRICITY AND MAGNETISM

BY

J. J. THOMSON, M.A., F.R.S.
HON. SC. D. DUBLIN
FELLOW OF TRINITY COLLEGE
PROFESSOR OF EXPERIMENTAL PHYSICS IN THE UNIVERSITY OF CAMBRIDGE

Oxford
AT THE CLARENDON PRESS
1893

PREFACE

In the twenty years which have elapsed since the first appearance of Maxwell's Treatise on *Electricity and Magnetism* great progress has been made in these sciences. This progress has been largely—perhaps it would not be too much to say mainly—due to the influence of the views set forth in that Treatise, to the value of which it offers convincing testimony.

In the following work I have endeavoured to give an account of some recent electrical researches, experimental as well as theoretical, in the hope that it may assist students to gain some acquaintance with the recent progress of Electricity and yet retain Maxwell's Treatise as the source from which they learn the great principles of the science. I have adopted exclusively Maxwell's theory, and have not attempted to discuss the consequences which would follow from any other view of electrical action. I have assumed throughout the equations of the Electromagnetic Field given by Maxwell in the ninth chapter of the second volume of his Treatise.

The first chapter of this work contains an account of a method of regarding the Electric Field, which is geometrical and physical rather than analytical. I have been induced to dwell on this because I have found that students, especially those who commence the subject after a long course of mathematical studies, have a great tendency to regard the whole of Maxwell's theory as a matter of the solution of certain differential equations, and to dispense with any attempt to form for themselves a mental picture of the physical processes which accompany the phenomena they are investigating. I think that this state of things is to be regretted, since it retards the progress of the science of

Electricity and diminishes the value of the mental training afforded by the study of that science.

In the first place, though no instrument of research is more powerful than Mathematical Analysis, which indeed is indispensable in many departments of Electricity, yet analysis works to the best advantage when employed in developing the suggestions afforded by other and more physical methods. One example of such a method, and one which is very closely connected with the initiation and development of Maxwell's Theory, is that of the 'tubes of force' used by Faraday. Faraday interpreted all the laws of Electrostatics in terms of his tubes, which served him in place of the symbols of the mathematician, while in his hands the laws according to which these tubes acted on each other served instead of the differential equations satisfied by such symbols. The method of the tubes is distinctly physical, that of the symbols and differential equations is analytical.

The physical method has all the advantages in vividness which arise from the use of concrete quantities instead of abstract symbols to represent the state of the electric field; it is more easily wielded, and is thus more suitable for obtaining rapidly the main features of any problem; when, however, the problem has to be worked out in all its details, the analytical method is necessary.

In a research in any of the various fields of electricity we shall be acting in accordance with Bacon's dictum that the best results are obtained when a research begins with Physics and ends with Mathematics, if we use the physical theory to, so to speak, make a general survey of the country, and when this has been done use the analytical method to lay down firm roads along the line indicated by the survey.

The use of a physical theory will help to correct the tendency —which I think all who have had occasion to examine in Mathematical Physics will admit is by no means uncommon—to look on analytical processes as the modern equivalents of the Philosopher's Machine in the Grand Academy of Lagado, and to regard as the normal process of investigation in this subject the manipulation of a large number of symbols in the hope that every now and then some valuable result may happen to drop out.

Then, again, I think that supplementing the mathematical theory by one of a more physical character makes the study of electricity more valuable as a mental training for the student. Analysis is undoubtedly the greatest thought-saving machine ever invented, but I confess I do not think it necessary or desirable to use artificial means to prevent students from thinking too much. It frequently happens that more thought is required, and a more vivid idea of the essentials of a problem gained, by a rough solution by a general method, than by a complete solution arrived at by the most recent improvements in the higher analysis.

The method of illustrating the properties of the electric field which I have given in Chapter I has been devised so as to lead directly to the distinctive feature in Maxwell's Theory, that changes in the polarization in a dielectric produce magnetic effects analogous to those produced by conduction currents. Other methods of viewing the processes in the Electric Field, which would be in accordance with Maxwell's Theory, could, I have no doubt, be devised; the question as to which particular method the student should adopt is however for many purposes of secondary importance, provided that he does adopt one, and acquires the habit of looking at the problems with which he is occupied as much as possible from a physical point of view.

It is no doubt true that these physical theories are liable to imply more than is justified by the analytical theory they are used to illustrate. This however is not important if we remember that the object of such theories is suggestion and not demonstration. Either Experiment or rigorous Analysis must always be the final Court of Appeal; it is the province of these physical theories to supply cases to be tried in such a court.

Chapter II is devoted to the consideration of the discharge of electricity through gases; Chapter III contains an account of the application of Schwarz's method of transformation to the solution of two-dimensional problems in Electrostatics. The rest of the book is chiefly occupied with the consideration of the properties of alternating currents; the experiments of Hertz and the development of electric lighting have made the use of these currents, both for experimental and commercial purposes, much more general than when Maxwell's Treatise was written; and though the principles which govern the action of these currents are clearly laid down by Maxwell, they are not developed to the extent which the present importance of the subject demands.

Chapter IV contains an investigation of the theory of such currents when the conductors in which they flow are cylindrical or spherical, while in Chapter V an account of Hertz's experiments on Electromagnetic Waves is given. This Chapter also contains some investigations on the Electromagnetic Theory of Light, especially on the scattering of light by small metallic particles; on reflection from metals; and on the rotation of the plane of polarization by reflection from a magnet. I regret that it was only when this volume was passing through the press that I became acquainted with a valuable paper by Drude (Wiedemann's *Annalen*, 46, p. 353, 1892) on this subject.

Chapter VI mainly consists of an account of Lord Rayleigh's investigations on the laws according to which alternating currents distribute themselves among a network of conductors; while the last Chapter contains a discussion of the equations which hold when a dielectric is moving in a magnetic field, and some problems on the distribution of currents in rotating conductors.

I have not said anything about recent researches on Magnetic Induction, as a complete account of these in an easily accessible form is contained in Professor Ewing's 'Treatise on Magnetic Induction in Iron and other Metals.'

I have again to thank Mr. Chree, Fellow of King's College, Cambridge, for many most valuable suggestions, as well as for a very careful revision of the proofs.

J. J. THOMSON.

CONTENTS

CHAPTER I.

ELECTRIC DISPLACEMENT AND FARADAY TUBES OF FORCE.

CHAPTER II.

PASSAGE OF ELECTRICITY THROUGH GASES.

CHAPTER III.

CONJUGATE FUNCTIONS.

CHAPTER IV.

ELECTRICAL WAVES AND OSCILLATIONS.

CHAPTER V.

ELECTROMAGNETIC WAVES.

CHAPTER VI.

DISTRIBUTION OF RAPIDLY ALTERNATING CURRENTS.

CHAPTER VII.

ELECTROMOTIVE INTENSITY IN MOVING BODIES.

APPENDIX.

ADDITIONS AND CORRECTIONS.

Page 67. For further remarks on electrification by incandescent bodies see Appendix, p. 569.

" 122. E. Wiedemann and Ebert have shown (*Wied. Ann.* 46, p. 158, 1892) that the repulsion between two pencils of negative rays is due to the influence which the presence of one cathode exerts on the emission of rays from a neighbouring cathode.

" 174. Dewar (*Proc. Roy. Soc.* 33, p. 262, 1882) has shown that the interior of the gaseous envelope of the electric arc always shows a fixed pressure amounting to about that due to a millimetre of water above that of the surrounding atmosphere.

" 182. Line 1, *for* 90°C. *read* 100°C.

NOTES ON ELECTRICITY AND MAGNETISM.

CHAPTER I.

ELECTRIC DISPLACEMENT AND FARADAY TUBES OF FORCE.

1.] THE influence which the notation and ideas of the fluid theory of electricity have ever since their introduction exerted over the science of Electricity and Magnetism, is a striking illustration of the benefits conferred upon this science by a concrete representation or '*construibar vorstellung*' of the symbols, which in the Mathematical Theory of Electricity define the state of the electric field. Indeed the services which the old fluid theory has rendered to Electricity by providing a language in which the facts of the science can be clearly and briefly expressed can hardly be over-rated. A descriptive theory of this kind does more than serve as a vehicle for the clear expression of well-known results, it often renders important services by suggesting the possibility of the existence of new phenomena.

The descriptive hypothesis, that of displacement in a dielectric, used by Maxwell to illustrate his mathematical theory, seems to have been found by many readers neither so simple nor so easy of comprehension as the old fluid theory; indeed this seems to have been one of the chief reasons why his views did not sooner meet with the general acceptance they have since received. As many students find the conception of 'displacement' difficult, I venture to give an alternative method of regarding the processes occurring in the electric field, which I have often found useful and which is, from a mathematical point of view, equivalent to Maxwell's Theory.

2.] This method is based on the conception, introduced by Faraday, of tubes of electric force, or rather of electrostatic induction. Faraday, as is well known, used these tubes as the language in which to express the phenomena of the electric field. Thus it was by their tendency to contract, and the lateral repulsion which similar tubes exert on each other, that he explained the mechanical forces between electrified bodies, while the influence of the medium on these tubes was on his view indicated by the existence of specific inductive capacity in dielectrics. Although the language which Faraday used about lines of force leaves the impression that he usually regarded them as chains of polarized particles in the dielectric, yet there seem to be indications that he occasionally regarded them from another aspect; i.e. as something having an existence apart from the molecules of the dielectric, though these were polarized by the tubes when they passed through the dielectric. Thus, for example, in § 1616 of the *Experimental Researches* he seems to regard these tubes as stretching across a vacuum. It is this latter view of the tubes of electrostatic induction which we shall adopt, we shall regard them as having their seat in the ether, the polarization of the particles which accompanies their passage through a dielectric being a secondary phenomenon. We shall for the sake of brevity call such tubes Faraday Tubes.

In addition to the tubes which stretch from positive to negative electricity, we suppose that there are, in the ether, multitudes of tubes of similar constitution but which form discrete closed curves instead of having free ends; we shall call such tubes 'closed' tubes. The difference between the two kinds of tubes is similar to that between a vortex filament with its ends on the free surface of a liquid and one forming a closed vortex ring inside it. These closed tubes which are supposed to be present in the ether whether electric forces exist or not, impart a fibrous structure to the ether.

In his theory of electric and magnetic phenomena Faraday made use of tubes of magnetic as well as of electrostatic induction, we shall find however that if we keep to the conception of tubes of electrostatic induction we can explain the phenomena of the magnetic field as due to the motion of such tubes.

The Faraday Tubes.

3.] As is explained in Art. 82 of Maxwell's *Electricity and Magnetism*, these tubes start from places where there is positive and end at places where there is negative electricity, the quantity of positive electricity at the beginning of the tube being equal to that of the negative at the end. If we assume that the tubes in the field are all of the same strength, the quantity of free positive electricity on any surface will be proportional to the number of tubes leaving the surface. In the mathematical theory of electricity there is nothing to indicate that there is any limit to the extent to which a field of electric force can be subdivided up into tubes of continually diminishing strength, the case is however different if we regard these tubes of force as being no longer merely a form of mathematical expression, but as real physical quantities having definite sizes and shapes. If we take this view, we naturally regard the tubes as being all of the same strength, and we shall see reasons for believing that this strength is such that when they terminate on a conductor there is at the end of the tube a charge of negative electricity equal to that which in the theory of electrolysis we associate with an atom of a monovalent element such as chlorine.

This strength of the unit tubes is adopted because the phenomena of electrolysis show that it is a natural unit, and that fractional parts of this unit do not exist, at any rate in electricity that has passed through an electrolyte. We shall assume in this chapter that in all electrical processes, and not merely in electrolysis, fractional parts of this unit do not exist.

The Faraday tubes either form closed circuits or else begin and end on atoms, all tubes that are not closed being tubes that stretch in the ether along lines either straight or curved from one atom to another. When the length of the tube connecting two atoms is comparable with the distance between the atoms in a molecule, the atoms are said to be in chemical combination; when the tube connecting the atoms is very much longer than this, the atoms are said to be 'chemically free'.

The property of the Faraday tubes of always forming closed circuits or else having their ends on atoms may be illustrated by the similar property possessed by tubes of vortex motion in a frictionless fluid, these tubes either form closed circuits or have their ends on the boundary of the liquid in which the vortex motion takes place.

The Faraday tubes may be supposed to be scattered throughout space, and not merely confined to places where there is a finite electromotive intensity, the absence of this intensity being due not to the absence of the Faraday tubes, but to the want of arrangement among such as are present: the electromotive intensity at any place being thus a measure, not of the whole number of tubes at that place, but of the excess of the number pointing in the direction of the electromotive intensity over the number of those pointing in the opposite direction.

4.] In this chapter we shall endeavour to show that the various phenomena of the electromagnetic field may all be interpreted as due to the motion of the Faraday tubes, or to changes in their position or shape. Thus, from our point of view, this method of looking at electrical phenomena may be regarded as forming a kind of molecular theory of Electricity, the Faraday tubes taking the place of the molecules in the Kinetic Theory of Gases: the object of the method being to explain the phenomena of the electric field as due to the motion of these tubes, just as it is the object of the Kinetic Theory of Gases to explain the properties of a gas as due to the motion of its molecules.

These tubes also resemble the molecules of a gas in another respect, as we regard them as incapable of destruction or creation.

5.] It may be asked at the outset, why we have taken the tubes of electrostatic induction as our molecules, so to speak, rather than the tubes of magnetic induction? The answer to this question is, that the evidence afforded by the phenomena which accompany the passage of electricity through liquids and gases shows that molecular structure has an exceedingly close connection with tubes of electrostatic induction, much closer than we have any reason to believe it has with tubes of magnetic induction. The choice of the tubes of electrostatic induction as our molecules seems thus to be the one which affords us the greatest facilities for explaining those electrical phenomena in which matter as well as the ether is involved.

6.] Let us consider for a moment on this view the origin of the energy in the electrostatic and electromagnetic fields. We suppose that associated with the Faraday tubes there is a distribution of velocity of the ether both in the tubes themselves and in the space surrounding them. Thus we may have rotation in the ether inside and around the tubes even when the tubes themselves have no translatory velocity, the kinetic energy due to this motion constituting the potential energy of the electrostatic field: while when the tubes themselves are in motion we have super-added to this another distribution of velocity whose energy constitutes that of the magnetic field.

The energy we have considered so far is in the ether, but when a tube falls on an atom it may modify the internal motion of the atom and thus affect its energy. Thus, in addition to the kinetic energy of the ether arising from the electric field, there may also be in the atoms some energy arising from the same cause and due to the alteration of the internal motion of the atoms produced by the incidence of the Faraday tubes. If the change in the energy of an atom produced by the incidence of a Faraday tube is different for atoms of different substances, if it is not the same, for example, for an atom of hydrogen as for one of chlorine, then the energy of a number of molecules of hydrochloric acid would depend upon whether the Faraday tubes started from the hydrogen and ended on the chlorine or vice versâ. Since the energy in the molecules thus depends upon the disposition of the tubes in the molecule, there will be a tendency to make all the tubes start from the hydrogen and end on the chlorine or vice versâ, according as the first or second of these arrangements makes the difference between the kinetic and potential energies a maximum. In other words, there will, in the language of the ordinary theory of electricity, be a tendency for all the atoms of hydrogen to be charged with electricity of one sign, while all the atoms of chlorine are charged with equal amounts of electricity of the opposite sign.

The result of the different effects on the energy of the atom produced by the incidence of a Faraday tube will be the same as if the atoms of different substances attracted electricity with different degrees of intensity: this has been shown by v. Helmholtz to be sufficient to account for contact and frictional electricity. It also, as we shall see in Chapter II, accounts for some of the effects observed when electricity passes from a gas to a metal or vice versâ.

7.] The Faraday tubes when they reach a conductor shrink to molecular dimensions. We shall consider the processes by which this is effected at the end of this chapter, and in the meantime proceed to discuss the effects produced by these tubes when moving through a dielectric.

8.] In order to be able to fix the state of the electric field at any point of a dielectric, we shall introduce a quantity which we shall call the 'polarization' of the dielectric, and which while mathematically identical with Maxwell's 'displacement' has a different physical interpretation. The 'polarization' is defined as follows: Let A and B be two neighbouring points in the dielectric, let a plane whose area is unity be drawn between these points and at right angles to the line joining them, then the polarization in the direction AB is the excess of the number of the Faraday tubes which pass through the unit area from the side A to the side B over those which pass through the same area from the side B to the side A. In a dielectric other than air we imagine the unit area to be placed in a narrow crevasse cut out of the dielectric, the sides of the crevasse being perpendicular to AB. The polarization is evidently a vector quantity and may be resolved into components in the same way as a force or a velocity; we shall denote the components parallel to the axes of x, y, z by the letters f, g, h; these are mathematically identical with the quantities which Maxwell denotes by the same letters, their physical interpretation however is different.

9.] We shall now investigate the rate of change of the components of the polarization in a dielectric. Since the Faraday tubes in such a medium can neither be created nor destroyed, a change in the number passing through any fixed area must be due to the motion or deformation of the tubes. We shall suppose, in the first place, that the tubes at one place are all moving with the same velocity. Let u, v, w be the components of the velocities of these tubes at any point, then the change in f, the number of tubes passing at the point x, y, z, through unit area at right angles to the axis of x, will be due to three causes. The first of these is the motion of the tubes from another part of the field up to the area under consideration; the second is the spreading out or concentration of the tubes due to their relative motion; and the third is the alteration in the direction of the tubes due to the same cause.

Let $\delta_1 f$ be the change in f due to the first cause, then in consequence of the motion of the tubes, the tubes which at the time $t+\delta t$ pass through the unit area will be those which at the time t were at the point

$$x-u\delta t, \quad y-v\delta t, \quad z-w\delta t,$$

hence $\delta_1 f$ will be given by the equation

$$\delta_1 f = -\left(u\frac{df}{dx}+v\frac{df}{dy}+w\frac{df}{dz}\right)\delta t.$$

In consequence of the motion of the tubes relatively to one another, those which at the time t passed through unit area at right angles to x will at the time $t+\delta t$ be spread over an area

$$1+\delta t\left\{\frac{dv}{dy}+\frac{dw}{dz}\right\};$$

thus $\delta_2 f$, the change in f due to this cause, will be given by the equation

$$\delta_2 f = \frac{f}{1+\delta t\left\{\frac{dv}{dy}+\frac{dw}{dz}\right\}}-f,$$

or

$$\delta_2 f = -\delta t f\left\{\frac{dv}{dy}+\frac{dw}{dz}\right\}.$$

In consequence of the deflection of the tubes due to the relative motion of their parts some of those which at the time t were at right angles to the axis of x will at the time $t+\delta t$ have a component along it. Thus, for example, the tubes which at the time t were parallel to y will after a time δt has elapsed be twisted towards the axis of x through an angle $\delta t\frac{du}{dy}$, similarly those parallel to z will be twisted through an angle $\delta t\frac{du}{dz}$ towards the axis of x in the time δt; hence $\delta_3 f$, the change in f due to this cause, will be given by the equation

$$\delta_3 f = \delta t\left\{g\frac{du}{dy}+h\frac{du}{dz}\right\}.$$

Hence if δf is the total change in f in the time δt, since

$$\delta f = \delta_1 f+\delta_2 f+\delta_3 f,$$

we have

$$\delta f = \left[-\left(u\frac{df}{dx}+v\frac{df}{dy}+w\frac{df}{dz}\right)-f\left(\frac{dv}{dy}+\frac{dw}{dz}\right)+\left(g\frac{du}{dy}+h\frac{du}{dz}\right)\right]\delta t,$$

which may be written as

$$\frac{df}{dt} = \frac{d}{dy}(ug-vf)-\frac{d}{dz}(wf-uh)-u\left(\frac{df}{dx}+\frac{dg}{dy}+\frac{dh}{dz}\right). \quad (1)$$

If ρ is the density of the free electricity, then since by the definition of Art. 8 the surface integral of the normal polarization taken over any closed surface must be equal to the quantity of electricity inside that surface, it follows that

$$\rho = \frac{df}{dx} + \frac{dg}{dy} + \frac{dh}{dz},$$

hence equation (1) may be written

$$\left.\begin{aligned} \frac{df}{dt} + u\rho &= \frac{d}{dy}(ug - vf) - \frac{d}{dz}(wf - uh). \\ \text{Similarly}\quad \frac{dg}{dt} + v\rho &= \frac{d}{dz}(vh - wg) - \frac{d}{dx}(ug - vf), \\ \frac{dh}{dt} + w\rho &= \frac{d}{dx}(wf - uh) - \frac{d}{dy}(vh - wg). \end{aligned}\right\} \quad (2)$$

If p, q, r are the components of the current parallel to x, y, z respectively, α, β, γ the components of the magnetic force in the same directions, then we know

$$\left.\begin{aligned} 4\pi p &= \frac{d\gamma}{dy} - \frac{d\beta}{dz}, \\ 4\pi q &= \frac{d\alpha}{dz} - \frac{d\gamma}{dx}, \\ 4\pi r &= \frac{d\beta}{dx} - \frac{d\alpha}{dy}. \end{aligned}\right\} \quad (3)$$

Hence, if we regard the current as made up of the convection current whose components are $u\rho$, $v\rho$, $w\rho$ respectively, and the polarization current whose components are $\frac{df}{dt}$, $\frac{dg}{dt}$, $\frac{dh}{dt}$, we see by comparing equations (2) and (3) that we may regard the moving Faraday tubes as giving rise to a magnetic force whose components α, β, γ are given by the equation

$$\left.\begin{aligned} \alpha &= 4\pi(vh - wg), \\ \beta &= 4\pi(wf - uh), \\ \gamma &= 4\pi(ug - vf). \end{aligned}\right\} \quad (4)$$

Thus a Faraday tube when in motion produces a magnetic force at right angles both to itself and to its direction of motion, whose magnitude is proportional to the component of the velocity at right angles to the direction of the tube. The magnetic force and the rotation from the direction of motion to that of the tube at any point are related like translation and rotation in a right-handed screw.

10.] The motion of these tubes involves kinetic energy, and this kinetic energy is the energy of the magnetic field. Now if μ is the magnetic permeability we know that the energy per unit volume is

$$\frac{\mu}{8\pi}(\alpha^2+\beta^2+\gamma^2),$$

or substituting the values of α, β, γ from equations (4),

$$2\pi\mu\left[(hv-gw)^2+(fw-hu)^2+(gu-fv)^2\right].$$

The momentum per unit volume of the dielectric parallel to x is the differential coefficient of this expression with regard to u, hence if U, V, W are the components of the momentum parallel to x, y, z, we have

$$\begin{aligned} U &= 4\pi\mu\{g(gu-fv)-h(fw-hu)\} \\ &= gc-hb, \end{aligned}$$

if a, b, c are the components of the magnetic induction parallel to x, y, z.

Similarly
$$\left.\begin{aligned} V &= ha-fc, \\ W &= fb-ga. \end{aligned}\right\} \qquad (5)$$

Thus the momentum per unit volume in the dielectric, which is due to the motion of the tubes, is at right angles to the polarization and to the magnetic induction, the magnitude of the momentum being equal to the product of the polarization and the component of the magnetic induction at right angles to it. We may regard each tube as having a momentum proportional to the intensity of the component of the magnetic induction at right angles to the direction of the tube. It is interesting to notice that the components of the momentum in the field as given by equations (5) are proportional to the amounts of energy transferred in unit time across unit planes at right angles to the axes of x, y, z in Poynting's theory of the transfer of energy in the electromagnetic field (*Phil. Trans.* 1884, Part II. p. 343); hence the direction in which the energy in Poynting's theory is supposed to move is the same as the direction of the momentum determined by the preceding investigation.

11.] The electromotive intensities parallel to x, y, z due to the motion of the tubes are the differential coefficients of the kinetic energy with regard to f, g, h respectively, hence we obtain the following expressions for X, Y, Z the components of the electromotive intensity,

$$\left.\begin{aligned} X &= wb - vc, \\ Y &= uc - wa, \\ Z &= va - ub. \end{aligned}\right\} \qquad (6)$$

Thus the direction of the electromotive intensity due to the motion of the tubes is at right angles both to the magnetic induction and to the direction of motion of the tubes.

From equations (6) we get

$$\frac{dZ}{dy} - \frac{dY}{dz} = v\frac{da}{dy} + w\frac{da}{dz} - u\left(\frac{db}{dy} + \frac{dc}{dz}\right)$$
$$+ a\left(\frac{dv}{dy} + \frac{dw}{dz}\right) - b\frac{du}{dy} - c\frac{du}{dz}.$$

But since the equation

$$\frac{da}{dx} + \frac{db}{dy} + \frac{dc}{dz} = 0$$

holds, as we shall subsequently show, on the view we have taken of the magnetic force as well as on the ordinary view, we have

$$\frac{dZ}{dy} - \frac{dY}{dz} = u\frac{da}{dx} + v\frac{da}{dy} + w\frac{da}{dz} + a\left(\frac{dv}{dy} + \frac{dw}{dz}\right) - b\frac{du}{dy} - c\frac{du}{dz}.$$

The right-hand side of this investigation is by the reasoning given in Art. 9 equal to $-\frac{da}{dt}$, the rate of diminution in the number of lines of magnetic induction passing through unit area at right angles to the axis of x: hence we have

$$\left.\begin{aligned} \frac{dZ}{dy} - \frac{dY}{dz} &= -\frac{da}{dt}. \\ \text{Similarly} \qquad \frac{dX}{dz} - \frac{dZ}{dx} &= -\frac{db}{dt}, \\ \frac{dY}{dx} - \frac{dX}{dy} &= -\frac{dc}{dt}. \end{aligned}\right\} \qquad (7)$$

Now by Stokes' theorem

$$\int(X\,dx + Y\,dy + Z\,dz)$$

taken round a closed circuit is equal to

$$\iint\left\{l\left(\frac{dZ}{dy}-\frac{dY}{dz}\right)+m\left(\frac{dX}{dz}-\frac{dZ}{dx}\right)+n\left(\frac{dY}{dx}-\frac{dX}{dy}\right)\right\}dS,$$

where l, m, n are the direction-cosines of the normal to a surface S which is entirely bounded by the closed circuit. Substituting the preceding values for $dZ/dy-dY/dz$, &c., we see that the line integral of the electromotive intensity round a closed circuit is equal to the rate of diminution in the number of lines of magnetic induction passing through the circuit. Hence the preceding view of the origin of magnetic force leads to Faraday's rule for the induction of currents by the alteration of the magnetic field.

12.] When the electromotive intensity is entirely due to the motion of the tubes in an isotropic medium whose specific inductive capacity is K, we have

$$f=\frac{K}{4\pi}X$$
$$=\frac{K}{4\pi}\{wb-vc\},$$

and since

$$b=4\pi\mu\{fw-hu\},\quad c=4\pi\mu\{gu-fv\},$$

we have $f=\mu K\{f(u^2+v^2+w^2)-u(fu+gv+hw)\}$;

similarly $g=\mu K\{g(u^2+v^2+w^2)-v(fu+gv+hw)\}$,

$$h=\mu K\{h(u^2+v^2+w^2)-w(fu+gv+hw)\},$$

hence $fu+gv+hw=0$,

and therefore $u^2+v^2+w^2=\frac{1}{\mu K}$.

Hence when the electromotive intensity is entirely due to the motion of the tubes, the tubes move at right angles to themselves with the velocity $1/\sqrt{\mu K}$, which is the velocity with which light travels through the dielectric. In this case the momentum is parallel to the direction of motion, and the electromotive intensity is in the direction of the polarization. In this case the polarization, the direction of motion and the magnetic force, are mutually at right angles; their relative disposition is shown in Fig. 1.

Collecting the preceding results, we see that when a Faraday tube is in motion it is accompanied by (1) a magnetic force at right angles to the tube and to the direction in which it is moving, (2) a momentum at right angles to the tube and to

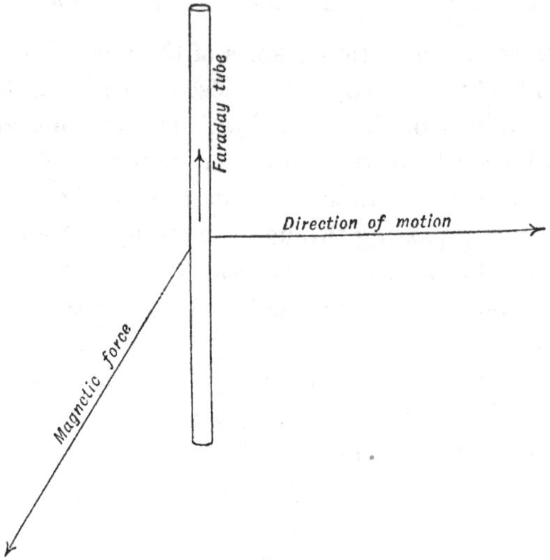

Fig. 1.

the magnetic induction, (3) an electromotive intensity at right angles to the direction of motion and to the magnetic induction; this always tends to make the tube set itself at right angles to the direction in which it is moving. Thus in an isotropic medium in which there is no free electricity and consequently no electromotive intensities except those which arise from the motion of the tubes, the tubes set themselves at right angles to the direction of motion.

13.] We have hitherto only considered the case when the tubes at any one place in a dielectric are moving with a common velocity. We can however without difficulty extend these results to the case when we have different sets of tubes moving with different velocities.

Let us suppose that we have the tubes f_1, g_1, h_1, moving with a velocity whose components are u_1, v_1, w_1, while the tubes f_2, g_2, h_2 move with the velocities u_2, v_2, w_2, and so on. Then the rate of increase in the number of tubes which pass through unit area at right angles to the axis of x is, by the same reasoning as before,

$$\frac{d}{dy}\Sigma(ug-vf)-\frac{d}{dz}\Sigma(wf-uh)-\Sigma(u\rho).$$

Hence we see as before that the tubes may be regarded as producing a magnetic force whose components α, β, γ are given by the equations

$$\left.\begin{aligned}\alpha &= 4\pi\Sigma(vh-wg),\\ \beta &= 4\pi\Sigma(wf-uh),\\ \gamma &= 4\pi\Sigma(ug-vf).\end{aligned}\right\} \qquad (8)$$

The Kinetic energy per unit volume, T, due to the motion of these tubes is given by the equation

$$T = \frac{\mu}{8\pi}\{\alpha^2+\beta^2+\gamma^2\},$$

or

$$T = 2\pi\mu\left[\{\Sigma(vh-wg)\}^2+\{\Sigma(wf-uh)\}^2+\{\Sigma(ug-vf)\}^2\right].$$

Thus dT/du_1, the momentum per unit volume parallel to x due to the tube with suffix 1, is equal to

$$\begin{aligned}&4\pi\mu\{g_1\Sigma(ug-vf)-h_1\Sigma(wf-uh)\},\\ &= g_1c-h_1b,\end{aligned}$$

where a, b, c are the components of the magnetic induction.

Thus U, V, W, the components of the momentum per unit volume parallel to the axes of x, y, z respectively, are given by the equations

$$\left.\begin{aligned}U &= c\Sigma g-b\Sigma h,\\ V &= a\Sigma h-c\Sigma f,\\ W &= b\Sigma f-a\Sigma g.\end{aligned}\right\} \qquad (9)$$

Thus when we have a number of tubes moving about in the electric field the resultant momentum at any point is perpendicular both to the resultant magnetic induction and to the resultant polarization, and is equal to the product of these two quantities into the sine of the angle between them.

The electromotive intensities X, Y, Z parallel to the axes of x, y, z respectively are equal to the mean values of dT/df, dT/dg, dT/dh, hence we have

$$\left.\begin{aligned}X &= b\bar{w}-c\bar{v},\\ Y &= c\bar{u}-a\bar{w},\\ Z &= a\bar{v}-b\bar{u};\end{aligned}\right\} \qquad (10)$$

where a bar placed over any quantity indicates that the mean value of that quantity is to be taken.

Thus when a system of Faraday tubes is in motion, the electromotive intensity is at right angles both to the resultant magnetic induction and to the mean velocity of the tubes, and is equal in magnitude to the product of these two quantities into the sine of the angle between them.

We see from the preceding equations that there may be a resultant magnetic force due to the motion of the positive tubes in one direction and the negative ones in the opposite, without either resultant momentum or electromotive intensity; for if there are as many positive as negative tubes passing through each unit area so that there is no resultant polarization, there will, by equations (9), be no resultant momentum, while if the number of tubes moving in one direction is the same as the number moving in the opposite, equations (10) show that there will be no resultant electromotive intensity due to the motion of the tubes. We thus see that when the magnetic field is steady the motion of the Faraday tubes in the field will be a kind of shearing of the positive past the negative tubes; the positive tubes moving in one direction and the negative at an equal rate in the opposite. When, however, the field is not in a steady state this ceases to be the case, and then the electromotive intensities due to induction are developed.

Mechanical Forces in the Field.

14.] The momentum parallel to x per unit volume of the medium, due to the motion of the Faraday tubes, is by equation (9)

$$c\Sigma g - b\Sigma h;$$

thus the momentum parallel to x which enters a portion of the medium bounded by the closed surface S in unit time is equal to

$$\iint [c\Sigma g(lu+mv+nw) - b\Sigma h(lu+mv+nw)]\, dS,$$

where dS is an element of the surface and l, m, n the direction-cosines of its inwardly directed normal.

If the surface S is so small that the external magnetic field may be regarded as constant over it, the expression may be written as

$$c\iint \Sigma g(lu+mv+nw)\,dS - b\iint \Sigma h(lu+mv+nw)\,dS.$$

Now $$\iint \Sigma g\,(lu + mv + nw)\,dS,$$
and $$\iint \Sigma h\,(lu + mv + nw)\,dS,$$
are the number of Faraday tubes parallel to y and z respectively which enter the element in unit time, that is, they are the volume integrals of the components q and r of the current parallel to y and z respectively: if the medium surrounded by S is a dielectric this is a polarization current, if it is a conductor it is a conduction current. Thus the momentum parallel to x communicated in unit time to unit volume of the medium, in other words the force parallel to x acting on unit volume of the medium, is equal to

$$cq - br;$$

similarly the forces parallel to y and z are respectively

$$\left.\begin{aligned} &ar - cp, \\ \text{and}\quad &bp - aq. \end{aligned}\right\} \qquad (11)$$

When the medium is a conductor these are the ordinary expressions for the components of the force per unit volume of the conductor when it is carrying a current in a magnetic field.

When, as in the above investigation, we regard the force on a conductor carrying a current as due to the communication to the conductor of the momentum of the Faraday tubes which enter the conductor, the origin of the force between two currents will be very much the same as that of the attraction between two bodies on Le Sage's theory of gravitation. Thus, for example, if we have two parallel currents A and B flowing in the same direction, then if A is to the left of B more tubes will enter A from the left than from the right, because some of those which would have come from the right if B had been absent will be absorbed by B, thus in unit time the momentum having the direction left to right which enters A will exceed that having the opposite direction; thus A will tend to move towards the right, that is towards B, while for a similar reason B will tend to move towards A.

15.] We have thus seen that the hypothesis of Faraday tubes in motion explains the properties and leads to the ordinary equations of the electromagnetic field. This hypothesis has the advantage of indicating very clearly why polarization and conduction currents produce similar mechanical and magnetic effects. For the mechanical effects and the magnetic forces at any point in the field are due to the motion of the Faraday tubes at that point, and any alteration in the polarization involves motion of these tubes just as much as does an ordinary conduction current.

16.] We shall now proceed to illustrate this method of regarding electrical phenomena by applying it to the consideration of some simple cases. We shall begin with the case which suggested the method; that of a charged sphere moving uniformly through the dielectric. Let us suppose the charge on the sphere is e and that it is moving with velocity w parallel to the axis of z. Faraday tubes start from the sphere and are carried along with it as it moves through the dielectric; since these tubes are moving they will, as we have seen, produce a magnetic field. We shall suppose that the system has settled down into a steady state, so that the sphere and its tubes are all moving with the same velocity w. Let f, g, h be the components of the polarization at any point, a, β, γ those of the magnetic force. The expressions for X, Y, Z, the components of the electromotive intensity, will consist of two parts, one due to the motion of the Faraday tubes and given by equations (6), the other due to the distribution of these tubes and derivable from a potential Ψ; we thus have, if the magnetic permeability is unity,

$$\left.\begin{aligned} X &= \quad w\beta - \frac{d\Psi}{dx}, \\ Y &= -wa - \frac{d\Psi}{dy}, \\ Z &= -\frac{d\Psi}{dz}. \end{aligned}\right\} \qquad (12)$$

By equations (4)

$$\begin{aligned} a &= -4\pi g w, \\ \beta &= \quad 4\pi f w, \\ \gamma &= \quad 0. \end{aligned}$$

If K is the specific inductive capacity of the medium, we have

$$X = \frac{4\pi}{K} f, \quad Y = \frac{4\pi}{K} g, \quad Z = \frac{4\pi}{K} h.$$

Since the magnetic permeability of the dielectric is taken as unity, we may put $1/K = V^2$, where V is the velocity of light through the dielectric.

Making these substitutions for the magnetic force and the electromotive intensity, equations (12) become

$$4\pi f(V^2 - w^2) = -\frac{d\Psi}{dx},$$

$$4\pi g(V^2 - w^2) = -\frac{d\Psi}{dy},$$

$$4\pi h V^2 \qquad = -\frac{d\Psi}{dz};$$

and since $$\frac{df}{dx} + \frac{dg}{dy} + \frac{dh}{dz} = 0,$$

we get $$\frac{d^2\Psi}{dx^2} + \frac{d^2\Psi}{dy^2} + \frac{V^2 - w^2}{V^2}\frac{d^2\Psi}{dz^2} = 0, \qquad (13)$$

or putting $$z' = \frac{V}{\{V^2 - w^2\}^{\frac{1}{2}}} z,$$

equation (13) becomes

$$\frac{d^2\Psi}{dx^2} + \frac{d^2\Psi}{dy^2} + \frac{d^2\Psi}{dz'^2} = 0,$$

a solution of which is

$$\Psi = \frac{A}{\{x^2 + y^2 + z'^2\}^{\frac{1}{2}}}$$

$$= \frac{A}{\left\{x^2 + y^2 + \frac{V^2}{V^2 - w^2} z^2\right\}^{\frac{1}{2}}}. \qquad (14)$$

To find A we notice that the normal polarization over any sphere concentric with the moving one must equal e, the charge on the sphere; hence if a is the radius of the moving sphere,

$$\iint \left\{\frac{x}{a} f + \frac{y}{a} g + \frac{z}{a} h\right\} dS = e.$$

Substituting for f, g, h their values, we find

$$\frac{Aa}{4\pi(V^2-w^2)}\iint\frac{dS}{\left\{x^2+y^2+\frac{V^2}{V^2-w^2}z^2\right\}^{\frac{3}{2}}}=e,$$

or
$$\frac{A}{2(V^2-w^2)}\int_0^\pi\frac{\sin\theta\,d\theta}{\left\{\sin^2\theta+\frac{V^2}{V^2-w^2}\cos^2\theta\right\}^{\frac{3}{2}}}=e.$$

The integral, if $V>w$, is equal to

$$\frac{2\{V^2-w^2\}^{\frac{1}{2}}}{V};$$

hence
$$A=eV\{V^2-w^2\}^{\frac{1}{2}},$$

so that
$$\left.\begin{aligned} f&=\frac{e}{4\pi}\frac{V}{\{V^2-w^2\}^{\frac{1}{2}}}\frac{x}{\left\{x^2+y^2+\frac{V^2}{V^2-w^2}z^2\right\}^{\frac{3}{2}}},\\ g&=\frac{e}{4\pi}\frac{V}{\{V^2-w^2\}^{\frac{1}{2}}}\frac{y}{\left\{x^2+y^2+\frac{V^2}{V^2-w^2}z^2\right\}^{\frac{3}{2}}},\\ h&=\frac{e}{4\pi}\frac{V}{\{V^2-w^2\}^{\frac{1}{2}}}\frac{z}{\left\{x^2+y^2+\frac{V^2}{V^2-w^2}z^2\right\}^{\frac{3}{2}}}.\end{aligned}\right\}\quad(15)$$

Thus
$$\frac{f}{x}=\frac{g}{y}=\frac{h}{z}.$$

The Faraday tubes are radial and the resultant polarization varies inversely as

$$r^2\left\{1+\frac{w^2}{V^2-w^2}\cos^2\theta\right\}^{\frac{3}{2}},$$

where r is the distance of the point from the centre, and θ the angle which r makes with the direction of motion of the sphere. We see from this result that the polarization is greatest where $\theta=\pi/2$, least where $\theta=0$; the Faraday tubes thus leave the poles of the sphere and tend to congregate at the equator. This arises from the tendency of these tubes to set themselves at right angles to the direction in which they are moving. The surface density of the electricity on the moving sphere varies inversely as

$$\left\{1+\frac{w^2}{V^2-w^2}\cos^2\theta\right\}^{\frac{3}{2}},$$

it is thus a maximum at the equator and a minimum at the poles.

The components a, β, γ of the magnetic force are given by the equations

$$\left.\begin{aligned} a &= -4\pi w g = -\frac{eVw}{\{V^2-w^2\}^{\frac{1}{2}}}\frac{y}{\left\{x^2+y^2+\frac{V^2}{V^2-w^2}z^2\right\}^{\frac{3}{2}}}, \\ \beta &= \quad 4\pi w f = \frac{eVw}{\{V^2-w^2\}^{\frac{1}{2}}}\frac{x}{\left\{x^2+y^2+\frac{V^2}{V^2-w^2}z^2\right\}^{\frac{3}{2}}}, \\ \gamma &= 0. \end{aligned}\right\} \quad (16)$$

These expressions as well as (15) were obtained by Mr. Heaviside by another method in the *Phil. Mag.* for April, 1889.

Thus the lines of magnetic force are circles with their centres in and their planes at right angles to the axis of z. When w is so small that w^2/V^2 may be neglected, the preceding equations take the simpler forms

$$f = \frac{e}{4\pi}\frac{x}{r^3}, \quad g = \frac{e}{4\pi}\frac{y}{r^3}, \quad h = \frac{e}{4\pi}\frac{z}{r^3},$$

$$a = -\frac{ewy}{r^3}, \quad \beta = \frac{ewx}{r^3}.$$

(See J. J. Thomson 'On the Electric and Magnetic Effects produced by the Motion of Electrified Bodies', *Phil. Mag.* April, 1881.)

The moving sphere thus produces the same magnetic field as an element of current at the centre of the sphere parallel to z whose moment is equal to ew. When as a limiting case $V = w$, that is when the sphere is moving with the velocity of light, we see from equations (15) and (16) that the polarization and magnetic force vanish except when $z = 0$ when they are infinite. The equatorial plane is thus the seat of infinite magnetic force and polarization, while the rest of the field is absolutely devoid of either. It ought to be noticed that in this case all the Faraday tubes have arranged themselves so as to be at right angles to the direction in which they are moving.

We shall now consider the momentum in the dielectric due to the motion of the Faraday tubes. Since the dielectric is non-magnetic the components U, V', W of this are by equations (9) given by the following expressions:

$$\left.\begin{aligned} U &= -\beta h &&= -\frac{e^2}{4\pi}\frac{V^2 w}{V^2-w^2}\frac{xz}{\left(x^2+y^2+\frac{V^2}{V^2-w^2}z^2\right)^3}, \\ V' &= \alpha h &&= -\frac{e^2}{4\pi}\frac{V^2 w}{V^2-w^2}\frac{yz}{\left(x^2+y^2+\frac{V^2}{V^2-w^2}z^2\right)^3}, \\ W &= \beta f-\alpha g &&= \frac{e^2}{4\pi}\frac{V^2 w}{V^2-w^2}\frac{(x^2+y^2)}{\left(x^2+y^2+\frac{V^2}{V^2-w^2}z^2\right)^3}. \end{aligned}\right\} \quad (17)$$

The resultant momentum at any point is thus at right angles to the radius and to the magnetic force; it is therefore in the plane through the radius and the direction of motion and at right angles to the former. The magnitude of the resultant momentum per unit volume at a point at a distance r from the centre of the sphere, and where the radius makes an angle θ with the direction of motion, is

$$\frac{e^2 w}{4\pi}\cdot\frac{V^2}{V^2-w^2}\,\frac{1}{r^4}\,\frac{\sin\theta}{\left\{1+\frac{w^2}{V^2-w^2}\cos^2\theta\right\}^3}.$$

Thus the momentum vanishes along the line of motion of the sphere, where the Faraday tubes are moving parallel to themselves, and continually increases towards the equator as the tubes get to point more and more at right angles to their direction of motion.

The resultant momentum in the whole of the dielectric is evidently parallel to the direction of motion; its magnitude I is given by the equation

$$\begin{aligned} I &= \frac{e^2 w}{4\pi}\frac{V^2}{V^2-w^2}\int_a^\infty\int_0^\pi\int_0^{2\pi}\frac{\sin^2\theta\, r^2 dr\sin\theta\, d\theta\, d\phi}{r^4\left\{1+\frac{w^2}{V^2-w^2}\cos^2\theta\right\}^3} \\ &= \frac{e^2 w}{a}\frac{V^2}{V^2-w^2}\int_0^1\frac{\sin^2\theta\, d(\cos\theta)}{\left\{1+\frac{w^2}{V^2-w^2}\cos^2\theta\right\}^3}, \end{aligned}$$

or putting

$$\frac{w}{\{V^2-w^2\}^{\frac{1}{2}}}\cos\theta = \tan\psi,$$

we see that

$$I = \frac{e^2 V^2}{a\{V^2 - w^2\}^{\frac{1}{2}}} \int_0^{\tan^{-1}\frac{w}{\{V^2-w^2\}^{\frac{1}{2}}}} \cos^2\psi \left(1 - \frac{V^2}{w^2}\sin^2\psi\right) d\psi;$$

or if

$$\tan^{-1}\frac{w}{\{V^2-w^2\}^{\frac{1}{2}}} = \vartheta,$$

$$I = \frac{e^2}{2a}\frac{V^2}{\{V^2-w^2\}^{\frac{1}{2}}}\left\{\vartheta\left(1 - \tfrac{1}{4}\frac{V^2}{w^2}\right) + \tfrac{1}{2}\sin 2\vartheta\left(1 + \tfrac{1}{4}\frac{V^2}{w^2}\cos 2\vartheta\right)\right\}.$$

Thus the momentum of the sphere and dielectric parallel to z is $mw + I$, where m is the mass of the sphere; so that the effect of the charge will be to increase the apparent mass of the sphere by I/w or by

$$\tfrac{1}{2}\frac{e^2}{a}\frac{V^2}{w\{V^2-w^2\}^{\frac{1}{2}}}\left\{\vartheta\left(1 - \tfrac{1}{4}\frac{V^2}{w^2}\right) + \tfrac{1}{2}\sin 2\vartheta\left(1 + \tfrac{1}{4}\frac{V^2}{w^2}\cos 2\vartheta\right)\right\}.$$

When the velocity of the sphere is very small compared to that of light,

$$\vartheta = \frac{w}{V}\left(1 + \tfrac{1}{6}\frac{w^2}{V^2}\right)$$

approximately, and the apparent increase in the mass of the sphere is

$$\frac{2}{3}\frac{e^2}{a}.$$

When in the limit $w = V$ the increase in mass is infinite, thus a charged sphere moving with the velocity of light behaves as if its mass were infinite, its velocity therefore will remain constant, in other words it is impossible to increase the velocity of a charged body moving through the dielectric beyond that of light.

The kinetic energy per unit volume of the dielectric is

$$\frac{1}{8\pi}(\alpha^2 + \beta^2),$$

and hence by equations (16) and (17) it is equal to

$$\frac{w}{2}W;$$

thus the total kinetic energy in the dielectric is equal to

$$\tfrac{1}{2}wI,$$

that is to

$$\frac{e^2}{4a} w \cdot \frac{V^2}{\{V^2-w^2\}^{\frac{1}{2}}} \left\{\vartheta\left(1-\tfrac{1}{4}\frac{V^2}{w^2}\right)+\tfrac{1}{2}\sin 2\vartheta\left(1+\tfrac{1}{4}\frac{V^2}{w^2}\cos 2\vartheta\right)\right\}.$$

We shall now proceed to investigate the mechanical forces acting on the sphere when it is moving parallel to the axis of z in a uniform magnetic field in which the magnetic force is everywhere parallel to the axis of x and equal to H.

If U, V', W are the components of the momentum,

$$U = gc - hb,$$
$$V' = ha - fc,$$
$$W = fb - ga.$$

In this case

$$c = 0, \qquad b = \beta, \qquad a = \alpha + H,$$

where α and β have the values given in equations (16).

The momentum transmitted in unit time across the surface of a sphere concentric with the moving one has for components

$$\iint wU\cos\theta\, dS, \qquad \iint wV'\cos\theta\, dS, \qquad \iint wW\cos\theta\, dS,$$

the integration being extended over the surface of the sphere. Substituting the values of U, V', W, we see that the first and third of these expressions vanish, while the second reduces to

$$\frac{e}{4\pi}\frac{VHw}{\{V^2-w^2\}^{\frac{1}{2}}}\iint\frac{\cos^2\theta\, dS}{r^2\left\{1+\dfrac{w^2}{V^2-w^2}\cos^2\theta\right\}^{\frac{3}{2}}},$$

or

$$\tfrac{1}{2}\frac{eVHw}{\{V^2-w^2\}^{\frac{1}{2}}}\int_0^\pi\frac{\cos^2\theta\sin\theta\, d\theta}{\left\{1+\dfrac{w^2}{V^2-w^2}\cos^2\theta\right\}^{\frac{3}{2}}},$$

which is equal to

$$-\frac{eHwV}{\{V^2-w^2\}^{\frac{1}{2}}}\left\{\frac{(V^2-w^2)^{\frac{3}{2}}}{Vw^2}-\left(\frac{V^2-w^2}{w^2}\right)^{\frac{3}{2}}\log\left(\frac{V+w}{V-w}\right)^{\frac{1}{2}}\right\},$$

or to

$$-eH\frac{(V^2-w^2)}{w}\left\{1-\tfrac{1}{2}\frac{V}{w}\log\left(\frac{V+w}{V-w}\right)\right\}.$$

When w/V is very small this expression reduces to

$$\tfrac{1}{3}eHw.$$

This is the rate at which momentum is communicated to the sphere, in other words it is the force on the sphere; hence the force on the charged sphere coincides in direction with the force on an element of current parallel to the axis of z, but the magnitude of the force on the moving sphere is only one-third that of the force on an element of current along z whose moment is ew. By the moment of an element of current we mean the product of the intensity of the current and the length of the element. When $w = V$, that is when the sphere moves through the magnetic field with the velocity of light, we see from the preceding expression that the force acting upon it vanishes.

We can get a general idea of the origin of the mechanical force on the moving sphere if we remember that the uniform magnetic field is (Art. 13) due to the motion of Faraday tubes, the positive tubes moving in one direction, the negative ones in the opposite, and that in their motion through the field these tubes have to traverse the sphere. The momentum due to these tubes when they enter the sphere is proportional to the magnetic force at the place where they enter the sphere, while their momentum when they leave the sphere is proportional to the magnetic force at the place of departure. Now the magnetic forces at these places will be different, because on one side of the sphere the magnetic force arising from its own motion will increase the original magnetic field, while on the other side it will diminish it. Thus by their passage across the sphere the tubes will have gained or lost a certain amount of momentum; this will have been taken from or given to the sphere, which will thus be subject to a mechanical force.

Rotating Electrified Plates.

17.] The magnetic effects due to electrified bodies in motion are more conveniently examined experimentally by means of electrified rotating plates than by moving electrified spheres. The latter have, as far as I know, not been used in any experiments on electro-convection, while most interesting experiments with rotating plates have been made by Rowland (*Berichte d. Berl. Acad.* 1876, p. 211), Rowland and Hutchinson (*Phil. Mag.* 27, p. 445, 1889), Röntgen (*Wied. Ann.* 35, p. 264, 1888; 40, p. 93, 1890), Himstedt (*Wied. Ann.* 38, p. 560, 1889). The general plan of these experiments is as follows: an air condenser with circular parallel plates is made to rotate about an axis through the centres of the plates and at right angles to their planes. To prevent induced currents being produced by the rotation of the plates in the earth's magnetic field, radial divisions filled with insulating material are made in the plates. When the plates are charged and set in rotation a magnetic field is found to exist in their neighbourhood similar to that which would be produced by electric currents flowing in concentric circular paths in the plates of the condenser, the centres of these circles being the points where the axis of rotation cuts the plates.

Let us now consider how these magnetic forces are produced. Faraday tubes at right angles to the plates pass from one plate to the other. We shall suppose when the condenser is rotating as a rigid body these tubes move as if they were rigidly connected with it. Then, taking the axis of rotation as the axis of z, the component velocities of a tube at a point whose coordinates are x, y are respectively $-\omega y$ and ωx, where ω is the angular velocity with which the plates are rotating.

If these were the only Faraday tubes in motion the components α, β, γ of the magnetic force would by equations (4) be given by the equations

$$\begin{aligned} \alpha &= 4\pi\sigma\omega x, \\ \beta &= 4\pi\sigma\omega y, \\ \gamma &= 0, \end{aligned} \tag{18}$$

where $\sigma\,(=h)$ is the surface-density of the electricity on either plate. These values for the components of the magnetic force do not however satisfy the relation

$$\frac{d\alpha}{dx} + \frac{d\beta}{dy} + \frac{d\gamma}{dz} = 0,$$

which must be satisfied since the value of

$$\frac{1}{8\pi}\iiint (\alpha^2 + \beta^2 + \gamma^2)\, dx\, dy\, dz$$

must, in a medium whose magnetic permeability is unity, be stationary for all values of α, β, γ which give assigned values to the currents, that is to

$$\frac{d\beta}{dx} - \frac{d\alpha}{dy}, \qquad \frac{d\alpha}{dz} - \frac{d\gamma}{dx}, \qquad \frac{d\gamma}{dy} - \frac{d\beta}{dz}.$$

For let α_0, β_0, γ_0 be any particular values of the components of the magnetic force which satisfy the assigned conditions, then the most general values of these components are expressed by the equations

$$\alpha = \alpha_0 + \frac{d\phi}{dx},$$

$$\beta = \beta_0 + \frac{d\phi}{dy},$$

$$\gamma = \gamma_0 + \frac{d\phi}{dz},$$

where ϕ is an arbitrary function of x, y, z.

Then if

$$\iiint (\alpha^2 + \beta^2 + \gamma^2)\, dx\, dy\, dz$$

is stationary,

$$\iiint (\alpha\delta\alpha + \beta\delta\beta + \gamma\delta\gamma)\, dx\, dy\, dz = 0. \tag{19}$$

Let the variations in α, β, γ be due to the increment of ϕ by an arbitrary function $\delta\phi$, then

$$\delta\alpha = \frac{d\delta\phi}{dx}, \qquad \delta\beta = \frac{d\delta\phi}{dy}, \qquad \delta\gamma = \frac{d\delta\phi}{dz}.$$

Substituting these values for $\delta\alpha$, $\delta\beta$, $\delta\gamma$, and integrating by parts, equation (19) becomes

$$\iint \delta\phi\, (\alpha\, dy\, dz + \beta\, dz\, dx + \gamma\, dx\, dy)$$
$$- \iiint \delta\phi \left\{ \frac{d\alpha}{dx} + \frac{d\beta}{dy} + \frac{d\gamma}{dz} \right\} dx\, dy\, dz = 0,$$

and therefore since $\delta\phi$ is arbitrary

$$\frac{d\alpha}{dx} + \frac{d\beta}{dy} + \frac{d\gamma}{dz} = 0.$$

The values of α, β, γ given by equation (18) cannot therefore be the complete expressions for the magnetic force, and since we regard all magnetic force as due to the motion of Faraday tubes, it follows that the tubes which connect the positive to the negative charges on the plates of the condenser cannot be the only tubes in the field which are in motion; the motion of these tubes must set in motion the closed tubes which, Art. 2, exist in their neighbourhood. The motion of the closed tubes will produce a magnetic field in which the forces can be derived from a magnetic potential Ω. When we include the magnetic field due to the motion of these closed tubes, we have

$$\alpha = 4\pi\sigma\omega x - \frac{d\Omega}{dx} = \frac{d\Omega'}{dx},$$

$$\beta = 4\pi\sigma\omega y - \frac{d\Omega}{dy} = \frac{d\Omega'}{dy},$$

$$\gamma = \qquad - \frac{d\Omega}{dz} = \frac{d\Omega'}{dz},$$

if $\Omega' = 2\pi\sigma\omega(x^2+y^2) - \Omega$;

and since $$\frac{d\alpha}{dx} + \frac{d\beta}{dy} + \frac{d\gamma}{dz} = 0,$$

we have $$\frac{d^2\Omega'}{dx^2} + \frac{d^2\Omega'}{dy^2} + \frac{d^2\Omega'}{dz^2} = 0.$$

The question now arises, does the motion of the tubes which connect the positive and negative electrifications on the plates only set those closed tubes in motion which are between the plates of the condenser, or does it affect the tubes outside as well? Let us examine the consequences of the first hypothesis. In this case, since the Faraday tubes outside the condenser are at rest, the magnetic force will vanish except between the plates of the condenser; it follows, however, from the properties of the magnetic potential that it must vanish inside as well, so that no magnetic force at all would be produced by the rotation of the plates. As this is contrary to the result of Rowland's experiments, the Faraday tubes stretching between the plates must by their rotation set in motion tubes extending far away from the region between the plates. The motion of these closed tubes must however be consistent with the condition that the magnetic force parallel to the plates due to the motion of the tubes must be continuous. Let us consider for a moment the radial magnetic force due to the closed tubes: this may arise either from the rotation round the axis of tubes which pass through the plates, or from the motion at right angles to the plates of tubes parallel to them. In the first case, the velocity tangential to the plates of the tubes must be continuous, otherwise the tubes would break, and since the tangential velocity is continuous, the radial magnetic force due to the motion of these tubes will be continuous also. In the second case, the product of the normal velocity of the tubes and their number per unit volume must be the same on the two sides of a plate, otherwise there would be an accumulation of these tubes in the plate. The product of the normal velocity into the number of the tubes is, however, equal to the tangential magnetic force due to the motion of the closed tubes, so that this must be continuous.

The open tubes which stretch from the positive electricity on one plate to the negative on the other will, however, by their motion produce a discontinuity in the radial magnetic force, since these tubes stop at the plates, and do not pass through them. The radial magnetic force at a point due to these tubes is $4\pi\sigma\omega r$, where r is the distance of the point from the axis of rotation. The conditions to determine the magnetic field are thus, (1) that except in the substance of the plates there must be a magnetic potential satisfying Laplace's equation, and (2) that at either plate the discontinuity in the radial magnetic force must be $4\pi\sigma\omega r$, where σ is the surface-density of the electricity on the plates. These conditions are, however, exactly those which determine the magnetic force produced by a system of electric currents circulating in circles in the plates of the condenser, the intensity of the currents at a distance r from the axis of rotation being $\sigma\omega r$ for the positive and $-\sigma\omega r$ for the negative plate. Hence the magnetic force due to rotating the plates will be the same as that produced by this distribution of electric currents.

This conclusion seems to be confirmed by the results of the experiments of Rowland and Hutchinson (*Phil. Mag.* 27, p. 445, 1889), as using this hypothesis they found a tolerably accurate value of 'v', the ratio of the electromagnetic to the electrostatic unit of electricity, by means of experiments on a rotating plate.

We can see by similar reasoning that if only one of the plates is rotating, the other being at rest, the magnetic effect will be the same as that due to a system of electric currents circulating in the rotating plate, the intensity of the current at a distance r from the axis being $\sigma\omega r$.

Some interesting experiments have been made by Röntgen (*Wied. Ann.* 35, p. 264, 1888), in which, while the plates of the condenser were at rest, a glass disc parallel to the plates and situated between them was set in rapid rotation and was found to produce a magnetic field. The rotation of the disc must thus have set in motion the Faraday tubes passing through it, and these in turn have affected closed tubes extending into the region beyond the condenser.

Experiments of this kind seem to open up a field of enquiry which will throw light upon a question which at present is one of the most obscure in electricity: that of the relation between the velocities of the dielectric and of the Faraday tubes passing through it. This question is one of great importance in the Electro-magnetic Theory of Light, as but little progress can be made in the Theory of Aberration until we have got an answer to it. Another question which we have not touched upon, but which is very important in this connexion, is whether the motion of the Faraday tubes through ether devoid of matter would produce magnetic force, or whether for this purpose it is necessary that the tubes should pass across ordinary matter as well as ether. The point may be illustrated by the following case. Suppose we have a plate of glass between two parallel charged plates rigidly electrified, whether uniformly or otherwise, and that the whole system is set in rotation and moves like a rigid body, then, is or is not the motion of the system accompanied by magnetic force? Here the Faraday tubes move through the ether (assuming that the velocity of the ether is not the same as that of the glass), but do not move relatively to the glass.

The motion of the tubes through the *dielectric* will be requisite for the production of magnetic effects if we suppose that there are no closed tubes in the electric field, and that all the tubes connect portions of ordinary matter. The recognition of closed tubes in the ether seems to be desirable in the present state of our electrical knowledge, as unless we acknowledge the existence of such tubes we have to suppose that light being an electro-magnetic phenomenon cannot traverse a region wholly devoid of ordinary matter, and further that the existence of magnetic force depends upon the presence of such matter in the field.

A Steady Magnetic Field.

18.] Magnetic force on the theory we are now discussing is due to the motion of the Faraday tubes. When the magnetic field is variable the presence of these tubes is rendered evident by the existence of electromotive intensities in the field: when however the field is steady, we have no direct electrical evidence of the presence of these tubes, and their disposition and velocities have to be deduced from the equations developed in the preceding pages. We shall now proceed to examine this very important case more in detail.

In a steady magnetic field in which there is no free electricity the Faraday tubes must be closed, exception being made of course of the short tubes which connect together the atoms in the molecules present in the field. Since in such a field there is no electromotive intensity, there must pass through each unit area of the field the same number of positive as of negative tubes, that is, there must be as many tubes pointing in one direction as the opposite. These tubes will (Art. 12) place themselves so as to be at right angles both to the direction in which they are moving and to the magnetic force. The distribution of the Faraday tubes and the directions in which they are moving cannot be determined solely from the magnetic force; but for the purpose of forming a clear conception of the way in which the magnetic force may be produced, we shall suppose that the positive tubes are moving with the velocity of light in one direction, the negative tubes with an equal velocity in the opposite, and that at any point the direction of a tube, its velocity, and the magnetic force are mutually at right angles.

In a steady magnetic field surfaces of equal potential exist which cut the lines of magnetic force at right angles, so that since both the Faraday tubes and the directions in which they are moving are at right angles to the lines of magnetic force we may suppose that the Faraday tubes form closed curves on the equipotential surfaces, a tube always remaining on one equipotential surface and moving along it at right angles to itself.

We shall now consider the motion of these tubes in a very simple magnetic field: that surrounding an infinitely long circular cylinder whose axis is taken as the axis of z, and which is uniformly magnetized at right angles to its axis and parallel to the axis of x.

The magnetic potential inside the cylinder is equal to

$$Hx,$$

where H is the magnetic force inside the cylinder.

The potential outside the cylinder, if a is the radius of the cylinder, is equal to

$$H\frac{a^2\cos\theta}{r},$$

where r is the distance from the axis of the cylinder of the point at which the potential is reckoned, and θ the azimuth of r measured from the direction of magnetization. Thus inside the cylinder the equipotential surfaces are planes at right angles to the

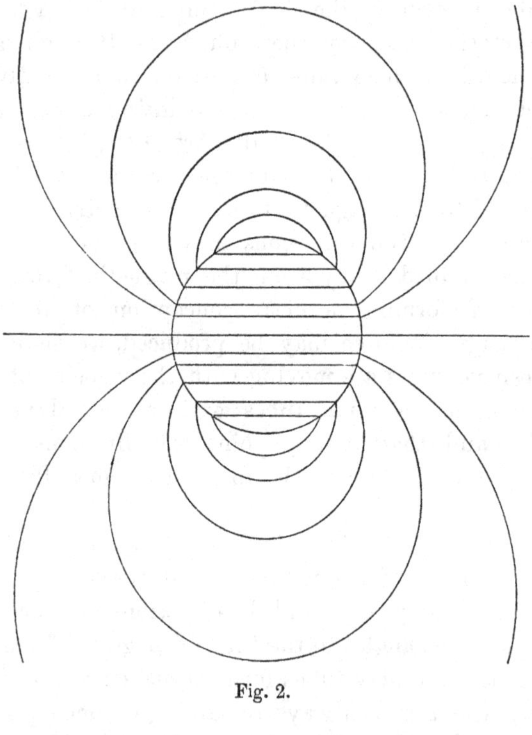

Fig. 2.

direction of magnetization, while outside they are a system of circular cylinders which if prolonged would pass through the axis of the magnet; the axes of all these cylinders are parallel to the axis of z and lie in the plane of xz. The cross sections of the original cylinder and the equipotential surfaces are represented in Fig. 2.

We shall suppose that the Faraday tubes are parallel to the axis of the cylinder; then we may regard the magnetic field as produced by such tubes travelling round the equipotential surfaces with uniform velocity, the positive tubes moving in one direction, the negative ones in the opposite. We shall show that the number of tubes passing through the area bounded by unit length of the cross-section of any equipotential surface, the normals and the consecutive equipotential surface will be constant. For since the magnetic force is at right angles both to the Faraday tubes and the direction in which they are moving, the magnetic force due to this distribution of Faraday tubes will be at right angles to the equipotential surfaces; and if N is the number of tubes of one sign between two consecutive equipotential surfaces per unit length of cross section of one of them, ds the length of a portion of such a cross section, $d\nu$ the normal distance between two consecutive equi-potential surfaces Ω_1 and Ω_2, then in the cylinder whose base is $ds\,d\nu$ the number of Faraday tubes of one sign will be $Nds(\Omega_2-\Omega_1)$; but since these tubes are distributed over an area $ds\,d\nu$, the number of tubes per unit area of the base of the cylinder is $N(\Omega_2-\Omega_1)/d\nu$. These tubes are however all moving at the same rate, so that the magnetic force due to them will be proportional to the number per unit area of the base of the cylinder, that is to $N(\Omega_2-\Omega_1)/d\nu$, so that since the magnetic force due to these tubes is proportional to $(\Omega_2-\Omega_1)/d\nu$, N will be constant. Thus the magnetic force due to the tubes moving in the way we have described coincides both in magnitude and direction with that due to the magnetized cylinder.

We see from Fig. 2 that the directions of motion of these tubes change abruptly as they enter the magnetized cylinder. The principles by which the amount of this bending of the direction of motion of the tubes may be calculated are as follows. If h_1 and h_2 are the densities of the tubes just inside and outside the cylinder, R_1, R_2 the corresponding velocities of these tubes along the normal to the cylinder, then since there is no accumulation of the tubes at the surface of the cylinder we must have

$$\Sigma R_1 h_1 = \Sigma R_2 h_2.$$

But since R is the radial velocity, $\Sigma\, 4\pi Rh$ is by (8), Art. 13, the tangential magnetic force: hence the preceding equation expresses the continuity of the tangential magnetic force as we cross the surface of the cylinder. Again, when a Faraday tube crosses the surface of the cylinder, the tangential component of its momentum will not change; but by equations (9) the tangential momentum of the tube is proportional to the normal magnetic induction, so that the continuity of the tangential momentum is equivalent to that of the normal component of the magnetic induction. We have thus deduced from this view of the magnetic field the ordinary boundary conditions (1) that the tangential component of the magnetic force is continuous, and (2) that the normal component of the magnetic induction is continuous.

The paths along which the tubes move coincide with the lines of flow produced by moving the cylinder uniformly at right angles to the direction of magnetization through an incompressible fluid.

Induction of Currents due to Changes in the Magnetic Field.

19.] Let Fig. 3 represent a section of the magnetized cylinder and one of its equipotential surfaces, the directions of the magnetic force round the cylinder being denoted by the dotted lines. We shall call those Faraday tubes which point upwards from the plane of the paper positive, the negative Faraday tubes of course pointing downwards. The positive and the negative tubes circulate round the equipotential surface in the directions marked in the figure. Let A and B represent the cross-sections of the wires of a circuit, the wires being at right angles to the plane of the paper. When the magnetic field is steady no current will be produced in this circuit, because there are as many positive as negative tubes at any point in the field. Let us now suppose that the magnetic field is suddenly destroyed; we may imagine that this is done by placing barriers across the equipotential surfaces in the magnetized cylinder so as to stop the circulation of the Faraday tubes. The inertia of these tubes will for a short time carry

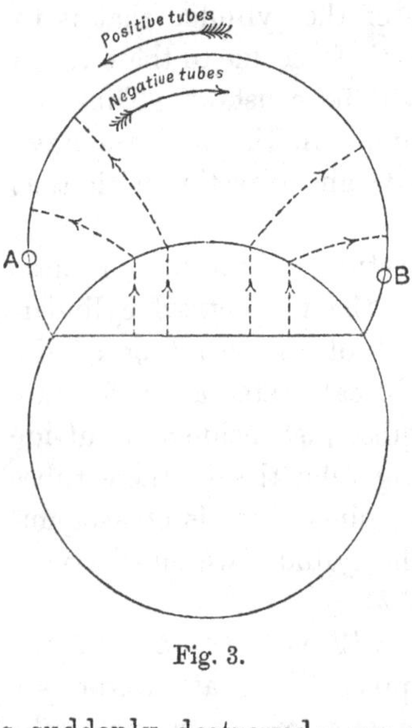

Fig. 3.

them on in the direction in which they were moving when the barrier was interposed, hence the positive tubes will run out on the right-hand side of the equipotential surfaces and accumulate on the left-hand side, while the negative tubes will leave the left-hand side and accumulate on the right. The equality which formerly existed between the positive and negative tubes will now be destroyed: there will be an excess of positive tubes in the neighbourhood of the conductor A, and an excess of negative ones round B. A current will therefore be started in the circuit running from A to B above the plane of the paper, and from B to A below it. We see in this way how the inertia of the Faraday tubes accounts for induced currents arising from variations in the intensity of the magnetic field.

Induction due to Motion of the Circuit.

20.] We can explain in a similar way the currents induced when a conductor is moved about in a magnetic field. Suppose we have a straight conductor moving about in the streams of Faraday tubes which constitute such a field, the Faraday tubes being parallel to each other and to the conductor: let the conductor be moved in the opposite direction to that in which the positive tubes are moving. This motion of the conductor will tend to stop the positive tubes in it and just in front of it; the inertia of the tubes further off will make them continue to move towards the conductor, and thus the density of the tubes in front (i. e. those entering the conductor) will increase, while the density of the tubes behind (i. e. those leaving the conductor) will diminish; the number of positive tubes in the conductor will thus be greater than the number which would have been present if the conductor had been at rest. Similar reasoning will show that there will be a decrease in the number of negative tubes in the conductor. Thus the positive tubes in the conductor will now outnumber the negative ones, and there will therefore be a positive current. The motion of the conductor in the direction opposite to that in which the positive Faraday tubes are moving will thus be accompanied by the production of a positive current. This current is the ordinary induction current due to the motion of a conductor in the magnetic field.

Effect of the Introduction of Soft Iron into a Magnetic Field.

21.] Another simple magnetic system which we shall briefly consider is that of an infinite cylinder of soft iron, whose axis is taken as that of z, placed in what was before its introduction a uniform magnetic field parallel to the axis of x. Before the cylinder was introduced into the field, the Faraday tubes, which we may suppose to be parallel to the axis of z, would all be moving parallel to the axis of y; as soon however as the cylinder is placed in the field, the tubes will turn so as to avoid as much as possible going through it, for since the tangential momentum is not altered the tangential velocity of the tubes must be smaller inside the cylinder than it is outside, as the effective inertia of a tube in a magnetic medium is greater than in a non-magnetic one (see Art. 10). The lines of flow of the Faraday tubes will thus be deflected by the cylinder in much the same way as a current of electricity flowing through a conducting field would be deflected by the introduction into the field of a cylinder made of a worse conductor than itself. The Faraday tubes bend away from the cylinder in the way shown in Figure 4. The paths of the Faraday tubes coincide however with equipotential surfaces; these surfaces therefore bend away from the cylinder, and the lines of magnetic force which are at right angles to the equipotential surface turn in consequence towards the cylinder as indicated in Fig. 4, in which the dotted lines represent lines of magnetic force.

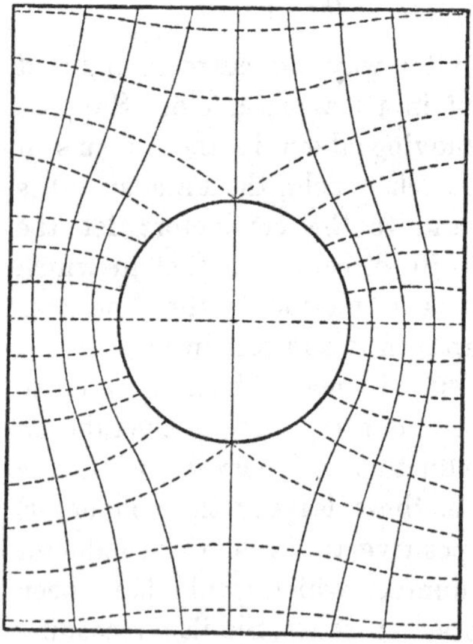

Fig. 4.

Permanent Magnets.

22.] In the interior of the magnet as well as in the surrounding magnetic field there is a shearing of the positive tubes past the negative ones. The magnet as it moves about carries this system of moving tubes with it, so that the motion of the tubes must in some way be maintained by a mechanism connected with the magnet: this mechanism exerts a fan-like action, driving the positive tubes in one direction, the negative ones in the opposite. This effect would be produced if the molecules of the magnet had the constitution described below and were in rapid rotation about the lines of magnetic force. Let the molecule ABC of a magnet consist of three atoms A, B, C, Fig. 5. Let one short tube go from B and end on A, another start from B and end on C, then if the molecule rotates in the direction of the arrow, about an axis through B perpendicular to the plane of the paper, since two like parallel Faraday tubes repel each other the rotation of the molecule will set the Faraday tubes in the ether surrounding the molecule in motion, the tubes going from left

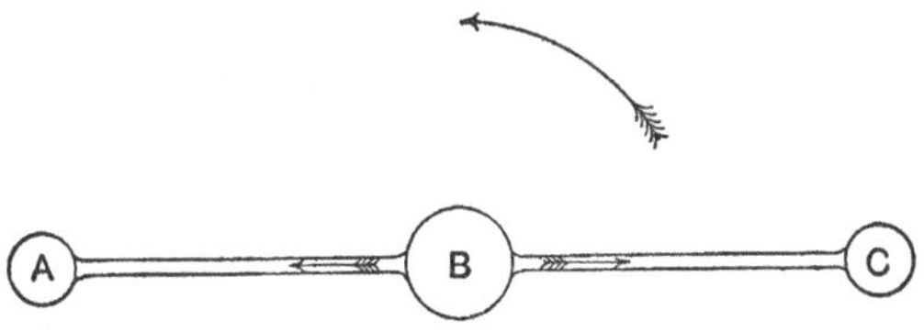

Fig. 5.

to right will move upwards in the plane of the paper, while those from right to left move downwards. This will produce a magnetic field in which, since the magnetic force is at right angles both to the moving tubes and the direction of motion, it will be at right angles to the plane of the paper and upwards; thus the magnetic force is parallel to the axis of rotation of the molecule. We notice that the atoms in the molecule are of different kinds with respect to the number of tubes incident upon them; thus B is the seat of two tubes, A and C of one each; in chemical language this would be expressed by saying that the valency of the atom B is twice that of either A or C.

This illustration is only intended to call attention to the necessity for some mechanism to be connected with a permanent magnet to maintain the motion of the Faraday tubes in the field, and to point out that the motion of molecular tubes is able to furnish such a mechanism.

Steady Current flowing along a Straight Wire.

23.] We shall now proceed to express in terms of the Faraday tubes the phenomena produced by a steady current flowing along an infinitely long straight vertical wire. We shall suppose that the circumstances are such that there is no free electricity on the surface of the wire, so that the Faraday tubes in its neighbourhood are parallel to its length. If we take the direction of the current as the positive direction, the positive tubes parallel to the wire will be moving in radially to keep up the current, and this inward radial flow of positive tubes will be accompanied by an outward radial flow of negative tubes, a positive tube when entering the wire displacing a negative tube which moves outward from the wire. This shearing of the positive and negative tubes past each other will give rise to a magnetic force which will be at right angles both to the direction of the tubes and the direction in which they are moving; thus the magnetic force is tangential to a circle whose plane is horizontal and whose centre is on the axis of the wire. When the positive tubes enter the wire they shrink to molecular dimensions in the manner to be described in Art. 31. At a distance r from the axis of the wire let N be the number of positive tubes passing through unit area of a plane at right angles to the wire, v the velocity of these tubes inwards, let N' be the number of negative tubes per unit area at the same point, v' their velocity outwards. The algebraical sum of the number of tubes which cross the circle whose radius is r and whose centre is on the axis of the wire is thus

$$(vN+v'N')\,2\pi r.$$

When the field is steady the value of this expression must be the same at all distances from the wire, because as many tubes must flow into any region as flow out of it. Hence when the field is steady this expression must equal the algebraical sum of the number of positive tubes which enter the wire in unit time; this number is however equal to i, the current through the wire; hence we have

$$(vN+v'N')\,2\pi r=i.$$

But by equations (4)

$$vN + v'N' = \frac{\gamma}{4\pi},$$

where γ is the magnetic force at a distance r from the axis. Substituting this value for $vN + v'N'$, we get

$$\gamma = \frac{2i}{r},$$

the usual expression for the magnetic force outside the wire produced by a straight current.

When the field is steady, there will be as many positive as negative tubes in each unit area, and therefore no electromotive intensity; if however the intensity of the current changes, this will no longer hold. To take an extreme case, let us suppose that the circuit is suddenly broken, then the inertia of the positive tubes will make them continue to move inwards; and since as the circuit is broken they can no longer shrink to molecular dimensions when they enter it, the positive tubes will accumulate in the region surrounding the wire: the inertia of the negative tubes carries them out of this region, so that now there will be a preponderance of positive tubes in the field around the wire. If any conductor is in this field these positive tubes will give rise to a positive current, which is the 'direct' induced current which occurs on breaking the circuit. When the field was steady no current would be produced in this secondary circuit, because there were as many positive as negative tubes in its neighbourhood.

The Faraday tubes have momentum which they give up when they enter the wire. If we consider a single wire where everything is symmetrical, the wire is bombarded by these tubes on all sides, so that there is no tendency to make it move off in any definite direction. Let us suppose, however, that we have *two* parallel wires conveying currents in the same direction, let A and B denote the cross-sections of these wires, B being to the right of A. Then some of the tubes which if B were absent would pass into A from the region on the right, will when B is present be absorbed by it, and so prevented from entering A. The supply of positive tubes to A will thus no longer be symmetrical; more will now come into A from the region on its left than from that on its right; hence since each of the tubes has momentum, more momentum will come to A from the left than from the right; thus A will be pushed from left to right or towards B. There will thus be an attraction between the parallel currents.

24.] It will be noticed that the tubes in the preceding case move radially in towards the wire, so that the energy which is converted into heat in the circuit comes from the dielectric sideways into the wire and is not transmitted longitudinally along it. This was first pointed out by Poynting in his paper on the Transfer of Energy in the Electromagnetic Field (*Phil. Trans.* 1884, Part. II. p. 343).

When however the current instead of being constant is alternating very rapidly, the motion of the tubes in the dielectric is mainly longitudinal and not transversal. We shall show in Chapter IV that if p is the frequency of the current, σ the specific resistance of the wire, a its radius, and μ its magnetic permeability, then when $4\pi\mu pa^2/\sigma$ is a large quantity the electromotive intensity outside the wire is normal to the wire and therefore radial. Thus in this case the Faraday tubes will be radial, and they will move at right angles to themselves parallel to the wire. There is thus a great contrast between this case and the previous one in which the tubes are longitudinal and move radially, while in this the tubes are radial and move longitudinally.

Discharge of a Leyden Jar.

25.] We shall now proceed to consider the distribution and motion of the Faraday tubes during the discharge of a Leyden jar.

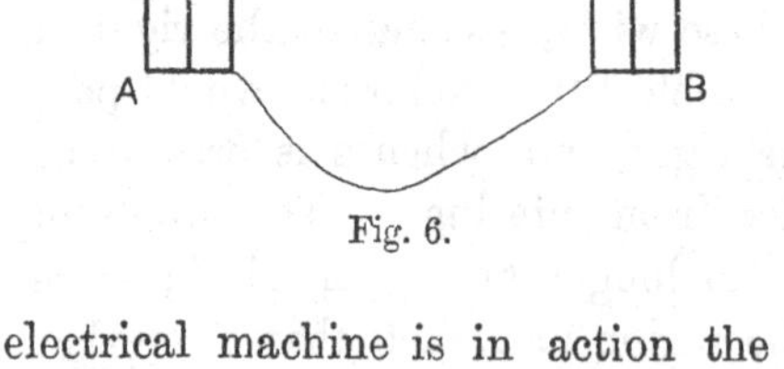

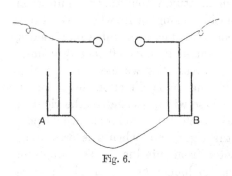

Fig. 6.

We shall take the symmetrical case in which the outside coatings of two Leyden jars A and B (Fig. 6) are connected by a wire, while the inside coating of A is connected to one terminal of an electrical machine, the inside coating of B to the other. When the electrical machine is in action the difference of potential between the inside coatings of the jars increases until a spark passes between the terminals of the machine and electrical oscillations are started in the jars.

Just before the passage of the spark the Faraday tubes will be arranged somewhat as follows. Some tubes will stretch from one terminal of the electrical machine to the other, others will go from these terminals to neighbouring conductors, such as the table on which the machine is placed, the floors and walls of the room. The great majority of the tubes will however be short tubes going through the glass from one coating to the other of the jars A and B.

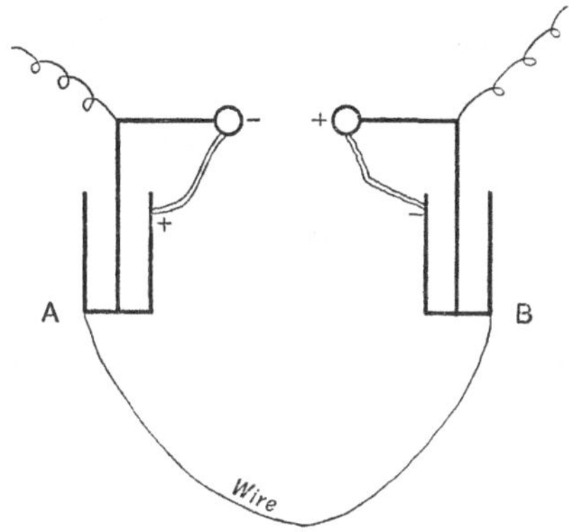

Fig. 7.

Let us consider the behaviour of two of these tubes, one from A, the other from B, when a spark passes between the terminals of the machine: while the spark is passing these terminals may be considered to be connected by a conductor. The tubes which before the spark passed stretched from one terminal of the machine to the other, will as soon as the air space breaks down shrink to molecular dimensions; and since the repulsion which these tubes exerted on those surrounding them is obliterated, the latter crowd into the space between the terminals. The short tubes which, before the spark passed, went from one coating of a jar to the other will now occupy some such positions as those shown in Fig. 7. These tubes being of opposite kinds tend to run together, they approach each other until they meet as in Fig. 8, the tubes now break up as in Fig. 9, the upper portion runs into the spark gap where it

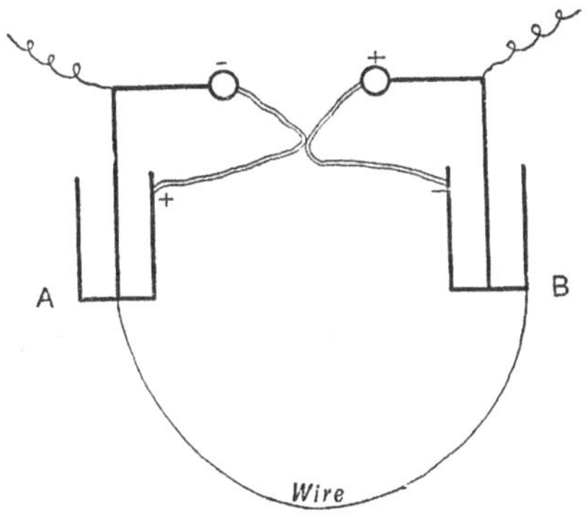

Fig. 8.

contracts, while the lower portion runs towards the wire connecting the outside coatings of the jars, Fig. 10. If this wire is a good conductor the tubes at their junction with the wire will be at right angles to it, and a tube will move somewhat as in Fig. 11. The inertia of the tube will carry the two sides past each other, until the tubes are arranged as in Fig. 12. The tube with its ends on the wire will travel backwards and will approach the positive tube which was emitted from the air gap when the negative tube (Fig. 9) entered it. The tubes then go through the processes illustrated in Figs. 9, 8, 7 in the reverse order, and the jars again get charged, but with electricity of opposite sign to that with which they started. After a time all the original Faraday tubes will be replaced by others of opposite sign, and the charges on the jars will be equal and opposite to the original charges. The new charge will then proceed to get reversed by similar processes to those by which the original charge was reversed, and thus the charges on the jar will oscillate from positive to negative and back again.

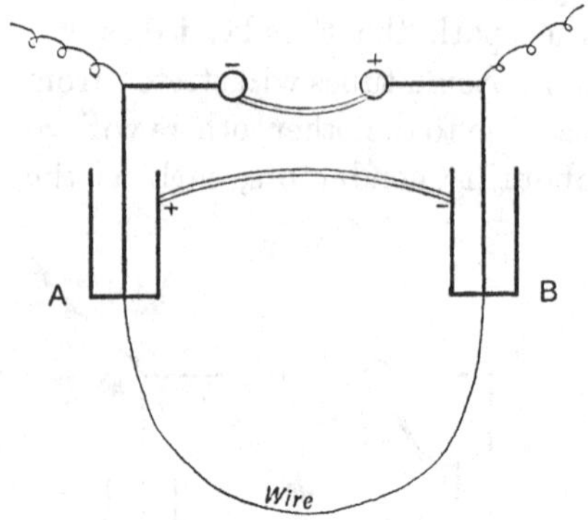

Fig. 9.

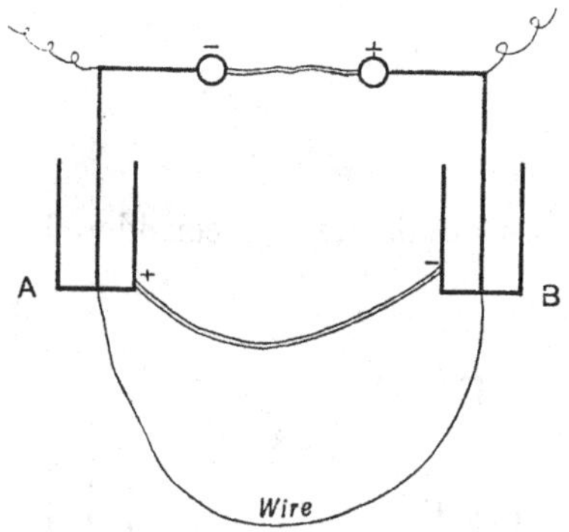

Fig. 10.

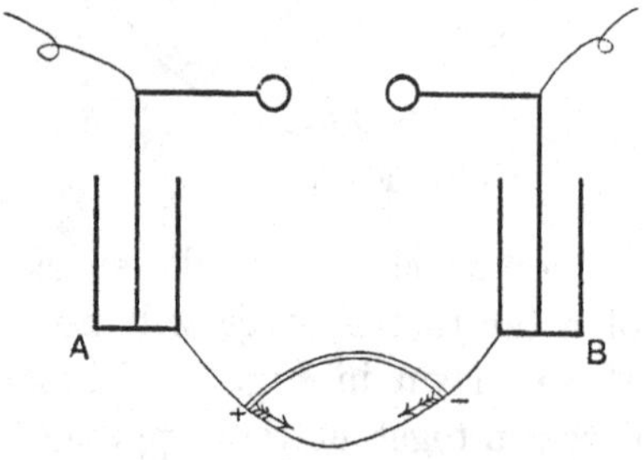

Fig. 11.

26.] When a conducting circuit is placed near the wire connecting the outer coatings of the jars, the Faraday tubes will strike against the circuit on their way to and from the wire. The passage of these tubes across the circuit will, since there is an excess of tubes of one name, produce a current in this circuit, which is the ordinary current in the secondary due to the variation of the intensity of the current in the primary circuit.

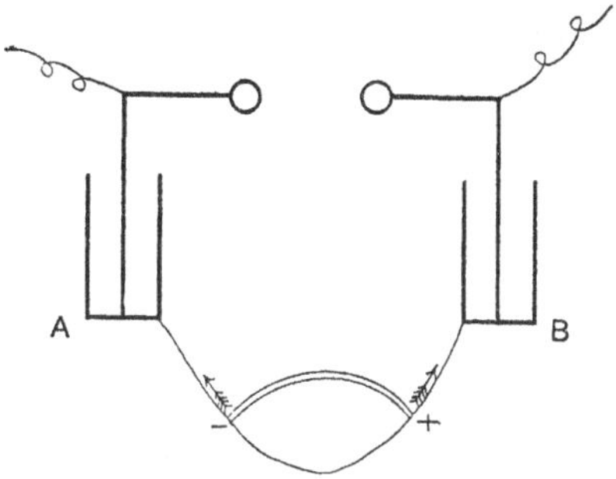

Fig. 12.

Some of the tubes as they rush from the jar to the wire connecting the outside coatings of the jar strike against the secondary circuit, break up into two parts, as shown in Fig. 13, the ends of these parts run along this circuit until they meet again, when the tube reunites and goes off as a single tube. The passage of the tube

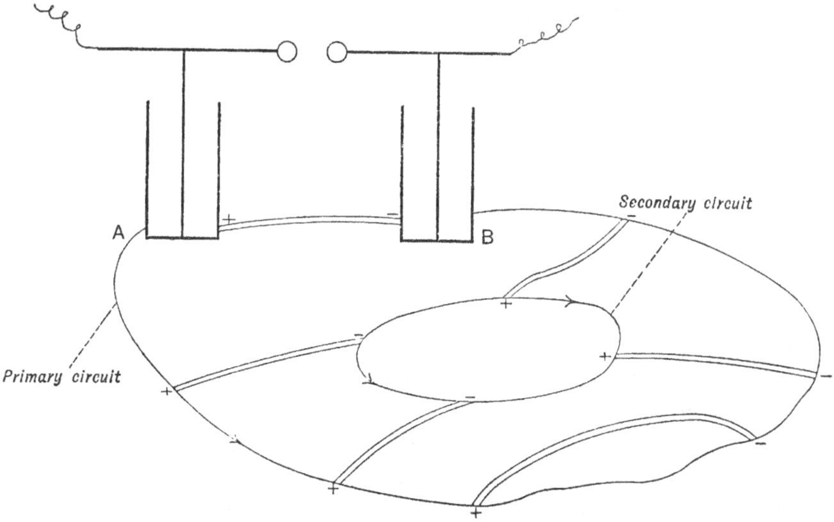

Fig. 13.

across the secondary circuit is thus equivalent to a current in the direction of rotation of the hands of a watch; this is opposite to that of the current in the wire connecting the outside coatings of the jars. The circuit by breaking up the tubes falling on it prevents them from moving across its interior, in other words, it tends to keep the number of lines of magnetic induction which pass through the circuit constant; this tendency gives the usual rule for finding the direction of the induced current. The introduction of magnetic force for the purpose of finding the currents in one circuit induced by alterations of the currents in another circuit seems however somewhat artificial.

Electromagnetic Theory of Light.

27.] We can by the aid of the Faraday tubes form a mental picture of the processes which on the Electromagnetic Theory accompany the propagation of light. Let us consider in the first place the uninterrupted propagation of a plane wave emitted from a plane source. Let z be the direction of propagation and let the wave be one of plane polarized light, the plane of polarization being that of yz. Then we may suppose that a bundle of Faraday tubes parallel to x are emitted from the plane source, and that either these, or other parallel tubes set in motion by them, travel at right angles to themselves and parallel to the axis of z with the velocity of light. By the principles we have been considering these tubes produce in the region through which they are passing a magnetic force whose direction is at right angles both to the direction of the tubes and that in which they are moving, the magnetic force is thus parallel to the axis of y. The magnitude of the magnetic force is by equations (4) equal to $4\pi v$ times the polarization, where v is the velocity of light, and since the electromotive intensity is $4\pi/K$, or, if the medium is non-magnetic, $4\pi v^2$ times the polarization, we see that the electromotive intensity is equal to v times the magnetic force. If there is no reflection the electromotive intensity and the magnetic force travel with uniform velocity v outwards from the plane of disturbance and always bear a constant ratio to each other. By supposing the number of tubes issuing from the plane source per unit time to vary harmonically we arrive at the conception of a divergent wave as a series of Faraday tubes travelling outwards with the velocity of light. In this case the places of maximum, zero and minimum electromotive intensity will correspond respectively to places of maximum, zero and minimum magnetic force.

The case is different, however, when light is reflected from a metallic surface. We shall suppose this surface plane and at right angles to the axis of z. In this case since the tangential electromotive intensity at the metallic surface vanishes, when a bundle of positive tubes enters the reflecting surface, an equal number of negative tubes are emitted from it; these travel backwards towards the source of light, moving in the opposite direction to the positive tubes. If we have a harmonic emission of tubes from the source of light we shall evidently also have a harmonic emission of tubes from the reflecting surface. Thus, at the various places in the path of the light, we may have positive tubes moving backwards or forwards accompanied by negative tubes moving in either direction. The magnetic effects of the positive tubes moving forwards are the same as those of the negative tubes moving backwards. Thus, when we have tubes of opposite signs moving in opposite directions, their magnetic effects conspire while their electromotive effects conflict; so that when, as in the case of reflection, we have streams of tubes moving in opposite directions the magnetic force will no longer be proportional to the electromotive intensity. In fact the places where the magnetic force is greatest will be places where the electromotive intensity vanishes, for such a place will evidently be one where we have the maximum density of positive tubes moving in one direction accompanied by the maximum density of negative tubes moving in the opposite, and since in this case there are as many positive as negative tubes the electromotive intensity will vanish. In a similar way we can see that the places where the electromotive intensity is a maximum will be places where the magnetic force vanishes.

This view of the Electromagnetic Theory of Light has some of the characteristics of the Newtonian Emission Theory; it is not, however, open to the objections to which that theory was liable, as the things emitted are Faraday tubes, having definite positions at right angles to the direction of propagation of the light. With such a structure the light can be polarized, while this could not happen if the things emitted were small symmetrical particles as on the Newtonian Theory.

28.] Before proceeding to interpret the production of a current by a galvanic cell in terms of Faraday tubes it is necessary to consider a little more in detail the process by which these tubes contract when they enter a conductor.

29.] When a Faraday tube is not closed its ends are places where electrification exists, and therefore are always situated on matter. Now the laws of Electrolysis show that the number of Faraday tubes which can fall on an atom is limited; thus only one can fall on an atom of a monad element, two on that of a dyad, and so on. The atoms in the molecule of a compound which is chemically saturated are already connected by the appropriate number of tubes, so that no more tubes can fall on such atoms. Thus on this view the ends of a tube of finite length are on free atoms as distinct from molecules, the atoms in the molecule being connected by short tubes whose lengths are of the order of molecular distances. Thus, on this view, the existence of free electricity, whether on a metal, an electrolyte, or a gas, always requires the existence of free atoms. The production of electrification must be accompanied by chemical dissociation, the disappearance of electrification by chemical combination; in short, on this view, changes in electrification are always accompanied by chemical changes. This was long thought to be a peculiarity attaching to the passage of electricity through electrolytes, but there is strong evidence to show that it is also true when electricity passes through gases. Reasons for this conclusion will be given in Chap. II, it will be sufficient here to mention one or two of the most striking instances, the details of which will be found in that chapter.

Perrot found that when the electric discharge passed through steam, oxygen came off in excess at the positive and hydrogen at the negative electrode, and that the excesses of oxygen at the positive and of hydrogen at the negative electrode were the same as the quantities of these gases set free in a water voltameter placed in series with the discharge through the steam. Grove found that when the discharge passed between a point and a silver plate through a mixture of hydrogen and oxygen, the plate was oxidised when it was the positive electrode, not when it was the negative. If the plate was oxidised to begin with, it was reduced by the hydrogen when it was the negative electrode, not when it was the positive. These and the other results mentioned in Chap. II seem to point unmistakably to the conclusion that the passage of electricity through gases is necessarily attended by chemical decomposition.

30.] Although the evidence that the same is true when electricity passes through metals is not so direct, it must be borne in mind that here, from the nature of the case, such evidence is much more difficult to obtain; there are, however, reasons for believing that the passage of electricity through metals is accomplished by much the same means as through gases or electrolytes. We shall return in Art. 34 to these reasons after considering the behaviour of the Faraday tubes when electricity is passing through an electrolyte, liquid or gaseous.

31.] To fix our ideas, let us take the case of a condenser discharging through the gas between its plates. Let us consider a Faraday tube which before discharge stretched from an atom O (Fig. 14) on the positive plate to another atom P on the negative

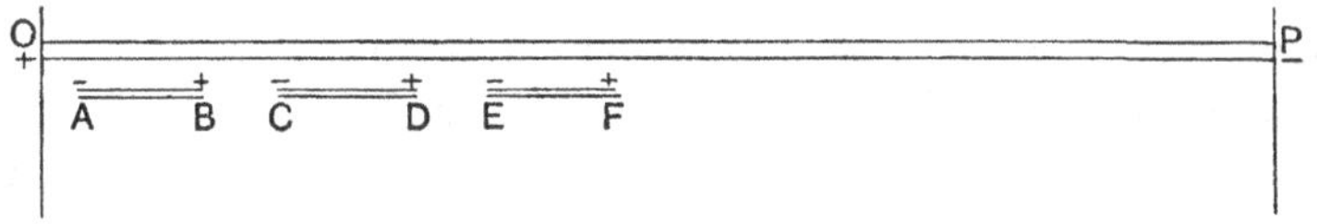

Fig. 14.

one. The molecules AB, CD, EF of the intervening gas will be polarized by induction, and the Faraday tubes which connect the atoms in these molecules will point in the opposite direction to the long tube OP. The tube in the molecule AB will lengthen and bend towards the tube OP (which is supposed to pass near to AB) since these are of opposite signs, until when the field is sufficiently strong the tube in the molecule AB runs up

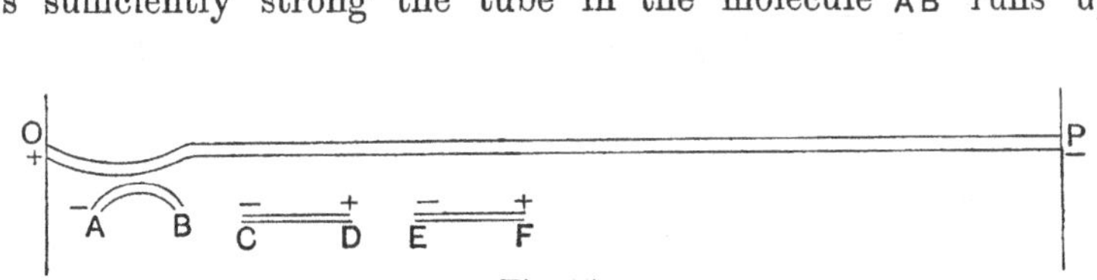

Fig. 15.

into the long tube OP as in Fig. 15. The long tube then breaks up into two tubes OA and BP as in Fig. 16, and the tube OA shortens to molecular dimensions. The result of these operations is that the tube OP has contracted to the tube BP, and the atoms O and A have formed a molecule. The process is then continued, until the tube OP has contracted into a tube of molecular dimensions at P. The above explanation is only intended to represent the general nature of the processes by which the Faraday tubes shorten; we must modify it a little in order to explain the very great velocity of the discharge along the positive column (see Chap. II, Art. 108). If the tubes shortened in the preceding manner, we see that the velocity of the

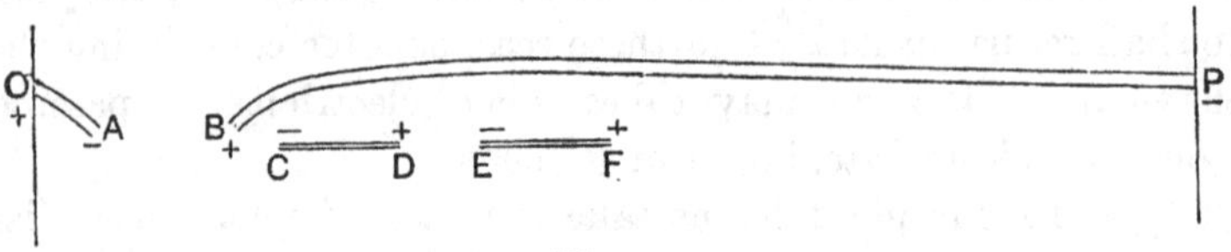

Fig. 16.

ends of the tube would only be comparable with the velocity of translation of the molecules of the gas, but the experiments alluded to above show that it is enormously greater than this. A very slight modification of the above process will, however, while keeping the essential features of the discharge the same, give a much greater velocity of discharge. Instead of supposing that the tube OP jumps from one molecule to the next, we may suppose that, under the induction in the field, several of the mole-

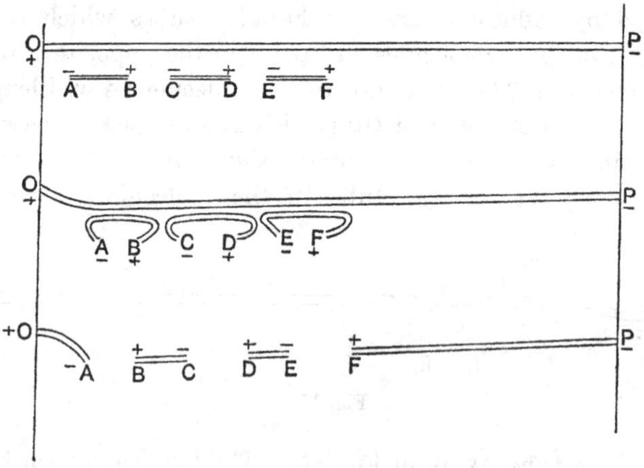

Fig. 17.

cules, say AB, CD, EF, form a chain, and that the tubes in these molecules instead of being successively affected by the long tube and by each other are simultaneously affected, so that the tube OP instead of merely jumping from one molecule to the next, moves as in Fig. 17 from one end of the chain AB, CD, EF to the other. In this case the long tube would shorten by the length of the chain in the same time as on the previous hypothesis it shortened by the distance between two molecules, so that on this view the velocity of discharge would be greater than that on the previous view in the proportion of the length of a chain to the distance between two molecules. We shall see in Chap. II that there is considerable evidence that in the electric field chains of molecules are formed having a structure much more complex than that of the molecules recognized in the ordinary Kinetic Theory of Gases.

32.] We can easily express the resistance of a conductor in terms of the time the Faraday tubes take to disappear (i.e. to contract to molecular dimensions). Let us for the sake of clearness take the case of a conducting wire, along which E is the electromotive intensity at any point, while K is the specific inductive capacity of the material of which the wire is made. Then the number of Faraday tubes passing through unit area of the cross-section of the wire is equal to

$$\frac{K}{4\pi}E.$$

Let T be the average life of a tube in the conductor, then the number of tubes which disappear from unit area in unit time is $KE/4\pi T$; and since the current c across unit area is equal to the number of tubes which disappear from unit area in unit time, we have

$$c=\frac{KE}{4\pi T}.$$

If σ be the specific resistance of the conductor measured in electromagnetic units

$$E=\sigma c,$$

hence

$$\sigma=\frac{4\pi T}{K},$$

or

$$T=\frac{K\sigma}{4\pi}.$$

Hence T has the same value as the quantity denoted by the same symbol in Maxwell's *Electricity and Magnetism* (Art. 325). It is often called the time of relaxation of the medium.

If $\{K\}$ be the value of K in electrostatic units,

then since $$\{K\} = \frac{K}{9 \times 10^{20}},$$

we have $$T = \frac{\{K\}}{4\pi} \frac{\sigma}{9 \times 10^{20}}.$$

The approximate values of $T/\{K\}$ for a few substances are given in the following table:—

	$T/\{K\}$
Silver	$1{\cdot}5 \times 10^{-19}$
Lead	$1{\cdot}8 \times 10^{-18}$
Mercury	$8{\cdot}7 \times 10^{-18}$
Water with 8·3 per cent of H_2SO_4	$3{\cdot}1 \times 10^{-13}$
Glass at 200° C	2×10^{-6}

Since the values of $\{K\}$ have not been determined for substances conducting anything like so well as those in the preceding list we cannot determine the value of T. Cohn and Arons have found however that the specific inductive capacity of distilled water is about 76. Cohn and Arons (*Wied. Ann.* 33, p. 13, 1888), and Cohn (*Berl. Ber.* p. 1037, 1891) found that the specific inductive capacity of a weak solution differs very little from that of the solvent, though the difference in the specific resistance is very great. If we suppose that the K for water mixed with sulphuric acid is the same as the K for water, we should find T for this electrolyte about 2×10^{-11}, which is about ten thousand times as long as the time of vibration of sodium light; hence this electrolyte when exposed to electrical vibration of this period will behave as if T were infinite or as if it were an insulator, and so will be transparent to electrical vibrations as rapid as those of light. We see too that if $\{K\}$ for the metals were as great as $\{K\}$ for distilled water, the values of T for these substances would not greatly exceed the time of vibrations of the rays in the visible spectrum: this result explains Maxwell's observation, that the opacity of thin metallic films is much less than the value calculated on the electromagnetic theory, on the assumption that the conductivity of the metals for the very rapidly alternating currents which constitute light is as great as that for steady currents.

Galvanic Cell.

33.] The production of a current by a cell is the reverse process to the decomposition of an electrolyte by a current; in the latter case the chemical processes make a long Faraday tube shrink to molecular dimensions, in the former they produce a long tube from short molecular tubes. Let A and B (Fig. 18) represent two metal plates immersed in an acid which combines chemically with A. Let a be a positive atom in the plate A connected by a

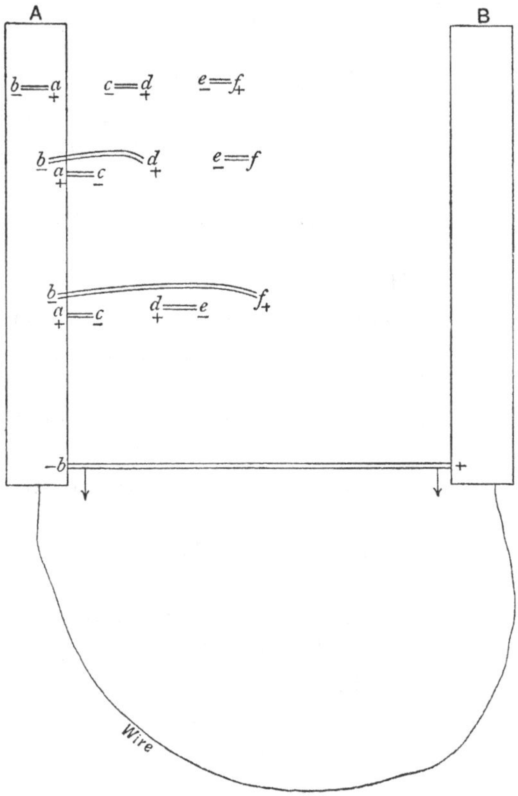

Fig. 18.

Faraday tube with a negative atom b, then if a enters into chemical combination with a molecule $c\,d$ of the acid, after the combination a and c will be connected by a Faraday tube, as will also b and d: it will be seen from the second line in the figure that the length of the tube bd has been increased by the chemical action. If now d enters into combination with another molecule $e\,f$, the result of this will be still further to increase the length of the tube, and this length will increase as the chemical combination progresses through the acid. In this way a long tube is produced, starting from the metal at which the chemical change occurs. This tube will rush to the wire connecting the plates, there shrink to molecular dimensions, and produce a current through the wire.

34.] The connection between electric conduction and chemical change is much more evident in the cases of liquid electrolytes and gases than it is in that of metals. There does not seem, however, to be sufficient difference between the *laws* of conduction through metals and electrolytes to make it necessary to seek an entirely different explanation for metallic conduction. The chief points in which metallic conduction differs from electrolytic are:—

1. The much greater ease with which electricity passes through metals than through electrolytes.

2. The difference of the effects of changes of temperature on the conductivity in the two cases. An increase of temperature generally diminishes the conductivity of a metal, while it increases that of an electrolyte.

3. The appearance of the products of chemical decomposition at the electrodes when electricity passes through an electrolyte, and the existence of polarization, while neither of these effects has been observed in metallic conduction.

With regard to the first of these differences, we may remark that though the conductivities of the best conducting metals are enormously greater than those of electrolytes, there does not seem to be any abrupt change in the values of the conductivities when we pass from cases where the conduction is manifestly electrolytic, as in fused lead or sodium chlorides, to cases where it is not recognised as being of this nature, as in tellurium or carbon. The following table, which contains the relative conductivities of a few typical substances, is sufficient to show this:—

Silver	63.
Mercury	1.
Gas Carbon	1×10^{-2}.
Tellurium	4×10^{-4}.
Fused Lead Chloride	2×10^{-4}.
Fused Sodium Chloride	$8{\cdot}6 \times 10^{-5}$.

With regard to the second difference between metallic and electrolytic conduction, viz. the effect of temperature on the conductivity, though it is true that in most cases the effect of an increase in temperature is to diminish the conductivity in one case and increase it in the other, this is a rule which is by no means without exceptions. There are cases in which, though the conduction is not recognised as being electrolytic, the conductivity increases as the temperature increases. Carbon is a striking instance of this, and Feussner* has lately prepared alloys of manganese, copper and nickel whose conductivities show the same peculiarity. On the other hand, Sack (*Wied. Ann.* 43, p. 212, 1891) has lately shown that above 95°C the conductivity of a ·5 per cent. solution of sulphate of copper decreases as the temperature increases, and in this respect resembles the conductivity of metals. These exceptions are sufficient to show that increase of conductivity with temperature is not a sufficient test to separate electrolytic from metallic conduction.

With regard to the third and most important point—the appearance of the products of chemical decomposition at the electrodes—it is evident that we could not expect to get any evidence of this in the case of the elementary metals. The case of alloys looks more hopeful. Roberts-Austen, however, who examined several alloys through which a powerful electric current had been passed, could not detect any difference in the composition of the alloy round the two electrodes. This result does not however seem conclusive against the conduction being electrolytic, for some alloys are little more than mixtures, while others behave as if they were solutions of one metal in another. In neither of these cases could we expect to find any separation of the constituents produced by the passage of the current; we could only expect to find this effect when the connection between the constituents was of such a nature that the whole alloy could be regarded as a chemical compound, in the molecule of which one metal could be regarded as the positive, the other as the negative element. The alloys investigated by Roberts-Austen do not seem to have been of this character.

One important respect in which metallic resembles electrolytic conduction is the way in which electrolytes and metals behave to the electrical vibrations which constitute light: an electrolyte, though a conductor for steady currents, behaves like an insulator to the rapidly alternating luminous electrical currents, and, as

* *Zeitschrift f. Instrumentenkunde*, 9, p. 233, 1889.

Maxwell's experiments on the transparency of their metallic films show, metals show an analogous effect, for their resistance for the light vibrations is enormously greater than their resistance to steady currents.

The theory of Faraday tubes which we have been considering is, as far as we have taken it, geometrical rather than dynamical; we have not attempted any theory of the constitution of these tubes, though the analogies which exist between their properties and those of tubes of vortex motion irresistibly suggest that we should look to a rotatory motion in the ether for their explanation.

Taking however these tubes for granted, they afford, I think, a convenient means of getting a vivid picture of the processes occurring in the electromagnetic field, and are especially suitable for expressing the relations which exist between chemical change and electrical action.

CHAPTER II.

THE PASSAGE OF ELECTRICITY THROUGH GASES.

35.] THE importance which Maxwell attached to the study of the phenomena attending the passage of electricity through gases, as well as the fact that there is no summary in English text books of the very extensive literature on this subject, lead me to think that a short account of recent researches on this kind of electric discharge may not be out of place in this volume.

Can the **Molecule** *of a Gas be charged with Electricity?*

36.] The fundamental question as to whether a body if charged to a low potential and surrounded by dust-free air at a low temperature will lose any of its charge, and the very closely connected one as to whether it is possible to communicate a charge of electricity to air *in this condition*, have occasioned considerable divergence of opinion among physicists.

Coulomb (*Mémoires de l'Académie des Sciences*, 1785, p. 612), who investigated the loss of electricity from a charged body suspended by insulating strings, thought that after allowing for the leakage along the supports there was a balance over, which he accounted for by a convective discharge through the air; he supposed that the particles of air when they came in contact with a charged body received a charge of electricity of the same sign as that on the body, and that they were then repelled by it. On this view the molecules of air, just like small pieces of metal, can be charged with electricity.

This theory of the loss of electricity from charged bodies has however not been confirmed by subsequent experiments, as Warburg (*Pogg. Ann.* 145, p. 578, 1872) and Nahrwold (*Wied. Ann.* 31, p. 448, 1887) have shown that the loss can be accounted for by the presence of dust in the air surrounding the bodies; and that it is the particles of dust striking against the bodies which carry off their electricity, and not the molecules of air.

This dust may either be present in the air originally, or it may consist of particles of metal given off from the charged conductors themselves, for, as Lenard and Wolf (*Wied. Ann.* 37, p. 443, 1889) have shown, metals either free from electrification or charged with negative electricity give off metallic dust when exposed to ultra-violet light. When the metals are positively electrified no dust seems to be given off.

The experiments of the physicists above mentioned point to the conclusion that the molecules of a gas at ordinary temperatures cannot receive a charge of electricity.

This view receives strong support from the results of Blake's experiments (*Wied. Ann.* 19, p. 518, 1883), which have been confirmed by Sohncke (*Wied. Ann.* 34, p. 925, 1888), which show that not only is there no electricity produced by the evaporation of an unelectrified liquid, but that the vapour arising from an electrified liquid is not electrified. If the molecules of a vapour were capable of receiving a charge of electricity under any circumstances we should expect them to do so in this case. This experiment is a striking example of the way in which important researches may be overlooked, for, as the following extract from Priestley's *History of Electricity*, p. 204, shows, Blake's experiment was made and the same result obtained more than one hundred years ago. 'Mr. Kinnersley of Philadelphia, in a letter dated March 1761, informs his friend and correspondent Dr. Franklin, then in England, that he could not electrify anything by means of steam from electrified boiling water; from whence he concluded, that, contrary to what had been supposed by himself and his friend, steam was so far from rising electrified that it left its share of common electricity behind.'

There does not seem to be any evidence that an electrified body can lose any of its charge by radiation through space without convection of electricity by charged particles.

Hot Gases.

37.] It is only at moderate temperatures that a conductor charged to a low potential retains its charge when surrounded by a gas, for Becquerel (*Annales de Chimie et de Physique* [3]

39, p. 355, 1853) found that air at a white heat would allow electricity to pass through it even though the potential difference was only a few volts. This result has been confirmed by Blondlot (*Comptes Rendus*, 104, p. 283, 1887), who found that air at a bright red heat was unable to insulate under potential differences as low as 1/1000 of a volt. He found, too, that the conduction through the hot gas did not obey Ohm's law.

From some experiments of my own (*Phil. Mag.* [5] 29, pp. 358, 441, 1890) I have come to the conclusion that hot gases conduct electricity with very different degrees of facility. Gases such as air, nitrogen, or hydrogen which do not experience any chemical change when heated conduct electricity only to a very small extent when hot, and in this case the conduction, as Blondlot supposed, appears to be convective. Gases, however, which dissociate at high temperatures, that is gases such as iodine, hydriodic acid gas, &c., whose molecules split up into atoms, conduct with very much greater facility, and the conduction does not exhibit that dependence on the material of which the electrodes are made which is found when the electricity is transmitted by convection.

A large number of gases were examined, and in every case where the hot gas possessed any considerable conductivity, I was able to detect by purely chemical means that chemical decomposition had been produced by the heat. In this connection it is necessary to distinguish between two classes of dissociation. The first kind is when the molecule is split up into atoms, as in iodine, hydriodic acid gas, hydrochloric acid gas (when the chlorine, though not the hydrogen, remains partly dissociated), and so on. In all cases when dissociation of this kind exists, the gas is a good conductor when hot. The second kind of dissociation consists in the splitting up of the molecules of the gas into simpler molecules but not into atoms. This kind of dissociation occurs when a molecule of ammonia splits up into molecules of nitrogen and hydrogen, or when a molecule of steam splits up into molecules of hydrogen and oxygen. In this case the gases only conduct on the very much lower scale of the non-dissociable gases.

The first of the following lists contains those gases which only conduct badly when heated, the second those which conduct comparatively well: chemical analysis showed that all the gases in the second list were decomposed when they were hot enough to conduct electricity:—

(1) Air, Nitrogen, Carbonic Acid, Steam, Ammonia, Sulphuric Acid gas, Nitric Acid gas, Sulphur (in an atmosphere of nitrogen), Sulphuretted Hydrogen (in an atmosphere of nitrogen).

(2) Iodine, Bromine, Chlorine, Hydriodic Acid gas, Hydrobromic Acid gas, Hydrochloric Acid gas, Potassium Iodide, Sal-Ammoniac, Sodium Chloride, Potassium Chloride.

The conductivities of the two classes of gases differ so greatly, both in amount and in the laws they obey, that the mechanism by which the discharge is effected is probably different in the two cases.

These experiments seem to show that when electricity passes through a gas otherwise than by convection, free atoms, or something chemically equivalent to them, must be present. It should be noticed that on this view the molecules even of a hot gas do not get charged, it is the *atoms* and not the molecules which are instrumental in carrying the discharge.

I also examined the conductivities of several metallic vapours, including those of Sodium, Potassium, Thallium, Cadmium, Bismuth, Lead, Aluminium, Magnesium, Tin, Zinc, Silver, and Mercury. Of these the vapours of Tin, Mercury, and Thallium hardly seemed to conduct at all, the vapours of the other metals conducted well, their conductivities being comparable with those of the dissociable gases.

The small amount of conductivity which hot gases, which are not decomposed by heat, possess, seems to be due to a convective discharge carried perhaps by dust produced by the decomposition of the electrodes: in some cases perhaps the electricity may be carried by atoms produced by the chemical action of the electrodes on the adjacent gas.

The temperature of the electrodes seems to exert great influence upon the passage of the electricity through the gas into which the electrodes dip. In the experiments described above I found it impossible to get electricity to pass through the gas, however hot it might be, unless the electrodes were hot enough to glow. A current passing through a hot gas was immediately stopped by placing a large piece of cold platinum foil between the electrodes —though a strong up-current of the hot gas was maintained to prevent the gas getting chilled by the cold foil. As soon as the foil began to glow, the passage of the electricity through the gas was re-established.

This is one among the many instances we shall meet with in this chapter of the difficulty which electricity has in passing from a gas to a cold metal.

Electric Properties of Flames.

38.] The case in which the passage of electricity through hot gases has been most studied is that of flames; here the conditions are far from simple, and the results that have been obtained are too numerous and intricate for us to do more than mention their main features. A full account of the experiments which have been made on this subject will be found in Wiedemann's *Lehre von der Elektricität*, vol. 4, B *.

A flame such as the oxy-hydrogen flame conducts electricity, the hotter parts conducting better than the colder: the conductivity of the flame is improved by putting volatile salts into it, and the increase in the conductivity is greater when the salts are placed near the negative electrode than when they are placed near the positive †.

The conduction through the flame exhibits polar properties, for if the electrodes are of different sizes the flame conducts better when the larger electrode is negative than when it is positive.

If wires made of different metals are connected together and dipped into the flame, there will be an electromotive force round the circuit formed by the flame and the wire; the flame apparently behaving in much the same way as the acid in a one-fluid battery; the electromotive force in some cases amounts to between three and four volts.

A current can also be obtained through a bent piece of wire if the ends of the wire are placed in different parts of the flame.

Escape of Electricity from a Conductor at Low Potential surrounded by Cold Gas.

39.] Though it seems to be a well-established fact that a conductor at a low potential, surrounded by cold air, may retain its charge for an indefinitely long time, recent researches have

* See also Giese, *Wied. Ann.* 38, p. 403, 1889.

† For an investigation on the effect of putting volatile salts in flames published subsequently to Wiedemann's work, see Arrhenius (*Wied. Ann.* 42, p. 18, 1891).

shown that when the conductor is exposed to certain influences leakage of the electricity may ensue.

One of the most striking of these influences is that of ultra-violet light. The effect of ultra-violet light on the electric discharge seems first to have been noticed by Hertz (*Wied. Ann.* 31, p. 983, 1887), who found that the disruptive discharge between two conductors is facilitated by exposing the air space, across which the discharge takes place, to the influence of ultra-violet light.

E. Wiedemann and Ebert (*Wied. Ann.* 33, p. 241, 1888) subsequently proved that the seat of this action is at the cathode; they showed that the light produces no effect when the cathode is shielded from its influence, however brightly the rest of the line of discharge may be illuminated.

They found that if the cathode is surrounded by air the effect of the ultra-violet light is greatest when the pressure is about 300 mm. of mercury: when the pressure is so low that the negative rays (see Art. 108) are visible, the effect of the ultra-violet light is not at all well marked.

They found also that the magnitude of the effects depends upon the gas surrounding the cathode; they tried the effect of immersing the cathode in carbonic acid, hydrogen and air, and found that for these three gases the effect is greatest in carbonic acid, least in air. In carbonic acid the effect is not confined to ultra-violet light, as the luminous rays when they fall on a cathode also facilitate the discharge.

Great light was thrown on the nature of this effect by an investigation made by Lenard and Wolf (*Wied. Ann.* 37, p. 443, 1889), in which it was proved that when ultra-violet light falls on a negatively electrified platinum surface, a steam jet in the neighbourhood of the surface shows by its change of colour that the steam in it has been condensed. This condensation always occurs when the negatively electrified surface on which the light falls is metallic, or that of a phosphorescent liquid, such as a solution of fuchsin or methyl violet. They found also that some, but much smaller, effects are produced when the surfaces are not electrified, but no effect at all can be detected when they are charged with positive electricity.

They attributed this condensation of the jet to dust emitted from the illuminated surface, the dust, in accordance with

Aitken's experiments (*Trans. Roy. Soc. Edinburgh*, 30, p. 337, 1881), producing condensation by forming nuclei round which the water drops condense.

The indications of a steam jet are not however free from ambiguity, as R. v. Helmholtz (*Wied. Ann.* 32, p. 1, 1887) has shown that condensation occurs in the jet when chemical reactions are going on in its neighbourhood, even though no dust is present. There is thus some doubt as to whether the condensation observed by Lenard and Wolf is due to disintegration of the illuminated surface or to chemical action taking place close to it. Taking however the interpretation which these observers givė to their own experiments, the effects observed by Hertz, E. Wiedemann and Ebert can easily be explained as due to the carrying of the discharge by particles disintegrated from the metallic surface by the action of the ultra-violet light.

40.] Closely connected with this effect is the discovery, made almost simultaneously by Hallwachs (*Phil. Mag.* [5], 26, p. 78, 1888) and Righi (*Phil. Mag.* [5], 25, p. 314, 1888), that a metallic surface, especially if the metal is zinc and freshly polished, becomes positively electrified when exposed to the action of ultra-violet light.

Lenard and Wolf's experiments suggest that this is probably due to the disintegration of the surface by the light, the metallic dust or vapour carrying off the negative electricity and leaving the positive behind.

Stoletow (*Phil. Mag.* [5], 30, p. 436, 1890) showed that a kind of voltaic battery might be made by taking two plates of different metals in metallic connection and exposing one of them to the action of ultra-violet light; the plate so exposed becoming the negative electrode of the battery. When ultra-violet light acts in this way, Stoletow found that, as we should expect, the light is powerfully absorbed by the surface on which it falls.

Probably another example of the same effect is the positive electrification observed by Crookes (*Phil. Trans.*, Part II. 1879, p. 647) on a plate placed inside an exhausted tube in full view of the negative electrode. We shall see, when we consider the discharges in such tubes, that something proceeds from the cathode which resembles ultra-violet light in its power of producing phosphorescence in bodies on which it falls. Crookes' experiment, which was made at Maxwell's suggestion, shows that the resemblance of the cathode discharge to ultra-violet light extends to its power of producing a positive charge on a metal plate exposed to its influence.

41.] A striking instance of the facility with which a negatively electrified surface disintegrates, whilst a positively electrified one remains intact, is afforded by the well-known 'spluttering' of the negative electrode in a vacuum tube. In such a tube the glass round the negative electrode is darkened by the deposition of a thin film of metal torn from the adjacent cathode; the glass round the positive electrode is, on the other hand, quite free from any such deposit. The amount of the disintegration of the cathode depends greatly upon the metal of which it is made. Crookes (*Proc. Roy. Soc.* 50, p. 88, 1891) has given the following table, which expresses the relative loss in weight in equal times of cathodes of the same size exposed to similar electrical conditions:—

Palladium	108·00
Gold	100·
Silver	82·68
Lead	75·04
Tin	56·96
Brass	51·58
Platinum	44·00
Copper	40·24
Cadmium	31·99
Nickel	10·99
Indium	10·49
Iron	5·50

The loss in weight of magnesium and aluminium electrodes was too small to be detected. In the same paper Crookes also describes an experiment which seems to show that the 'spluttering' at the negative electrode exists in water even when surrounded by air at atmospheric pressure.

42.] Since a metal surface when exposed to the action of sunlight emits negative electricity and retains positive, we should expect positively electrified bodies when exposed to light to behave differently from negatively electrified ones. This has been found to be the case. The first observations on this subject seem to have been made by Hoor (*Repertorium d. Physik.* 25, p. 105, 1889), who found that freshly prepared surfaces of zinc, copper, and brass quickly lost a negative charge when exposed to the action of ultra-violet light, while the same surfaces retained a positive charge.

The subject was afterwards taken up by Elster and Geitel (*Wied. Ann.* 38, pp. 40, 497, 1889; 41, p. 161, 1890; 42, p. 564, 1891), who verified Hoor's result for zinc, but could not detect any loss of negative electricity from freshly prepared surfaces of brass or copper. They also established the interesting fact that the effect is most marked in the case of the electro-positive metals, zinc or amalgamated zinc, aluminium, and magnesium. For the still more electro-positive metals, potassium and sodium, or rather for their amalgams, since the pure metals are difficult to work with on account of the tarnishing of their surfaces, they found that the effect is so strong that it can readily be observed even when the amalgams are enclosed in glass tubes, though glass, as is well known, absorbs most of the ultra-violet rays. When they succeeded subsequently in working with surfaces of potassium and sodium instead of their amalgams, they found that these substances are sensitive not merely to the ultra-violet rays but even to those emitted by an ordinary petroleum lamp (*Wied. Ann.* 43, p. 225, 1891).

Thus when the surface of some metals is negatively electrified and exposed to the action of light, and especially of ultra-violet light, we have an exception to the general rule that a charged body surrounded by cold air can retain its charge, for an indefinite time, provided the charge is not large enough to produce a spark. For as Elster and Geitel proved, the smallest negative charge rapidly disappears from the illuminated surface.

The order of sensitiveness of metals to this effect is given by Elster and Geitel as

Potassium,
Alloy of Sodium and Potassium,
Sodium,
Amalgams of Rubidium, Potassium, Sodium and Lithium,
Magnesium, Aluminium,
Zinc,
Tin.

It is interesting to note that this is roughly the order of the metals in Volta's contact electricity series, as each metal is positive to the one after it. Elster and Geitel found that the effect is too small to be measured in Cadmium, Lead, Copper, Iron, Platinum, Mercury, and Carbon. They also found no clear indications of it with water. It is well marked, however, in phosphorescent substances such as Balmain's luminous paint (sulphide of calcium), and Elster and Geitel (*Wied. Ann.* 44, p. 722, 1891) have quite recently shown that it is exhibited by Fluor Spar and other phosphorescent minerals.

Another way of observing this effect is to place the illuminated body without a charge and in connection with the earth in the neighbourhood of a charged body, when the latter will lose its charge if it is positively electrified, while it will not lose its charge if it is negatively electrified; the positive charge induces a negative one on the illuminated body, this negative electricity escapes, travels up to and neutralises the positive electricity which induced it. When the pressure of the gas surrounding the body is less than 1 mm., the escape of the negative electricity from the illuminated surface is considerably checked by placing it in a strong magnetic field (Elster and Geitel, *Wied. Ann.* 41, p. 166, 1890).

Discharge of Electricity caused by Glowing Bodies.

43.] Somewhat similar differences between the discharge of positive and negative electricity are observed when the charged body, instead of being illuminated, is raised to so high a temperature that it becomes luminous itself. Elster and Geitel (*Wied. Ann.* 38, p. 27, 1889) found that when a platinum wire is heated to a bright red heat in an atmosphere of air or oxygen at a low pressure, a cold metal plate in its neighbourhood discharges negative electricity with much greater ease than positive. If, on the other hand, a thin platinum wire or carbon-filament is heated to incandescence in an atmosphere of hydrogen at a low pressure, the cold plate discharges positive electricity more easily than negative. Guthrie, who (*Phil. Mag.* [4] 46, p. 257, 1873) was the first to call attention to phenomena of this kind, observed that an iron sphere in air when white hot cannot retain a charge either of positive or of negative electricity, and that as it cools it acquires the power of retaining a negative charge before it can retain a positive one. If the sphere is connected to the earth and held near a charged body, then, when the sphere is white hot, the body soon loses its charge whether this be positive or negative; when the sphere is somewhat colder, the body is discharged if negatively electrified but not if positively.

The converse problem of the production of electrification by a glowing wire has been studied in great detail by Elster and Geitel, a summary of whose results is given in *Wied. Ann.* 37, p. 315, 1889. The conclusions they have come to are that when an insulated plate is placed near an incandescent platinum wire, the plate becomes positively electrified in air and oxygen, negatively electrified in hydrogen. It thus appears that incandescent wires discharge most easily the electricity of opposite sign to that which they produce on plates placed in their neighbourhood. If the incandescence is continued for a long time, then if the wire is thin and the pressure low, a plate in the neighbourhood of the wire receives a negative charge, whatever be the gas by which it is surrounded. Elster and Geitel seem to ascribe this to the action of gases driven out of the electrodes. Nahrwold, who also observed this effect (*Wied. Ann.* 35, 107, 1888), regards it as the normal one, and ascribes the positive electrification observed when the wire first begins to glow to the action of dust in the gas. It is noteworthy that hydrogen, which in Elster and Geitel's experiments behaved with platinum electrodes oppositely to the other gases, is the only gas in which, according to Nahrwold, a platinum wire does not disintegrate when heated. With carbon filaments, Elster and Geitel found that the neighbouring plate is always negatively electrified, but so much gas is given off from these filaments that the interpretation of these results is ambiguous.

Elster and Geitel have also observed that the ease with which electricity is produced in a plate near a glowing wire is diminished if the gas is hydrogen by placing the wire in a magnetic field, increased if it is oxygen or air.

44.] The investigations we have just described show clearly that metallic surfaces have in general a much greater tendency to attract a positive than a negative charge. Thus, for example, we have seen that when originally uncharged they become positively charged when exposed to the action of ultra-violet light, and if charged to begin with, then under the influence of the light they lose a negative charge much more rapidly than a positive one, indeed there seems no evidence to show that there is any loss of a small positive charge from this effect.

The phenomena depending on the action of ultra-violet light and of incandescent surfaces can be co-ordinated by the conception introduced by v. Helmholtz (*Erhaltung der Kraft, Wissenschaftliche Abhand.* vol. 1. p. 48), that bodies attract electricity with different degrees of intensity. This conception was shown by him to be able to explain electrification by friction, and the difference of potential produced by the contact of metals. Thus, for example, the difference of potential produced by the contact of zinc and copper is explained on this hypothesis by saying that the positive electricity is attracted more strongly by the zinc than it is by the copper.

Instead of considering the specific attraction of different bodies for electricity directly, it is equivalent in theory and generally more convenient in practice to regard the potential energy possessed by a body charged with electricity as consisting of two parts, (1) the part calculated by the ordinary rules of electrostatics, and (2) a part proportional to the charge and equal to σQ, where Q is the charge and σ a quantity which we shall call the 'Volta potential' of the body, and which varies from one substance to another.

To investigate the nature of the effects produced by the presence of this second term, let us consider the case of two parallel plates A and B made of different metals and connected electrically with each other.

Let Q be the charge on the plate A, $-Q$ that on the plate B, σ_A, σ_B the values of the co-efficient σ for the plates A and B respectively, then if C is the capacity of the condenser formed by the two plates, the potential energy of the system will be given by the equation

$$V = \tfrac{1}{2}\frac{Q^2}{C} + \sigma_A Q - \sigma_B Q.$$

The system will be in equilibrium when the potential energy is a minimum, i. e. when $dV/dQ = 0$, or

$$\frac{Q}{C} + \sigma_A - \sigma_B = 0.$$

Thus, by the contact of the metals the potential of the plate A is raised above that of B by $\sigma_B - \sigma_A$.

It is worthy of notice that on this view the medium separating the plates does not affect the value of the potential difference between them, however great the value of σ for this medium may be, provided that, as in the case of cold air, the medium is incapable of receiving a charge of electricity.

The idea of the possession by a charged body of a quantity of energy proportional to the first power of the charge is involved in the well-used phrase 'specific heat of electricity'; for if we regard electricity as having a specific heat which varies from one substance to another, a body charged with electricity will in conquence of this specific heat possess some energy proportional to the charge. The electromotive forces which occur in unequally heated bodies may be explained as due to the tendency of the electricity to adjust itself so that the potential energy is a minimum; if the quantity σ is a function of the temperature, the energy will not be a minimum when the body is devoid of electrification.

The existence of the term σQ in the expression for the energy of a charged body, since the electrification is on the surface, makes the energy per unit area of the surface depend upon whether the electrification is positive, negative, or zero. Now since the apparent surface tension of a liquid is equal to the energy per unit area of surface, it may be objected that if this view were true the surface tension of such liquids as are conductors ought to be changed by electrification, the change being in one direction when the electrification is positive and in the opposite when it is negative. A short calculation will show however that this change in the surface tension is so small that it might easily have escaped detection. We have seen that $\sigma_B - \sigma_A$ is the potential difference produced by the contact of two metals A and B, we know from observation that this difference, and therefore presumably σ_A and σ_B, is of the order of a volt, or in electromagnetic units 10^8. Now the greatest electrification which can exist on the surface without discharge when the metal is surrounded by air at the atmospheric pressure is such as to produce an electromotive intensity equal approximately to 10^2 in electrostatic measure; thus the greatest surface density is in electrostatic units about $10^2/4\pi$, or in electromagnetic units $10^{-8}/12\pi$. Hence σQ, the energy of the kind we are considering, will at the most be of the order $1/(12\pi)$ ergs per square centimetre.

This is so small compared with the energy due to the surface tension that it would require very careful observations to detect it.

45.] When a conductor, which does not disintegrate, is surrounded by air in its normal state, or by some other dielectric incapable of receiving a charge of electricity, the conductor cannot get charged, however much the σ for the conductor may differ from that for the dielectric; for the electricity of opposite sign to that which would be left on the conductor has no place to which it can go.

The case is however different when the conductor is exposed to the action of ultra-violet light, for then, as Lenard and Wolf's experiments prove, one or both of the following effects must take place: (1) disintegration of the conductor, (2) chemical changes in the gas in the neighbourhood of the conductor which put the gas in a state in which it can receive a charge of electricity. If either of these effects takes place it is possible for the conductor to be electrified, for the electricity of opposite sign to that left on the conductor may go to the disintegrated metal or the gas. The experiments hitherto made leave undecided the question which of these bodies serves as the refuge of the electricity discarded from the metal.

The researches of Hallwachs and Righi on electrification by ultra-violet light can be explained on either hypothesis, if we assume that σ_1, the value of σ for the metallic vapour or for the dissociated gas, is greater than σ_2, the value of σ for the solid metal. For when negative electricity $-Q$ escapes from the metal and positive electricity equal to $+Q$ remains behind, the diminution in the part of the potential energy due to the Volta potential is $\sigma_1 Q - \sigma_2 Q$ or $(\sigma_1 - \sigma_2)Q$. Thus, since σ_1 is by hypothesis greater than σ_2, the departure of the negative electricity from the metal will be accompanied by a diminution in the potential energy, and will therefore go on until the increase in the ordinary potential energy due to the new distribution of electricity is sufficient to balance the diminution in the part of the energy due to the Volta potential. The positive electrification of the plate produced by ultra-violet light can thus be accounted for.

Again, if the metal were initially positively electrified it would not be so likely to lose its charge as if it were initially charged with negative electricity, for the passage of positive electricity from the metal to its vapour or to the dissociated gas would involve an increase in the energy depending upon the Volta potential, and so would be much less likely to occur than an escape of negative electricity, which would produce a diminution in this energy. We can thus explain the observations of Elster and Geitel on the difference in the rates of escape of positive and negative electricity from illuminated surfaces. The causes of the electrification by incandescence observed by Elster and Geitel (l.c.) are more obscure. Thus if we take the case when a plate receives a positive charge in air owing to the presence of a neighbouring incandescent platinum wire, the most obvious interpretation would be that the incandescence produces electrical separation, the wire getting negatively and the adjacent gas positively electrified. This view is however open to the very serious objection that in the other cases of the electrification of a metal in contact with a gas the metal receives the positive charge and not the negative one, as it would have to do if the preceding explanation were correct.

The plate is exposed to the radiation from the incandescent wire and may perhaps under the influence of this radiation become a cathode, i. e. give out negative electricity and thus become positively electified, just as it would if, as in Hallwach's and Righi's experiments, it were exposed to the action of ultra-violet light, or as in Crookes' experiment (Art. 40) to the emanations from a negative electrode. It seems however difficult to explain the anomalous behaviour of hydrogen on this view, and Nahrwold's discovery of the absence of 'spluttering' in platinum wires heated to incandescence in an atmosphere of hydrogen seems to suggest that the charge on the plate may possibly arise in some such way as the following, even though the first effect of the incandescence is to produce a positive electrification over the wire and a negative one over the adjacent gas. When a metallic wire is heated, disintegration may take place in two ways, the metal may go off as vapour, or it may be torn off in solid lumps or dust. Now there seems to be no reason why σ for these lumps should differ from σ for the wire, for both the lump and the wire consist of the same substance in the same state of aggregation; but if the σ's were the same there would be no separation of electricity between the two. On the contrary, if the wire were charged with positive electricity, the lump, when it broke away, would carry positive electricity off with it. The case is however different when the metal goes off as vapour, or when it dissociates the gas in its neighbourhood: here the wire and the vapour or gas are in different states of aggregation, for which the values of σ are probably different, so that there may now be a separation of electricity, the wire getting the positive and the vapour or gas the negative.

In air there is such an abundant deposition of platinum on a glass tube surrounding an incandescent platinum wire that the latter in all probability gives off dust as well as either dissociating the surrounding gas or giving off platinum vapour ; while Nahrwold (*Wied. Ann.* 35, 107, 1888) has shown that the deposition of platinum is so small in hydrogen that very little can be given off as dust in this gas.

Let us now consider what will happen in air. When the platinum becomes incandescent there is a separation of electricity, the positive remaining on the wire, the negative going to the metallic vapour or dissociated gas. Since the wire has got a positive charge, any lumps that break away from it will be positively electrified. If the positive electricity given by these lumps to the plate, which in Elster and Geitel's experiments was held above the glowing wire, is greater than the negative charge given to it by such vapour or gas as may come in contact with it, the charge on the plate will be positive, as in Elster and Geitel's experiments. In hydrogen however, where the lumps are absent, there is nothing to neutralize the negative electricity on the metallic vapour or dissociated gas, so that the charge on the plate will, as Elster and Geitel found, be negative.

Spark Discharge.

Electric Strength of a Gas.

46.] In Art. 51 of the first volume of the *Electricity and Magnetism* Maxwell defines the *electric strength* of a gas as the greatest electromotive intensity it can sustain without discharge taking place. This definition suggests that the electric strength is a definite specific property of a gas, otherwise the introduction of this term would not be of much value. If discharge through a gas at a definite pressure and temperature always began when the electromotive intensity reached a certain value, then this value, which is what Maxwell calls the electric strength of the gas, would have a perfectly definite meaning. The term 'electric strength of the gas' would however be misleading if it were found to depend on such things, for example, as the materials of which the electrodes are made, the state of their surface, their shape, size, or distance apart, or on whether the electric field was uniform or variable either with regard to time or space. It has been found that the 'electric strength' does depend upon some, perhaps even upon all, of the preceding conditions.

47.] Righi (*Nuovo Cimento*, [2] 16, p. 97, 1876) made some experiments with electrodes of carbon, bismuth, lead, zinc, tin and copper, but found that the substance of which the electrodes are made has little effect on the electromotive intensity necessary for discharge. Mr. Peace, who made careful experiments in the Cavendish Laboratory on this point, could not detect any difference in the electromotive intensity required to spark across electrodes made of brass and those made of zinc. De la Rue and Hugo Müller (*Phil. Trans.* 169, Pt. 1. p. 93, 1878) came to the conclusion that sparks pass more easily between aluminium terminals than between terminals of other metals, but that with this exception the nature of the electrodes has no influence upon the spark length.

Jaumann has shown (*Wien. Berichte*, 97, p. 765, 1888) that the spark discharge is very much facilitated by making small but rapid changes in the potential of one of the electrodes.

48.] The reduction by Schuster (*Phil. Mag.* [5] 29, p. 182, 1890) of the experiments of Baille, Paschen, and Gaugain on the spark discharge shows that with spherical electrodes of different sizes (1 cm., ·5 cm., and ·25 cm. in radius respectively) the maximum electromotive intensity when the spark just passes through air at atmospheric pressure varies from 142 to 372, the maximum intensity for small spheres being greater than for large ones. Schuster sums up the conclusions he draws from these experiments as follows, l. c. p. 192:—

(1) 'For two similar systems of two equal spheres in which only the linear dimensions vary, the breaking-stress is greater the greater the curvature of the spheres.'

(2) 'If the distance between the spheres is increased, the breaking-stress at first diminishes.'

(3) 'There is a certain distance for which the breaking-stress is a minimum.'

We shall find too when we consider the relation between spark length and potential difference that the distance between the electrodes may have an enormous effect on the electromotive intensity required to produce discharge.

The 'electric strength' as defined by Maxwell seems to depend upon so many extraneous circumstances that there does not appear to be any reason for regarding it as an intrinsic property of the gas.

Connection between Spark Length and Potential Difference, when the Field is approximately uniform.

49.] This subject has been investigated by a large number of physicists. We have however only space to consider the most recent investigations on this subject. Baille (*Annales de Chimie et de Physique*, [5] 25, p. 486, 1882) has made an elaborate investigation of the potential difference required to produce in air at atmospheric pressure sparks of varying lengths, between planes, cylinders, and spheres of various diameters. The method he used was to charge the conductors between which the sparks passed by a Holtz machine, the potential between the electrodes being measured by an attracted disc electrometer provided with a guard ring: this method is practically the same as that employed by Lord Kelvin (*Reprint of Papers on Electrostatics and Magnetism*, p. 247), who in 1860 made the first measurements in absolute units of the electromotive intensity required to produce a spark.

For very short sparks between two planes Baille (l. c., p. 515) found the results given in the following table:—

Potential Difference and Spark Length; (temperature 15° to 20° C, pressure 760 mm.)

Spark Length in Centimetres.	Potential Difference in Electrostatic Units.	Surface Density in Electrostatic Units.
·0015	1·42	75·4
·0020	1·62	64·5
·0025	1·90	60·5
·0050	2·51	39·9
·0075	2·81	29·8
·0100	3·15	25·1
·0125	3·48	22·1
·0150	3·80	20·1

In another series of experiments where the sparks were slightly longer, Baille, p. 515, found the following results:—

Potential Difference and Spark Length.

Spark Length.	Potential Difference.	Surface Density.	Spark Length.	Potential Difference.	Surface Density.
·01	3·17	25·2	·08	12·38	12·3
·02	4·51	17·9	·09	13·44	11·9
·03	6·22	16·5	·10	14·67	11·7
·04	7·32	14·6	·11	15·75	11·4
·05	8·71	13·8	·12	16·84	11·1
·06	9·84	13·2	·13	17·94	11·0
·07	11·20	12·7	·14	19·00	10·8
			·15	20·16	10·7

For spark lengths between ·025 cm. and ·5 cm. the following results were obtained, p. 516, in a different series of experiments :—

Potential Difference and Spark Length.

Spark Length.	Potential Difference.	Surface Density.	Spark Length.	Potential Difference.	Surface Density.
·025	5·94	18·86	·275	32·69	9·46
·050	8·68	13·76	·300	35·35	9·37
·075	11·87	12·57	·325	37·83	9·25
·100	14·79	11·76	·350	39·95	9·08
·125	17·45	11·06	·375	42·17	8·94
·150	20·29	10·76	·400	44·74	8·90
·175	22·94	10·43	·425	47·30	8·86
·200	25·51	10·15	·450	49.70	8·79
·225	28·17	9·96	·475	52·18	8·75
·250	30·47	9·70	·500	54·48	8·67

For longer sparks Baille, l. c., p. 517, got the numbers given in the two following Tables, which represent the results of different sets of experiments :—

Potential Difference and Spark Length.

TABLE (I).

Spark Length.	Potential Difference.	Surface Density.	Spark Length.	Potential Difference.	Surface Density.
·40	44·80	8·90	·60	63·82	8·47
·45	49·63	8·78	·65	68·75	8·42
·50	54·36	8·65	·70	74·09	8·42
·55	59·09	8·55	·75	79·02	8·39

TABLE (II).

Spark Length.	Potential Difference.	Surface Density.	Spark Length.	Potential Difference.	Surface Density.
·70	73·48	8·84	·90	94·72	8·38
·75	80·13	3·55	·95	100·16	8·38
·80	84·80	8.40	1.00	105·50	8·39
·85	89·89	8·42			

50.] We may compare with these results those obtained by Liebig (*Phil. Mag.* [5], 24, p. 106, 1887), who used a similar method, but whose electrodes were segments of spheres 9·76 cm. in radius. Liebig's results are as follows:—

Potential Difference and Spark Length.

Spark Length in centimetres.	Potential Difference.	Electromotive Intensity.	Spark Length.	Potential Difference.	Electromotive Intensity.
·0066	2·630	398·5	·2398	30·622	127·7
·0105	3·357	319·7	·2800	35·196	125·7
·0143	4·017	280·9	·3245	39·816	122·7
·0194	4·573	235·7	·3920	47·001	119·9
·0245	5·057	206·4	·4715	55·165	117·0
·0348	7·190	206·6	·5588	63·703	114·0
·0438	8·863	195·5	·6226	69·980	112·4
·0604	10·866	179·9	·7405	82·195	111·0
·0841	13·548	161·1	·8830	95·540	108·2
·0903	13·816	153·0	·9576	102·463	107·0
·1000	15·000	150·0	1·0672	110·775	103·8
.1520	20·946	137·8	1·1440	117·489	102·7
·1860	24·775	133·2			

The potential difference and electromotive intensity are measured in electrostatic units.

Liebig's results for hydrogen, coal gas and carbonic acid as well as air are exhibited graphically in Fig. 19, where the nearly straight curve represents the relation between potential difference and spark length, and the other the relation between electromotive intensity and spark length. The abscissae are the spark lengths, the ordinates, the potential difference or electromotive intensity. It will be seen that Liebig's values for the potential difference required to produce a spark of given length are about 8 per cent. higher than Baille's. It also appears from any of the preceding tables that the electromotive intensity required to spark across a layer of air varies very greatly with the thickness of the layer. Thus from Baille's result we see that the electromotive intensity required to spark across a layer ·0015 cm. thick is about nine times that required to spark across a layer

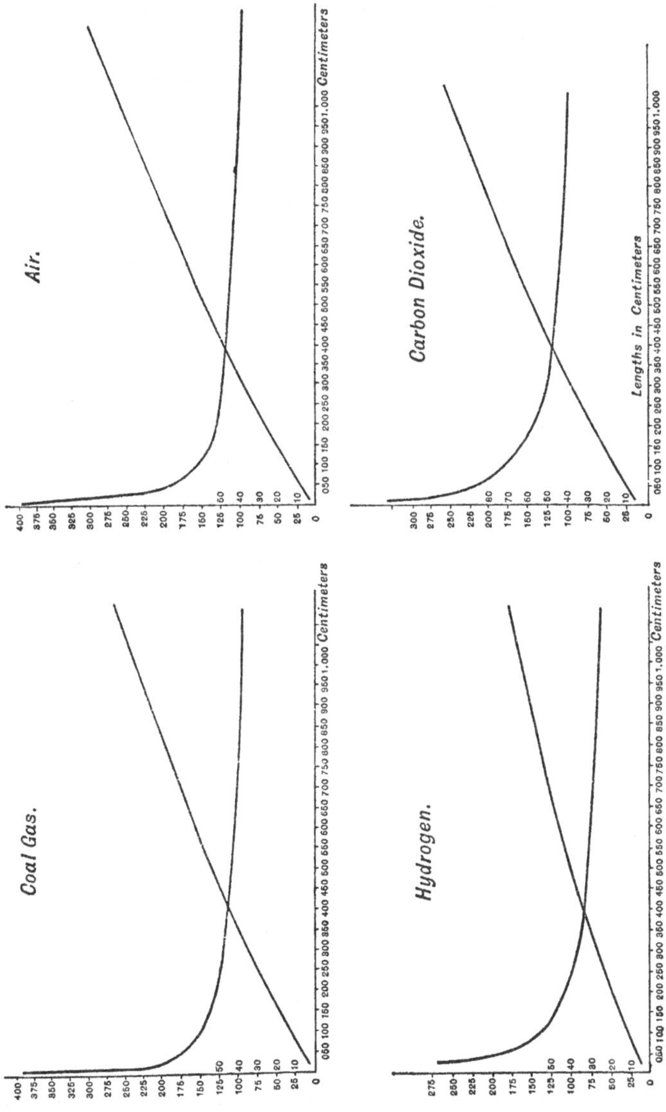

Fig. 19.

1 cm. thick. The fact that a greater electromotive intensity is required to spark across a thin layer of air than a thick one was discovered by Lord Kelvin (l. c.) in 1860.

51.] With regard to the relation between the potential difference V and spark length l, Baille deduced from his experiments the relation

$$V^2 = 10500\,(l + 0{\cdot}08)\,l.$$

The agreement between the numbers calculated by this formula and those found by experiment is not very close, and Chrystal (*Proc. Roy. Soc. Edin.* vol. 11. p. 487, 1882) has shown that for spark lengths greater than 2 millimetres the linear relation

$$V = 4{\cdot}997 + 99{\cdot}593\,l$$

represents Baille's results within experimental errors. This linear relation is confirmed by Liebig's results, as the curves, Fig. 19, are nearly straight when the spark length is greater than one millimetre.

Carey Foster and Pryson (*Chemical News*, 49, p. 114, 1884) found that the linear relation $V = a + \beta\,l$ was the one which represented best the results of their experiments on the discharge through air at atmospheric pressure.

52.] When the spark length in air at atmospheric pressure is less than about a millimetre, the curve which expresses the relation between potential difference and spark length gets concave to the axis along which the spark lengths are measured ; that is, for a given small increase in the spark length the increase in the corresponding potential difference is greater when the sparks are short than when they are long. For exceedingly short sparks there seems to be considerable evidence that when the spark length is reduced to a certain critical value there is a point of inflexion in the potential difference curve, and that when the spark length is reduced below this value the previous concavity is replaced by convexity, the curve for very small spark lengths taking somewhat the shape of the one in Fig. 20. This indicates that the potential difference required to produce a spark however short cannot be less than a certain finite value, which for air at ordinary temperatures is probably between 300 and 400 volts. If a curve similar to Fig. 20 represents the relation between potential difference and spark length, we see that it would be possible under certain conditions to start a spark by pulling two plates maintained at a constant potential difference further apart, and to stop the spark by pushing the plates nearer together.

53.] At atmospheric pressure the spark length at which the potential difference is a minimum must, if such a length exist at all, be so small, that it would be very difficult to measure the spark lengths with sufficient accuracy to investigate this point completely; when however the air is at a lower pressure the critical spark length is longer, and the investigation of this problem easier. The evidence to which I have alluded in Art. 52 comes indirectly from an investigation (which we shall have to

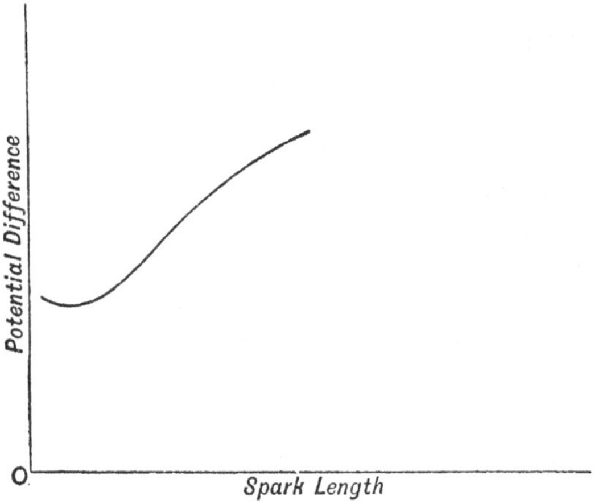

Fig. 20.

consider later) made by Mr. Peace in the Cavendish Laboratory, Cambridge. Mr. Peace's experiments were made with the view of finding the relation between the potential difference in air and the pressure when the spark length is kept constant, but as experiments were made on this relation for sparks of many different lengths, they furnish material for drawing the curve expressing the relation between potential difference and spark length at constant pressure. Such curves are given in Fig. 27, and it will be seen that at lower pressures they exhibit the peculiarities referred to. The discharge took place between very large electrodes, one of which was plane and the other a segment of a sphere about 20 cm. in radius, and as the difference of potential was produced by a large number of storage cells, the equality of whose E. M. F. was very carefully tested, the measurements of the potential difference could be made with great accuracy. It must be remembered, however, that the apparatus used was designed for the purpose of determining the relation between potential difference and pressure for constant spark length, and not for the relation between potential difference and spark length for constant pressure, so that its indications on this point are somewhat indirect. The conclusion that with very short sparks the potential difference increases as the spark length diminishes was, however, borne out to some extent by the observation that when the voltage was just not sufficient (i. e. was about two volts too small) to spark across ·002 of an inch at a pressure of 20 mm. of mercury, the same voltage would not send a spark between the plates when the distance was reduced to ·001 or even to ·0004 of an inch. Mr. Peace also found that when he removed the electrodes from the apparatus after sparks had passed between them when they were very close together, the part of the electrodes most affected by breathing upon them formed an annulus at some little distance from the centre, indicating that discharge had taken place most freely at distances which were slightly greater than the shortest distance between the electrodes, which was along the line joining their centres. Mr. Peace has more recently tested this result directly by placing two spark gaps in parallel, the electrodes being parallel plane plates. One pair of these electrodes were separated by a single thickness of thin pieces of glass such as are used for cover slips, while the other pair of electrodes were kept at a greater distance apart by placing between them two or more of the pieces of glass piled one on the top of the other. At atmospheric pressure the spark passed across the short gap rather than the long one, but when the pressure was reduced the reverse effect took place, the spark going across the longer air gap before any discharge could be detected across the shorter, and after the spark had first passed across the longer path it required in some cases an additional potential difference of more than 100 volts to make it go across the shorter as well. When in Art. 170 we consider discharge at very low pressures we shall find that in some experiments of Hittorf's a long spark passed much more easily than a very much shorter one between the same electrodes; in this case however the electrodes were wires, and the field before discharge was not uniform as in the case under consideration.

Discharge when the Electric Field is not uniform.

54.] In the experiments tabulated above the electrodes were so large that the electric field between them might be considered as uniform before the spark passed. Baille and Paschen have however made some very interesting experiments on the potential differences required to spark between spheres small enough to make the variations in the electric field considerable. Baille's results (*Annales de Chimie et de Physique* (5), 25, p. 531, 1882) are given in the following table, the potential difference being measured in absolute electrostatic units:—

Potential Differences: pressure 760 mm., *temperature* 15° *to* 20° C.

Spark Length in cm.	Planes.	Spheres 6 cm. in diameter.	Spheres 3 cm. in diameter.	Spheres 1 cm. in diameter.	Spheres ·6 cm. in diameter.	Spheres ·35 cm. in diameter.	Spheres ·1 cm. in diameter.
·05	8·94	8·96	9·18	9·18	9·26	9·30	**9·63**
·10	14·70	14·78	14·99	15·25	15·53	16·04	**16·10**
·15	20·20	20·31	20·47	21·28	21·24	**21·87**	19·58
·20	25·42	25·59	25·95	26·78	26·82	**27·13**	21·91
·25	30·38	30·99	31·33	32·10	**32·33**	31·96	23·11
·30	35·35	36·12	36·59	37·32	**37·38**	36·29	24·12
·35	40·45	41·45	41·47	**42.48**	42·16	39·39	25·34
·40	45·28	46·34	46·77	**47·62**	46·34	41·77	26·03
·45	50·48	51·46	**51·60**	**51·56**	50·44	43·76	26·62
·40	44·80	45·00	45·00	**45·50**	44·80	41·07	26·58
·45	49·63	50·33	49·63	**52·04**	48·42	43·29	28·49
·50	54·35	**55·06**	**54·96**	**54·66**	53·25	47·21	30·00
·60	63·82	**65·23**	**65·23**	**65.23**	59·69	53·75	31·51
·70	74·09	**75·40**	73·79	72·28	64·22	56·47	32·92
·80	84·83	**87·98**	84·76	77·61	67·75	58·79	33·82
·90	94·72	**97·44**	94·62	80·13	70·56	59.09	34·93
1·00	105·49	**112·94**	104·69	83·05	72·38	59·49	36·24

From this table Baille concludes that for a given length of spark between two equal spheres, one charged and insulated and the other put to earth, the potential difference varies with the diameter of the sphere; starting from the plane the potential difference at first increases with the curvature, and attains a maximum when the sphere has a certain diameter. This critical diameter of the sphere depends upon the spark length, the shorter the spark the smaller the critical diameter. In the preceding table the maximum potential differences have been printed in bolder type.

The two parts into which the table is divided by the horizontal line correspond to two different sets of experiments.

Paschen's results (*Wied. Ann.* 37, p. 79, 1889) are given in the following table:—

Potential Difference at first Spark: pressure 756 mm. *mean temperature* 15° C.

SHORT SPARKS.

Spark Length in centimetres.	Spheres 1 cm. radius.	Spheres ·5 cm. radius.	Spheres ·25 cm. radius.
·01	3·38	3·42	3·61
·02	5·04	5·18	5·58
·03	6·62	6·87	6·94
·04	8·06	8·22	8·43
·05	9·56	9·75	9·86
·06	10·81	10·87	11·19
·07	11·78	12·14	12·29
·08	13·40	13·59	13·77
·09	14·39	14·70	14·89
·10	15·86	15·97	16·26
·11	16·79	17·08	17·26
·12	18·28	18·42	18·71
·14	20·52	20·78	21·26

LONG SPARKS.

Spark Length in centimetres.	Spheres 1 cm. radius.	Spheres ·5 cm. radius.	Spheres ·25 cm. radius.
·10	15·96	16·11	**16·45**
·15	21·94	22·17	**22·59**
·20	27·59	27·87	**28·18**
·25	32·96	33·42	**33·60**
·30	38·59	**39·00**	38·65
·35	43·93	**44·32**	43·28
·40	49·17	**49·31**	47·64
·45	**54·37**	54·18	51·56
·50	**59·71**	59·03	54·67
·55	**64·60**	63·35	57·27
·60	**69·27**	67·80	59·95
·70	**78·51**	75·04	63·14
·80	**87·76**	81·95	66·39
·90			68·65
1·00			70·68
1·20			74·94
1·50			79·42

Here the heavy type again denotes the maximum potential differences.

These results are represented graphically in Fig. 21. They confirm Baille's conclusion that for a spark of given length the

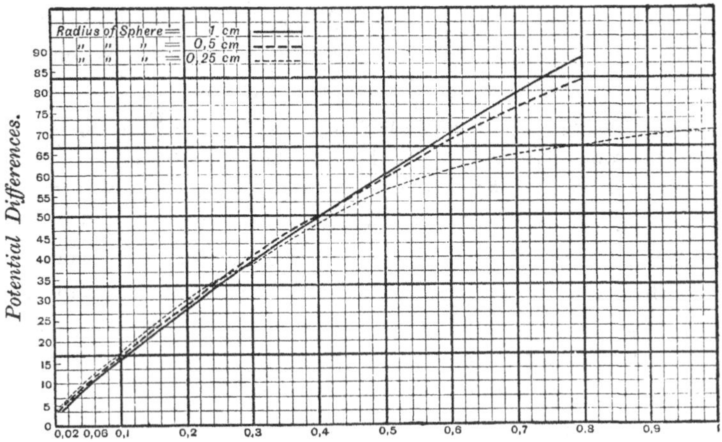

Spark Length.

Fig. 21.

potential difference is a maximum when the spheres have a cer-

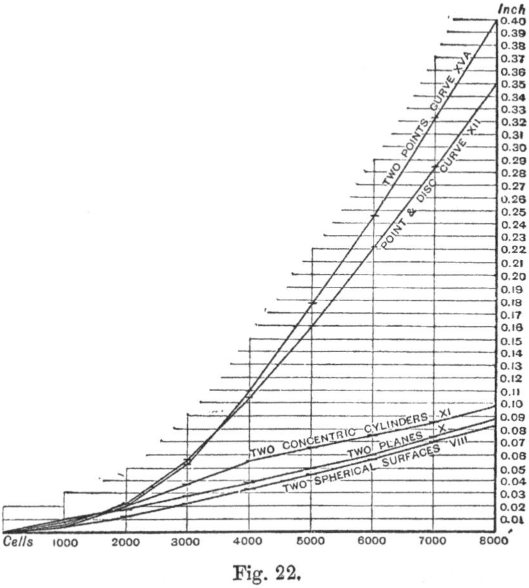

Fig. 22.

tain critical diameter, the critical diameter increasing with the length of the spark.

Both Baille's and Paschen's measurements show that when the spheres are very small, the potential difference required to produce a spark of given length is, if the spark length is not too small, much less than the potential difference required to produce the same length of sparks between parallel plates. When the spark passes between pointed electrodes the potential differences are still smaller. This effect is clearly shown in Fig. 22, which is taken from a paper by De la Rue and Hugo Müller (*Phil. Trans.* 1878, Pt. 1. p. 55), and which contains curves representing the relation between potential difference and spark length when the electrodes are (i) two plates, (ii) two spheres, one 3 cm. in radius the other 1·5 cm. in diameter, (iii) two concentric cylinders, (iv) a plane and a point, (v) two points. It will be noticed that the two points, which give the greatest striking distance for long sparks, give the least for short sparks.

55.] If the spark length between parallel plates is taken as unity, the spark length corresponding to various potential differences for different kinds of electrodes was found by De la Rue and Müller to be as follows (*Proc. Roy. Soc.* 36, p. 157, 1883):—

Number of cells, each cell having an E. M. F. of 1·03 volts . . .	1000	3000	6000	9000	12,000	15,000
Striking distance for point and plane	·60	2·09	3·82	3·89	3·58	3·30
Striking distance for two points . .	·84	1·94	4·65	4·65	4·18	3·68

This table would appear to indicate that the ratio of the striking distance for pointed electrodes to that of planes attains a maximum. It must however be remembered that when the sparks are long the conditions are not the same in the two cases; in the case of the plates the discharge takes place abruptly, while when the electrodes are pointed a brush discharge starts long before the spark passes, and materially modifies the conditions.

56.] Schuster (*Phil. Mag.* [5] 29, p. 182, 1890) has, by the aid of Kirchhoff's solution of the problem of the distribution of electricity over two spheres, calculated from Baille's and Paschen's experiments the maximum electromotive intensity in the field when the spark passed. The results for Baille's experiments are given in Table 1, for Paschen's in Table 2.

TABLE 1.

Value of Maximum Electromotive Intensity in Electrostatic Units.

Spark Length in cm.	Planes.	Spheres, diameter 6 cm.	Spheres, diameter 3 cm.	Spheres, diameter 1 cm.	Spheres, diameter ·6 cm.	Spheres, diameter ·35 cm.	Spheres, diameter ·1 cm.
·05	179	180	186	190	197	206	292
·10	147	149	153	163	176	198	376
·15	135	138	141	157	170	206	425
·20	127	131	137	154	170	219	460
·25	122	127	134	154	180	236	478
·30	118	124	130	156	189	253	494
·35	116	122	129	159	197	263	516
·40	113	122	129	164	204	272	528
·45	112	120	127	166	214	278	540
·40	112	118	124	157	197	268	539
·45	110	119	122	167	206	275	578
·50	109	117	125	166	218	296	608
·60	106	116	125	181	233	327	639
·70	106	117	126	188	234	339	667
·80	106	123	130	192	250	349	685
·90	105	120	132	191	255	349	708
1·00	106	128	133	194	258	349	733

TABLE 2.

Maximum Electromotive Intensity in Electrostatic Units.

Spark Length in cm.	Spheres, diameter 2 cm.	Spheres, diameter 1 cm.	Spheres, diameter ·5 cm.	Spark Length.	Spheres, diameter 2 cm.	Spheres, diameter 1 cm.	Spheres, diameter ·5 cm.
·01	336	347	372	·10	166	175	[illegible]
·02	258	262	277	·15	155	165	190
·03	224	236	240	·20	148	162	198
·04	206	213	222	·25	145	161	204
·05	194	202	215	·30	143	163	215
·06	184	190	202	·35	143	166	226
·07	175	183	193	·40	142	170	236
·08	172	179	192	·45	142	174	249
.09	165	174	187	·50	144	180	256
·10	164	171	187	·55	145	184	265
·11	160	167	183	·60	145	190	272
·12	159	167	185	·70	148	196	281
·14	154	164	187	·80	151	205	288
				·90			293
				1·00			301
				1·20			312
				1·50			327

57.] It will be seen from these tables that the smaller the spheres, or in other words the more irregular the electric field, the greater the value of the maximum electromotive intensity. This is sometimes expressed by saying that the curvature of the electrodes increases the electric strength of the gas, and Gaugain (*Annales de Chimie et de Physique*, [iv] 8, p. 75, 1866) has found that when the spark passes between two coaxial cylinders, the maximum value R of the electromotive intensity can be expressed by an equation of the form

$$R = a + \beta r^{-\frac{1}{2}},$$

where a and β are constants and r is the radius of the inner cylinder.

58.] The variations in the value of the electromotive intensity are so great that they prove that it is not the value of the electromotive intensity which primarily determines whether or not discharge must take place; and it is probable that the use of this quantity as the measure of the electric strength has retarded the progress of this subject by withdrawing attention from the most important cause of the discharge to this which is probably merely secondary.

59.] The following results taken from Paschen's experiments show that when the sparks are not too long the variations in the electromotive intensity are very much greater than the variations in the potential difference; suggesting that for such sparks the potential difference is the most important consideration.

Radius of Electrodes in cm.	1·	·5	·25	
Potential Difference Maximum Intensity	13·4 172	13·6 179	13·8 192	Spark length ·08 cm.
Potential Difference Maximum Intensity	20·5 154	20·8 164	21·3 187	Spark length ·14 cm.
Potential Difference Maximum Intensity	49·2 142	49·3 170	47·6 236	Spark length ·40 cm.
Potential Difference Maximum Intensity	87·8 151	81·9 205	66·4 288	Spark length ·80 cm.

60.] We can explain by the following geometrical illustration the two effects produced by the irregularity of the field—the diminution in the potential difference, and the increase in the maximum electromotive intensity. When a discharge is passing through gas, we shall see later on, from the consideration of the discharge at low pressures, reasons for believing that the distribution of potential during discharge may be approximately represented by the equation

$$V = \alpha + \beta l,$$

where α and β are constants and l the distance from the negative electrode. If the curve representing the distribution of potential before discharge cuts the curve representing the distribution after discharge, a spark will pass, while if it does not cut it, no discharge can take place.

In Fig. 23, A, B represent the electrodes, CD the distribution of potential during the discharge. If the electric field is uniform the curve which represents the distribution of potential before the spark passes is a straight line such as AE, as the intensity of the field increases E moves higher and higher, the first point at which it intersects the curve representing the distribution of potential after discharge being D. In this case the difference of potential between the electrodes when the spark passes is BD, so that the relation between the potential difference V and the spark length l is

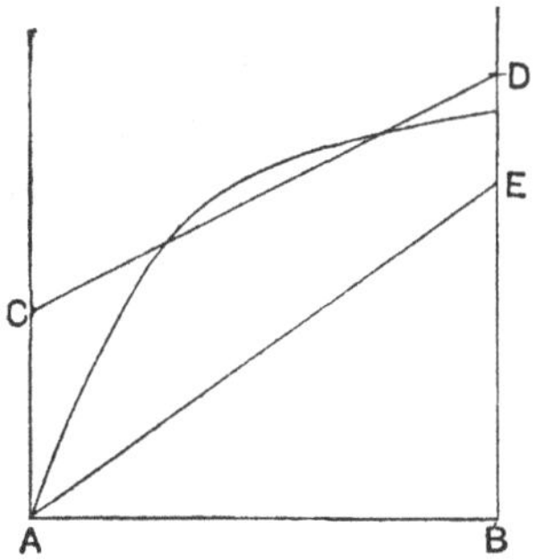

Fig. 23.

$$V = \alpha + \beta l.$$

When however the electric field is not uniform it is possible for the curve representing the potential before discharge to intersect the potential curve after discharge, even though the difference of potential before discharge is less than B D. This will be evident from Fig. 23, where the curved line represents the distribution of potential in an irregular field. Here we have a very rapid change in potential in the neighbourhood of one of the electrodes, followed by a comparatively slow rate of change midway between them. In this case the curves intersect and a discharge would take place, though the difference of potential between the electrodes is less than that required for sparking in a uniform field. Thus for equal spark lengths the potential difference may be less when the field is variable than when it is uniform. Again, we notice that the slope of this curve in the neighbourhood of the electrode A is steeper than that of a line joining A and D, in other words the maximum electromotive intensity when discharge takes place is greater when the field is variable than when it is uniform. Both these results are confirmed by Baille's and Paschen's observations.

For a theory of the spark discharge the reader is referred to the discussion at the end of this chapter.

61.] It is sometimes said that the reason a thin layer of gas is electrically stronger than a thick one is, that a film of condensed gas is spread over the surface of the electrodes, and that this film is electrically stronger than the free gas. This consideration however, as Chrystal (*Proc. Roy. Soc. Edin.*, 11, 1881–2, p. 487) has pointed out, is quite incapable of explaining the variation in electric strength, for it is evident that if this were all that had to be taken into account the discharge would pass whenever the electromotive intensity was great enough to break through this film of condensed gas, so that this intensity would be constant when the spark passed whatever the thickness of the layer of free gas.

Connection between Spark Potential and the Pressure of the Gas.

62.] The general nature of this connection is as follows: as the pressure of the gas diminishes the difference of potential required to produce a spark of given length also diminishes, until the pressure falls to a critical value depending upon the length of the spark, the nature of the gas, the shape and size of the electrodes and of the vessel in which the gas is contained; at this pressure the potential difference is a minimum, and any further diminution in the pressure is accompanied by an increase in the potential difference. The critical pressure varies very greatly with the length of the spark; in Mr. Peace's experiments, which we shall consider later, when the spark length was about 1/100 of a millimetre, the critical pressure was that due to about 250 mm. of mercury, while for sparks several millimetres long the critical pressure was less than that due to 1 mm. of mercury.

63.] At pressures considerably greater than the critical pressure, the curve which represents the relation between potential difference and pressure, the spark length being constant, approximates to a straight line, or more accurately to a slightly curved hyperbola concave with respect to the axis along which the pressures are measured. Thus Wolf, who has determined (*Wied. Ann.* 37. 306, 1889) the potential difference required to produce a spark through air, hydrogen, carbonic acid, oxygen and nitrogen at pressures varying from 1 to 5 atmospheres, found that the electromotive intensity, y, required to produce a spark across a length of 1 mm. between electrodes 5 cm. in radius when the pressure was x atmospheres, could be expressed by the following equations:—

For hydrogen	$y = 65{\cdot}09x + 62.$
For oxygen	$y = 96{\cdot}0x + 44.$
For air	$y = 107x + 39.$
For nitrogen	$y = 120{\cdot}8x + 50.$
For carbonic acid . . .	$y = 102{\cdot}2x + 72.$

64.] For pressures less than one atmosphere the connection between spark length and pressure has been investigated by Baille (*Annales de Chimie et de Physique*, [5] 29, p. 181, 1883),

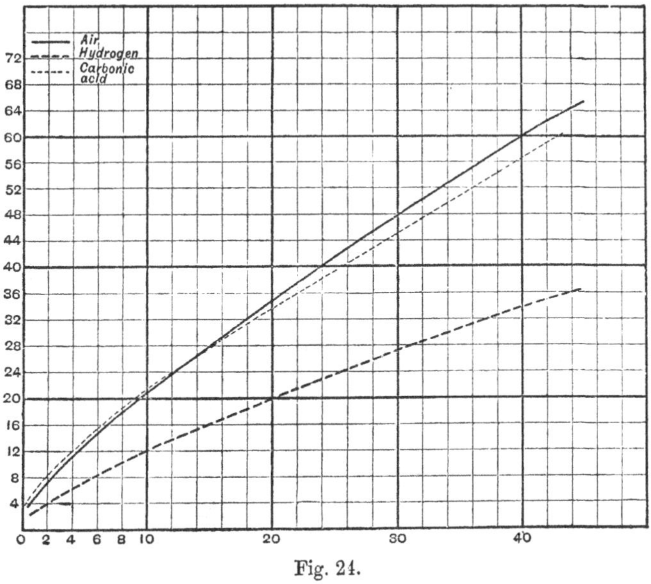

Fig. 24.

Macfarlane (*Phil. Mag.* [5] 10, p. 389, 1880), and Paschen (*Wied. Ann.* 37, p. 69, 1889), who have found that the relation is graphically represented by very slightly curved portions of a hyperbola. Paschen (*l. c.* p. 91) made the interesting observation that as long as the product of the density and spark length is constant the sparking potential is for a considerable range of pressure constant for the same gas. This result can also be expressed by saying that the sparking potential for a gas can be expressed in terms of the ratio of the spark length to the mean free path of the molecules of the gas. The curves given in Fig. 24, which represent for air, hydrogen and carbonic acid the relation between the spark potential in electrostatic units as ordinates, and the products of the pressure of the gas in centimetres of mercury and the spark length in centimetres as abscissæ, seem to show that this relation is approximately a linear one.

65.] The preceding experiments were made at pressures much greater than the critical pressure. A series of very interesting experiments has lately been made by Mr. Peace in the

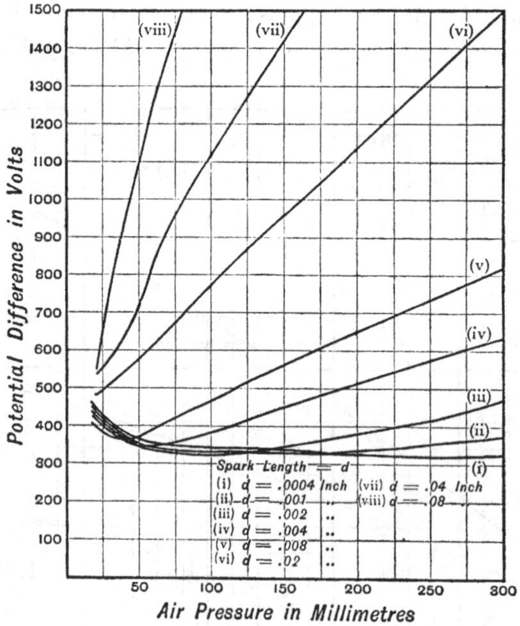

Fig. 25.

Cavendish Laboratory, Cambridge, on the shape of these curves in the neighbourhood of the critical pressure. In these experiments the potential difference could be determined with great accuracy, as it was produced by a large number of small storage cells whose E. M. F. could very easily be determined. Mr. Peace's curves are represented in Figs. 25, 26, 27, 28. Fig. 25 represents the relation between potential difference in air and pressure for spark lengths varying from ·0010 cm. to ·2032 cm. Fig. 26 represents the relation between electromotive intensity and pressure for the same spark lengths, and Fig. 27 the relation between potential difference and spark length for a series of different pressures: the curve representing the relation between electromotive

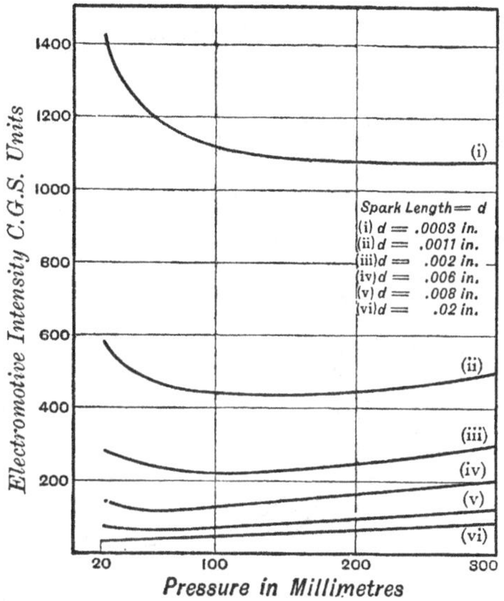

Fig. 26.

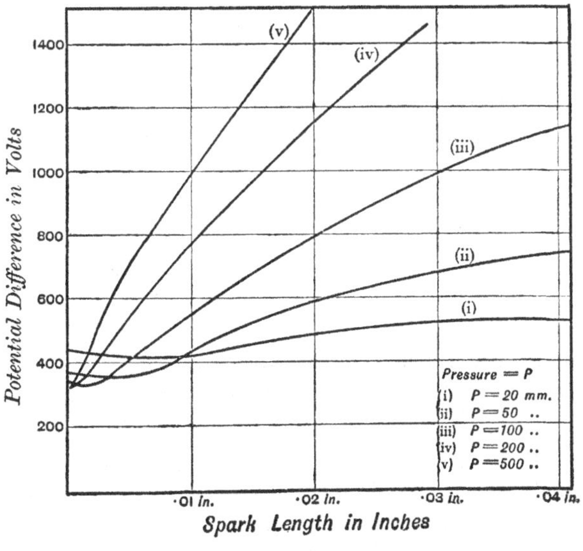

Fig. 27.

intensity and spark length is given in Fig. 28. These curves will be seen to present several points of great interest. In the first place, Fig. 25 shows how much the critical pressure depends

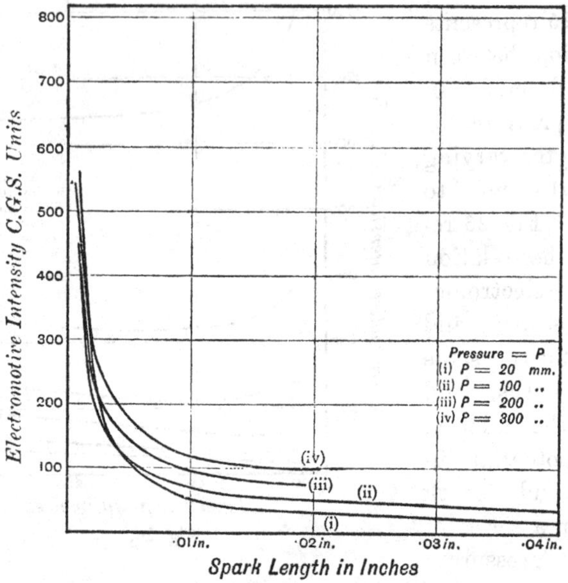

Fig. 28.

upon the spark length; this will also be seen from the following table:—

Spark Length.	Minimum Potential Difference.	Critical Pressure.
·0010 cm.	326 volts.	250 mm.
·00254 cm.	330 volts.	150 mm.
·00508 cm.	333 volts.	110 mm.
·01016 cm.	354 volts.	55 mm.
·02032 cm.	370 volts.	35 mm.

Thus when the spark length was increased twenty-fold the critical pressure was reduced from 250 mm. to 35 mm. Another very remarkable feature is the small variation in the minimum potential difference required to produce the spark. In the preceding table there is a very considerable range of pressure, but the variation in the potential difference is comparatively small. Mr. Peace too made the interesting observation that he could not produce a spark however near he put the electrodes together, or however the pressure was altered, if the potential difference was less than something over 300 volts. Gases in this respect seem to resemble electrolytes which require a finite difference of potential to produce a steady current through them. This constancy in the minimum value of the potential required to produce a spark seems additional evidence that the passage of the spark is regulated more by the value of the potential difference than by that of the electromotive intensity. Another thing to be remarked about the curves in Fig. 25 is the way in which they get flatter and flatter as the spark length diminishes: the flatness of the curve corresponding to the spark length ·0010 cm., or ·0004 inch, is so remarkable that I give the numbers from which it was drawn:—

Spark Length ·00101 cm.

Pressure in mm. of Mercury.	Potential Difference in Volts.	Electromotive Intensity.
20	433	1420
30	398	1310
40	380	1245
50	370	1215
60	357	1170
70	353	1160
80	349	1145
90	346	1135
100	343	1125
120	337	1105
140	332	1090
160	330	1085
180	329	1080
200	328	1075
240	326	1070
280	327	1072
300	328	1075

The curves representing the relation between potential difference and pressure for different lengths of spark cut each other; this indicates that at a pressure lower than that where the curves cut it requires a greater potential difference to produce the short spark than it does the long one. This point has already been considered in Art. 53.

66.] The connection between the critical pressure and the spark length proves that the gas at the critical pressure when conveying the electric discharge has a structure of which the linear measure of the coarseness is comparable with the spark length. This spark length is very much greater than the mean free path of the molecules, and thus these experiments show that a gas conveying electrical discharge possesses a much coarser structure than that recognized by the ordinary Kinetic Theory of Gases. For the nature of this structure we must refer to the general theory of the electrical discharge given at the end of this chapter.

67.] Although the magnitude of the critical pressure depends, as we have seen, to a very great extent on the distance between the electrodes, the actual existence of a critical pressure does not seem to depend on the presence of electrodes. In Art. 74 a method is described by which an endless ring discharge can be produced in a bulb containing gas at a low pressure; in this case the discharge is in the gas throughout the whole of its course, and there are no electrodes. If in such an experiment the bulb is connected to an air pump it will be found that when the pressure of the gas in the bulb is high no discharge at all is visible; as however the pressure is reduced a discharge gradually appears and increases in brightness until the pressure is reduced to a small fraction of a millimetre, when the brightness is a maximum; when the pressure is reduced below this value the discharge has greater difficulty in passing, it gets dimmer and dimmer, and finally stops altogether when the exhaustion is very great. This experiment shows that there is a critical pressure even when there are no electrodes, but that it is very much lower than in an ordinary sized tube when electrodes are used.

68.] De la Rue and Hugo Müller (*Proc. Roy. Soc.* 35, p. 292, 1883), using the ordinary discharge with electrodes, found that the critical pressure depends on the diameter of the tube in which the rarefied gas is confined, the critical pressure getting lower as the diameter of the tube is increased.

Potential Difference required to produce Sparks through various Gases.

69.] The potential difference required to send a spark between the same electrodes, separated by the same distance, depends, as Faraday found, on the nature of the gas surrounding the electrodes: thus, for example, the potential difference required to produce a spark of given length in hydrogen is much less than in air. Measurements of the potential differences required to produce discharge through a series of gases have been made by, among others, Faraday, Baille (*Annales de Chimie et de Physique*, [5] 29, p. 181, 1883), Liebig (*Phil. Mag.* [5] 24, p. 106, 1887), Paschen (*Wied. Ann.* 37, p. 69, 1889). The results obtained by different observers seem to differ very largely. This will be seen from the following table, in which Paschen gives the ratio of the potential difference required to spark across hydrogen or carbonic acid, to the potential difference required to spark across a layer of air of the same thickness, the pressure for all the gases being 750 mm. of mercury.

Spark length in centimetres.	Hydrogen.			Carbonic Acid.		
	Baille.	Liebig.	Paschen.	Baille.	Liebig.	Paschen.
·1	·49	·873	·639	1·67	1·20	1·05
·2	·49	·787	·578	1·24	1·16	·988
.3	.50	·753	·560	·94	1·07	·962
·4	·50	·704	·553	·76	1·03	·930
·5	·50	·670	·548		·994	·910
·6		·656	·555		·974	·940

It will be seen that, though the numbers got by different observers differ very widely, they all agree in making carbonic acid stronger than air for short sparks and weaker than it for long. This would indicate that in the formula

$$V = a + \beta l,$$

which gives the spark potential V in terms of the spark length l, a for carbonic acid is greater than a for air, while β for carbonic acid is less than β for air.

It will be seen from Fig. 24, which contains Paschen's curves showing the relation between potential difference and pressure for air, hydrogen and oxygen, that these curves cut each other; thus the relation between their 'electric strengths' depends to a large extent upon the pressure. Liebig's curves for air, hydrogen, carbonic oxide and coal gas were given in Fig. 19.

70.] Röntgen (*Göttinger Nachrichten*, 1878, p. 390) arrived at the conclusion that the potential difference required to produce a spark of given length in different gases was, approximately, inversely proportional to the mean free path of the molecules of the gas. This approximation, if it exists at all, must be exceedingly rough, for we have seen that the relation between the potential differences required to spark through different gases depends on the spark length and the pressure of the gases. If the result found by Mr. Peace for air (Art. 65),—that the minimum potential difference required to produce a spark varied very little with the spark length,—were to hold for other gases, there would be much more likelihood of this minimum potential difference being connected with some physical or chemical property of the gas, than the potential difference required to produce a spark of arbitrary length at a pressure chosen at random being so connected.

71.] If a permanent gas in a closed vessel be heated up to 300°C, the discharge potential does not change (see Cardani, *Rend. della R. Acc. dei Lincei*, 4, p. 44, 1888 ; J. J. Thomson, *Proc. Camb. Phil. Soc.*, vol. 6, p. 325, 1889): if however the vessel be open so that the pressure remains constant, there will be a diminution in the discharge potential due to the diminution in density. When the temperature gets so high that chemical changes such as dissociation take place in the gas the discharge potential may fall to zero.

A great number of experiments have been made on the relative 'electric strengths' of damp and dry air. The only observer who seems to have found any difference is Baille, and in his case the difference was so large as to make it probable that some of the water vapour had condensed into drops.

Phenomena accompanying the Electric Discharge at Low Pressures.

72.] When the discharge passes between metallic electrodes sealed into a tube filled with gas at a low pressure, the appearance it presents is very complicated: many of the effects observed in the tube are however evidently due to the action of the electrodes, as the phenomena at the anode are very different from those at the cathode; it therefore appears desirable to begin the study of the phenomena shown in vacuum tubes by investigating the discharge when no electrodes are present.

73.] If we wish to produce the endless discharge in a closed vessel without electrodes, we must produce in some way or another round a closed curve in the vessel an electromotive force large enough to break down the insulation of the gas. Since, for discharge to take place, the electromotive force round a closed curve must be finite, it cannot be produced electrostatically, we must use the electromotive forces produced by electromagnetic induction, and make the closed curve in the exhausted vessel practically the secondary of an induction coil. As the primary of this induction coil I have used a wire connecting the inside and outside coatings of a Leyden jar; when the jar is discharged through the wire enormous currents pass for a short time backwards and forwards along the wire, the currents when the wire is short and the jar small reversing their directions millions of times in a second. We thus have here all the essentials for producing a very large electromotive force round the secondary, viz. a very intense current in the primary and an exceedingly rapid rate of alternation of this current; and though the electromotive force only lasts for an exceedingly short time, it lasts long enough to produce the discharge through the gas and to enable us to study its appearance.

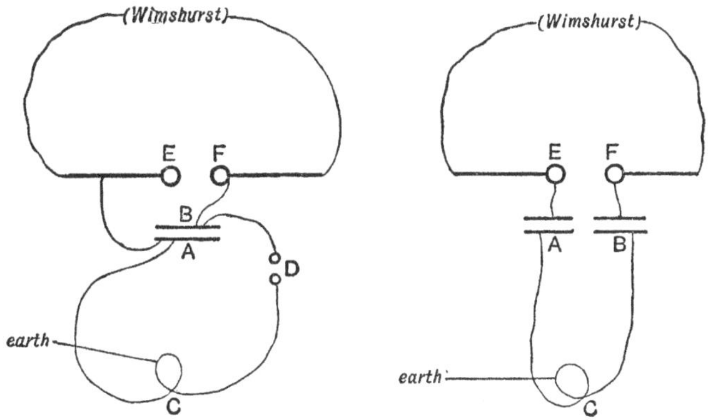

Fig. 29.

74.] Two convenient methods of producing the discharge are shown in Fig. 29: in the one on the right two jars are used, the outside coatings of which (A and B) are connected by a wire in which a few turns C are made; C forms the primary coil. The inside coatings of these jars are connected, one to one terminal E of a Wimshurst electrical machine or of an induction coil, the other coating to F, the other terminal of such a machine. If the tubes in which the discharge is to be observed are spherical bulbs, they are placed inside the coil C; if they are endless tubes, they are placed just outside it. When the difference of potential between E and F becomes great enough to spark across E F, the jars are discharged and electrical oscillations set up in the wire ACB. The oscillating currents in the primary produce a large electromotive intensity in its neighbourhood, sufficient under favourable conditions to cause a bright discharge to pass through the rarefied gas in the bulb placed inside the coil.

We have described in Art. 26 the way in which the Faraday tubes, which before the spark took place were mainly in the glass between the two coatings of the jars, spread through the region outside the jars, as soon as the discharge passes, keeping their ends on the wire ACB. They will pass in their journey through the bulb in the coil C, and if they congregate there in sufficient numbers the electromotive force will be sufficient to cause a discharge to pass through the gas. Anything which concentrates the Faraday tubes in the bulb will increase the brightness of the discharge through it.

75.] It is necessary to prevent the coil C getting to a high potential before the spark passes, otherwise it may induce a negative electrification on the parts of the inside of the glass bulb nearest to it and a positive electrification on the parts more remote: when the potential of the coil suddenly falls in consequence of the passage of the spark, the positive and negative electricities will rush together, and in so doing may pass through the rarefied gas in the bulb and produce luminosity. This luminosity will spread throughout the bulb and will not be concentrated in a well-defined ring, as it is when it arises from the electromotive force due to the alternating currents passing along the wire ACB. This effect may explain the difference in the appearance presented by the discharge in the following experiments, where the discharge passes as a bright ring, from that observed by Hittorf (*Wied. Ann.* 21, p. 138, 1884), who obtained the discharge in a tube by twisting round it a wire connecting the two coatings of a Leyden jar: in Hittorf's experiment the luminosity seems to have filled the tube and not to have been concentrated in a bright ring. To prevent these electrostatic effects, due to causes which operate before the electrical oscillations in the wires begin, the coil C is connected to earth, and as an additional precaution the discharge tube may be separated from the coil by a screen of blotting paper moistened with dilute acid. The wet blotting paper is a sufficiently good conductor to screen off any purely electrostatic effect, but not a good enough one to interfere to any appreciable extent with the electromotive forces arising from the rapidly alternating currents.

76.] If C is the capacity of the jars, L the coefficient of self-induction of the discharging circuit, then if the difference of potential between the terminals of the electric machine is initially V_0, γ the current through the wire at a time t after the spark has passed will (Chap. IV) be given by the equation

$$\gamma = \frac{CV_0}{(LC)^{\frac{1}{2}}} \sin \frac{t}{(LC)^{\frac{1}{2}}},$$

supposing as a very rough approximation that there is no decay either from resistance or radiation in the vibrations.

The rate of variation of the current, $\dot{\gamma}$, is thus given by the equation

$$\dot{\gamma} = \frac{V_0}{L} \cos \frac{t}{(LC)^{\frac{1}{2}}}.$$

Thus if M is the coefficient of mutual induction between the primary and a secondary circuit, the maximum electromotive force round the secondary will be MV_0/L, which for a given spark length is independent of the capacity of the jars. But though the maximum electromotive force does not depend upon the capacity of the jars, the oscillations will last longer when the jars have a large capacity than when they have a small one, as the energy to begin with is greater; hence, though it is possible to get the discharge with jars whose capacity is not more than 70 or 80 in electrostatic measure, it is not nearly so bright as when larger capacities are used. The best number of turns to use in the coil is that which makes M/L a maximum. If n is the number of turns, then M and L will be respectively of the forms βn and $L_0 + an^2$, where a and β are constants and L_0 the self-induction of the part of the wire ACB not included in the coil; thus M/L will be of the form

$$\frac{\beta n}{L_0 + an^2},$$

and this is a maximum when $L_0 = an^2$, that is when the self-induction in the coil is equal to that in the rest of the circuit. Though the electromotive force is greatest in this case, in practice it is found to be better to sacrifice a little of the electromotive force for the sake of prolonging the vibrations;

this can be done by increasing the self-induction of the coil. It is thus advisable to use rather more turns in the coil than is indicated by the preceding rule.

Appearance of the Discharge.

77.] Let us suppose that a bulb fused on to an air pump is placed within the coil C, and that the jars are kept sparking while the bulb is being exhausted. When the pressure is high, no discharge at all is to be seen inside the bulb; but when the exhaustion has proceeded until the pressure of the air has fallen to a millimetre of mercury or thereabouts, a thin thread of reddish light is seen going round the bulb in the zone of the coil. As the exhaustion proceeds still further, the brightness of this thread rapidly increases as well as its thickness; it also changes its colour, losing the red tinge and becoming white. Continuing the exhaustion, the luminosity attains a maximum and the discharge passes as a very bright and well-defined ring. When the pressure is still further diminished, the luminosity also diminishes, until when an exceedingly good vacuum is reached no discharge at all passes. The pressure at which the luminosity is a maximum is very much less than the pressure at which the electric strength is a minimum in a tube provided with electrodes and comparable in size with the size of the bulb; the former pressure is in air less than 1/200 of a millimetre of mercury, while the latter is about half a millimetre.

78.] We see from this result that the difficulty which is experienced in getting the discharge to pass through an ordinary vacuum tube when the pressure is very low is not altogether due to the difficulty of getting the electricity to pass from the electrodes into the gas, but that it also occurs in tubes without electrodes, though in this case the critical pressure is very much lower.

79.] The existence of a critical pressure can also be easily shown by putting some mercury in the bulb, and, when the bulb has been well exhausted, driving out the remainder of the air by heating the mercury and filling the bulb with mercury vapour. After this process has been repeated two or three times, the bulb should be fused off from the pump when full of mercury vapour. It will only be found possible to get a discharge through this bulb within a narrow range of temperature, between about 70° and

160°C; when the bulb is colder than this, the pressure of the mercury vapour is too small to allow the discharge to pass; when it is hotter, the vapour pressure is too great.

The critical pressure can also be proved by using the principle that a conductor screens off the electromotive intensities due to rapidly alternating currents while an insulator does not. For this purpose we use two glass bulbs one inside the other, the inner bulb containing gas at such a pressure that the discharge can pass freely through it. The outer bulb contains nothing but mercury and mercury vapour, and is prepared in the way just described. If the primary coil is placed round the outer bulb, then, when the bulb is cold, the discharge passes through the inner bulb, but not through the outer, showing that at this low pressure the conductivity of the vapour in the outer bulb is not great enough for the vapour to act as an electrical screen to the inner bulb. If, however, the outer bulb is warmed, the vapour pressure of the mercury increases, and with it the conductivity; a discharge now passes through the outer bulb but not through the inner, the mercury vapour acting as a screen. When the temperature of the outer bulb is still further increased, the pressure of the mercury vapour gets so great that it ceases to conduct, and the discharge, as at first, passes through the inner bulb but not through the outer.

80.] These experiments show that after a certain exhaustion has been passed the difficulty of getting a discharge to pass through a highly exhausted tube increases as the exhaustion is increased. This result is in direct opposition to a theory which has found favour with some physicists, viz. that a vacuum is a conductor of electricity. The reason advanced for this belief is that when the discharge passes through highly exhausted tubes provided with electrodes, the difficulty which it experiences in getting through such a tube, though very great, seems to be almost as great for a short tube as for a long one; from this it has been concluded that the resistance to the discharge is localised at the electrodes, and that when once the electricity has succeeded in escaping from the electrode it has no difficulty in making its way through the rare gas. But although there is no doubt that in a highly exhausted tube the rise in potential close to the cathode is great compared with the rise in unit length of the gas elsewhere, it does not at all follow that the latter vanishes or that it continually diminishes as the pressure is diminished. The experiment we have just described on the bulb without electrodes shows that it does not. Numerous other experiments of very different kinds point to the conclusion that a vacuum is not a conductor. Thus Worthington (*Nature*, 27, p. 434, 1883) showed that electrostatic attraction was exerted across the best vacuum he could produce, and that a gold-leaf electroscope would work inside it. Ayrton and Perry (*Ayrton's Practical Electricity*, p. 310) have determined the electrostatic capacity of a condenser in a vacuum in which they estimated the pressure to be only ·001 mm. of mercury. If the air at this pressure had been a good conductor the electrostatic capacity would have been infinite, instead of being, as they found, less than at atmospheric pressure. Again, if we accept Maxwell's Electromagnetic Theory of Light, a vacuum cannot be a conductor or it would be opaque, and we should not receive any light from the sun or stars.

81.] The discharge has considerable difficulty in passing across the junction of a metal and rarefied gas. This can easily be shown by placing a metal diaphragm across the bulb in which the discharge takes place, care being taken that the diaphragm extends right up to the surface of the glass. In this case the discharge does not cross the metal plate, but forms two separate closed circuits, one circuit being on one side of the diaphragm, the other on the other. The nature of the discharge is shown in Fig. 30, in which it is seen that it travels through a comparatively long distance in the rarefied gas to avoid the necessity of crossing a thin plate of a very good conductor. If the bulb, instead of merely being bisected by one diaphragm, is divided into six or more regions by a suitable number of diaphragms, it will be found a matter of great difficulty to get any discharge at all through it. The metal plate in fact behaves in this case almost

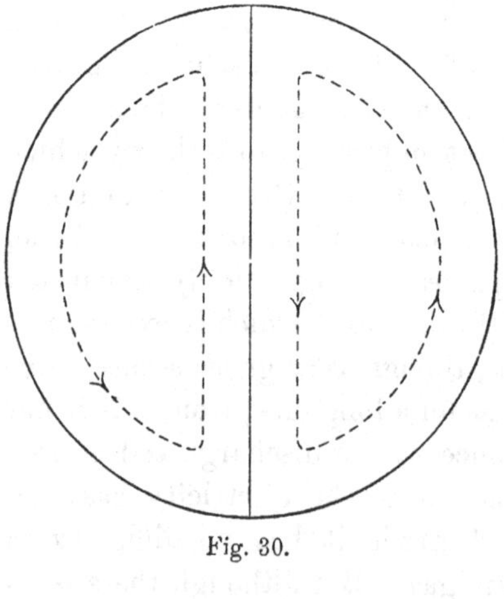

Fig. 30.

exactly like a plate of an insulating substance such as mica, which when continuous also breaks the discharge up into as many circuits as there are regions formed by the mica diaphragms. When however small holes are bored through the mica diaphragms the discharge will not be split up into separate circuits, but will pass through these holes. By properly choosing the position of the holes relative to that of the primary coil, we can get an undivided discharge in part of the circuit branching in the neighbourhood of the diaphragm into as many separate discharges as there are holes through either side of the mica plate. The appearance presented by the discharge when there are two holes on each side of the mica plate is shown in Fig. 31.

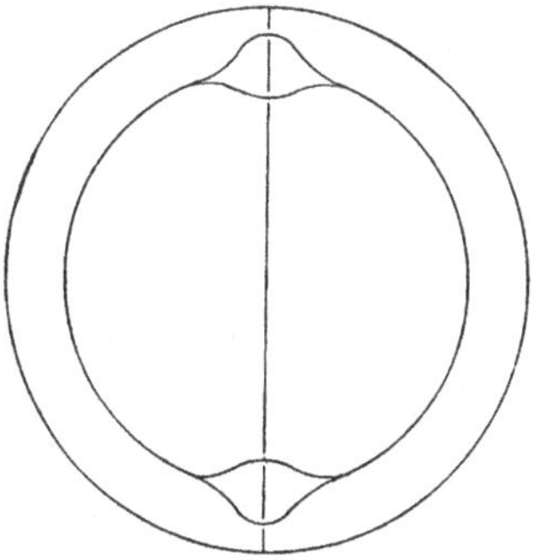

Fig. 31.

82.] A rarefied gas is usually regarded as an exceedingly bad conductor, and the experiments of many observers, such as those of Hittorf, De la Rue and Hugo Müller, have shown that when a tube provided with electrodes in the usual way and filled with such a gas is placed in a circuit round which there is a given electromotive force, it produces as great a diminution in the intensity of the current as a resistance of several million ohms would produce. This great apparent resistance, when the pressure of the gas is not too low, is principally due however to the difficulty which the discharge has in passing from the electrodes into the gas. If we investigate the amount of current sent by a given electromotive force round a circuit exclusively confined to the rarefied gas, we find that, instead of being exceedingly bad conductors, rarefied gases (at not too low a pressure) are, on the contrary, surprisingly good ones, having molecular conductivities—that is specific conductivities divided by the number of molecules in unit volume—enormously greater than those of any electrolytes with which we are acquainted.

83.] We cannot avail ourselves of any of the ordinary methods of measuring resistances to measure the resistance of rarefied gases to these electrodeless discharges; but while the very high frequency of the currents through our primary coil makes the ordinary methods of measuring resistances impracticable, it at the same time makes other methods available which would be useless if the currents were steady or only varied slowly. One such method, which is very easily applied, is based on the way in which plates made of conductors screen off the action of rapidly alternating currents. If a conducting plate be placed between a primary circuit conveying a rapidly alternating current and a secondary coil, the electromagnetic action of the currents induced in the plate will be opposed to that of the currents in the primary, so that the interposition of the plate diminishes the intensity of the currents induced in the secondary. When we are dealing with currents through the primary with frequencies as high as those produced by the discharge of a Leyden jar, the thinnest plate of any metal is sufficient to entirely screen off the primary from the secondary, and no currents at all are produced in the latter when a metal plate is interposed between it and the primary; we could not therefore use this method conveniently to distinguish between the conductivities of different metals. If however instead of a metal plate we use a layer of an electrolyte, the conductivity of the electrolyte is not sufficient to screen off from the secondary the effect of the primary unless the layer is some millimetres in thickness, and the worse the conductivity of the electrolyte the thicker will be the layer of it required to reduce the action of the primary on the secondary to a given fraction of its undisturbed value. By comparing the thicknesses of layers of different electrolytes which produce the same effect when interposed between the primary and the secondary we can, since this thickness is proportional to the specific resistance, determine the conductivity of electrolytes for very rapidly alternating currents (see J. J. Thomson, *Proc. Roy. Soc.* 45, p. 269, 1889).

84.] The conductivity of a rarefied gas can on this principle be compared with that of an electrolyte in the following way: A, B, C, Fig. 32, represents the section of a glass vessel shaped something like a Bunsen's calorimeter; in the inner portion ABC of this vessel, which is exposed to the air, an exhausted tube E is placed. A tube from the outer vessel leads to a mercury pump which enables us to alter its pressure at will. The primary coil LM is wound round the outer tube. When the air in the outer tube is at atmospheric pressure, a discharge caused by the action of the primary passes through the tube E; but when the pressure of the gas in the outer tube is reduced until a discharge passes through it, the discharge in E stops, showing that the currents induced in the gas in the outer vessel have been sufficiently intense to neutralise the direct action of the primary coil on the tube in E.

In order to compare the intensity of the currents through the rarefied gas with those produced under similar circumstances in an electrolyte, the outer vessel ABC, Fig. 32, through which the discharge has passed is disconnected from the pump, and the portion which has previously been occupied by the rarefied gas is filled with water, to which sulphuric acid is gradually added. Pure water does not seem to produce any effect on the brightness of the discharge in E, but as more and more sulphuric acid is added to the water the discharge in E gets fainter and fainter, until when about 25* per cent. by volume of sulphuric acid has been added the effect produced by the electrolyte seems to be as nearly as possible the same as that produced by the rarefied gas. Thus the currents through the rarefied gas must, since they produced the same shielding effect, be as intense as those through a 25 per cent. solution of sulphuric acid. The conductivity of the gas must therefore be as great as that of the mixture of sulphuric acid and water, which is one of the best liquid conductors we know. This shielding effect can be produced by the rarefied gas when its pressure is as low as that due to 1/100 mm. of mercury, while the number of molecules of sulphuric acid in a 25 per cent. solution is such as would, if the sulphuric acid were in a gaseous state, produce a pressure of about 100 atmospheres. Thus, comparing the conductivity per molecule of the gas and of the electrolyte, the molecular conductivity of the gas is about seven and a-half million times that of sulphuric acid. The relation which the molecular conductivity of the gas bears to that of an electrolyte, which produces the same effect in shielding

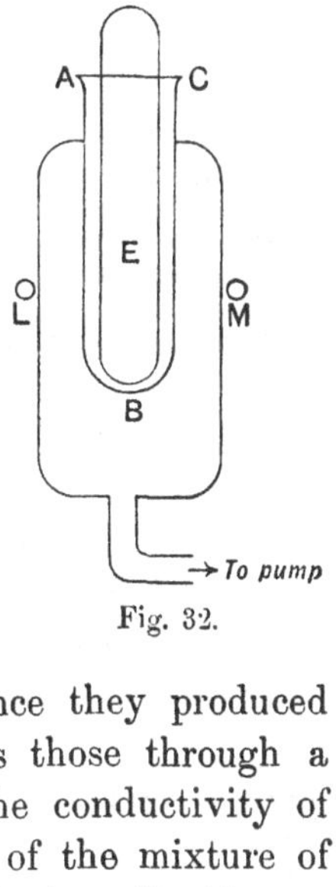

Fig. 32.

* The actual percentage depends on the pressure of the gas as well as on what kind of gas it is; the figures given above refer to an actual experiment.

off the effects of the primary, depends upon the length of spark passing between the jars, and so upon the electromotive intensity acting on the gas: in other words, conduction through these gases does not obey Ohm's law: the conductivity instead of being constant increases with the electromotive intensity. This is what we should expect if we regard the discharge through the gas as due to the splitting up of its molecules: the greater the electromotive intensity the greater the number of molecules which are split up and which take part in the conduction of the electricity.

85.] Another method by which we can prove the great conductivity of these rarefied gases at the pressures when they conduct best, is by measuring the energy absorbed by a secondary circuit made of the rarefied gas when placed inside a primary circuit conveying a rapidly alternating current. We shall see, Chapter IV, that when a conductor, whose conductivity is comparable with that of electrolytes, is placed inside the primary coil, the amount of energy absorbed per unit time is proportional to the conductivity of the conductor; so that if we measure the absorption of energy by equal and similar portions of two electrolytes we can find the ratio of their conductivities. In the case of these electrodeless discharges we can easily compare the absorption of energy by two different secondary circuits in the following manner. In the primary circuit connecting the outside coatings of two jars, two loops, A and B, Fig. 33, are made, a standard bulb is placed in A and the substance to be examined in B. When a large amount of energy is absorbed by the secondary in B, the brightness of the discharge through the bulb placed in A is diminished, and by observing the brightness of this discharge we can estimate whether the absorption of energy by two different secondaries placed in B is the same. If, now, an exhausted bulb be placed in B, the brightness of the discharge of the A bulb is at once diminished; indeed it is not difficult so to adjust the spark by which the jars are discharged, that a brilliant discharge passes in A when the B bulb is out of its coil, and no visible discharge when it is inside the coil. To compare the absorption of energy

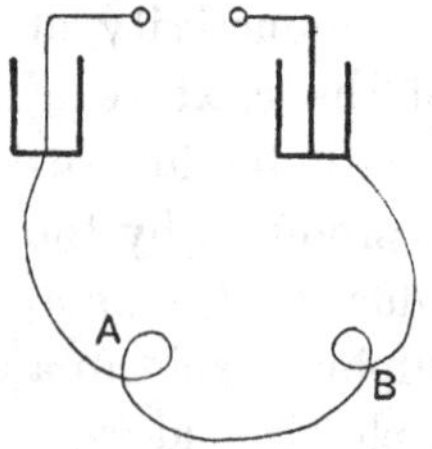

Fig. 33.

by the rarefied gas with that by an electrolyte we have merely to fill the bulb with an electrolyte, and alter the strength of the electrolyte until the bulb when filled with it produces the same effect as when it contained the rarefied gas. It will be found that in order to produce as great an absorption of energy as that due to a comparatively inefficient bulb filled with rarefied air, a very strong solution of an electrolyte must be put into the bulb; while a bulb which is exhausted to the pressure at which it produces its maximum effect absorbs a greater amount of energy than when filled even with the best conducting electrolyte we can obtain. We conclude from these experiments that the very large electromotive intensities which are produced by the discharge of a Leyden jar can, when no electrodes are used, send through a rarefied gas when the pressure is not too low much larger currents than the same electromotive intensities could send through even the best conducting mixture of water and sulphuric acid.

The results just quoted show that the conductivity, if estimated per molecule taking part in the discharge, is much higher for rare gases than even for metals such as copper or silver.

86.] The large values of the conductivities of these rarefied gases when no electrodes are used are in striking contrast to the almost infinitesimal values which are obtained when electrodes are present. This illustrates the reluctance which the discharge has to pass across the junction of a rarefied gas and a metal: the experiments described in Art. 81 are a very direct proof of this peculiarity of the discharge. It seems also to be indicated, though perhaps not quite so directly, by some experiments made by Liveing and Dewar (*Proc. Roy. Soc.* 48, p. 437, 1890) on the spectrum of the discharge. They found that the spectrum of a discharge passing through a gas which holds in suspension a considerable quantity of metallic dust does not show any of the lines of the metal. This is what we should expect from the experiments described in Art. 81, as these show that the discharge would take a very round-about course to avoid passing through the metal.

87.] There seem some indications that this reluctance of the discharge to pass from one substance to another extends also to the case when both substances are in the gaseous state, and that when the discharge passes through a mixture of two gases

A and B, the discharges through A and through B respectively are in parallel rather than in series: in other words, that the polarized chains of molecules, which are formed before the discharge passes, consist some of A molecules and some of B molecules, but that the chains conveying the discharge do not consist partly of A and partly of B molecules. Thus, if the discharge is passing through a mixture of hydrogen and nitrogen, the chains in which the molecules split up and along which the electricity passes may be either hydrogen chains or nitrogen chains, but not chains containing both hydrogen and nitrogen. This seems to be indicated by the fact that when the discharge passes through a mixture of hydrogen and nitrogen, the spectrum of the discharge may, though a considerable quantity of nitrogen is present, show nothing but the hydrogen lines.

Crookes' observations on the striations in a mixture of gases (*Presidential Address to the Society of Telegraph Engineers*, 1891) seem also to point to the conclusion that the discharges through the different gases in the mixture are separate; for he found that when several gases are present in the discharge tube, different sets of striations, Art. 99, are found when the discharge passes through the tube, the spectrum of the bright portions of the striae in one set showing the lines of one, and only one, of the gases in the mixture; the spectrum of another set showing the lines of another of the gases and so on, indicating that the discharges through the components of the mixture are distinct.

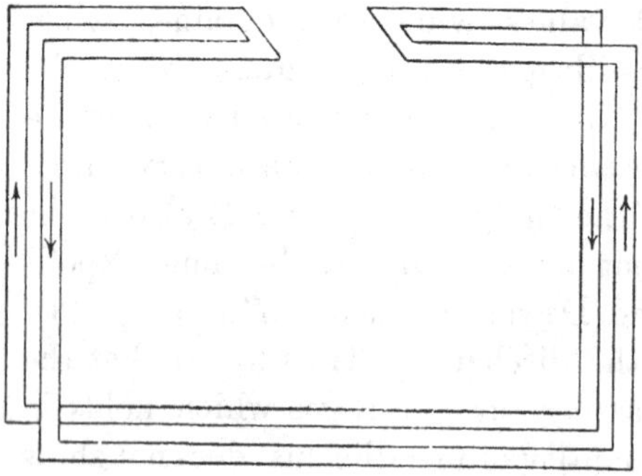

Fig. 34.

88.] When the discharge can continue in the same medium all the way it can traverse remarkably long distances, even though the greater portion of the secondary may be of such a shape as not to add anything to the electromotive force acting round it. Thus, for example, the discharge will pass through a very long secondary, even though the tube of which this secondary is made is bent up so that the greater part of it is at right angles to the electromotive intensity acting upon it. By using square coils with several turns for the primaries, I have succeeded in sending discharges through tubes of this kind over 12 feet in length. On the other hand, there will be no discharge through a rarefied gas if the shape of the tube in which it is contained is such that the electromotive force round it is either zero or very small: it is impossible, for example, to get a discharge of this kind through a tube shaped like the one shown in Fig. 34.

Action of a Magnet on the Electrodeless Discharge.

89.] A magnet deflects the discharge through a rarefied gas in much the same way as it does a flexible wire carrying a current which flows in the same direction as the one through the gas. As the electrodeless discharges through the rarefied gas are oscillatory, they are when under the action of a magnet separated into two distinct portions, the magnet driving the discharge in one direction one way and that in the opposite direction the opposite way. Thus, when a bulb in which the discharge passes as a ring in a horizontal plane is placed between the poles of an electromagnet arranged so as to produce a horizontal magnetic field, those parts of the ring which are at right angles to the lines of magnetic force are separated into two portions, one being driven upwards, the other downwards. The displacement of the discharge is not however the only effect observed when the discharge bulb is placed in a magnetic field, for the difficulty which the discharge experiences in getting through the rarefied gas is very much increased when it has to pass across lines of magnetic force. This effect, which is very well marked, can perhaps be most readily shown when the discharge passes as a bright ring through a spherical bulb. If such a bulb is placed near a strong electromagnet it is easy to adjust the length of spark in the primary circuit, so that when the magnet is 'off' a brilliant discharge passes through the bulb, while when the magnet is 'on' no discharge at all can be detected.

90.] The explanation of this effect would seem to be somewhat as follows. The discharge through the rarefied gas does not rise to its full intensity quite suddenly, but, as it were, feels its way. The gas first breaks down along the line where the electromotive intensity is a maximum, and a small discharge takes place along this line. This discharge produces a supply of dissociated molecules along which subsequent discharges can pass with greater ease. The gas is thus in an unstable state with regard to the discharge, since as soon as any small discharge passes through it, it becomes electrically weaker and less able to resist subsequent discharges. When, however, the gas is in a magnetic field, the magnetic force acting on the discharge produces a mechanical force which displaces the molecules taking part in the discharge from the line of maximum electromotive intensity; thus subsequent discharges will not find it any easier to pass along this line in consequence of the passage of the previous discharge. There will not therefore be the same unstability in this case as there is in the one where the gas is free from the action of the magnetic force. A confirmation of this view is afforded by the appearance presented by the discharge when the intensity of the magnetic field is reduced until the discharge just, but only just, passes when the magnetic field is on: in this case the discharge instead of passing as a steady fixed ring, flickers about the tube in a very undecided way. Unless some displacement of the line of easiest discharge is produced by the motion of the dissociated molecules under the action of the magnetic force, it is difficult to understand why the magnet should displace the discharge at all, unless the Hall effect in rarefied gases is very large.

91.] In the preceding case the discharge was retarded because it had to flow across the lines of magnetic force, when however the lines of magnetic force run along the line of discharge the action of the magnet facilitates the discharge instead of retarding it. This effect is easily shown by an arrangement of the following kind. A square tube ABCD, Fig. 35, is placed outside the primary EFGH, the lower part of the discharge tube being situated between the poles L, M of an electromagnet. By altering the length of the spark between the jars, the electromotive intensity acting on the secondary circuit can be adjusted until no discharge passes round the tube ABCD when the magnet is off, whilst a bright discharge occurs as long as the magnet is on. The two effects of the magnet on the discharge, viz. the stoppage of the discharge across the lines of force and the help given to it along these lines, may be prettily illustrated by placing in this experiment an exhausted bulb N inside the primary. The spark length can be adjusted so that when the magnet is 'off' the discharge passes through the bulb and not in the square tube; while when the magnet is 'on' the discharge passes in the square tube and not in the bulb.

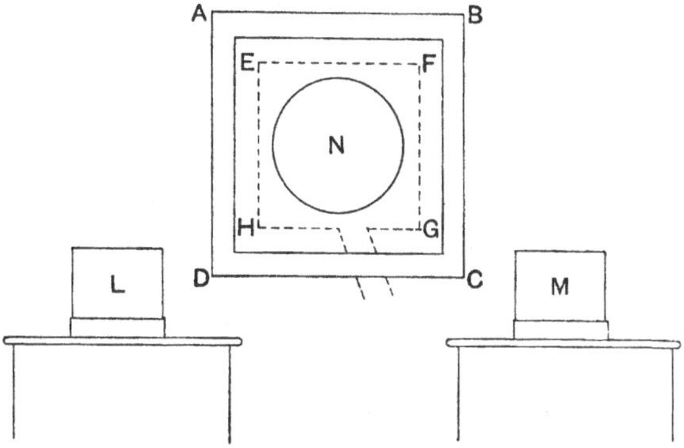

Fig. 35.

92.] The explanation of the longitudinal effect of magnetic force is more obscure than that of the transverse effect, it is possible however that both are due to the same cause. For if the feeble discharge with which we suppose the total discharge to begin branches away at all from the main line, these branches will, when the magnetic force is parallel to the line of discharge, be brought into this line by the action of the magnetic force; there will thus be a larger supply of dissociated molecules along the main line of discharge, and therefore an easier path for subsequent discharges when the magnetic force is acting than when it is not.

This action of the magnet is not confined to this kind of discharge; in fact I observed it first for a glow discharge, which took place more easily from the pole of an electromagnet when the magnet was 'on' than when it was 'off'.

93.] Professor Fitzgerald has suggested that this effect of the magnetic field on the discharge may be the cause of the streamers which are observed in the aurora, the rare air, since it is electrically weaker along the lines of magnetic force than at right angles to them, transmitting brighter discharges along these lines than in any other direction.

Electric discharge through rarefied Gases when Electrodes are used.

94.] When the discharge passes between electrodes through a rare gas, the appearance of the discharge at the positive and negative electrodes is so strikingly different that the discharge loses all appearance of uniformity. Fig. 36, which is taken from a paper by E. Wiedemann (*Phil. Mag.* [5], 18, p. 35, 1884), represents the appearance presented by the discharge when it passes through a gas at a pressure comparable with that due to half a millimetre of mercury. Beginning at the negative electrode k we meet with the following phenomena. A velvety glow runs often in irregular patches over the surface of the negative electrode; a wire placed inside this glow casts a shadow towards the negative electrode (Schuster, *Proc. Roy. Society*, 47, p. 557, 1890).

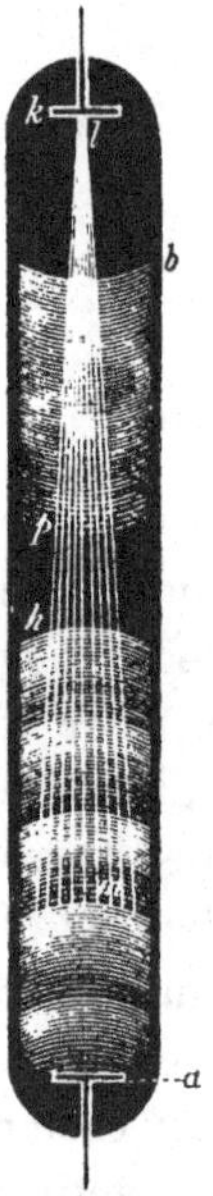

Fig. 36.

Next to this there is a comparatively dark region lb, called sometimes 'Crookes' space' and sometimes the 'first dark space;' the length of this region depends on the density of the gas, it gets longer as the density diminishes. Puluj's experiments (*Wien. Ber.* 81 (2), p. 864, 1880) show that the length does not vary directly as the reciprocal of the density, in other words, that it is not proportional to the mean free path of the molecules.

The luminous boundary b of this dark space is approximately such as could be got by tracing the locus of the extremities of normals of constant length drawn from the nega-

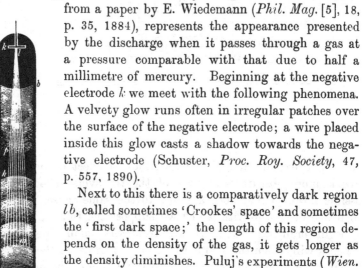

Fig. 37.

tive electrode: thus, if the electrode is a disc, the luminous boundary of the dark space is over a great part of its surface nearly plane as in Fig. 37, which is given by Crookes; while if it is a circular ring of wire, the luminous boundary resembles that shown in Fig. 38 (De la Rue). The length of the dark space also

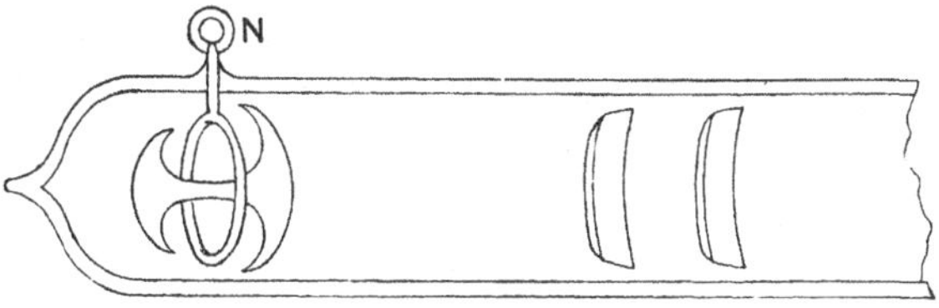

Fig. 38.

depends to some extent on the current passing through the gas, an increase of current producing (see Schuster, *Proc. Roy. Society*, 47, p. 556, 1890) a slight increase in the length of the dark space. Some idea of the length of the dark space at different pressures may be got from the following table of the results of some experiments made by Puluj (*Wien. Ber.* 81 (2), p. 864, 1880) with a cylindrical discharge tube and disc electrodes:—

Pressure in millimetres of mercury.	Length of dark space in air in mm.
1·46	2·5
.66	4·5
·51	5·8
.30	7·8
·24	9·5
.16	14·0
·12	15·5
·09	19·5
·06	22·0

The mean free path of the molecules is very much smaller than the length of the dark space; thus at a pressure of 1·46 mm. of mercury, the mean free path is only ·04 mm. Crookes found (*Phil. Trans.* Part I, 1879, pp. 138–9) that the dark space is longer in hydrogen than in air at the same pressure, but that in carbonic acid it is considerably shorter.

95.] Crookes' theory of the dark space is that it is the region which the negatively electrified particles of gas shot off from the cathode (see Art. 108) traverse before making an appreciable number of collisions with each other, and that the brightly luminous boundary of this space is the region where the collisions occur, these collisions exciting vibrations in the particles and so making them luminous. It is an objection, though perhaps not a fatal one, to this view, that the thickness of the dark space is very much greater than the mean free path of the molecules. We shall see later on that if the luminosity is due to gas shot from the negative electrode, this gas must be in the atomic and not in the molecular condition; in the former condition its free path would be greater than the value calculated from the ordinary data of the Molecular Theory of Gases, though if we take the ordinary view of what constitutes a collision we should not expect the difference to be so great as that indicated by Puluj's experiments.

96.] The size of the dark space does not seem to be much affected by the material of which the negative electrode is made, as long as it is metallic. It is however considerably shorter over sulphuric acid electrodes than over aluminium ones (Chree, *Proc. Camb. Phil. Soc.* vii, p. 222, 1891). Crookes (*Phil. Trans*, 1879, p. 137) found that if a metallic electrode is partly coated with lamp black the dark space is longer over the lamp-blacked portion than over the metallic. Lamp black however absorbs gases so readily that this effect may be due to a change in the gas and not to the change in the electrode. The dark space is also, as Crookes has shown (loc. cit.), independent of the position of the positive electrode. When the cathode is a metal wire raised to a temperature at which it is incandescent, Hittorf (*Wied. Ann.* 21, p. 112, 1884) has shown that the changes in luminosity which with cold electrodes are observed in the neighbourhood of the cathode disappear. There is a difference of opinion as to whether the dark space exists when the discharge passes through mercury vapour, Crookes maintaining that it does, Schuster that it does not.

97.] Adjoining the 'dark space' is a luminous space, *bp* Fig. 36, called the 'negative column,' or sometimes the 'negative glow;' the length of this is very variable even though the pressure is constant. The spectrum of this part of the discharge exhibits peculiarities which are not in general found in that of the other luminous parts of the discharge. Goldstein (*Wied. Ann.* 15, p. 280, 1882) however has found that when very intense discharges are used, the peculiarities in the spectrum, which are usually confined to the negative glow, extend to the other parts of the discharge.

98.] The negative glow is independent of the position of the positive electrode; it does not bend round, for example, in a tube shaped as in Fig. 41, but is formed in the part of the tube away from the positive electrode. This glow is stopped by any substance, whether a conductor or an insulator, against which it strikes. The development of the negative glow is also checked when the space round the negative electrode is too much restricted by the walls of the discharge tube. Thus Hittorf (*Pogg. Ann.* 136, p. 202, 1869) found that if the discharge took place in a tube shaped like Fig. 39, when the wire c

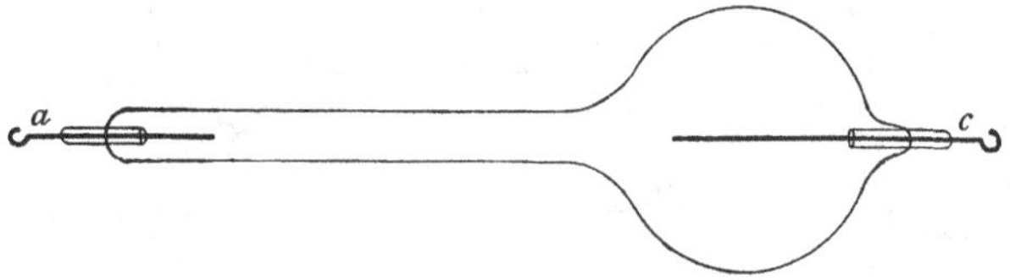

Fig. 39.

in the bulb was made the negative electrode, the negative glow spread over the whole of its length, while if the wire a in the neck was used as the negative electrode the glow only occurred at its tip.

99.] Next after the negative glow comes a second comparatively non-luminous space, *ph* Fig. 36, called the 'second negative dark space,' or by some writers the 'Faraday space;' this is of very variable length and is sometimes entirely absent. Next after this we have a luminous column reaching right up to the positive electrode, this is called the 'positive column.' Its luminosity very often exhibits remarkable periodic alterations in intensity such as those shown in Fig. 40, which is taken from a paper by De la Rue and Hugo Müller (*Phil. Trans.*, 1878, Part I, p. 155); these are called 'striations,' or 'striae;' under favourable circumstances they are exceedingly regular and constitute the most striking feature of the discharge. The bright parts of the striations are slightly concave to the positive electrode. The distance between the bright parts depends upon the pressure of the gas and the diameter of the discharge tube. The distance increases as the density of the gas diminishes.

According to Goldstein (*Wied. Ann.* 15, p. 277, 1882), if d is the distance between two striations and ρ the density of the gas, d varies as ρ^{-n}, where n is somewhat less than unity. The distance between the bright parts of successive striations increases as the diameter of the discharge tube increases, provided the striations reach to the sides of the tube. Goldstein (l. c.)

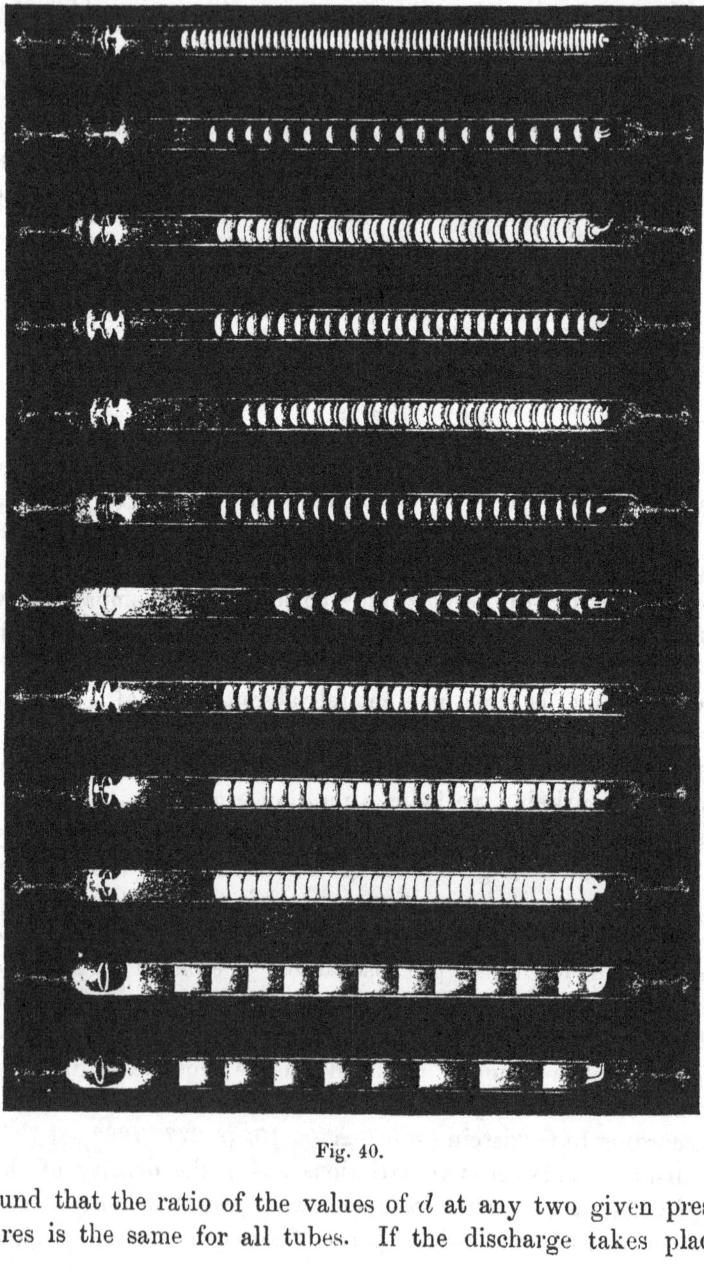

Fig. 40.

found that the ratio of the values of d at any two given pressures is the same for all tubes. If the discharge takes place in a tube which is wider in some places than in others, the striations are more closely packed in the narrow parts of the tube than they are in the wide.

The striations have very often a motion of translation along the tube; this motion is quite irregular, being sometimes towards the positive electrode and sometimes away from it. This can easily be detected by observing, as Spottiswoode did, the discharge in a rapidly rotating mirror. These movements of the striae tend to make the striated appearance somewhat indistinct, and if the movements are too large may obliterate it altogether; thus many discharges which show no appearance of striation when examined in the ordinary way, are seen to be striated when looked at in a revolving mirror. The difficulty of detecting whether a discharge is striated or not is, in consequence of the motion of the striae, very much greater when the striae are near together than when they are far apart, so that it is quite possible that discharges are striated at pressures much greater than those at which striations are usually observed.

Goldstein, using a tube with moveable electrodes, showed (*Wied. Ann.* 12, p. 273, 1881) that when the cathode is moved the striae move as if they were rigidly connected with it, while when the anode is moved the position of the striae is not affected except in so far as they may be obliterated by the anode moving past them.

100.] The striations are not confined to any one particular method of producing the discharge, they occur equally well whether the discharge is produced by an induction coil or by a very large number of galvanic cells. They do not, however, occur readily in the electrodeless discharge; indeed I have never observed them when a considerable interval intervened between consecutive sparks. By using an induction coil large enough to furnish a supply of electricity sufficient to produce an almost continuous torrent of sparks between the jars, I have been able to get striations in exhausted bulbs containing hydrogen or other gases.

101.] The striations are influenced by the quantity of current flowing through the tube; this can easily be shown by putting a great external resistance in the circuit, such as a wet string. The changes produced by altering the current are complex and irregular: there seems to be a certain intensity of current for which the steadiness of the striations is a maximum (De la Rue and Hugo Müller, *Comptes Rendus*, 86, p. 1072, 1878). Crookes has found (*Presidential Address to the Society of Telegraph Engineers*, 1891) that when the discharge passes through a mixture of different gases there is a separate set of striations for each gas: the colour of the striations in each set being different. Crookes proved this by observing the spectra of the different striae. A full account of the different coloured striations observed in air is given by Goldstein (*Wied. Ann.* 12, p. 274, 1881).

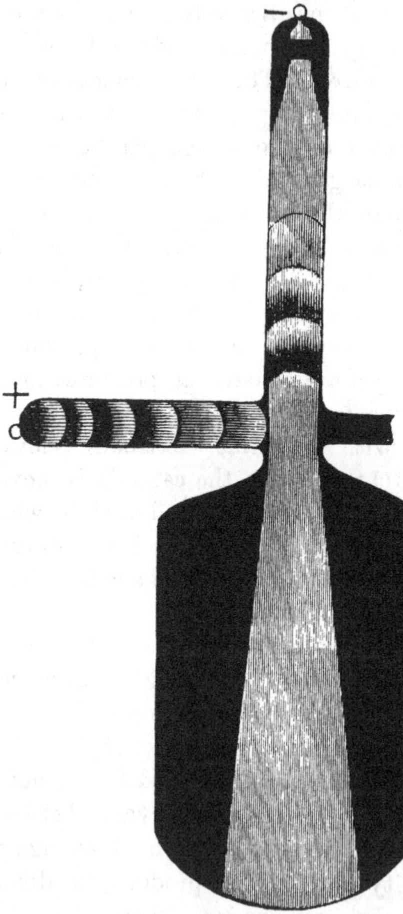

Fig. 41.

102.] When we consider the action of a magnet on the striated positive column we shall see reasons for thinking that any portion of the positive column between the bright parts of consecutive striations constitutes a separate discharge, and that the discharges in the several portions do not occur simultaneously, but that the one next the anode begins the discharge, and the others follow on in order.

103.] The positive column bears a very much more important relation to the discharge than either the negative dark space or the negative glow. The latter effects are merely local, they do not depend upon the position of the positive electrode, nor do they increase when the length of the discharge tube is increased. The positive column, on the other hand, takes the shortest route through the gas to the negative electrode. Thus, if, for example, the discharge takes place in a tube like Fig. 41, the positive column bends round the corner so as to get to the negative electrode, while the negative glow goes straight down the vertical tube, and is not affected by the position of the positive electrode. Again, if the length of the tube is increased the size of the negative dark space and of the negative glow is not affected, it is only the positive column which lengthens out. I have, for example, obtained the discharge through a tube 50 feet long, and this tube, with the exception of a few inches next the cathode, was entirely filled by the positive column, which was beautifully striated. These examples show that it is the positive column which really carries the discharge through the gas, and that the negative dark space and the negative glow are merely local effects, depending on the peculiarities of the transference of electricity from a gas to a cathode.

104.] By the use of long discharge tubes such as those mentioned above, it is possible to determine the direction in which the luminosity in the positive column travels and to measure its rate of progression. The first attempt at this seems to have been made by Wheatstone, who, in 1835, observed the appearance presented in a rotating mirror by the discharge through a vacuum tube 6 feet long; he concluded from his observations that the velocity with which the flash went through the tube could not have been less than 8×10^7 cm. per second. This great velocity is not accompanied by a correspondingly large velocity of the luminous molecules, for von Zahn (*Wied. Ann.* 8, p. 675, 1879) has shown that the lines of the spectrum of the gas in the discharge tube are not displaced by as much as $\frac{1}{40}$ of the distance between the D lines when the line of sight is in the direction of the discharge. It follows from this by Döppler's principle, that the particles when emitting light are not travelling at so great a rate as a mile a second, proving, at any rate, that the luminous column does not consist of a wind of luminous particles travelling with the velocity of the discharge.

105.] Wheatstone's observations only give an inferior limit to the velocity of the discharge; they do not afford any information as to whether the luminous column travels from the anode to the cathode or in the opposite direction. To determine this, as well as measure the velocity of the luminosity in the positive column, I made the following experiment. ABCDEFG, Fig. 42, is a glass tube about 15 metres long and 5 millimetres in diameter, which, with the exception of two horizontal pieces of BC and GH, is

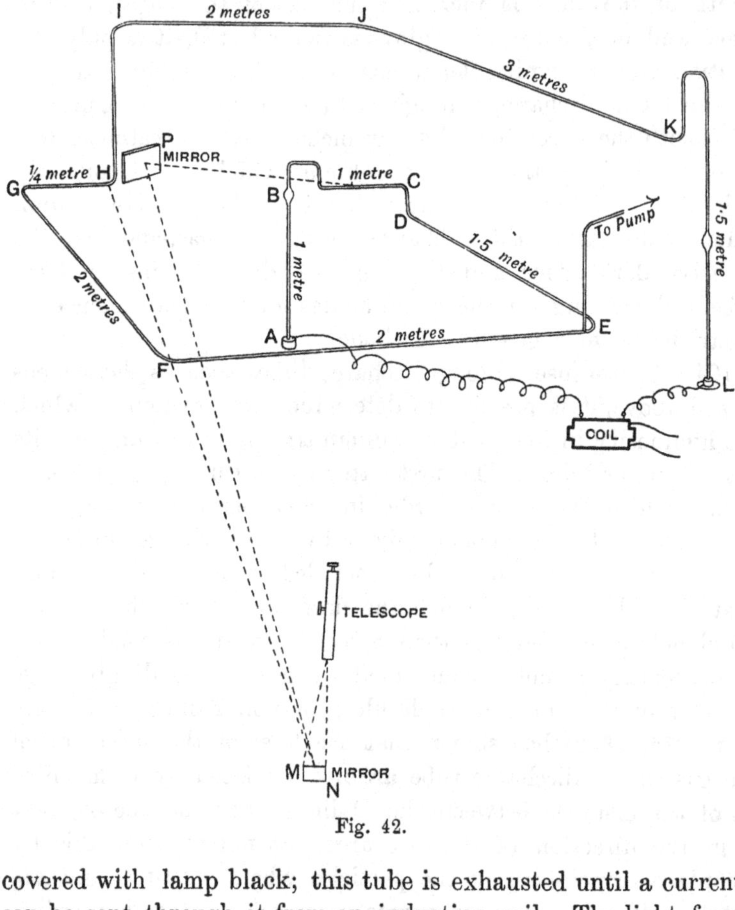

Fig. 42.

covered with lamp black; this tube is exhausted until a current can be sent through it from an induction coil. The light from the uncovered portions of the tube falls on a rotating mirror MN, placed at a distance of about 6 metres from BC; the light from GH falls on the rotating mirror directly, that from BC after reflection from the plane mirror P. The images of the bright portions of the tube after reflection from the mirror are viewed through a telescope, and the mirrors are so arranged that when the revolving mirror is stationary the images of the bright portions GH and BC of the tube appear as portions of the same horizontal straight line. The terminals of the long vacuum tube are pushed through mercury up the vertical tubes AB KL. This arrangement was adopted because by running sulphuric acid up these tubes the terminals could easily be changed from pointed platinum wires to flat liquid surfaces, and the effect of very different terminals on the velocity and direction of the discharge readily investigated. The bulbs on the tube are also useful as receptacles of sulphuric acid, which serves to dry the gas left in the tube. The rotating mirror was driven at a speed of from 400 to 500 revolutions per second by a Gramme machine. It was not found possible to make any arrangement work well which would break the primary circuit of the induction coil when the mirror was in such a position that the images of the luminous portions of the tube would be reflected by it into the field of view of the telescope. The method finally adopted was to use an independent slow break for the coil and look patiently through the telescope at the rotating mirror until the break happened to occur just at the right moment. When the observations were made in this way the observer at the telescope saw, on an average about once in four minutes, sharp bright images of the portions BC and GH of the tube, not sensibly broadened but no longer quite in the same straight line. The relative displacement of those images was reversed when the poles of the coil were reversed, and also when the direction of rotation of the mirror was reversed. This displacement of the images of BC and GH from the same straight line is due to the finite velocity with which the luminosity is propagated: for, if the mirror can turn through an appreciable angle while the luminosity travels from BC to GH or from GH to BC, these images of BC and GH, when seen in the telescope after reflection from the revolving mirror, will no longer be in the same straight line. If the mirror is turning so that on looking through the telescope the images seem to come in at the top and go out at the bottom of the field of view, the image of that part of the tube at which the luminosity first appears will be raised above that of the other part. If we know the rate of rotation of the mirror, the vertical displacement of the images and the distance between BC and GH, the rate of propagation of the luminosity may be calculated. The displacement of the images showed that the luminosity always travelled from the positive to the negative electrode. When AB was the negative electrode, the luminous discharge arrived at GH, a place about 25 feet from the positive electrode, before it reached BC, which was only a few inches from the cathode, and as the interval between its appearance at these places was about the same as when the current was reversed, we may conclude that when AB is the cathode the luminosity at a place BC, only a few inches from it, has started from the positive electrode and traversed a path enormously longer than its distance from the cathode. The velocity of the discharge through air at the pressure of about $\frac{1}{2}$ a millimetre of mercury in a tube 5 millimetres in diameter was found to be rather more than half the velocity of light.

106.] The preceding experiment was repeated with a great variety of electrodes; the result, however, was the same whether the electrodes were pointed platinum wires, carbon filaments, flat surfaces of sulphuric acid, or the one electrode a flat liquid surface and the other a sharp-pointed wire. The positive luminosity travels from the positive electrode to the negative, even though the former is a flat liquid surface and the latter a pointed wire. The time taken by the luminosity to travel from BC to GH was not affected to an appreciable extent by inserting between BC and GH a number of pellets of mercury, so that the discharge had to pass from the gas to the mercury several times in its passage between these places: the intensity of the light was however very much diminished by the insertion of the mercury.

107.] The preceding results bear out the conclusion which Plücker (*Pogg. Ann.* 107, p. 89, 1859) arrived at from the consideration of the action of a magnet on the discharge, viz. that the positive column starts from the positive electrode; they also confirm the result which Spottiswoode and Moulton (*Phil. Trans.* 1879, p. 165) deduced from the consideration of what they have termed 'relief' effects, that the time taken by the negative electricity to leave the cathode is greater than the time taken by the positive luminosity to travel over the length of the tube.

Negative Rays or Molecular Streams.

108.] Some of the most striking of the phenomena shown by the discharge through gases are those which are associated with the negative electrode. These effects are most conspicuous at low pressures, but Spottiswoode and Moulton's experiments (*Phil. Trans.* 1880, pp. 582, 85 *seq.*) show that they exist over a wide range of pressure. The sides of the tube exhibit a brilliant phosphorescence, behaving as if something were shot out at right angles, or nearly so, to the surface of the cathode, which had the power of exciting phosphorescence on any substance on which it fell, provided that this substance is one which becomes phosphorescent under the action of ultra-violet light. The portions of the tube enclosed within the surface formed by the normals to the cathode will, when the pressure of the gas is low, show a bright green phosphorescence if the tube is made of German glass, while the phosphorescence will be blue if the tube is made of lead glass. Perhaps the easiest way of describing the general features of this effect is to say that they are in accordance with Mr. Crookes' theory, that particles of gas are projected with great velocities at right angles, or nearly so, to the surface of the cathode, and that these particles in a highly exhausted tube strike the glass before they have lost much momentum by collision with other molecules, and that the bombardment of the glass by these particles is intense enough to make it phosphoresce. The following extract from Priestley's *History of Electricity*, p. 294, 1769, is interesting in connection with this view: 'Signior Beccaria observed that hollow glass vessels, of a certain thinness, exhausted of air, gave a light when they were broken in the dark. By a beautiful train of experiments, he found, at length, that the luminous appearance was not occasioned by the breaking of the glass, but by the dashing of the external air against the inside, when it was broke. He covered one of these exhausted vessels with a receiver, and letting the air suddenly on the outside of it, observed the very same light. This he calls his *new invented phosphorous.*'

109.] If a screen made either of an insulator or a conductor is placed between the electrode and the walls of the tube, a shadow of the screen is thrown on the walls of the tube, the shadow of the screen remaining dark while the glass round the shadow phosphoresces brightly. In this way many very beautiful and brilliant effects have been produced by Mr. Crookes and Dr. Goldstein, the two physicists who have devoted most attention to this subject. One of Mr. Crookes' experiments in which the shadow of a Maltese cross is thrown on the walls of the tube is illustrated in Fig. 43.

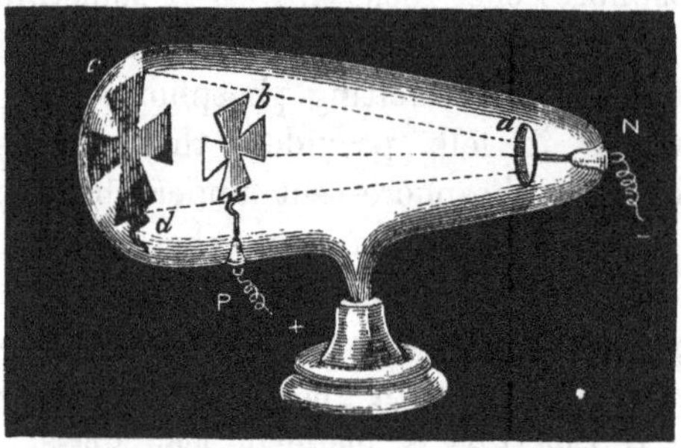

Fig. 43.

110.] As we have already mentioned, the colour of the phosphorescence depends on the nature of the phosphorescing substance; if this substance is German glass the phosphorescence

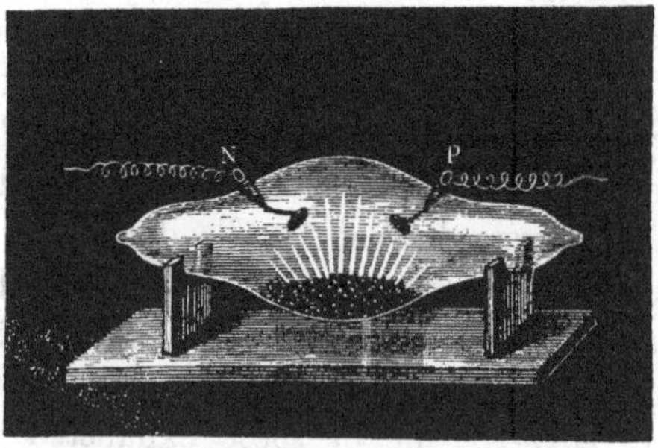

Fig. 44.

is green, if it is lead glass the phosphorescence is blue. Crookes found that bodies phosphorescing under this action of the negative electrode give out characteristic band spectra, and he has developed this observation into a method of the greatest importance for the study of the rare earths: for the particulars of this line of research we must refer the reader to his papers 'On Radiant Matter Spectroscopy,' *Phil. Trans.* 1883, Pt. III, and 1885, Pt. II.

The way the spectrum is produced is represented in Fig. 44, the substance under examination being placed in a high vacuum in the path of the normals to the cathode.

111.] Crookes also found that some substances, when submitted for long periods to the action of these rays, undergoes remarkable modifications, which seems to suggest that the phosphorescence is attended (or caused by?) chemical changes slowly taking place in the phosphorescent body. He also observed that glass which has been phosphorescing for a considerable time seems to get tired, and to respond less readily to this action of the cathode. Thus, for example, if after the experiment in Fig. 44 has been proceeding for some time the cross is shaken down, or a new cathode used whose line of fire does not cut the cross, the pattern of the cross will still be seen on the glass, but now it will be brighter than the adjacent parts instead of darker. The portions outside the pattern of the cross have got tired by their long phosphorescence, and respond less vigorously to the stimulus than the portions forming the cross which were previously shielded. Crookes found this 'exhaustion' of the glass could survive the melting and reblowing of the bulb.

By using a curved surface for the negative electrode, such as a portion of a hollow cylinder or of a spherical shell, this effect of the negative rays may be concentrated to such an extent that a platinum wire placed at the centre of the cylinder or sphere becomes red hot.

112.] The negative rays are deflected by a magnet in the same way as they would be if they consisted of particles moving away from the negative electrode and carrying a charge of negative electricity. This deflection is made apparent by the movement of the phosphorescence on the glass when a magnet is brought near the discharge tube.

On the other hand they are not deflected when a charged body is brought near the tube; this does not prove, however, that the rays do not consist of electrified particles, for we have seen that gas conveying an electric discharge is an extremely good conductor, and so would be able to screen the inside of the tube from any external electrostatic action. Crookes (*Phil. Trans.* 1879, Pt. II, p. 652) has shown, moreover, that two pencils of these rays repel each other, as they would do if each pencil consisted of particles charged with the same kind of electricity. The experiment by which this is shown is represented in Fig. 45; *a*, *b* are metal discs either or both of which may be made into cathodes, a diaphragm with two

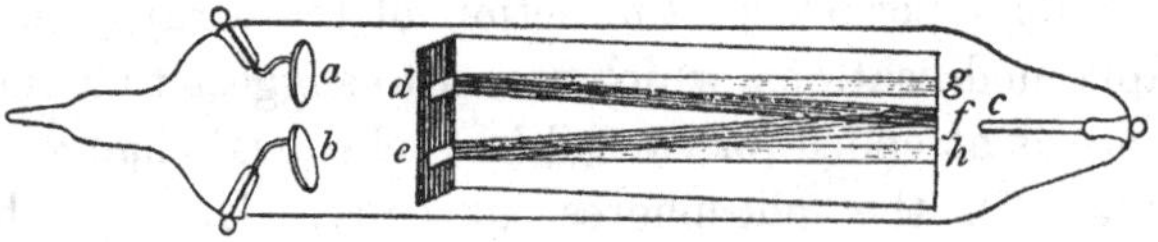

Fig. 45.

openings *d* and *e* is placed in front of the disc, and the path of the rays is traced by the phosphorescence they excite in a chalked plate inclined at a small angle to their path. When *a* is the cathode and *b* is idle, the rays travel along the path *d f*, and when *b* is the cathode and *a* idle they travel along the path *e f*, but when *a* and *b* are cathodes simultaneously the paths of the rays are *dg* and *eh* respectively, showing that the two streams have slightly repelled each other.

113.] Crookes (*Phil. Trans.* 1879, Part II, p. 647) found that if a disc connected with an electroscope is placed in the full line of fire of these rays it receives a charge of *positive* electricity. This is not, however, a proof that these rays do not consist of negatively electrified particles, for the experiments described in Art. 81 show that electricity does not pass at all readily from a gas to a metal, and the positive electrification of the disc may be a secondary effect arising from the same cause as the positive electrification of a plate when exposed to the action of ultra-violet light. For since the action of these rays is the same as that of ultra-violet light in producing phosphorescence in the bodies upon which they fall, it seems not unlikely that the rays may resemble ultra-violet light still further and make any metal plate on which they fall a cathode.

Hertz (*Wied. Ann.* 19, p. 809, 1883) was unable to discover that these rays produced any magnetic effect.

The paths of the negative rays are governed entirely by the shape and position of the cathode, they are quite independent of the shape or position of the anode. Thus, if the cathode and anode are placed at one end of an exhausted tube, as in Fig. 46, the cathode rays will not bend round to the anode, but will go straight down the tube and make the opposite end phosphoresce.

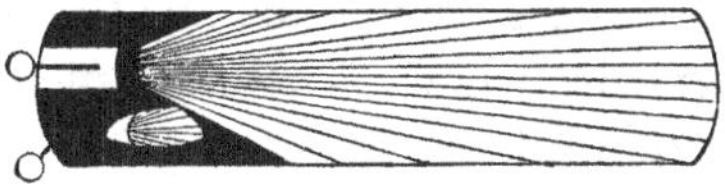

Fig. 46.

Any part of the tube which is made to phosphoresce by the action of these rays seems to acquire the power of sending out such rays itself, or we may express the same thing by saying that the rays are diffusely reflected by the phosphorescent body (Goldstein, *Wied. Ann.* 15, p. 246, 1882). Fig. 47 represents the appearance presented by

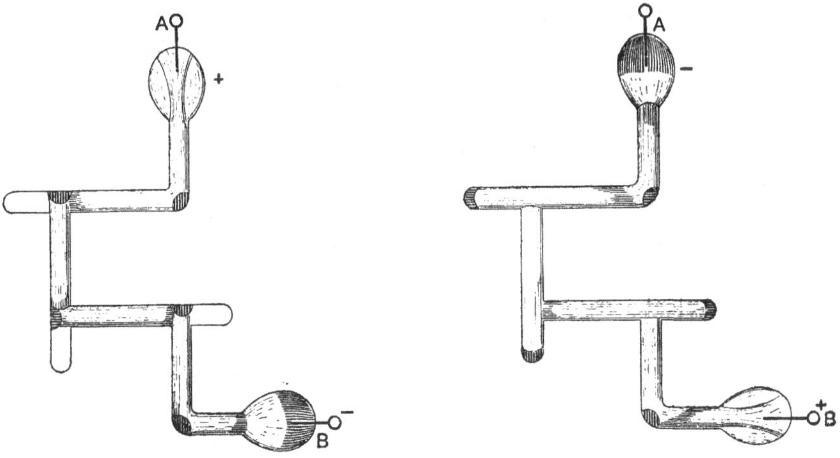

Fig. 47.

a bent tube when traversed by such rays, the darkly shaded places being the parts of the tube which show phosphorescence.

114.] These rays seem to be emitted by any negative electrode, even if this be one made by putting the finger on the glass of the tube near the anode. This produces a discharge of negative electricity from the glass just underneath the finger, and the characteristic green phosphorescence (if the tube is made of German glass) appears on the opposite wall of the tube; this phosphorescence is deflected by a magnet in exactly the same way as if the rays came from a metallic electrode. This experiment is sufficient to show the inadequacy of a theory that has sometimes been advanced to explain the phosphorescence, viz. that the particles shot off from the electrode are not gaseous particles, but bits of metal torn from the cathode; the phosphorescence being thus due to the disintegration of the negative electrode, which is a well-known feature of the discharge in vacuum tubes. The preceding experiment shows that this theory is not adequate, and Mr. Crookes has still further disproved it by obtaining the characteristic effects in tubes when the electrodes were pieces of tinfoil placed *outside* the glass.

115.] Goldstein (*Wied. Ann.* 11, p. 838, 1880) found that a sudden contraction in the cross section of the discharge tube produces on the side towards the anode the same effect as a cathode. These quasi-cathodes produced by the contraction of the tube are accompanied by all the effects which are observed with metallic cathodes, thus we have the dark space, the phosphorescence, and the characteristic behaviour of the glow in a magnetic field.

116.] Spottiswoode and Moulton (*Phil. Trans.* 1880, pp. 615–622) have observed a phosphorescence accompanying the positive column. They found that in some cases when this strikes the gas the latter phosphoresces. They ascribe this phosphorescence to a negative discharge called from the sides of the tube by the positive electricity in the positive column.

Mechanical Effects produced by the Negative Rays.

117.] Mr. Crookes (*Phil. Trans.* 1879, Pt. I, p. 152) has shown that when these rays impinge on vanes mounted like those in a radiometer the vanes are set in rotation. This can be shown by making the axle of the vanes run on rails as in Fig. 48. When the discharge passes through the tube, the vanes travel from the negative to the positive end of the tube. It is not clear, however, that this is a purely mechanical effect; it may, as suggested by Hittorf, be due to secondary thermal effects making the vanes act like those of a radiometer. In another experiment the vanes are suspended as in Fig. 49, and can be screened from the negative rays by the screen e; by tilting the tube the vanes can be brought wholly or partially out of the shadow of the screen. When the vanes are completely out of the shade they do not rotate as the bombardment is symmetrical; when, however, they are half in and half out of the shadow, they rotate in the same direction as they would if exposed to a bombardment from the negative electrode. The deflection of the negative

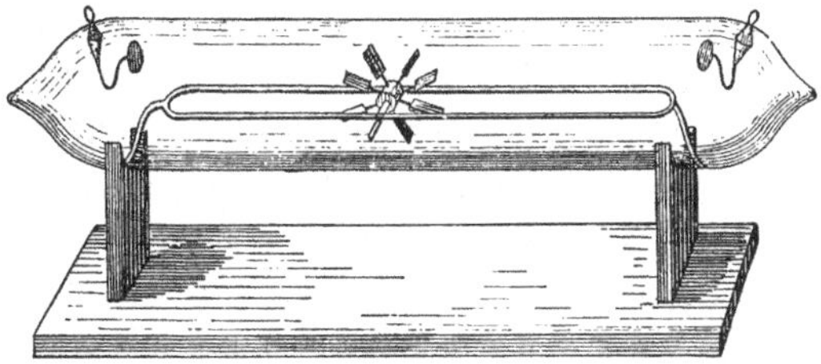

Fig. 48.

rays by a magnet is well illustrated by this apparatus. Thus, if the vanes are placed wholly within the shadow no rotation takes place; if, however, the south pole of an electro-magnet is brought to S, the shadow is deflected from the former position

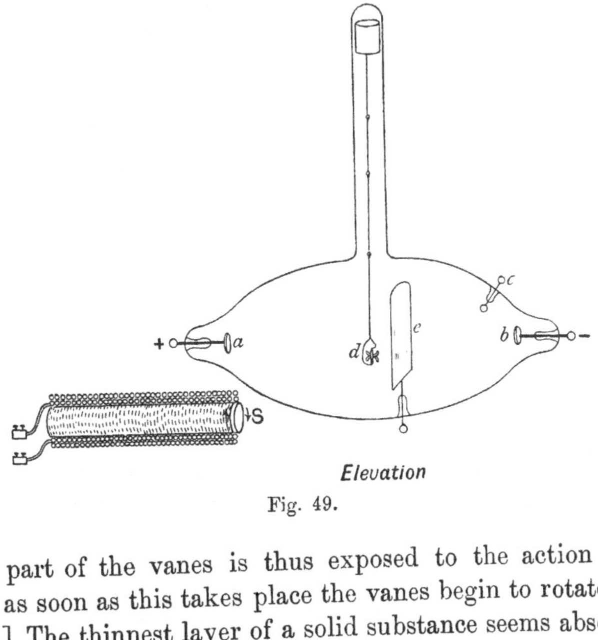

Fig. 49.

and a part of the vanes is thus exposed to the action of the rays; as soon as this takes place the vanes begin to rotate.

118.] The thinnest layer of a solid substance seems absolutely opaque to these radiations. Thus Goldstein (*Phil. Mag.* [5] 10, p. 177, 1880) found that a thin layer of collodion placed on the glass gave a perfectly black shadow, and Crookes (*Phil. Trans.*

1879, Part I, p. 151) found that a thin film of quartz, which is transparent to ultra-violet light, produced the same effect. This last result is of great importance in connection with a theory which has received powerful support, viz. that these 'rays' are a kind of ethereal vibration having their origin at the cathode. If this view were correct we should not expect to find a thin quartz plate throwing a perfectly black shadow, as quartz is transparent to ultra-violet light. To make the theory agree with the facts we have further to assume that no substance has been discovered which is appreciably transparent to these vibrations *. The sharpness and blackness of these shadows are by far the strongest arguments in support of the impact theory of the phosphorescence.

119.] Though Crookes' theory that the phosphorescence is due to the bombardment of the glass by gaseous particles projected from the negative electrode is not free from difficulties, it seems to cover the facts better than any other theory hitherto advanced. On one point, however, it would seem to require a slight modification: Crookes always speaks of the *molecules* of the gas receiving a negative charge. We have, however (see Art. 3), seen reasons for thinking that a molecule of a gas is incapable of receiving a charge of electricity, and that free electricity must be on the *atoms* as distinct from the molecules. If this view is right, we must suppose that the gaseous particles projected from the negative electrode are atoms and not molecules. This does not introduce any additional difficulty into the theory, for in the region round the cathode there is a plentiful supply of dissociated molecules or atoms; of these, those having a negative charge may under the repulsion of the negative electricity on the cathode be repelled from it with considerable violence.

120.] An experiment which I made in the course of an investigation on discharge without electrodes seems to afford considerable evidence that there is such a projection of atoms from the

* Since the above was written, Hertz (*Wied. Ann.* 45, p. 28, 1892) has found that thin films of gold leaf do not cast perfectly dark shadows but allow a certain amount of phosphorescence to take place behind them, which cannot be explained by the existence of holes in the film. It seems possible, however, that this is another aspect of the phenomenon observed by Crookes (Art. 113) that a metal plate exposed to the full force of these rays becomes a cathode; in Hertz's experiments the films may have been so thin that each side acted like a cathode, and in this case the phosphorescence on the glass would be caused by the film acting like a cathode on its own account.

cathode. The interpretation of this evidence depends upon the fact that the presence in a gas of atoms, or the products of a previous discharge through the gas, greatly facilitates the passage of a subsequent discharge. The experiment is represented in Fig. 50: the discharge tube A was fused on to the pump, and two terminals *c* and *d* were fused through the glass at an elbow of the tube. These terminals were connected with an induction coil, and the pressure in the discharge tube was such that the electrodeless discharge would not pass. When the induction coil was turned on in such a way that *c* was the negative electrode the electrodeless discharge at once passed through the tube, but no effect at all was produced when *c* was positive and *d* negative.

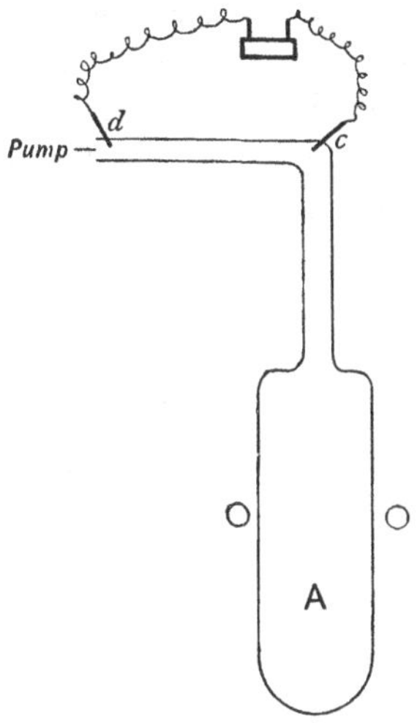

Fig. 50.

121.] Assuming with Mr. Crookes that it is the impact of particles driven out of the region around the negative electrode which produces the phosphorescence, it still seems an open question whether the luminosity is due to the mechanical effect of the impulse or whether the effect is wholly electrical. For since these particles are charged, their approach, collision with the glass, and retreat, will produce much the same electrical effect as if a body close to the glass were very rapidly charged with negative electricity and then as rapidly discharged. Thus the glass in the neighbourhood of the point of impact of one of these particles is exposed to a very rapidly changing electric polarization, the effect of which, according to the electromagnetic theory of light, would be much the same as if light fell on the glass, in which case we know it would phosphoresce.

The sharpness of the shadows cast by these rays shows that the phosphorescence cannot be due to what has been called a 'lamp action' of the particles, each particle acting like a lamp, radiating light, and causing the glass to phosphoresce by the light it emits.

122.] The distance which these particles travel before losing their power of affecting the glass is surprising, amounting to a large multiple of the mean free path of the molecules of the gas when in a molecular condition; it is possible, however, that they travel together, forming something analogous to the 'electrical wind,' and that their passage through the gas resembles the passage of a mass of air by convection currents rather than a process of molecular diffusion. We must remember, too, that since atoms are smaller than molecules, the mean free path of a gas in the atomic condition would naturally be greater than when in the molecular.

123.] Strikingly beautiful as the phenomena connected with these 'negative rays' are, it seems most probable that the rays are merely a local effect, and play but a small part in carrying the current through the gas. There are several reasons which lead us to come to this conclusion: in the first place, we have seen that the great mass of luminosity in the tube starts from the anode and travels down the tube with an enormously greater velocity then we can assign to these particles; again, this discharge seems quite independent of the anode, so that the rays may be quite out of the main line of the discharge. The exact function of these rays in the discharge is doubtful, it seems just possible that they may constitute a return current of gas by which the atoms which carry the discharge up to the negative electrode are prevented from accumulating in its neighbourhood.

124.] These rays have been used by Spottiswoode and Moulton (*Phil. Trans.* 1880, p. 627) to determine a point of fundamental importance in the theory of the discharge, viz. the relative magnitudes of the following times:—

(1) The period occupied by a discharge.

(2) The time occupied by the discharge of the positive electricity from its terminal.

(3) The time occupied by the discharge of the negative electricity from its terminal.

(4) The time occupied by molecular streams in leaving a negative terminal.

(5) The time occupied by positive electricity in passing along the tube.

(6) The time occupied by negative electricity in passing along the tube.

(7) The time occupied by the particles composing molecular streams in passing along the tube.

(8) The time occupied by electricity in passing along a wire of the length of the tube.

The phenomenon which was most extensively used by Spottiswoode and Moulton in investigating the relative magnitude of these times was the repulsion of one negative stream by another in its neighbourhood. This effect may be illustrated in several ways: thus, if the finger or a piece of tin-foil connected to earth be placed on the discharge tube, not too far away from the anode, the portion of the glass tube immediately underneath the finger becomes by induction a cathode and emits a negative stream; this stream produces a phosphorescent patch on the other side of the tube, diametrically opposite to the finger. If two fingers or two pieces of tin-foil are placed on the

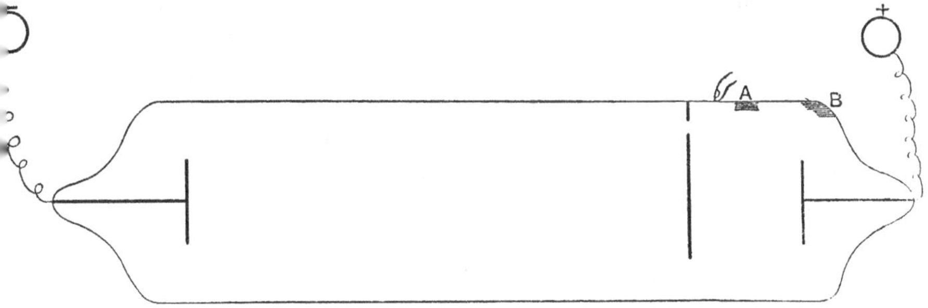

Fig. 51.

tube two phosphorescent patches appear on the glass, but neither of these patches occupies quite the position it would if the other patch were away. Another experiment (see Spottiswoode and Moulton, *Phil. Trans.* 1880, Part II, p. 614) which also illustrates the same effect is the following. A tube, Fig. 51, was taken, in which there was a flat piece of aluminium containing a small hole; when the more distant terminal was made negative, a bright image A of the hole appeared on the side of the tube in the midst of the shadow cast by the plate. When the tube was touched on the side on which this image appeared, but at a point on the negative side of the image, it was found that the image was splayed out to B, part of it moving down the tube away from the negative terminal. This seems to show that the negative electrode formed by the finger pushes away from it the rays forming the image. From this case Spottiswoode and Moulton reasoned as follows (*Phil. Trans.* 1880, Part II, p. 632):

This image was *splayed out* by the finger being placed on the tube. Now a magnet displaced it as a whole without any splaying out. This then pointed to a variation in the relative strength of the interfering stream and the stream interfered with, and such variation must have occurred during the period that they were encountering one another, and were moving in the ordinary way of such streams, for it showed itself in a variation in the extent to which the streams from the negative terminal were diverted. We may hence conclude that the time requisite for the molecules to move the length of the tube was decidedly less than that occupied by the discharge, but was sufficiently comparable with it to allow the diminution of intensity of the streams from the sides of the tube to make itself visible before the streams from the negative terminal experienced a similar diminution.'

125.] This may serve as an example of the method used by Spottiswoode and Moulton in comparing the time quantities enumerated in Art. 124. We regret that we have not space to describe the ingenious methods by which they brought other time quantities into comparison, for these we must refer to their paper; we can only quote the final result of their investigation. They arrange (l. c. pp. 641–642) the time quantities in groups which are in descending order of magnitude, the quantities in any group are exceedingly small compared with those in any group above them, while the quantities in the same group are of the same order of magnitude.

A. The interval between two discharges.

B. The time occupied by the discharge of the negative electricity from its terminal.

The time occupied by negative streams in leaving a negative terminal.

The time occupied by the particles composing molecular streams in passing along the tube.

C. The time occupied by positive electricity in passing along the tube.

The time occupied by negative electricity in passing along the tube.

D. The time occupied by positive discharge.

The time required for the formation of positive luminosity at the seat of positive discharge.

The time required for the formation of the dark space at the seat of negative discharge.

E. The time occupied by either electricity in passing along a wire of the length of the tube.

The time of a complete discharge is of the order B.

It will be seen that one of the conclusions given above, viz. that the time required for the positive luminosity to travel the length of the tube is very small compared with the time occupied by the negative discharge, is confirmed by the experiments with the rotating mirror described in Art. 104. According to these experiments however C and E are of the same order.

Action of a Magnet upon the Discharge when Electrodes are used.

126.] The appearance and path of the discharge in a vacuum tube are affected to a very great extent by the action of magnetic force. We may roughly describe the effect produced by a magnet by saying that the displacement of the discharge is much the same as that of a perfectly flexible wire conveying a current in the direction of that through the tube, the position of the wire coinciding with the part of the luminous discharge under consideration. This statement, which at first sight seems to bring the behaviour of the discharge under magnetic force into close analogy with that of ordinary currents, is apt, however, to obscure an essential difference between the two cases. A current through a wire is displaced by a magnetic force because the wire itself is displaced, and there is no other path open to the current. If, however, the current were flowing through a large mass of metal, if, for example, the discharge tube were filled with mercury instead of with rarefied gas, there would (excluding the Hall effect) be no displacement of the current through it. In the case of the rarefied gas, however, we have, what we do not have in the metal to any appreciable extent, a displacement of the lines of flow through the conductor—the rarefied gas. Thus the effects of the magnetic force on currents through wires, and on the discharge through a rarefied gas, instead of being, as they seem at first sight, the same, are apparently opposed to each other.

127.] The explanation which seems the most probable is that

by which we explained the effect of a magnet on the discharge without electrodes: viz. that when an electric discharge has passed through a gas, the supply of dissociated molecules, or of molecules in a peculiar condition, left behind in the line of the discharge, has made that line so much better a conductor than the rest of the gas, that when the particles composing it are displaced by the action of the magnetic force, the discharge continues to pass through them in their displaced position, and maintains by its passage the high conductivity of this line of particles. On this view the case would be very similar to that of a current along a wire, the line of particles along which the discharge passes being made by the discharge so much better a conductor than the rest of the gas, that the case is analogous to a metal wire surrounded by a dielectric.

128.] This view seems to be confirmed by the behaviour of a spark between electrodes when a blast of air is blown across it; the spark is deflected by the blast much as a flexible wire would be if fastened at the two electrodes. On the preceding view the explanation of this would be, that by the passage of the spark through the gas, the electric strength of the gas along the line of discharge is diminished, partly by the lingering of atoms produced by the discharge, partly perhaps by the heat produced by the spark. When a blast of air is blowing across the space between the electrodes, the electrically weak gas will be carried with it, so that the next spark, which will pass through the weak gas, will be deflected. Feddersen's observations (*Pogg. Ann.* 103, p. 69, 1858) on the appearance presented by a succession of sparks in a revolving mirror when a blast of air was directed across the electrodes, seem to prove conclusively that this explanation is the true one, for he found that the first spark was quite straight, while the successive sparks got, as shown in Fig. 52, gradually more and more bent by the blast.

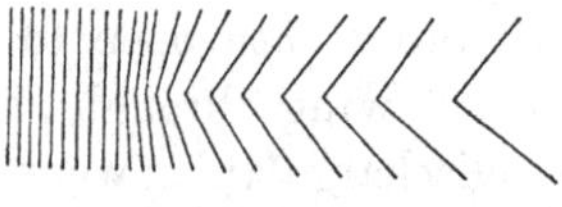

Fig. 52.

129.] The effects produced by a magnet show themselves in different ways, at different parts of the discharge. Beginning with the negative glow, Plücker (*Pogg. Ann.* 103, p. 88, 1858), who was the first to observe the behaviour of this part of the discharge when under the action of a magnet, found that the appearance of the glow in the magnetic field could be described by saying that the negative glow behaved as if it consisted of a

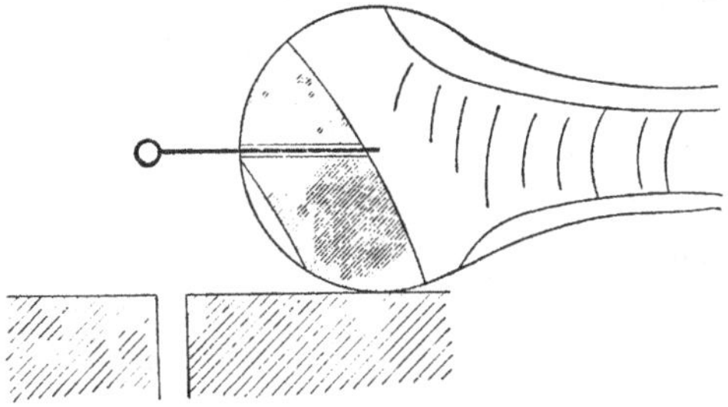

Fig. 53.

paramagnetic substance, such as iron filings without weight and with perfect freedom of motion. He found that the bright

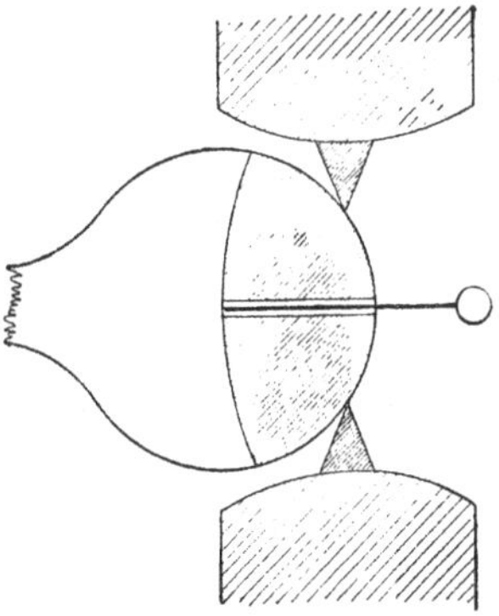

Fig. 54.

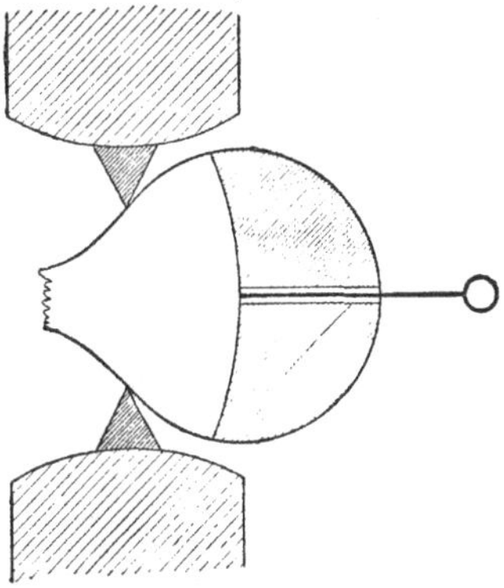

Fig. 55.

boundary of the negative glow coincided with the line of magnetic force passing through the extremity of the negative elec-

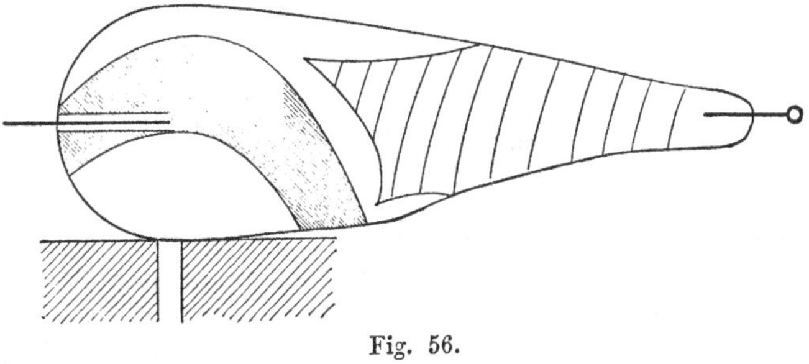

Fig. 56.

trode. Figs. 53, 54, 55, 56, which are taken from Plücker's

paper, show the shape taken by the glow when placed in the magnetic field due to a strong electro-magnet, the tube being placed in Figs. 54, 55 so that the lines of magnetic force are transverse to the line of discharge; while in Figs. 53 and 56 the line of discharge is more or less tangential to the direction of the magnetic force.

130.] Hittorf (*Pogg. Ann.* 136, p. 213 *et seq.*, 1869) found that when the negative rays were subject to the action of magnetic

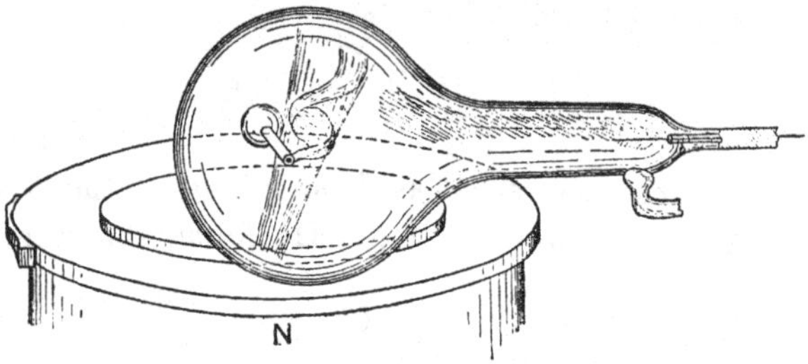

Fig. 57.

force, they were twisted into spirals and sometimes into circular rings. In his experiments the negative electrode was fused into a small glass tube fused into the discharge tube, the open end of the small tube projecting beyond the electrode. The negative rays were by this means limited to those which were approxi-

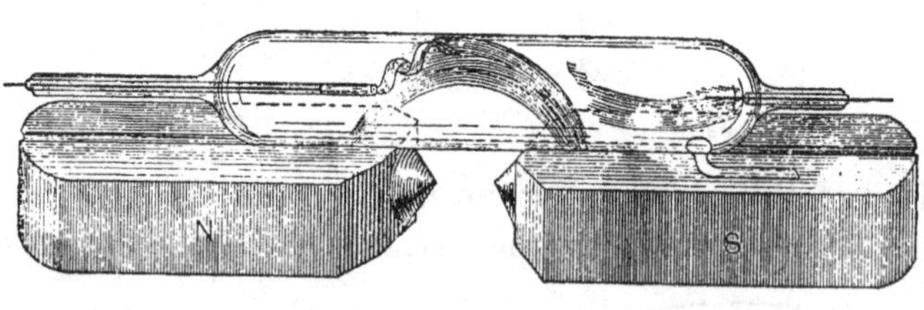

Fig. 58.

mately parallel to the axis of the small tube, so that it was easy to alter the angle which these rays made with the lines of magnetic force either by moving the discharge tube or altering the position of the electro-magnet. The discharge tube was shaped so that the walls of the tube were at a considerable distance from the negative electrode. Hittorf found that when the direction of the negative rays was tangential to the line of magnetic force passing through the extremity of the cathode, the rays continued to travel along this line; that when the rays were initially at right angles to the lines of magnetic force they curled up into circular rings; and that when the rays were oblique to the direction of the magnetic force they were twisted into spirals of which

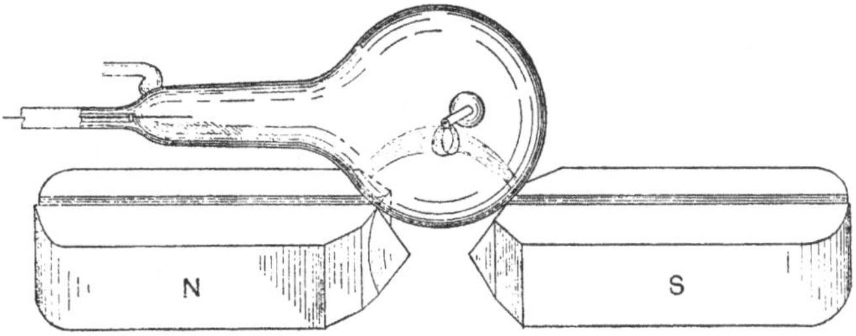

Fig. 59.

two or three turns were visible, the axis of the spiral being parallel to the direction of the magnetic force. These effects are illustrated in Figs. 57, 58, 59, and 60. In Figs. 57 and 58 the rays are at right angles to the lines of magnetic force, while in Figs. 59 and 60 they are oblique to them.

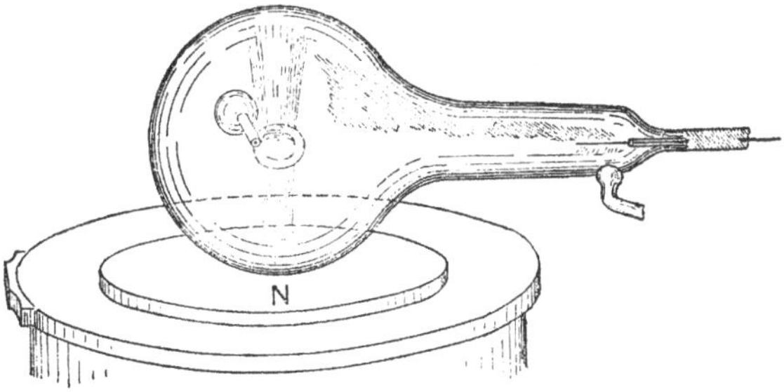

Fig. 60.

131.] This spiral form is the path which would be traversed by a negatively charged particle moving away from the cathode. To prove this, let us assume that the magnetic field is uniform, and that the axis of z is parallel to the lines of magnetic force. Let e be the charge on the particle, v its velocity. Then if we regard the particle as a small conducting sphere, the mechanical force on it in the magnetic field is, if v is small compared with the velocity of light, the same (see Art. 16) as that which would be exerted on unit length of a wire carrying a current whose components parallel to the axes of x, y, z are respectively

$$\tfrac{1}{3}ev\frac{dx}{ds}, \quad \tfrac{1}{3}ev\frac{dy}{ds}, \quad \tfrac{1}{3}ev\frac{dz}{ds},$$

where ds is an element of the path of the particle. Thus, if m is the mass of the particle, Z the magnetic force, the equations of motion of the particle are

$$m\frac{d^2x}{dt^2} = \tfrac{1}{3}evZ\frac{dy}{ds}, \qquad (1)$$

$$m\frac{d^2y}{dt^2} = -\tfrac{1}{3}evZ\frac{dx}{ds}, \qquad (2)$$

$$m\frac{d^2z}{dt^2} = 0. \qquad (3)$$

Since the force on the particle is at right angles to its direction of motion, the velocity v of the particle will be constant, and since by (3) the component of the velocity parallel to the axis of z is constant, the direction of motion of the particle must make a constant angle, α say, with the direction of the magnetic force. Since ds/dt is constant, equations (1)—(3) may be written

$$mv^2\frac{d^2x}{ds^2} = \tfrac{1}{3}evZ\frac{dy}{ds},$$

$$mv^2\frac{d^2y}{ds^2} = -\tfrac{1}{3}evZ\frac{dx}{ds},$$

$$mv^2\frac{d^2z}{ds^2} = 0.$$

If ρ is the radius of curvature of the path, λ, μ, ν its direction cosines,

$$\frac{d^2x}{ds^2} = \frac{\lambda}{\rho}, \quad \frac{d^2y}{ds^2} = \frac{\mu}{\rho}, \quad \frac{d^2z}{ds^2} = \frac{\nu}{\rho}.$$

Hence from the preceding equations

$$\frac{\lambda}{\rho} = \tfrac{1}{3}\frac{Ze}{mv}\frac{dy}{ds},$$

$$\frac{\mu}{\rho} = -\tfrac{1}{3}\frac{Ze}{mv}\frac{dx}{ds},$$

$$\frac{\nu}{\rho} = 0.$$

Squaring and adding, we get

$$\frac{1}{\rho^2} = \left(\tfrac{1}{3}\frac{Ze}{mv}\right)^2\left\{\left(\frac{dx}{ds}\right)^2 + \left(\frac{dy}{ds}\right)^2\right\}.$$

But

$$\left(\frac{dx}{ds}\right)^2 + \left(\frac{dy}{ds}\right)^2 = \sin^2 a.$$

So that

$$\frac{1}{\rho} = \tfrac{1}{3}\frac{Ze}{mv}\sin a.$$

Hence the radius of curvature of the path of the particle is constant, and since the direction of motion makes a constant angle with that of the magnetic force, the path of the particle is a helix of which the axis is parallel to the magnetic force; the angle of the spiral is the complement of the angle which the direction of projection makes with the magnetic force. If a is the radius of the cylinder on which the spiral is wound, $a = \rho \sin^2 a$, so that

$$a = 3\frac{mv}{Ze}\sin a.$$

If $a = \pi/2$, the spiral degenerates into a circle of which the radius is $3\,mv/Ze$.

Let the particle be an atom of hydrogen charged with the quantity of electricity which we find always associated with the atom of hydrogen in electrolytic phenomena: then since the electro-chemical equivalent of hydrogen is about 10^{-4}, we have, if N is the number of hydrogen atoms in one gramme of that substance, $Ne = 10^4$ and $Nm = 1$; hence when the ray is curled up into a ring of radius a,

$$a = 10^{-4}3\frac{v}{Z},$$

or $3v = 10^4 aZ$ in hydrogen.

132.] In one of Hittorf's experiments, that illustrated in Fig. 60, he estimated the diameter of the ring as less than 1 mm.: the gas in this case was air, which is not a simple gas; we shall assume, however, that m/e is the same as for oxygen, or eight times the value for hydrogen. Putting

$a = 5 \times 10^{-2}$, and $m/e = 8 \times 10^{-4}$, we get

$$v = \frac{5}{24}10^2 Z.$$

The value of Z is not given in Hittorf's paper; we may be sure, however, that it was considerably less than 10^4, and it follows that v must have been less than 2×10^5; this superior limit to the value of v is less than six times the velocity of sound. Hence the velocity of these particles must be infinitesimal in comparison with that of the positive luminosity which, as we have seen, is comparable with that of light.

133.] A magnet affects the disposition of the negative glow over the surface of the electrode as well as its course through the gas. Thus Hittorf (*Pogg. Ann.* 136, p. 221, 1869) found that when the negative electrode is a flat vertical disc, and the discharge tube is placed horizontally between the poles of an electromagnet, with the disc in an axial plane of the electromagnet; the disc is cleared of glow by the magnetic force except upon the highest point on the side most remote from the positive electrode, or the lowest point on the side nearest that electrode according to the direction of the magnetic force. In another experiment Hittorf, using as a cathode a metal tube about 1 cm. in diameter, found that when the discharge tube is placed so that the axis of the cathode is at right angles to the line joining the poles of the electromagnet the cathode is cleared of glow in the neighbourhood of the lines where the normals are at right angles to the magnetic force. These experiments show that the action of a magnet on the glow is the same as its action on a system of perfectly flexible currents whose ends can slide freely over the surface of the negative electrode.

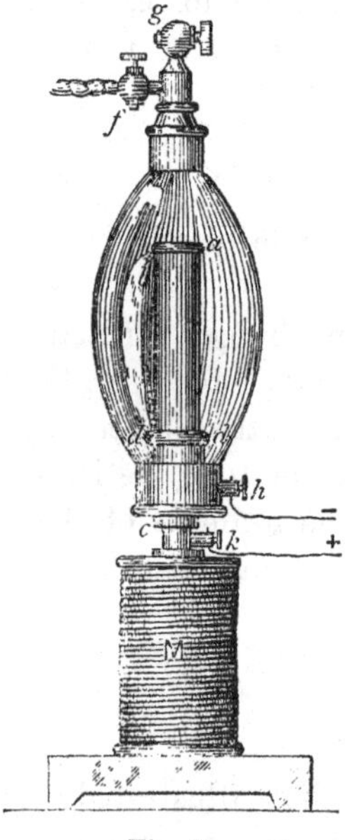

Fig. 61.

134.] The positive column is also deflected by a magnet in the same way as a perfectly flexible wire carrying a current in the direction of that through the discharge tube. This is beautifully illustrated by an experiment due to De la Rive in which the discharge through a rarefied gas is set in continuous rotation by the action of a magnet. The method of making this experiment is shown in Fig. 61; the two terminals *a* and *d* are metal rings separated from each other by an insulating tube which fits over a piece of iron resting on one of the poles of an electromagnet *M*. This arrangement is placed in an egg-shaped vessel from which the air can be exhausted. To make the experiment successful it is advisable to introduce a small quantity of the vapour of alcohol or turpentine. The terminals *a* and *d* are connected with an induction coil, which, when the pressure in the vessel is sufficiently reduced, produces a discharge through the gas between the terminals *a* and *d*, which rotates under the magnetic force with considerable velocity. The rotation of the discharge through the gas is probably due, as we have seen, to the displacement of the particles through which one discharge has already passed; the displaced particles form an easier path for a subsequent discharge than the original line of discharge along which none of the dissociated molecules have been left. The new discharge will thus not be along the same line as the old one, and the luminous column will therefore rotate. We can easily see why a simple gas like hydrogen should not show this effect nearly so well as a complicated one like the vapour of alcohol or turpentine. For the discharges of the induction coil are intermittent, so that to produce this rotation the dissociated molecules produced by one discharge must persist until the arrival of the subsequent one. Now we should expect to find that when a molecule of a stable gas like hydrogen is dissociated by the discharge, the recombination of its atoms will take place in a much shorter time than similar recombination for a complex gas like turpentine vapour; thus we should expect the effects of the discharge to be more persistent, and therefore the rotation more decided in turpentine vapour than in hydrogen.

135.] Crookes (*Phil. Trans.* 1879, Part II, p. 657) has produced somewhat analogous rotations of the negative rays in a very highly exhausted tube. The shape of the tube he employed is shown in Figure 62. When the discharge went through this tube, the neck surrounding the negative pole was covered with two or three bright patches which rotated when the tube was placed over an electromagnet. Crookes found that the direction of rotation was reversed when the magnetic force was reversed, but that if the magnetic force were not altered the direction of rotation was not affected by reversing the poles of the discharge tube. This is what we should expect if we remember that the bright spots on the glass are due to the negative rays, and that these will be at right angles to the negative electrode; thus the reversal of the poles of the tube does not reverse the direction of these rays; it merely alters their distance from the pole of the electromagnet. The curious thing about the rotation was that it had the *opposite* direction to that which would have been produced by the action of a magnet on a current carrying electricity in the same direction as that carried by the negative rays, showing clearly that this rotation is due to some secondary effect and not to the primary action of the magnetic force on the current.

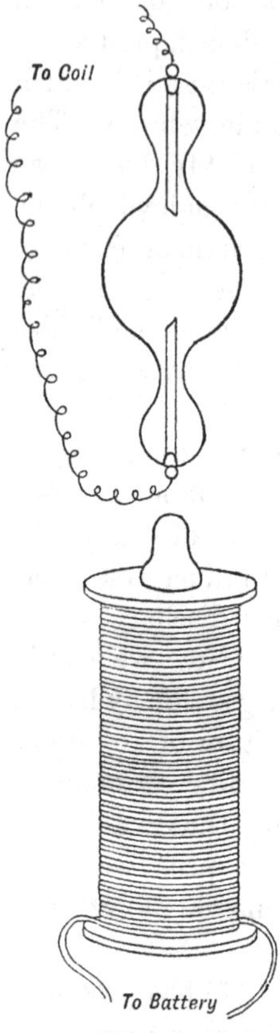

Fig. 62.

136.] An experiment due to Goldstein, which may seem inconsistent with the view we have taken, viz. that the deflection of the discharge is due to the deflection of the line of least electric strength, should be mentioned here. Goldstein (*Wied. Ann.* 12, p. 261, 1881) took a large discharge tube, 4 cm. wide by 20 long, the electrodes being at opposite ends of the tube. A piece of sodium was placed in the tube which was then quickly filled with dry nitrogen, the tube was then exhausted until a discharge passed freely through the tube, and the sodium heated until any hydrogen it might have contained had been driven off. When this had been done the tube was refilled with nitrogen and then exhausted until the positive column filled the tube with a reddish purple light. The sodium was then slowly heated until its vapour began to come off, when the discharge in the lower part of the tube over the sodium became yellow as it passed through sodium vapour, while the discharge at the top of the tube remained red as the sodium vapour did not extend all the way across the tube. The positive discharge was now deflected by a magnet and driven to the top of the tube out of the region occupied by the sodium vapour, the discharge was now entirely red and showed no trace of sodium light. The experiment does not seem inconsistent with the view we have advocated, as we cannot suppose that more than an infinitesimal quantity of sodium vapour travelled across the tube under the action of the magnetic force, and it does not follow that because we suppose the line of discharge to be weakened by the presence of the dissociated molecules that these molecules are the only ones affected by the discharge; it seems much more probable that they serve as nuclei round which the chemical changes which transmit the discharge take place.

137.] The striations are affected by magnetic force; in Figs. 53 and 56 may be seen the distortion of striae when the discharge tube is placed in a magnetic field. If the negative glow is driven away from the line joining the terminals by magnetic force, the positive column lengthens and fills part of the space previously occupied by the negative glow; if the positive column is striated new striae appear, so that in this case we have a creation of striae by the action of magnetic force. The most remarkable effect of a magnet on the striated discharge, however, is that discovered by Spottiswoode and Fletcher Moulton, and Goldstein; Spottiswoode and Moulton (*Phil. Trans.* 1879, Part I, p. 205) thus describe the effect: 'If a magnet be applied to a striated column, it will be found that the column is not simply thrown up or down as a whole, as would be the case if the discharge passed in direct lines from terminal to terminal, threading the striae in its passage. On the contrary, each stria is subjected to a rotation or deformation of exactly the same character as would be caused if the stria marked the termination of flexible currents radiating from the bright head of the stria behind it and terminating in the hazy inner surface of the stria in question. An examination of several cases has led the authors of this paper to conclude that the currents do thus radiate from the bright head of a stria to the inner surface of the next, and that there is no direct passage from one terminal of the tube to the other.' Goldstein (*Wied. Ann.* 11, p. 850, 1880) found that the striated column could by the action of magnetic force be broken up into a number of bright curves, of the same kind as those observed by Hittorf in the negative rays (see Art. 130), the number of bright curves being the same as the number of striae which had disappeared; each striation was transformed by the magnetic force into a separate curve, and these curves were separated from each other by dark spaces. We may conclude from these experiments that the positive column does not consist of a current of electricity traversing the whole of its length in the way that such a current would traverse a metal cylinder coincident with the positive column, but that it rather consists of a number of separate currents, each striation corresponding to a current which is to a certain extent independent of those which precede or follow. The discharge along the positive column might perhaps be roughly illustrated by placing pieces of wire equal in length to the striae and separated by very minute air spaces along the line of discharge.

138.] Goldstein found that the rolling up of the striae by the magnetic force was most marked at the end of the positive column nearest the negative electrode: the following is a translation of Goldstein's description of this process (l. c. p. 852). The appearance is very characteristic when in the unmagnetized condition the negative glow penetrates beyond the first striation into the positive column. The end of the negative glow is then further from the cathode than the first striation or, even if the rarefaction is suitable, than the second or third. Nevertheless the end of the negative glow rolls itself under the magnetic action up to the cathode in the magnetic curve which passes through the cathode. Then separated from this by a dark space follows on the side of the anode a curve in which all the rays of the first striation are rolled up, then a similar curve for the second striation, and so on.

We shall have occasion to refer to these experiments again in the discussion of the theory of the discharge.

On the Distribution of Potential along an Exhausted Tube through which an Electric Discharge is passing.

139.] The changes which take place in the potential as we pass along the discharge tube are extremely interesting, as they present a remarkable contrast to those which take place along a metal wire through which a steady uniform current is passing; in this case the potential-gradient is uniform along the wire, but changes when the current changes, being by Ohm's law proportional to the intensity of the current; in the exhausted tube, on the other hand, the potential-gradient varies greatly in different parts of the tube, but in the positive column is almost independent of the intensity of the current passing through the gas. The potentials measured are those of wires immersed in the rarefied gas, and the question arises, whether the potentials of these wires are constant, as they would be if the wires were in a steady current, or whether they are variable, the potentials determined in these experiments being the mean values about which the potentials of the wires fluctuate? This question is the same as, whether the current through the gas is continuous or intermittent? On this point considerable difference of opinion has existed among physicists. There is no doubt that by the aid of a battery consisting of a large number of cells a discharge can be got, which, if not continuous, has such a high rate of intermittence that no unsteadiness can be detected when it is observed in a rotating mirror making 100 revolutions per second; this is sufficient to prove that if the intermittence exists at all it must be exceedingly rapid. As long, however, as the discharge retains the property of requiring a large potential difference to exist between the electrodes, this difference varying continuously with the pressure, while the latter varies from that of an atmosphere to the pressure in the discharge tube, we should expect the electrodes to act like condensers continually being charged and discharged as they are at atmospheric pressures, in other words we should expect the discharge to be intermittent. When, however, the discharge passes as the 'arc discharge,' see Art. 169, the potential difference falls to a comparatively small value, and it is probable that this discharge is much more nearly continuous than the striated one.

It ought also to be remembered that the current through the gas may be interrupted even though that through the leads is continuous. For since the current through the gas does not obey the same laws as when it goes through a metallic conductor, the current across a section of the discharge tube need not at any specified instant be the same as that across the section of one of the leads. The average current must of course be the same in the two cases, but only the *average* current and not that at any particular instant. To quote an illustration given by Spottiswoode and Moulton, the discharge tube may act like the air vessel of a fire engine; all the electricity that goes in comes out again, but no longer with the same pulsation. The tube may sometimes contain more and sometimes less free electricity, and may act as an expansible vessel would act if it formed part of the path of an incompressible fluid.

The rapidity of the intermittence can to some extent be tested by observing whether or not the discharge is deflected by the approach of a conductor. When the discharge is intermittent and the interval between the discharges so long that the intermittence of the discharge can be detected either by the eye or by a slowly rotating mirror, the discharge is deflected when a conductor is brought near it; when however the intermittence is very rapid, the discharge is not affected by the approach of the conductor. This effect has been very completely investigated by Spottiswoode and Moulton (*Phil. Trans.* 1879, Part I, p. 166; 1880, Part II, p. 564).

140.] We shall begin by considering Hittorf's experiments on the potential gradient (*Wied. Ann.* 20. p. 705, 1883). The discharge tube, Fig. 63, which was 5·5 cm. in diameter and 33·7 cm.

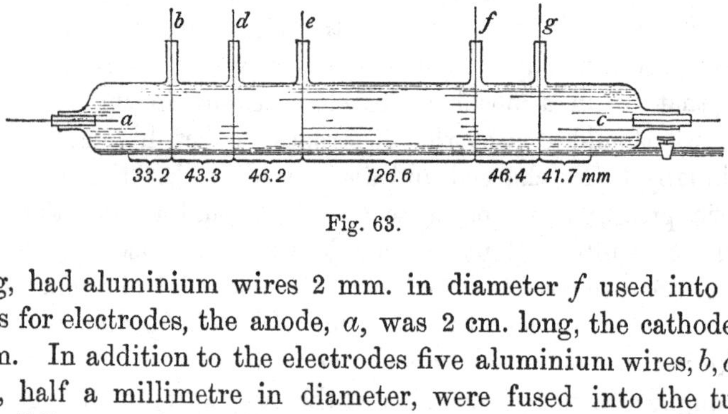

Fig. 63.

long, had aluminium wires 2 mm. in diameter *f* used into the ends for electrodes, the anode, *a*, was 2 cm. long, the cathode, *c*, 7 cm. In addition to the electrodes five aluminium wires, *b*, *d*, *e*, *f*, *g*, half a millimetre in diameter, were fused into the tube. The difference of potential between any two of these wires could be determined by connecting them to the plates of a condenser, and then discharging the condenser through a galvanometer. The deflection of the galvanometer was proportional to the charge in the condenser, which again was proportional to the difference of potential between the wires. The discharge was produced by means of a large number of cells of Bunsen's chromic acid battery, and the intensity of the current was varied by inserting in the circuit a tube containing a solution of cadmium iodide, which is a very bad conductor. No intermittence in the discharge could be detected either by a mirror rotating 100 times a second or a telephone. The tube was filled with nitrogen, as this gas has the advantage of not attacking the electrodes and of not being absorbed by them so greedily as hydrogen. The results of some of the measurements are given in the following Table, l. c. p. 727:—

Pressure of Nitrogen ·6 *mm.*

Number fixing the experiment	1	2	3	4	5	6	7	8
Number of cells	500	500	500	600	700	800	900	1000
Intensity of current in millionths of an Ampère . . .	244	814	1282	3175	5189	7000	8791	11192
Kick of galvanometer due to the charging of the condenser to the potential difference between—								
ac . . .	133	132	133·5	141·5	150	157	165	173
ab . . .	22	22·5	22	21·5	21	21	21	21
bd . . .	14	13	13	12	12·5	12	12	12·25
de . . .	13	13	13	14	14	13·5	12	12·5
ae . . .	52	50	49	47	47	47	47	47
fg . . .	—	2·25	3	4	3·75	4	3·25	3

The difference of potential in volts can be approximately got by multiplying the galvanometer deflection by 6. In experiment 1 the negative glow covered about 1·5 cm. of the cathode, and the positive light extended to *f*. In experiment 2 the negative glow covered 6 cm. of the cathode, and in 3 and the following experiments the whole of the cathode. In experiments 1, 2, 3 the thickness of the negative glow remained the same; in the later experiments where the negative glow covered the whole of the cathode its thickness increased as the intensity of the current increased, and in 7 and 8 it extended to the walls of the tube. The table shows that no changes in the potential differences occurred until the negative glow began to increase in thickness. We see that by far the greatest fall in the potential occurs in the immediate neighbourhood of the cathode, the rise in potential from the negative electrode to the outside of the negative glow being far greater than the rise in all the rest of the tube; we also see that the changes which take place when the thickness of the negative glow alters take place in this part of the tube, and that the potential differences in the positive column are *independent of the strength of the current.* The portions *bd* and *de* of the positive column, which are very nearly equal in length, have also practically the same potential differences; and these are each less than that of the portion *ab* which contains the anode, although the latter portion is considerably shorter. The wires *f*; *g* were in all these experiments in the dark space between the negative glow and the positive column. The small difference of potential between these wires is very noteworthy.

141.] Hittorf also investigated the potential differences for lower pressures of the gas than that used in the last experiment; for this purpose the tube in Fig. 63 was not suitable, as the negative glow was very much interfered with by the walls of the tube, he therefore used a tube shaped like that in Fig. 64, which

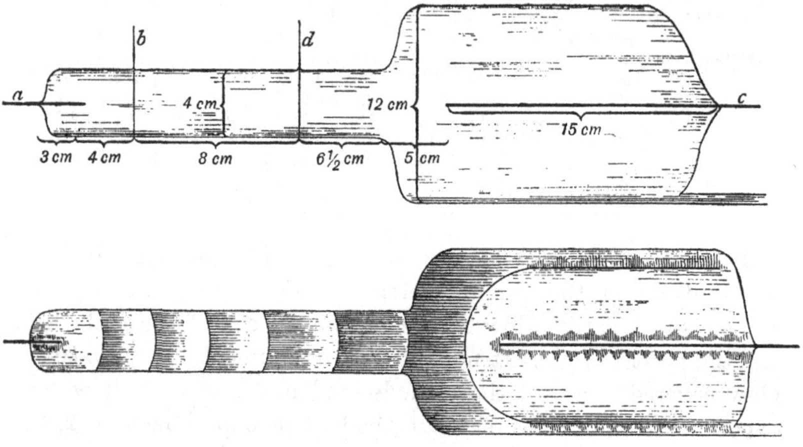

Fig. 64.

was purposely made wide in the region round the negative electrode. The diameter of the positive part of the tube was 4 cm., that of the negative 12 cm. The length of the negative electrode was 15 cm., that of the positive 3 cm. In this case only two wires, *b* and *d*, were inserted in the tube. The results of experiments with this tube are given in the following table:—

Number fixing the experiment	1	2	3	4	5	6
Pressure of the nitrogen in millimetres of mercury . .	·70	·35	·175	·088	·044	·022
Number of cells	600	600	600	600	600	600
Strength of current in millionths of an Ampère . .	2870	2076	1791	1360	916	488
Kick of galvanometer due to the charging of the condenser to the potential difference between—						
1 *ac* . . .	151	140	145	157	168	178
2 *ab* . . .	21	15	12	9·5	8	7
3 *bd* . . .	30	19	12	8	5	4·25
4 *ad* . . .	51	34	24	17·5	13	11·25

Number fixing the experiment	7	8	9	10	11
Pressure of the nitrogen in millimetres of mercury . .	·011 ?	·0055 ?	·0029 ?	.0014 ?	·0007 ?
Number of cells	600	800	1000	1200	1400
Strength of current in millionths of an Ampère . .	326	610	814	814	1100
Kick of galvanometer due to the charging of the condenser to the potential difference between—					
1 *ac* . . .	184	242	298	352	422
2 *ab* . . .	7	7	8	8·5	8·75
3 *bd* . . .	4	4	3·75	2·5	2·25
4 *ad* . . .	11	11·5	12	11·5	10·5

The negative glow in all these experiments covered the cathode, and in all but the first three it extended to the walls of the tube. The appearance of the glow at the higher exhaustions is shown in Fig. 64, where the shaded portions represent the bright parts of the discharge; it will be seen from the figure that the positive column was striated.

142.] The table shows that at high exhaustions the potential difference between the electrodes increases as the density of the gas diminishes, but that this increase is confined to the neighbourhood of the cathode; the ratio of the change in potential near the cathode to that in the rest of the tube increases as the pressure of the gas diminishes. The potential difference in the positive light diminishes as the pressure is reduced, but the diminution in the potential difference is not so rapid as the diminution in the pressure. The table seems to suggest that the potential gradient in the positive column tends towards a constant value which is independent of the density. We must remember however that Hittorf's experiments do not give the potential difference required to initiate the discharge through the gas, but the distribution of potential which accompanies the passage of electricity through the gas when the discharge has been established for some time, and where there are a plentiful supply of dissociated molecules produced by the passage of previous discharges. Hittorf found that the number of cells which would maintain a discharge after it was once started was frequently quite insufficient to initiate it, and the gas had to be broken through by a discharge from another source.

143.] The experiments described in Art. 79 on the discharge

without electrodes, when the interval between two discharges was long enough to give the gas through which the discharge had passed an opportunity of returning to its normal condition before the passage of the next discharge, show that even when no electrodes are used the electromotive intensity required to start the discharge has a minimum value at a particular pressure, and that when the pressure is reduced below this value the electromotive intensity required for discharge increases.

144.] The supply of dissociated molecules furnished by previous discharges also explains another peculiarity of these experiments. It will be seen from the table that at a pressure of ·0007 mm. of mercury, a potential difference which gave a galvanometer deflection of 10·5, corresponding to about 63 volts, was all that occurred in a length of 12 cm. of the positive light; it does not follow however that a potential gradient of about 5 volts per centimetre would be sufficient to *initiate* the discharge even if the great change in potential at the cathode were absent. In fact the experiments previously described on the discharge without electrodes show that it requires a very much greater electromotive intensity than this even when the cathode is entirely done away with.

The table shows that the potential difference between *a* and *b*, a space which includes the anode, has at the higher exhaustions passed its minimum value and commenced to increase.

145.] Though the potential differences between wires immersed in the positive column is independent of the strength of the current passing through the tube, yet in such a tube as Fig. 63 the potential differences between wires in the middle of the tube may be affected by variations in the current if these variations are accompanied by changes in the appearance of the discharge.

Let us suppose, for example, that the tube is filled with nitrogen at a pressure of from 2 to 3 mm. of mercury, then when the intensity of the current is very small the tube will appear to be dark throughout almost the whole of its length, the positive column and negative glow being reduced to mere specks in the neighbourhood of the electrodes; when however the intensity of the current increases the positive column increases in length, and if the increase is great enough to make it envelop two wires which were previously in the dark Faraday space, the difference of potential between these wires will be found to be very much greater than when the gas round them was non-luminous. This is illustrated for lower pressures by the table in Art. 140, which shows that the potential gradient between f and g, the wires in the dark space between the positive column and the negative glow, was very much less than the potential gradient in the positive column. It is shown however still more clearly in the following set of experiments made with the tube shown in Fig. 63 (l. c. p. 739).

Pressure of Nitrogen 3·95 *mm. of Mercury.*
Temperature 12° C.

Number fixing the experiment . . .	1	2	3	4	5	6	7
Number of cells . .	700	700	700	800	900	1000	1200
Intensity of the current in millionths of an Ampère . .	1465	2035	2391	2483	2830	3541	5820
Kick of the galvanometer from the charge in the condenser due to the potential difference between—							
1 *ac* . .	166–168	175–168	190–188	212–208	238–232	255	292–285
2 *ab* . .	—	—	—	—	63	60	79
3 *bd* . .	16·5	18	18·5	25	43	61	56·5
4 *de* . .	17·5	18	17	18	20	26	62
5 *fg* . .	10	10·5	11·5	12	13	13	12·5

146.] In experiments 1–3 the tube was quite dark, except quite close to the electrodes; the anode had a thin coating of positive light. The negative glow extended in experiment 1 over 1 cm. of the cathode, in experiment 2 over 3 cm., and in experiment 3 over 3 cm. In experiment 3 the beginning of a brush discharge was discernible at the anode. In consequence of the wires being in the dark Faraday space instead of the positive column, it will be noticed that the potential difference between b and d is very little greater than in the experiments described in Art. 140, though the pressure is more than six times greater.

147.] In experiment 4 the positive column reached past b; it will be seen that the potential difference between b and d rose to 25, while the differences between d and e and between *f* and *g*, which were still in the dark, remained unaltered. In experiment 5 the positive column reached past the middle of *bd*; the potential difference in *bd* rose from 25 to 43, the potential differences between the wires in the dark still being unaltered. In experiment 6 the positive light filled the whole space *ad*; the potential difference between *b* and *d* rose to 61, and that between *d* and *e* also began to rise as *d* was now in the positive column; this difference increased very much in experiment 7, when the positive column reached to *e*.

148.] We now pass on to the effect of an alteration in the strength of the current on the potential difference at the cathode. We have already remarked that if the negative glow does not spread over the whole of the cathode, the only effect of an increase in the intensity of the current is to make the negative glow spread still further over the cathode, without altering the potential difference. Until the glow has covered the electrode, there is, according to Hittorf, no considerable increase in temperature at the cathode: when however the intensity of the current is increased beyond the point at which the whole of the cathode is covered by the glow, the temperature of the cathode begins to increase; when the current through the gas is very strong, the cathode, and sometimes even the anode, becomes white hot. When this is the case the character of the discharge changes in a remarkable way, all luminosity disappears from the gas, which when examined by the spectroscope does not show any trace of the lines of its spectrum. The tube with its white hot electrodes surrounded by the dark gas presents a remarkable appearance, and it is especially to be noted that the electrodes are raised to incandescence by a current, which if it passed through them when they formed part of a metallic circuit, would hardly make them appreciably hot.

Hittorf also found (*Wied. Ann.* 21. p. 121, 1884) that if in a vacuum tube conveying an ordinary luminous discharge, a platinum spiral which could be raised by a battery to a white heat was placed so as to be in the path of the discharge, the latter lost all luminosity in the neighbourhood of the spiral when this was white hot. If the spiral was allowed to cool, the luminosity appeared again before the spiral had cooled below a bright red heat.

149.] For experiments of this kind aluminium electrodes melt too easily. Hittorf used in most of his experiments iridium electrodes, which can be raised to a very high temperature without melting. These were raised to a white heat before any measurements were made, so as to get rid of any gas they might have occluded. The length of the electrodes was 48 mm. The result of some experiments on nitrogen is given in the following table (*Wied. Ann.* 21, p. 111, 1884); in this, when the number of cells is given as $600 \times x$, it means that x sets of cells, each containing 600 elements, were connected up in parallel.

Experiments with Nitrogen. Iridium electrodes at a distance of 15 *mm.*

Number fixing the experiment	1	2	3	4	5
Pressure of the nitrogen in millimetres of mercury . .	19·65	31·9	53·1	53·1	52·4
Number of cells	600×3	600×3	600×3	600×4	400×6
Strength of current in Ampères	·535	1·225	1·4	2·0	2·1
Kick of the galvanometer due to the charge in the condenser produced by the potential difference between the electrodes	75–82	25–32	25–32	15–20	17–20

In the first experiment a reddish-yellow positive column stretched at first from the anode to an intensely bright patch on the cathode; the cathode however soon became white hot along the whole of its length, and then showed no trace of the negative glow, nor were any nitrogen lines detected when the region round the cathode was examined by the spectroscope. The tip of the anode was white hot.

From the second experiment we see that though the density of the nitrogen was much greater, the potential difference was less than half what it was in the first experiment. This is due to the electrodes being hotter in this experiment than in the preceding one. In the third experiment only half of the cathode was white hot, but the length of the anode which was incandescent was greater than in the preceding experiment. In the fourth experiment, in which a current of 2 Ampères passed through the gas, the end of the anode was hotter than that of the cathode, in fact with this current the anode, though made of iridium, began to melt. In the ordinary arc lamp, in which we have probably a discharge closely resembling that in this experiment, the anode is also hotter than the cathode when the current is intense.

In this case the gas was quite dark. A very remarkable feature shown by it is the smallness of the potential difference between the electrodes, not amounting to more than 100 volts, though the gas was at the pressure of 53·1 millimetres, and the distance between the electrodes 15 mm. When the electrodes were cold, the battery power used, about 1200 volts, was not sufficient to break down the gas: the discharge had to be started by sending a spark from a Leyden jar through the tube. The conduction through the gas in this case is of the same character as that described in Art. 169.

150.] Hittorf also made experiments on hydrogen and carbonic oxide; the results for hydrogen are given in the following table (*Wied. Ann.* 21, p. 113, 1884):—

Experiments with Hydrogen. Distance of the Iridium electrodes 15 *mm.*

Number fixing the experiment . .	1	2	3	4	5	6
Pressure of hydrogen in millimetres of Hg.	20	33·8	47·05	47·05	47·05	68·55
Number of cells. .	400 × 6	400 × 6	400 × 6	600 × 4	800 × 3	800 × 3
Intensity of current in Ampères . .	·5465	·3415	·3074	·9222	·9905	·8197
Kick of the galvanometer due to the charge in the condenser produced by the potential difference between the electrodes. .	100	107–108	110	100–110	107–110	110

In experiment 1 the pressure and the current were almost the same as for experiment 1, Art. 149, in nitrogen; the potential difference between the electrodes was however much greater in hydrogen than in nitrogen, though the potential difference required to *initiate* a discharge in hydrogen is considerably less than in nitrogen. In these experiments the potential difference between the electrodes for this dark discharge seems almost independent of the current and of the density of the gas.

Experiments with Carbonic Oxide Gas. Distance between Iridium electrodes 15 *mm.*

Number fixing the experiment	1	2	3	4
Pressure of CO in millimetres of mercury	13·1	22·75	51·7	75·85
Number of cells	800 × 3	800 × 3	800 × 3	800 × 3
Intensity of current in Ampères	·8880	·9734	1·3662	1·2978
Kick of the galvanometer due to the charge in the condenser produced by the potential difference between the electrodes	92–100	89–92	40	42

The great fall in potential, which occurs between experiments 2 and 3 on CO, was accompanied by a loss of luminosity; in 1 and 2 there was a little positive blue light at the anode, but in 3 this had disappeared, and the discharge was quite dark and showed in the spectroscope no trace of the carbonic oxide bands.

151.] When repeating these experiments with carbon electrodes instead of iridium ones, Hittorf found that with strong currents and at pressures between 10 mm. and 2 mm. the discharge through hydrogen took a very peculiar form, it consisted of ring-shaped striae, the insides of which were dark. These rings extended through the tubes and encircled both the anode and the cathode, as shown in Fig. 65.

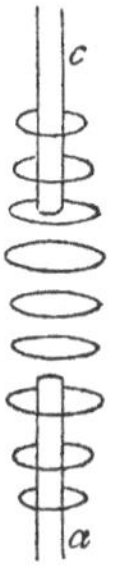

Fig. 65.

152.] The preceding experiments show that when the electrodes are white hot, the negative glow disappears, and the potential difference between the electrodes when a current is passing through the gas sinks to a fraction of the value it has when the electrodes are cold and the negative glow exists. Hittorf (*Wied. Ann.* 21, p. 133) has shown by a direct experiment that when the cathode is white hot a very small electromotive force is sufficient to maintain the discharge. The arrangement he used is shown in Fig. 66. A thin carbon filament which serves as a cathode is stretched between two conductors *mn*, and can be raised to a white heat by a current passing through it and these conductors; the anode *a* is vertically below the cathode and remains cold. When the pressure was very low, Hittorf found that 1 cell of his battery, equivalent to about 2 volts, would maintain a current between the anode and cathode when they were separated by 6 cm.: in this case the discharge was quite dark. When ten or more cells were used a pale bluish light spread over the anode. It should be noticed that the single cell does not *start* the current, it only maintains it: the current has previously to be started by the application of a much greater potential difference. Hittorf generally started the current by discharging a Leyden jar through the tube. No current at all will pass if the poles are reversed so that the anode is hot and the cathode cold. In these experiments it is necessary for the cathode to be at a white heat for an appreciable current to pass between the electrodes; very little effect seems to be produced on the potential difference at the cathode until the latter is hotter than a bright red heat. The current produced by a given electromotive force is greater at higher exhaustions than at low ones, but Hittorf found he could get appreciable effects at pressures up to 9 or 10 mm.

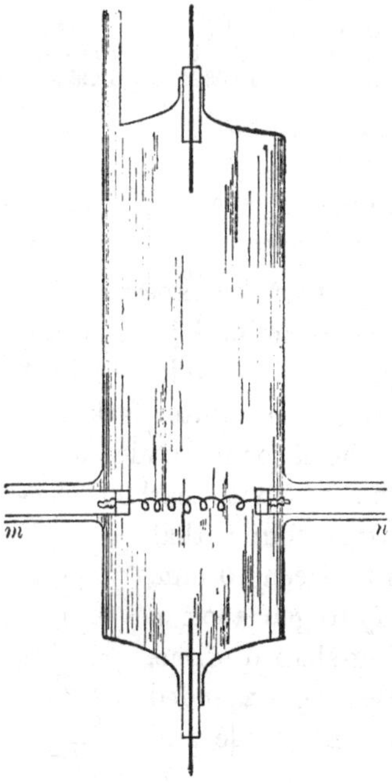

Fig. 66.

153.] In considering the results of experiments in which carbon filaments or platinum wires are raised to incandescence, we must remember that, as Elster and Geitel have shown (Art. 43), electrification is produced by the incandescent body, the region around which receives a charge of electricity; though whether the carrier of this charge is the disintegrated particles of the incandescent wire, or the dissociated molecules of the gas itself, is not clear. This electrification often makes the interpretation of experiments in which incandescent bodies are used ambiguous. Thus for example, Hittorf in one experiment (*Wied. Ann.* 21, p. 137, 1884) used a U-shaped discharge tube, in one limb of which a carbon filament was raised to incandescence; the other limb of the tube contained a small gold leaf electroscope; when the pressure of the gas in the tube was very low, Hittorf found that the electroscope would retain a charge of negative electricity but immediately lost a positive charge. This experiment does not however show conclusively that positive electricity escapes more easily than negative from a metal into a gas which is in the condition in which it conducts electricity, because the same effect would occur if the incandescent carbon filament produced a negative electrification in the gas around it.

154.] The way in which the passage of electricity from metal to gas, or *vice versâ*, is facilitated by increasing the temperature of the metal to the point of incandescence is illustrated by an effect observed in the experiments on hot gases described in Art. 37. It was found that when a current was passing between electrodes immersed in a platinum tube at a bright yellow heat and containing some gas, such as iodine, which conducts well, the current was at once stopped if a large piece of cold platinum foil was lowered between the electrodes, although there was a strong up-current of gas in the tube which prevented a cold layer of gas being formed against the platinum foil: as soon, however, as the foil became incandescent the current from one or two Leclanché cells passed freely. It would appear, therefore, that even when the gas is in the condition in which it conducts electricity freely, some of the cathode potential difference will remain as long as the cathode itself is not incandescent.

155.] The passage of electricity from a gas to a negative electrode seems, as we shall see later, to require something equivalent to chemical combination between the charged atoms of the metal and the atoms of the gas which carry the discharge; and the reason for the disappearance of the fall in potential at the cathode when the latter is incandescent is probably due to this combination taking place under these circumstances much more easily than when the electrode is cold.

156.] Warburg (*Wied. Ann.* 31, p. 545, 1887: 40. p. 1, 1890) has made a valuable series of experiments on the circumstances which influence the fall of potential at the cathode. He has investigated the effect produced on this fall by altering the gas, the size and material of the electrodes, and the amount of impurity in the gas. Hittorf, as we have seen, had already shown that as long as there is room for the negative glow to spread over the surface of the cathode, the cathode fall in potential is approximately independent of the intensity of the current.

In Warburg's experiments, the fall in potential at the cathode, by which is meant the potential difference between the cathode and a wire at the luminous boundary of the negative glow, was measured by a quadrant electrometer. Warburg found that, so long as the whole of the cathode was not covered by the negative glow, the fall in potential at the cathode was nearly independent of the density of the gas: this is shown by the following table (l. c. p. 579), in which E represents the potential difference between the electrodes, which were made of aluminium, e the potential fall at the cathode, E and e being measured in volts, p the pressure of the gas, dry hydrogen, measured in millimetres of mercury, i the current through the gas in millionths of an Ampère.

p.	e.	$E-e$.	i.
9·5	191	139	6140
6·4	190	103	4740
4·4	190	70	4810
3·0	189	50	2640
1·79	191	40	1730
1·20	192	39	1360
·80	191	39	508

This table shows that though the fall in potential in the positive light decreased as the pressure diminished, the fall in potential at the cathode remained almost constant.

157.] In imperfectly dried nitrogen, which contained also a trace of oxygen, the cathode potential difference depended to some extent on the metal of which the electrode was made; platinum, zinc, and iron electrodes had all practically the same potential fall; for copper electrodes the fall was about 3 per cent. and for aluminium electrodes about 15 per cent. less than for platinum. In hydrogen which contained a trace of oxygen, the potential fall for platinum, silver, copper, zinc, and steel was practically the same, about 300 volts. In the case of the last three metals, however, the value of the cathode potential fall at the beginning of the experiment was much lower than 300 volts, and it was not until after long sparking that it rose to its normal value; Warburg attributed this to the presence at the beginning of the experiment of a thin film of oxide which gradually got dissipated by the sparking; he found by direct experiment that the potential fall of a purposely oxidised steel electrode was less than the value reached by a bright steel electrode after being used for some time. The potential fall for aluminium and magnesium electrodes was about 180 volts, and was thus considerably smaller than for platinum electrodes (cf. Art. 47); these metals, however, are easily oxidised; and as, unlike other metals, they do not disintegrate when used as cathodes, the film of oxide would not get removed by use.

158.] The fact that a large number of metals give the same potential fall, while others give a varying one, seems to indicate that this potential fall depends upon whether the electrodes do or do not take part in some chemical change occurring at the cathode; and the connection between this fall in potential and the chemical changes which take place near the cathode seems still more clearly shown by the surprisingly large effects produced by a small quantity of impurity in the gas. Warburg found that the fall of potential at the cathode in nitrogen which contained traces both of moisture and oxygen was 260 volts, while the same nitrogen, after being very carefully dried, gave a cathode fall of 343 volts: thus, in this case, a mere trace of moisture had diminished the cathode fall by 25 per cent., the removal of the trace of oxygen produced equally remarkable effects, see Art. 160. This points clearly to the influence exerted by chemical actions at the cathode on the fall of potential in that region; since a mere trace of a substance is often sufficient to start chemical reactions which would be impossible without it: thus, for example, Pringsheim (*Wied. Ann.* 32, p. 384, 1887) found that unless traces of moisture were present, hydrogen and chlorine gas would not combine to form hydro-chloric acid under the action of sunlight unless it was very intense.

159.] The fall of potential at the cathode seems to be lowered as much by a trace of moisture as by a larger quantity, as long as the total quantity of moisture in the nitrogen remains small; if, however, the amount of aqueous vapour is considerable, the fall in potential is greater than for pure nitrogen; thus in a mixture of nitrogen and aqueous vapour, in which the pressure due to the nitrogen was 3·9 mm., that due to the aqueous vapour 2·3 mm., Warburg found that the fall in potential was about 396 volts, as against about 343 volts for nitrogen containing a trace of oxygen; the increase in the fall of potential at the cathode was, however, not nearly so great comparatively as the increase in the potential differences along the positive column.

In hydrogen, Warburg found that a trace of aqueous vapour increased the potential difference at the cathode instead of diminishing it as in nitrogen.

160.] Warburg (*Wied. Ann.* 40, p. 1, 1890) also investigated the effects produced by removing from the nitrogen or hydrogen any trace of oxygen that might have been present. This was done by placing sodium in the discharge tube, and then after the other gas had been let into the tube, heating up the sodium, which combined with any oxygen there might be in the tube. The effect of removing the oxygen from the nitrogen was very remarkable: thus, in nitrogen free from oxygen, the fall of potential at the cathode when platinum electrodes were used was only 232 volts as against 343 volts when there was a trace of oxygen present; when magnesium electrodes were used the fall in potential was 207 volts; in hydrogen free from oxygen the fall of potential was 300 volts with platinum electrodes, and 168 volts with magnesium electrodes; thus with platinum electrodes the potential fall in hydrogen is greater than in nitrogen, while with magnesium electrodes it is less.

161.] Warburg also investigated a case in which the conditions for chemical change at the cathode were as simple as possible, one in which the gas was mercury vapour (with possibly a trace of air) and the cathode a mercury surface; he found that the negative dark space was present, and that the cathode fall was very considerable, amounting to about 340 volts; this, at the pressures used in these experiments between 3·5 mm. and 14·0 mm., was much greater than the potential difference in a portion of the positive light about half as long again as the piece at the cathode, for which the potential fall was measured.

162.] In air free from carbonic acid, but containing a little moisture, Warburg (*Wied. Ann.* 31, p. 559, 1887) found that the potential fall was about 340 volts: this is very nearly the value found by Mr. Peace for the smallest potential difference which would send a spark between two parallel plates. When we consider the theory of the discharge we shall see that there are reasons for concluding that it is impossible to produce a spark by a smaller potential difference than the cathode fall of potential in the gas through which the spark has to pass.

The researches made by Hittorf on the distribution of potential along the tube show, as we have seen, Art. 140, that the potential gradient is by no means constant; to produce the changes in this gradient which occur in the neighbourhood of the cathode, there must in that region be a quantity of free electricity in the tube. Schuster (*Proc. Roy. Soc.* 47, p. 542, 1890) concludes from his measurements of the potential in the neighbourhood of the cathode that if ρ is the volume density of the free positive electricity at a distance x from the cathode, ρ varies as $\epsilon^{-\kappa x}$.

163.] The measurements of potential along the positive column have been less numerous than those of the negative dark space. Hittorf, De la Rue and Hugo Müller concur in finding that the potential gradient close to the anode is, though not comparable with that at the cathode, greater than that in the middle of the tube.

164.] The potential gradient in the positive column is not like the fall in potential at the cathode approximately independent of the density, it diminishes as the pressure of the gas diminishes: but as the pressure of the gas diminishes, the distance between two consecutive striations increases, and though I can find no experiments bearing on this point, it would be a matter of great interest to know whether or not the potential difference along a length of the positive column equal to the distance between two striations, where these are regular, is approximately independent of the density of the gas.

165.] De la Rue and Hugo Müller (*Phil. Trans.* 1878, Part I, p. 159) measured the potential gradients along a tube in which two wide portions were connected by a piece of capillary tubing, narrow enough to constrict the striae; they found the potential gradient much greater along the capillary portion than along the wide one. Thus the potential difference along 4·25 inches of the positive column in the wide tube, which was about $\frac{15}{16}$ of an inch in diameter, was, on an arbitrary scale, 75, while the potential difference along a portion of the positive column, which included 2 inches of the wide tube and 3·75 inches of the capillary tube ($\frac{1}{8}$ of an inch in diameter), was 138; the potential gradients along the wide and narrow portions are thus in the proportion of 1 to 1·55.

In this case the cathode was in the wide part of the tube; when the tube round the cathode is so narrow that it restricts the negative glow, the increase in the potential difference at the cathode produced by this restriction makes it very much more difficult to get a discharge to pass through the narrow tube than through a wider one. An experiment due to Hittorf (*Wied. Ann.* 21, p. 93, 1884) illustrates this effect in a very remarkable way; at a pressure of ·03 mm. of mercury, it took 1100 of his cells to force the discharge through a tube 1 cm. in diameter, while 300 cells were sufficient to force it between similar electrodes the same distance apart in a tube 11 cm. in diameter, filled with the same kind of gas at the same pressure.

166.] When the electrodes are placed so near together that the dark space round the cathode extends to the anode, the appearance of the discharge is completely changed: this is very well shown in an experiment due to Hittorf (*Pogg. Ann.* 136, p. 213, 1869) represented in Fig. 67; the electrodes were parallel to each other, and the pressure of the gas in the discharge tube was so low that the dark space round the cathode extended beyond the anode; the positive discharge in this case, instead of turning towards the cathode, started from the bend in the anode on the side furthest away from the cathode, and then crept along the surface of the glass until it reached the boundary of the negative dark space. I observed a similar effect in the course of some experiments on the discharge between large parallel plates (*Proc. Camb. Philos. Soc.* 5, p. 395, 1886); when the pressure of the gas was very small, the positive column, instead of passing between the plates, went, as in Fig. 68, from the under

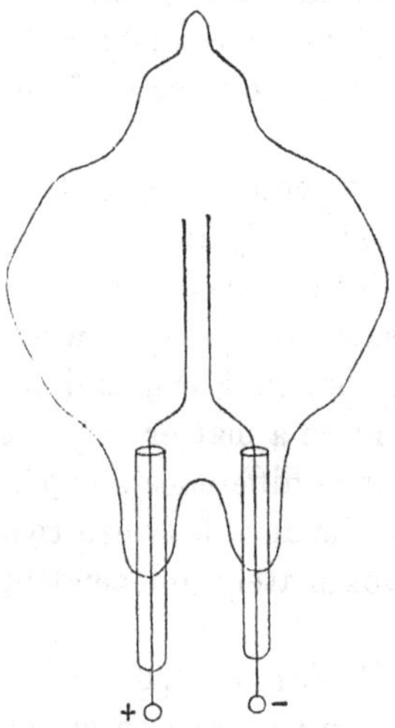

Fig. 67.

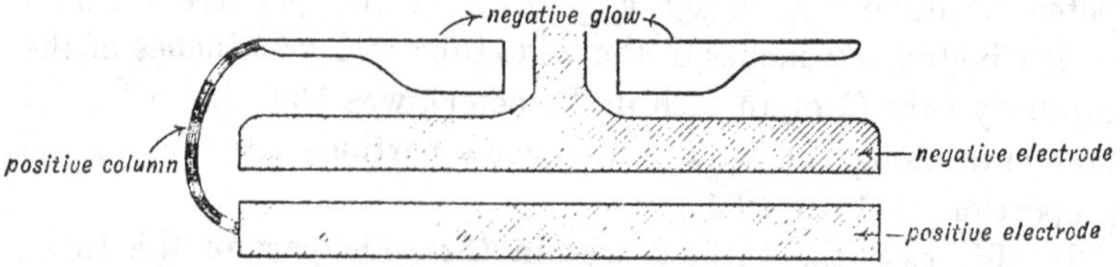

Fig. 68.

side of the lower plate which was the positive electrode, and after passing between the glass and the plates reached right up to the negative glow, which was above the negative plate: the space between the plates was quite dark and free from glow.

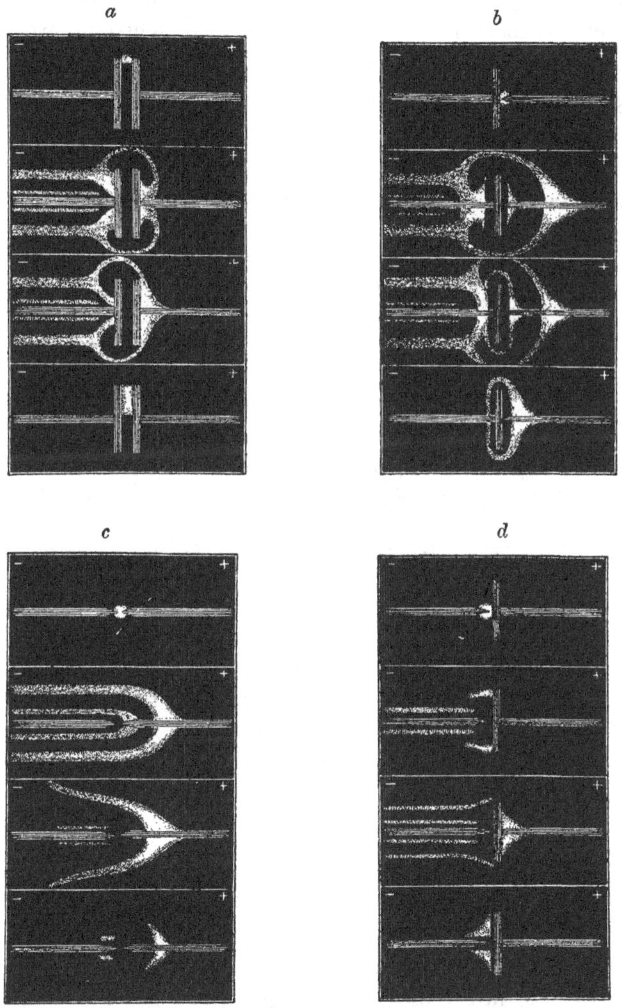

Fig. 69.

Lehmann (*Molekularphysik*, bd. 2, p. 295) has observed with a microscope the appearance of the discharge passing between electrodes of different shapes, placed very close together; they exhibit in a very beautiful way the same peculiarities as those just described; Lehmann's figures are represented in Fig. 69.

When the distance between the electrodes is less than the thickness of the dark space, it is very difficult to get the discharge to pass between them; this is very strikingly illustrated by another experiment of Hittorf's (*Wied. Ann.* 21, p. 96, 1884) which is represented in Fig. 70. The two electrodes were only 1 mm. apart, but the regions surrounding them were connected by a long spiral tube $3\frac{3}{4}$ m. long; in spite of the enormous difference between the lengths of the two paths, the discharge, when the pressure was very low, all went round through the spiral, and the space between the electrodes remained quite dark.

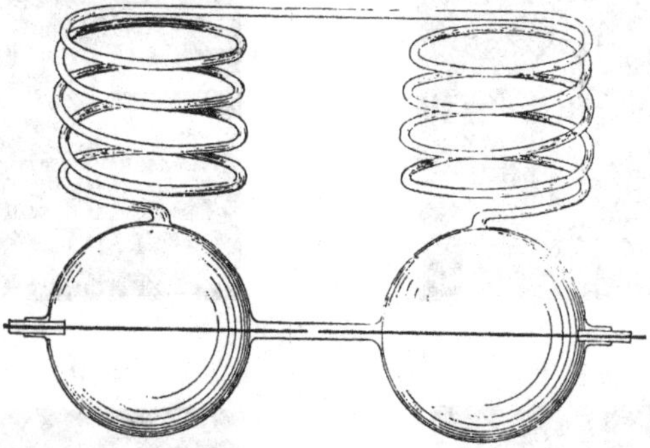

Fig. 70.

167.] In cases of this kind the potential difference required to produce discharge between two electrodes must be *diminished* by increasing the distance between them. For in Hittorf's experiments, the potential difference between the electrodes was equal to the potential fall at the cathode, plus the change in potential due to the $3\frac{3}{4}$ m. of positive light in the spiral, while if the shortest distance between the electrodes had been increased until it was just greater than the thickness of the negative dark space, the potential difference between the electrodes when the discharge passed would only have amounted to the cathode fall, plus the potential difference due to a short positive column instead of to one $3\frac{3}{4}$ metres long, so that the potential difference would have been less than when the electrodes are nearer together. Peace's experiment described in Art. 53. is a direct proof of the truth of this statement for higher pressure, and is free from the objection to which the preceding deduction from Hittorf's experiment is liable, that the cathode fall may not be the same when the discharge starts in the large vessel when the negative glow is unrestricted, as it is when the discharge passes through the narrow tubes, the walls of which constrict the negative glow.

168.] These results explain a peculiar effect which is observed when the discharge passes between slightly curved electrodes at not too great a distance apart; until the pressure is very low the discharge passes across the shortest distance between the electrodes, but after a very low pressure is reached the discharge leaves the centre of the field, and in order to get a longer spark length departs further and further from it as the pressure of the gas is reduced.

The Arc Discharge.

169.] The 'arc discharge,' of which the well-known arc lamp is a familiar example, is characterised by the passage of a large current and the incandescence of both the terminals, as well as by the comparatively small potential difference between them; we considered a case of this discharge in Art. 148, the gas was, however, in that case, at a low pressure; the cases when the gas is at higher pressures are of special interest, on account of the extensive use made of this form of discharge for lighting purposes.

If the current through a vacuum tube with electrodes is gradually increased, the discharge, as Gassiot found in 1863, gradually changes from the ordinary type of the vacuum tube discharge with the negative space and a striated positive column to the arc discharge, in which there is comparatively little difference between the appearances at the terminals. The terminals are brilliantly incandescent while the gas remains comparatively dark, being probably in the state in which it has a large supply of dissociated molecules by means of which it can transmit the current even though the potential gradient is small.

The connection between spark length, potential difference and current in the arc discharge, has been investigated by many physicists, who have all found that the potential difference V is almost independent of the current and can be expressed by the formula

$$V = a + bl,$$

where l is the spark length and a and b are constants. Ayrton and Perry (*Phil. Mag.* [5] 15, p. 346, 1883), using a formula which is identical with the preceding one if the sparks are not very short, found that for carbon electrodes $a = 63$ volts and $b = 21{\cdot}6$ volts, if l is measured in centimetres. The value of a probably depends on the quality of the carbon of which the electrodes are made, as other observers, who have also used carbon electrodes, have found considerably smaller values for a. When more volatile substances than carbon are used the values of a are smaller, the more volatile the substance the smaller in general being the value of a. This is borne out by the following determinations made by Lecher (*Wied. Ann.* 33, p. 625, 1888); the length l in these equations is measured in centimetres, and V in volts:—

Horizontal Carbon Electrodes . .	$V = 33 + 45l$.
Vertical Carbon Electrodes . . .	$V = 35{\cdot}5 + 57l$.
Platinum Electrodes, (·5 cm. in diameter)	$V = 28 + 41l$.
Iron Electrodes, (·55 cm. in diameter) .	$V = 20 + 50l$.
Silver Electrodes, (·49 cm. in diameter) .	$V = 8 + 60l$.

170.] The form of the expression for V shows that the potential required to maintain the current between two incandescent electrodes cannot fall short of a certain minimum value, however short the arc may be. The preceding measurements for a show that this potential difference, though small compared with the 'cathode fall' when the electrodes are cold, is much greater than that which Hittorf in his experiments (see Art. 152) found necessary to maintain a constant current when the cathode was incandescent; we must remember, however, that in Lecher's experiments the gas was at atmospheric pressure, while in Hittorf's the pressure was very low.

171.] Lecher (l. c.) investigated the potential gradient in the arc by inserting a spare carbon electrode, and found that it was far from uniform: thus when the difference of potential between the anode and the cathode was 46 volts, there was a fall of 36 volts close to the anode, and a smaller fall of ten volts near the cathode. The result that the great fall of potential in the arc discharge occurs close to the anode is confirmed by an experiment made by Fleming (*Proc. Roy. Soc.* 47, p. 123, 1890), in which a spare carbon electrode was put into the arc; when this electrode was connected with the anode sufficient current went round the new circuit to ring an electric bell, but when it was connected to the cathode the current which went round the circuit was not appreciable.

172.] The term in the expression for the potential in Art. 169, which is independent of the length of the arc, and which involves an expenditure of energy when electricity travels across an infinitesimally small air space, is probably connected with the work required to disintegrate the electrodes, since the more volatile are the electrodes the smaller is this term.

173.] The disintegration of the electrodes is a very marked feature of the arc discharge, and it is not, as in the case when small currents pass through a highly exhausted gas, confined to the negative electrode; in fact, when carbon electrodes are used, the loss in weight of the anode is greater than that of the cathode, the anode getting hollowed out and taking a crater-like form.

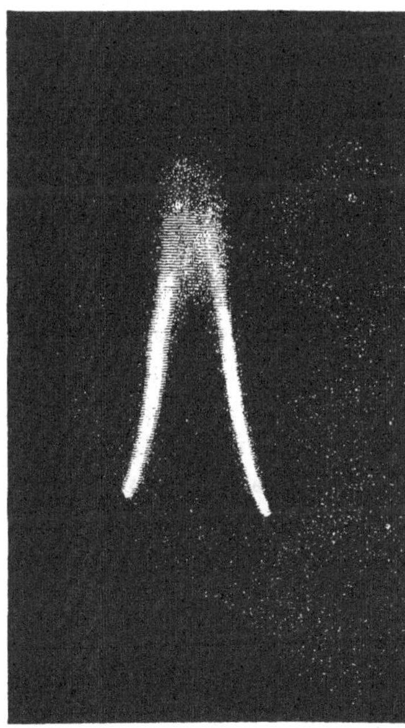

Fig. 71.

174.] Perhaps the most interesting examples of the arc discharge are those which occur when we are able by means of transformers to produce a great difference of potential, say thirty or forty thousand volts between two electrodes, and also to transmit through the arc a very considerable current. In this case the arc presents the appearance illustrated in Fig. 71. The discharge, instead of passing in a straight line between the electrodes, rises from the electrodes in two columns which unite at the top, where striations are often seen though these do not appear in the photograph from which Fig. 71 was taken. The vertical columns are sometimes from eighteen inches to two feet in length, they flicker slowly about and are very easily blown out, a very slight puff of air being sufficient to extinguish them. The air blast apparently breaks the continuity of the belt of dissociated molecules along which the current passes, and the current is stopped just as a current through a wire would be stopped if the wire were cut. The discharge is accompanied by a crackling sound, as if a number of minute sparks were passing between portions of the arc temporarily separated by very short intervals of space.

175.] The relation between the losses of weight of the anode and the cathode in the arc discharge depends however very much on the material of which the electrodes are made; thus Matteucci (*Comptes Rendus*, 30, p. 201, 1850) found that for copper, silver and brass electrodes the cathode lost more than the anode, while for iron the loss in weight of the anode was greater than that of the cathode.

The electrodes in the arc discharge are at an exceedingly high temperature, in fact probably the highest temperatures we can produce are obtained in this way. With carbon electrodes the anode is much hotter than the cathode (compare Art. 149). Since the temperature of the electrodes is so high, it is probable that they are disintegrated partly by the direct action of the heat and not wholly by purely electrical processes such as those which occur in electrolysis; for this reason, we should not expect to find any simple relation between the loss in weight of the electrode and the quantity of electricity which has passed through the arc. Grove (*Phil. Mag.* [3] 16, p. 478, 1840), who used a zinc anode sufficiently large for the temperature not to rise about its melting point, came to the conclusion that the amounts of zinc lost and oxygen absorbed by the electrode were chemically equivalent to the oxygen liberated in a voltameter placed in the circuit. On the other hand, Herwig, (*Pogg. Ann.* 149, p. 521, 1873), who investigated the relation between the loss of weight of a silver electrode in the arc and the amount of chemical decomposition in a voltameter placed in the same circuit, was however unable to find any simple law connecting the two. The brightness of the light given by carbon electrodes is much increased by soaking them in a solution of sodium sulphate.

176.] The particles projected from the electrodes in the arc discharge are presumably charged with electricity, since they are deflected by a magnet; thus some of the electricity passing between the electrodes will be carried by these particles. Comparatively few experiments bearing on this point have, however, been made on the arc discharge, and we have not the information which would enable us to estimate how much of the current is carried by the disintegrated electrodes and how much by the gas.

Fleming (*Proc. Roy. Soc.* 47, p. 123, 1890) has suggested that *all* the current is carried by particles torn off the electrodes, that these particles are projected (chiefly from the cathode) with enormous velocities, and that the incandescence of the electrodes is due to the heat developed by their bombardment by these particles; the hollowing out of the anode is on this theory supposed to be due to a kind of sand blast action exerted by the particles coming from the negative electrode.

On this theory, if I understand it rightly, the gas by which the electrodes are enveloped plays no part in the discharge. I do not think that the theory is consistent with Hittorf's and Gassiot's observations on the continuity of the arc discharge with the ordinary striated discharge produced in a vacuum tube through which only a very small current is passing, nor does it seem in accordance with what we know about the high conductivity of gases which are at a high temperature or through which an electric discharge has recently passed.

The Heat produced by the Discharge.

177.] Though the electric discharge is generally accompanied by intense light, the average temperature of the molecules of the gas through which it passes is often by no means high. Thus E. Wiedemann (*Wied. Ann.* 6, p. 298, 1879) has found that the average temperature of a column of air at a pressure of about 3 mm. made luminous by the passage of the discharge can be under 100° C. As, however, any instrument which we may use to measure the temperature of the gas merely measures the average temperature of molecules filling a considerable space, the fact that this temperature is low does not preclude the existence of a small number of molecules moving with velocities immensely greater than the mean velocity corresponding to the temperature indicated by the thermometer.

On the other hand, the fact that the gas is luminous during the discharge does not afford conclusive evidence of the existence of molecules in a state comparable with that of the majority of the molecules in a gas at a very high temperature, for mere increase of temperature unaccompanied by chemical changes seems to have little effect in increasing the luminosity of a gas; thus in one of Hittorf's experiments already mentioned, where the temperature of the electrodes was great enough to melt iridium, the gas surrounding them when examined by the spectroscope did not show any spectroscopic lines. It would seem that the interchange of atoms between the molecules which probably goes on when the discharge passes through the gas is much more effective in making it luminous than mere increase in temperature unaccompanied by chemical changes.

178.] Many experiments have been made by G. and E. Wiedemann, Hittorf, and others on the distribution along the line of discharge of the heat produced by the spark. Hittorf's experiments are the easiest to interpret, since by means of a large battery he produced through the discharge tube a current which, if not absolutely continuous, was so nearly so, that no want of continuity could be detected either by a revolving mirror or by a telephone; the gas had therefore a much better chance of getting into a steady state than if intermittent discharges such as those produced by an induction coil had been used.

Hittorf (*Wied. Ann.* 21, p. 128, 1884) inserted three thermometers in the discharge tube, one close to the cathode, another in the bright part of the negative glow, and the third in the positive column. He found, using small currents and low gaseous pressures, that the temperature of the thermometer next the cathode was the highest, that of the one in the negative glow the next, and that of the one in the positive column the lowest.

The distribution of temperature depends very much upon the intensity of the current. Hittorf found that when the strength was increased the difference between the temperatures of his thermometers increased also. When however the increase in the current is so great that the discharge becomes an arc discharge, then, at any rate when carbon electrodes are used, the temperature at the anode is higher than that of the cathode; with weak currents we have seen that it is lower.

E. Wiedemann (*Wied. Ann.* 10, p. 225 et seq., 1880) found that the distribution of temperature along the discharge depended on the pressure. In his experiments the temperature at the anode was slightly higher than that at the cathode when the pressure was about 26 mm. of mercury, at lower pressures the cathode was the hotter, and the difference between the temperatures of the cathode and the anode increased as the pressure diminished.

Differences between the Phenomena at the Positive and Negative Electrodes.

179.] We have seen already that when the pressure of the gas is small the two electrodes present very different appearances, there are however many differences between an anode and a cathode even at atmospheric pressure.

The appearance of the spark discharge at the two electrodes is different. The following figure is from a photograph of the spark in air at atmospheric pressure. It will be noticed that

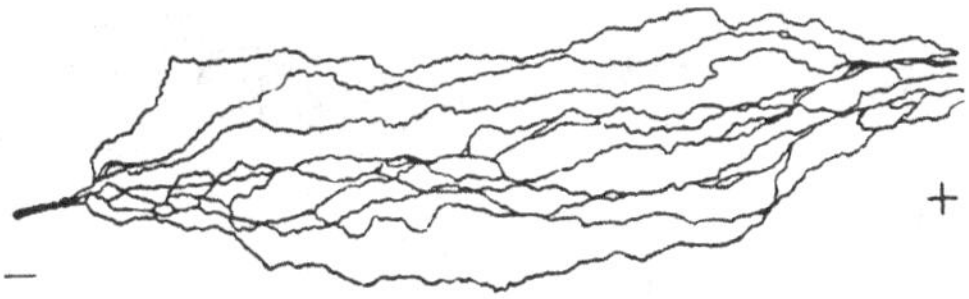

Fig. 72.

the sparks seem to reach a definite point on the negative electrode, but to spread over a considerable area of the positive. Bright dots of light are often to be seen on the positive electrode but not on the negative, these are still more striking at lower pressures. When the spark is branched as in Fig. 73, the branches point to the negative electrode.

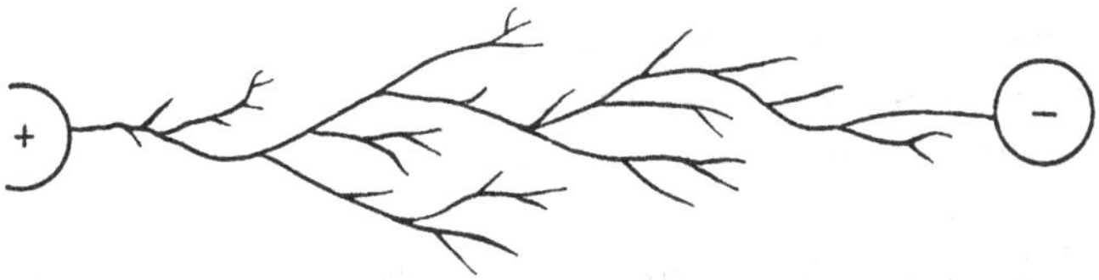

Fig. 73.

If the electrodes are not of the same size, the spark length for the same potential difference seems to depend upon whether the larger or smaller electrode is used as the cathode, though it is a disputed question whether this difference exists if the spark is not accompanied by some other form of discharge.

Thus, if for example the electrodes are spheres of different sizes, Faraday (*Experimental Researches*, § 1480) found that the spark length was greater when the smaller sphere was positive than when it was negative. We may express this result by saying that when the electric field is not uniform the gas does not break down so easily when the greatest electromotive intensity is at the cathode as it does when it is at the anode.

Macfarlane's measurements (*Phil. Mag.* [5] 10, p. 403, 1880) of the potential difference required to start a discharge between a ball and a disc are in accordance with this result, as he found that for a given length of spark the potential difference between the electrodes was smaller when the ball was positive than when it was negative.

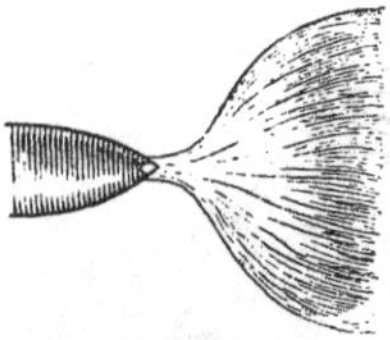

Fig. 74.

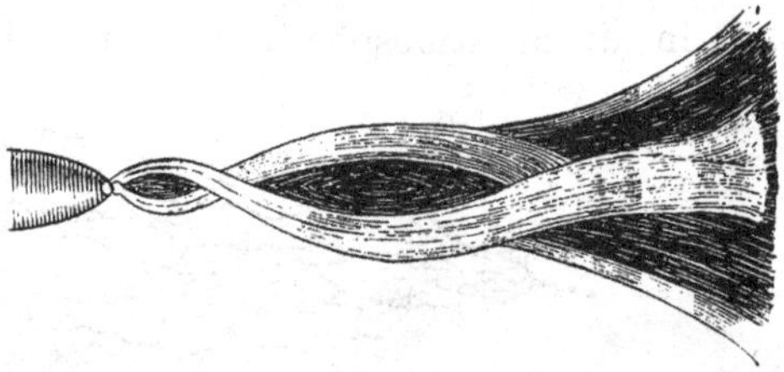

Fig. 75.

180.] De la Rue and Hugo Müller (*Phil. Trans.* 1878, Part I, p. 55) observed analogous effects in the experiments they made with their large chloride of silver battery on the sparking distance between a point and a disc. They found that for potential differences between 5000 and 8000 volts the sparking distance was greatest when the point was positive and the disc negative, while for smaller potential differences they found that the opposite result was true. The appearance of the discharge at the positive point they found was different from that at the negative. The discharge at the negative point is represented in Fig. 74, that at the positive in Fig. 75.

181.] Wesendonck (*Wied. Ann.* 38, p. 222, 1889), however, concludes from his experiments that there are no polar differences of this kind when the discharge passes entirely as a spark, and that the differences which have been observed are due to the coexistence of other kinds of discharge such as a brush and glow.

The existence of this kind of discharge would put the gas into a condition in which it is electrically weak and thus ill-fitted to resist the passage of the spark. This explanation does not seem inconsistent with Faraday's experiment, for, as we shall see in the next paragraph, the negative brush is formed more easily than the positive one. Thus if the sparks in his experiments only passed when they were preceded by the formation of brushes at both the electrodes, it might be produced if the greatest electromotive intensity was at the place where the brush was formed with the greatest difficulty—the anode—while it might not be produced if the smallest intensity was at the anode, thus the gas would be electrically weaker in the first case than in the second.

182.] Considerable polar differences seem undoubtedly to occur in the brush and glow discharges. Thus Faraday (*Experimental Researches*, § 1501) found that if two equal spheres were electrified until they discharged their electricity by a brush discharge into the air, the discharge occurred at a lower potential for the negative ball than for the positive; more electricity thus accumulates on the positive ball than on the negative before the brush occurs, so that when the positive brush does take place it is finer than the negative one.

The brush discharge is also intermittent, and since the positive brush requires a greater accumulation of electricity than the negative one, the interval between consecutive discharges is greater for the positive than for the negative brush.

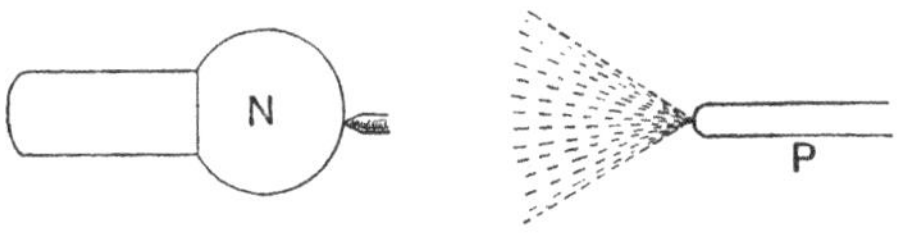

Fig. 76.

The positive and negative brushes are represented in Fig. 76, copied from a figure given by Faraday.

In the brush discharge the electricity seems to be carried partly by particles of metal torn from the electrodes. Nahrwold (*Wied. Ann.* 31, p. 473, 1887) has confirmed the conclusion that the negative brush is more easily formed than the positive.

Wesendonck (*Wied. Ann.* 39, p, 601, 1890) has shown that when the discharge passes as a glow discharge from a point into air, hydrogen, or nitrogen, the potential at which the discharge begins is less when the point is negative than when it is positive.

Lichtenberg's Figures and Kundt's Dust Figures.

183.] Very tangible differences between the discharges from the positive and negative electrodes at ordinary pressures are obtained if we allow the discharge from one or other of the electrodes to pass on to a non-conducting plate covered with

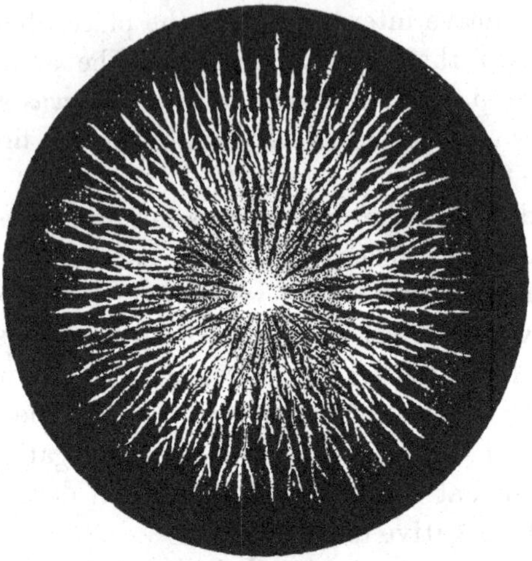

Fig. 77.

some badly conducting powder. If, for example, we powder a plate with a mixture of red minium and yellow sulphur and then cause a discharge from a positively electrified point to pass to the plate, the sulphur, which by friction against the minium is negatively electrified, adheres to the positively electrified parts of the plate, and will be found to be arranged in a star-like form like that represented in Fig. 77. If, on the other hand, the discharge is taken from a negatively electrified body the appearance of the minium on the plate is that represented in Fig. 78. These are known as Lichtenberg's figures; the positive ones are larger than the negative.

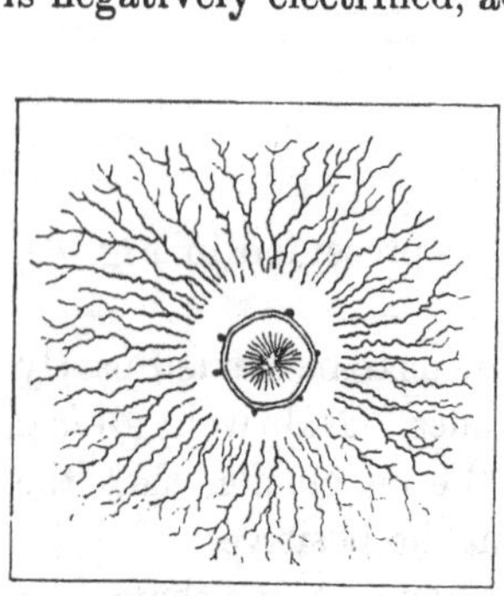

Fig. 78.

If the electrodes are made of very bad conductors, such as wood, there is no difference between the positive and the negative figures.

184.] Very beautiful figures are obtained if a plate of glass covered with a non-conducting powder, such as lycopodium, is placed on a metal plate, and two wires connected with the poles of an induction coil made to touch the powdered surface of the glass. When the discharge passes the powder arranges itself in patterns which are finely branched and have a moss-like appearance at the anode and a more feathery or lichenous appearance at the cathode. The accompanying figure is from a paper by Joly (*Proc. Roy. Soc.* 47, p. 84, 1890); the negative electrode is on the left.

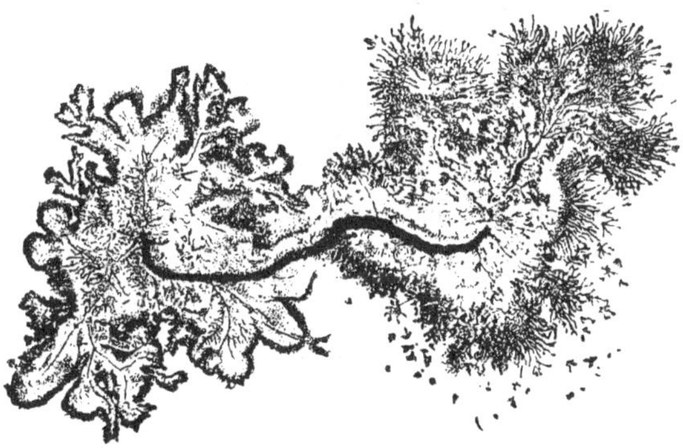

Fig. 79.

185.] As Lehmann has remarked (*Molekularphysik*, b. 11, p. 303), the differences between the positive and negative figures are what we should expect if the discharge passed as a brush from the positive electrode and as a glow from the negative one. He has verified by direct observation that this is frequently the case.

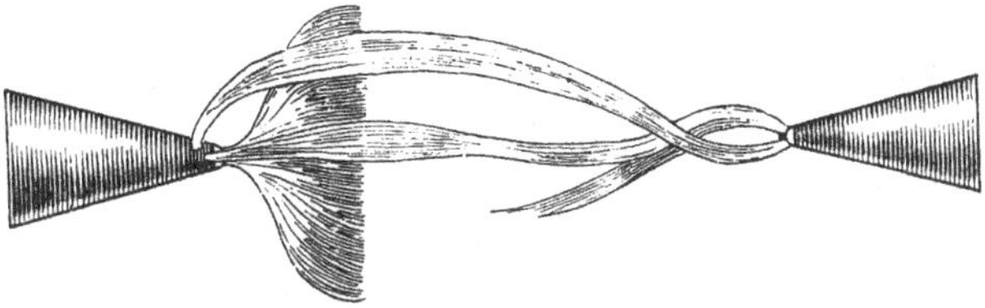

Fig. 80.

A good deal of light is also, I think, thrown on the difference between the positive and negative figures by Fig. 80, which is given by De la Rue and Hugo Müller (*Phil. Trans.* 1878, Part I, p. 118) as the discharge produced by 11,000 of their chloride of silver cells in free air. It will be noticed that there is at the negative electrode a continuous discharge superposed on the streamers which are the only form of discharge at the positive, this continuous discharge will fully account for the comparative want of detail in the negative figure.

186.] Kundt's figures are obtained by scattering non-conducting powders over a horizontal metal plate, instead of, as in Lichtenberg's figures, over a non-conducting one. If the plate be shaken after a discharge has passed from a negative point to the positive plate, it will be found that the powder will fall from every part of the plate except a small circle under the negative electrode, where the powder sticks to the plate and forms what is called Kundt's 'dust figure.' The dimensions of this circle are very variable, ranging in Kundt's original experiments (*Pogg. Ann.* 136, p. 612, 1869) from 10 to 200 mm. in diameter. If the point is positive and the plate negative Kundt's figures are only formed with great difficulty.

Mechanical Effects produced by the Discharge.

187.] We have already considered the mechanical effects produced by the projection of particles from the cathode: many other such effects are however produced by the electric discharge. One of the most interesting of these is that described by De la Rue and Hugo Müller (*Phil. Trans.* 1880, p. 86): they found that when the discharge from their large chloride of silver battery passed through air at the pressure of 53 mm. of mercury, the pressure of the air was increased by about 30 per cent., and they proved, by measuring the temperature, that the increase in pressure could not be accounted for by the heat produced by the spark.

This effect can easily be observed if a pressure gauge is attached to any ordinary discharge tube, the gas inside being most conveniently at a pressure of from 2 to 10 mm. of mercury. At the passage of each spark there is a quick movement of the liquid in the gauge as if it had been struck by a blow coming from the tube; immediately after the passage of the spark the liquid in the gauge springs back to within a short distance of its position of equilibrium, and then slowly creeps back the rest of the way. This creeping effect is probably due to the slow escape of the heat produced by the passage of the spark.

The gauge behaves as if a wave of high pressure rushed through the tube when the spark passed.

188.] Meissner, *Abhand. der König. Gesellschaft, Gottingen*, 16, p. 98 et seq., 1871 (who seems to have been the first to observe this effect, though in his experiments it was not developed to such an extent as in De la Rue's and Müller's), found that if a tube provided with a gauge were placed between the plates of a condenser there was an increase of pressure when the plates were charged or discharged, and no effect as long as the charge on the condenser remained constant. In this case there was no spark between the plates of the condenser, and the effect must have been due to the passage through the gas of the electricity which, when it was in equilibrium before the spark passed, was spread over the glass of the tube.

Meissner observed this effect when the tube was filled with oxygen, hydrogen, carbonic acid, and nitrogen, though it was very small when the tube was filled with hydrogen.

189.] The effect seems too great to be accounted for merely by the increased statical pressure due to the decomposition of the molecules of the gas by the discharge, for in De la Rue's experiment, where the gas was contained in a large vessel and the discharge passed as a narrow thread between the electrodes, the pressure was increased by about 30 per cent. Now if this increase of pressure was due to the splitting up of the molecules into atoms it would require about one-third of the molecules to be so split up by a discharge which only occupied an infinitesimal fraction of the volume of the gas.

190.] It would seem more probable that in this case we had something analogous to the driving off of particles from an electrified point, as in the ordinary phenomenon of the 'electrical wind,' or that of the projection of particles from the cathode which occurs when the discharge passes through a gas at a very low pressure; the difference between this case and the one we are considering being that in the latter, since the pressure is greater, the molecules shot off from the cathode communicate their momentum to the surrounding gas instead of retaining it until they strike against the walls of the discharge tube. This would have the effect of diminishing the density of the gas in the neighbourhood of the line of discharge, and would therefore increase the density and pressure in other parts of the tube.

191.] Töpler (*Pogg. Ann.* 134, p. 194, 1868) has investigated by means of a stroboscopic arrangement the disturbance in the air produced by the passage of a spark. The following figures taken

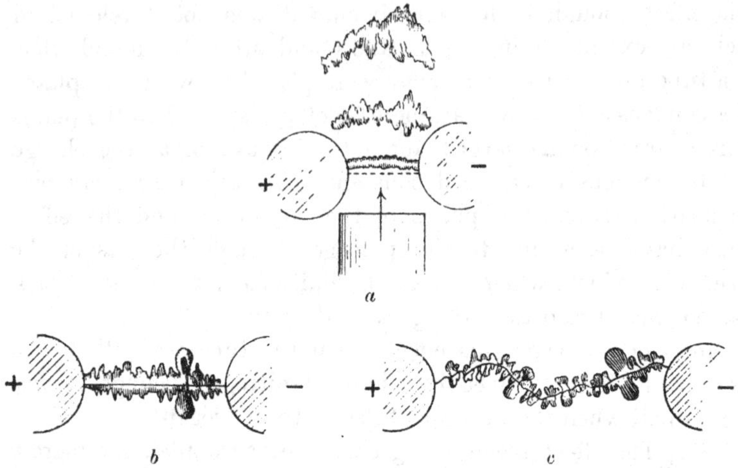

Fig. 81.

from his paper show the regions when the gas is expanded in the neighbourhood of the spark line at successive small intervals of time after the passage of the spark. It will be noticed that these regions show periodic swellings and contractions as if the centres of greatest disturbance were distributed at regular and finite intervals along the line of discharge. A similar appearance was observed by Antolik (*Pogg. Ann.* 154, p. 14, 1875) when the discharge passed over a plate covered with fine powder; the powder placed itself in ridges at regular intervals along the line of discharge.

192.] This effect is also beautifully illustrated in an experiment made by Joly (*Proc. Roy. Soc.* 47, p. 78 et seq., 1890), in which the discharge passed from one strip of platinum to another between plates of glass placed so close together that they showed Newton's rings; it was only with difficulty that the discharge could be got through this narrow space at all, it declined to go through the centre of the rings, and went out of its way to get through the places where the distance between the plates was greatest. Where it passed it made furrows on the glass at right angles to the line of discharge and separated by regular intervals; a magnified representation of these is shown in Fig. 82, taken from Joly's paper. When the air between the plates was replaced by hydrogen these furrows had a tendency to be more widely separated.

193.] The explosive effects produced by the spark are well illustrated by an experiment due to Hertz (*Wied. Ann.* 19, p. 87, 1883), in which the anode was placed at the bottom of a glass tube with a narrow mouth, while the cathode was placed

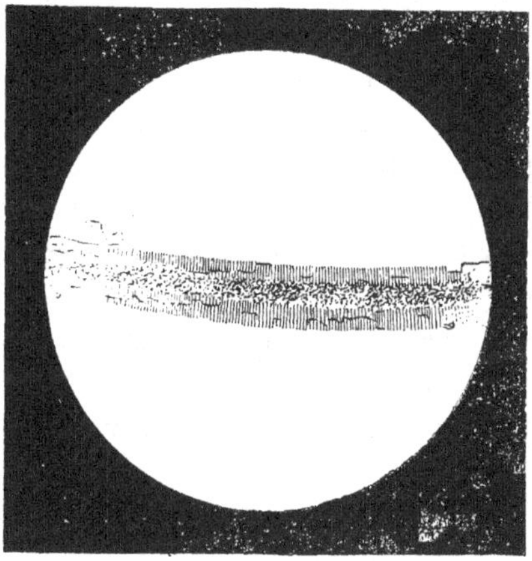

Fig. 82.

outside the tube and close to the open end. The tube and the electrodes were in a bell jar filled with dry air at a pressure of 40–50 mm. of mercury. When the discharge from a Leyden jar charged by an induction-coil passed, the glow accompanying it was blown out of the tube and extended several centimetres from the open end. In this experiment, as in the well-known 'electric wind,' the explosive effects seem to be more vigorous at the anode than they are at the cathode.

Chemical Action of the Electric Discharge.

194.] When the electric discharge passes through a gas, it produces in the majority of cases perceptible chemical changes, though whether these changes are due to the electrical action of the spark, or whether they are secondary effects due to a great increase of temperature occurring either at the electrodes or along the path of the discharge, is very difficult to determine when the discharge takes the form of a bright spark.

195.] For this reason we shall mainly consider the chemical changes produced by those forms of discharge in which the thermal effects are as small as possible, though even in these cases, since we can only measure the average temperature of a large number of molecules, it is always possible to account for any chemical effect by supposing that although the average temperature is not much increased by the discharge, a small number of molecules have their kinetic energy so much increased that they can enter into fresh chemical combinations.

The thermal explanation of the chemical changes requires that they should be subsequent to, and not contemporaneous with the passage of the discharge; on the view adopted in this book chemical changes of some kind are necessary before the discharge can pass at all, though it by no means follows that the chemical changes which are instrumental in carrying the current are those which are finally apparent. When electricity passes through a liquid electrolyte the substances liberated at the electrodes are in consequence of secondary chemical actions frequently different from the ions which carry the current.

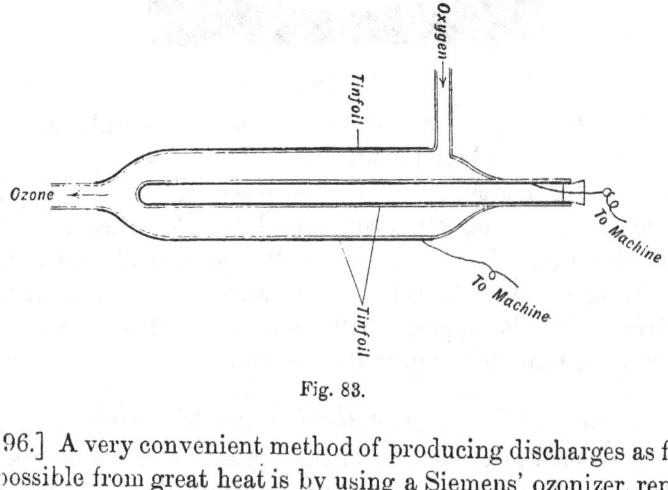

Fig. 83.

196.] A very convenient method of producing discharges as free as possible from great heat is by using a Siemens' ozonizer, represented in Fig. 83. Two glass tubes are fused together, and the gas through which the discharge takes place circulates between them, entering by one of the side tubes and leaving by the other; the inside of the inner tube and the outside of the outer are coated with tin-foil, and are connected with the poles of an induction-coil. When the coil is working a quiet discharge passes as a series of luminous threads between the surfaces of the glass opposed to each other. This form of discharge is often called the 'silent discharge,' and by French writers *l'effluve electrique.*

When air or oxygen is sent through a tube of this kind when the coil is working a considerable amount of ozone is produced.

Ozone is not produced by the action of a steady electric field on oxygen or air unless the field is intense enough to produce a discharge through the gas (see J. J. Thomson and R. Threlfall. *Proc. Roy. Soc.* 40, p. 340, 1886).

Meissner (*Abhandlungen der König. Gesell. Göttingen*, 16, p. 3, 1871) found that ozone was produced in tubes placed between the plates of a condenser when the condenser was charged or discharged, although no sparks passed between the plates, but that no ozone was produced when the charges on the plates of the condenser were kept constant. This was probably due to the passage through the gas of electricity which had distributed itself over the walls of the tube under the inductive action of the charged plates of the condenser.

Bichat and Guntz (*Annales de Chimie et de Physique* [6], 19, p. 131, 1890) ascribe the formation of ozone, even by the silent discharge, to purely thermal causes. They regard the bright thread-like discharge surrounded by the non-luminous gas as a column of very hot oxygen surrounded by a cold atmosphere, and consider the conditions analogous to those which obtain in a St. Claire Deville 'chaud froid' tube, by the aid of which they state that Troost and Hautefeuille have produced ozone from oxygen without the use of the electric discharge.

197.] By the aid of the silent discharge a great many chemical changes are produced, of which the following are given by Lehmann, *Molekularphysik,* (bd. 2, p. 328.) Carbonic acid is split up by the discharge into carbonic oxide, oxygen, and ozone: water vapour into hydrogen and oxygen: when the discharge passes through acetylene a solid and a liquid are produced: phosphoretted hydrogen yields under similar circumstances a solid: methyl hydride gives marsh gas, hydrogen, and an acid: nitrous oxide splits up into nitrogen and oxygen: nitric oxide into nitrous oxide, nitrogen and oxygen.

A mixture of carbonic acid and marsh gas gives a viscous

fluid; nitrogen partly combines with ammonia: carbonic oxide and hydrogen give a solid product: carbonic oxide and marsh gas a resinous substance: nitrogen and hydrogen ammonia.

Dextrine, benzine, and sodium absorb nitrogen under the influence of the discharge, and enter into chemical combination with it. Hydrogen forms with benzine and turpentine resinous compounds.

198.] Perthelot (*Annales de Chimie et de Physique*, [5], 10, p. 55, 1877) has shown that the absorption of nitrogen by dextrine takes place under very small electromotive intensities; he showed this by connecting the inside and the outside coatings of the ozonizer to points at different heights above the surface of the ground, and found that this difference of potential, which varied in the course of the experiments from $+60$ to -180 volts, was sufficient to produce in the course of a few weeks an appreciable absorption of nitrogen by a solution of dextrine in contact with it. The potential differences in these experiments were so small, and their rate of variation so slow, that it seems improbable that any discharge could have passed through the nitrogen, and the experiments suggest that chemical action between a gas and a substance with which it is in contact can be produced by the action of a variable electric field without the passage of electricity through the bulk of the gas. Berthelot suggests that plants may, under the influence of atmospheric electricity, absorb nitrogen by an action of this kind. This suggestion also raises the very important question as to whether the chemical changes which accompany the growth of plants can have any influence on the development of atmospheric electricity.

199.] We must now consider the relation between the quantity of electricity which passes through a gas and the amount of chemical action which takes place in consequence. It is necessary here to make a distinction, which has been too much neglected, between the part of this action which occurs at the electrodes and the part which occurs along the length of the spark. When a current of electricity passes through a liquid electrolyte the only evidence of chemical decomposition is to be found at the electrodes. When, however, the electric discharge passes through a gas the chemical changes are not confined to the electrodes but occur along the line of the discharge as well. This is proved by the fact that when the electrodeless discharge passes through oxygen ozone is produced, as is testified by the existence for several seconds after the discharge has passed of a beautiful phosphorescent glow: the same thing is also proved by the behaviour of the discharge when it passes through acetylene; the first two or three sparks are of a beautiful light green colour, while all subsequent discharges are a kind of whitish pink, showing that the first two or three sparks have decomposed the gas.

200.] Since chemical decomposition is not confined to the electrodes its amount must depend upon the length of the spark; this has been proved by Perrot (*Annales de Chimie et de Physique* [3], 61, p. 161, 1861), who compared the amounts of water vapour decomposed in the same time in a number of discharge tubes placed in series, the spark lengths in the tubes ranging from two millimetres to four centimetres; he found that the volumes of gas decomposed varied from 2 c.c. to 52 c.c., and that neither the longest nor the shortest spark produced the maximum effect. By placing a voltameter in the circuit Perrot found that in one of his tubes the amount of water vapour decomposed by the sparks was about 20 times the amount of water decomposed in the voltameter. It is evident from this that if we wish to arrive at any simple relation between the quantity of electricity passing through the gas and the amount of chemical decomposition produced we must separate the part of the latter which occurs along the length of the spark from that which takes place at the electrodes.

201.] This seems to have been done in a remarkable investigation made more than thirty years ago by Perrot (l.c.), which does not seem to have attracted the attention it merits, and which would well repay repetition. The apparatus used by Perrot in his experiments is represented in Fig. 84 from his paper. The spark passed between two platinum wires sealed into glass tubes, *cfg*, *dfg*, which they did not touch except at the places where they were sealed: the open ends, *c*, *d*, of these tubes were about 2 mm. apart, and the wires terminated inside the tubes at a distance of about 2 mm. from the ends. The other ends of these tubes were inserted under test tubes *e e*, in which the gases which passed up the tubes were collected. The air was exhausted from the vessel A and the water vapour through which the discharge passed was obtained by heating the water in the vessel to about 90°C.: special precautions were taken to free this water from any dissolved gas. The stream of vapour arising from this water drove up the tubes the gases produced by the passage of the spark; part of these gases was produced along the length of the spark, but in this case the hydrogen and

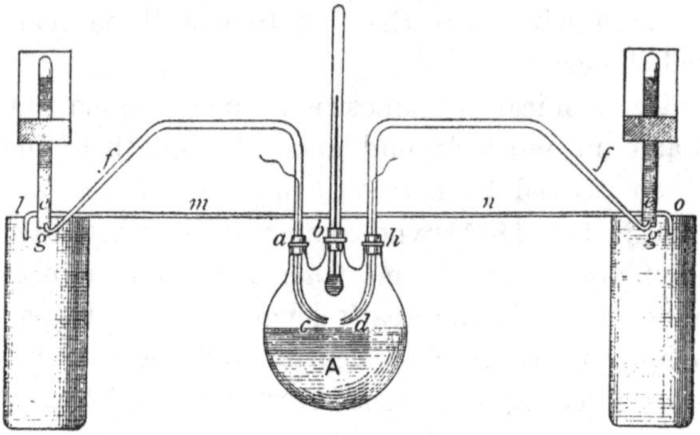

Fig. 84.

oxygen would be in chemically equivalent proportions; part of the gases driven up the tubes would however be liberated at the electrodes, and it is this part only that we could expect to bear any simple relation to the quantity of electricity which had passed through the gas.

When the sparking had ceased, the gases which had collected in the test tubes *e* and *e* were analysed; in the first place they were exploded by sending a strong spark through them, this at once got rid of the hydrogen and oxygen which existed in chemically equivalent proportions and thus got rid of the gas produced along the length of the spark. After the explosion the gases left in the tubes were the hydrogen or oxygen in excess, together with a small quantity of nitrogen, due to a little air which had leaked into the vessel in the course of the experiments, or which had been absorbed by the water. The results of these analyses showed that there was always an excess of oxygen in the test tube in connection with the positive electrode, and an excess of hydrogen in the test tube connected with the negative electrode, and also that the amounts of oxygen and hydrogen in the respective tubes were very nearly chemically equivalent to the amount of copper deposited from a solution of copper sulphate in a voltameter placed in series with the discharge tube.

These results are so important that I shall quote one of Perrot's experiments in full (l.c. pp. 182–3).

Duration of experiment 4 hours. 8·5 milligrammes of copper deposited in the voltameter from copper sulphate; this amount of copper is chemically equivalent to 3 c.c. of hydrogen and 1·5 c.c. of oxygen at atmospheric pressure.

In the test tube over the negative electrode there were at the end of the experiment 37·5 c.c. of gas, after the explosion by the spark this was reduced to 3·1 c.c., so that by far the greater part of the gas collected consisted of hydrogen and oxygen in chemically equivalent proportions, produced not at the electrodes but along the line of the spark. 5·3 c.c. of oxygen were added to the original gas, which was again exploded and the contraction was 4·5 c.c.; in the original gas in the test tube there was therefore an excess of 3 c.c. of hydrogen and ·1 c.c. of something besides hydrogen and oxygen, probably nitrogen. In the test tube over the positive electrode there were 35·8 c.c. of gas at the end of the experiment, after the explosion by the spark this was reduced to 1·6 c.c. 1·8 c.c. of oxygen were added, but there was no explosion when the spark passed; 8·7 c.c. of hydrogen were added and the mixture exploded when the spark passed; the contraction produced was 9·6 c.c., showing that the excess of oxygen originally present was 1·4 c.c. and that ·2 c.c. of nitrogen were mixed with it. Thus the excesses of hydrogen and oxygen in the tubes were very nearly chemically equivalent to the amount of copper deposited in the voltameter. This is also borne out by the following results of other experiments made by Perrot (l.c. p. 183).

2nd experiment. Duration of experiment 4 hours. Copper deposited in voltameter 6 milligrammes, chemically equivalent to 2·12 c.c. of hydrogen and 1·06 c.c. of oxygen.

Gas in the test tube over the positive electrode 35·10 c.c.; excess of oxygen ·95 c.c.; nitrogen ·2 c.c.

Gas in the test tube over the negative electrode 32·40 c.c.; excess of hydrogen 2·10 c.c.; nitrogen ·1 c.c.

4th experiment. Duration of experiment 3 hours. Copper deposited in voltameter 5·5 milligrammes, chemically equivalent to 1·94 c.c. of hydrogen and to ·97 c.c. of oxygen.

Gas in the test tube over the positive electrode 25·10 c.c.; excess of oxygen ·85 c.c.; nitrogen ·15 c.c.

Gas in the test tube over the negative electrode 27·70 c.c.; excess of hydrogen 1·8 c.c.; nitrogen ·21 c.c.

6th experiment. Duration of experiment $3\frac{1}{2}$ hours. Copper deposited in voltameter 6 milligrammes, chemically equivalent to 2·12 c.c. of hydrogen and to 1·06 c.c. of oxygen.

Gas in the test tube over the positive electrode 30·20 c.c.; excess of oxygen ·90 c.c.; nitrogen ·2 c.c.

Gas in the test tube over the negative pole 32·50 c.c.; excess of hydrogen 2·05 c.c.; nitrogen ·2 c.c.

These results seem to prove conclusively (assuming that the discharge passed straight between the platinum wires and did not pass through a layer of moisture on the sides of the tubes) that the conduction through water vapour is produced by chemical decomposition, and also that in a molecule of water vapour the atoms of hydrogen and oxygen are associated with the same electrical charges as they are in liquid electrolytes.

202.] Another way in which the chemical changes which accompany the passage of the spark through a gas manifest themselves is by the production of a phosphorescent glow, which often lasts for several seconds after the discharge has ceased. In a great many gases this glow does not occur, it is however extremely bright in oxygen. A convenient way of producing the glow is to take a tube about a metre long filled with oxygen at a low pressure, and produce an electrodeless discharge at the middle of the tube. From the bright ring produced by the discharge a phosphorescent haze will spread through the tube moving sufficiently slowly for its motion to be followed by the eye. The haze seems to come from the ozone, and the phosphorescence to be due to the gradual reconversion of the ozone into oxygen. This view is borne out by the fact that if the tube is heated the glow is not formed by the discharge, but as soon as the tube is allowed to cool down the glow is again produced: thus the glow, like ozone, cannot exist at a high temperature.

The spectrum of this glow in oxygen is a continuous one, in which, however, a few bright lines can be observed if very high dispersive power is used. The glow is also formed in air, though not so brightly as in pure oxygen. When electrodes are used it seems to form most readily over the negative electrode, especially if this is formed of a flat surface of sulphuric acid.

I have experimented with a large number of gases in order to see whether or not the glow was formed when the electrodeless discharge passed through them. I have never detected any glow in a single gas (as distinct from a mixture) unless that gas was one which formed polymeric modifications, but all the gases I examined which do polymerize have shown the after-glow. The gases in which I have found the glow are oxygen; cyanogen (in which it is extremely persistent, though not so bright as in oxygen), acetylene, and vinyl chloride, all of which polymerize.

A bulb filled with oxygen seems to retain its power of glowing unimpaired, however much it may be sparked through. In bulbs filled with the other gases, however, the glow after long sparking is not so bright as it was originally. This seems to suggest that the polymeric modification produced by the sparking does not get completely reconverted into the original form.

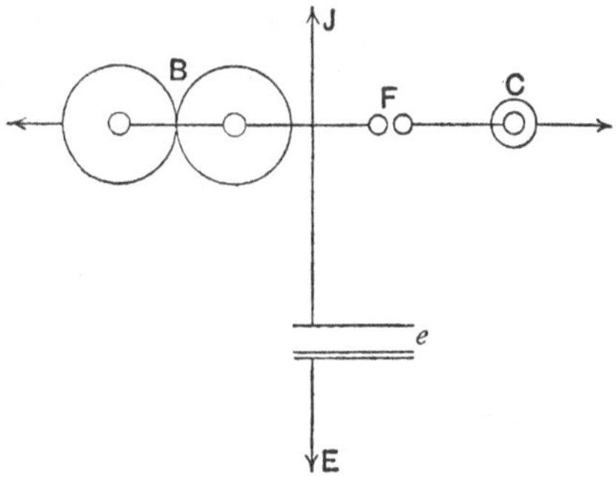

Fig. 85.

Spark facilitated by rapid changes in the intensity of the Electric Field.

203.] Jaumann (*Sitzb. d. Wien Akad.* 97, p. 765, 1888) has made some interesting experiments on the effect on the spark length of small but rapid changes in the electrical condition of the electrodes. The arrangement used for these experiments is represented in Fig. 85, which is taken from Jaumann's paper.

The main current from an electrical machine charged the condenser B, while a neighbouring condenser C could be charged through the air-space F; C was a small condenser whose capacity was only ·55 m., while B was a battery of Leyden Jars whose capacity was 1000 times that of C. Another circuit connected with the machine led to a thin wire placed about 5 mm. above a plate *e* which was connected to the earth. A glow discharge passed between the wire and the plate, and the difference of potential between the inside and outside coatings of the jar B was constant and equal to about 12 electrostatic units. When the knobs of the air-break F were pushed suddenly together a spark about ·5 mm. in length was produced at F, and in addition a bright spark 5 mm. long jumped across the air space at *e* where there was previously only a glow. The passage of the spark at F put the two condensers B and C into electrical communication, and this was equivalent to increasing the capacity of B by about one part in a thousand; this alteration in the capacity produced a corresponding diminution in the potential difference between its coatings. This disturbance of the electrical equilibrium would give rise to small but very rapid oscillations in the potential difference between the wire and the plate *e*, and this variable field seemed able to send a spark across *e*, where when the potential was steady nothing but a glow was to be seen.

204.] It thus appears that a gas is electrically weaker under oscillating electric fields than under steady ones, for it is not apparent why the addition of the capacity of the small condenser to that of B should produce any considerable difference in the electromotive intensity at *e*. It is true that while the discharge is oscillating the tubes of electrostatic induction are not distributed in the same way as they are when the field is steady, and some concentration of these tubes may very likely take place, but it does not seem probable that the disturbance produced by so small a condenser would be sufficient to account for the large effects observed by Jaumann, unless, as he supposes, the gas is electrically weaker in variable electric fields.

Another point which might affect the electromotive intensity at *e* is the following: the comparatively small difference of potential between the wire and the plate is partly due to the glowing air-space at *e* acting as a conductor, this conductivity is due to dissociated molecules produced by the discharge, and it is likely that this would exhibit what are called 'unipolar' properties, that is, that its conductivity for a current in one direction would not be the same as for one in the opposite. Even when the change produced in the distribution of electricity is not so great as that due to an actual reversal of the current it is conceivable that the conductivity of the space at *e* might depend upon the way the electricity was distributed over the wire and plate. Thus when this distribution of electricity was altered, the air, by becoming a worse conductor, might cause the electricity to accumulate on the wire and thus increase the electromotive intensity at *e*. Since, however, there is a condenser of large capacity in electrical connection with the wire any increase in its electrification would be slow, whereas the spark observed by Jaumann seems to have followed that across F without the lapse of any appreciable interval.

205.] The observations of other physicists seem to afford confirmatory evidence of the way in which electric discharge is facilitated by rapid alterations in the electromotive intensity. Thus Meissner (*Abhand. der König. Gesell. Göttingen*, 16, p. 3, 1871; see also Art. 196) found that ozone was produced in a tube placed between the plates of a condenser when these were suddenly charged or discharged, while none was produced when the charges on the plates were kept constant; the potential difference in this experiment was not sufficient to cause a spark to pass between the plates. Again, R. v. Helmholtz and Richarz (*Wied. Ann.* 40, p. 161, 1890) using an induction coil that would give sparks in air about 4 inches long, found that when the electrodes were separated by about a foot and encased in wet linen bags to stop any particles of metal that might be given off from them, a steam jet some distance away from the electrodes showed very distinct signs of condensation whenever the current in the primary of the coil was broken. A steam jet is a very sensitive detector of chemical decomposition, free atoms producing condensation of the steam even when no particles of dust are present.

If we suppose that the electric field produces a polarized arrangement of the molecules of the gas, then considering the case when the left-hand electrode is the negative one, the right-hand the positive, there will be between the electrodes a chain of molecules arranged as in the first line in Fig. 86, the positively charged atoms being denoted by *A*, the negatively charged ones by *B*. If the field is now reversed, the molecules will be arranged as in the second line in Fig. 86. If the reversal takes place very slowly, the molecules will reverse their polarity by swinging round, but if the rate of reversal is very rapid the resistance offered by the inertia of the molecules to this rotation will give rise to a tendency to produce the reversal of polarity of the molecules by chemical decomposition without rotation.

A——*B* *A*——*B* *A*——*B* *A*——*B*

B——*A* *B*——*A* *B*——*A* *B*——*A*

A *B*——*A* *B*——*A* *B*——*A* *B*

Fig. 86.

This may be done by the molecules splitting up and rearranging themselves as in the third line of Fig. 86.

I have observed the effect of the reversal of the electric field when experimenting on the discharge produced in hydrogen at low pressures by a battery consisting of a large number of storage cells. I found that when the electromotive force was insufficient to produce continuous discharge, a momentary discharge occurred when the battery was reversed; this discharge merely flashed out for an instant, and took place when no discharge could be obtained by merely making or breaking the circuit without reversing the battery. A momentary discharge, however, occurred on making the circuit long before the electromotive force was sufficient to maintain a permanent discharge.

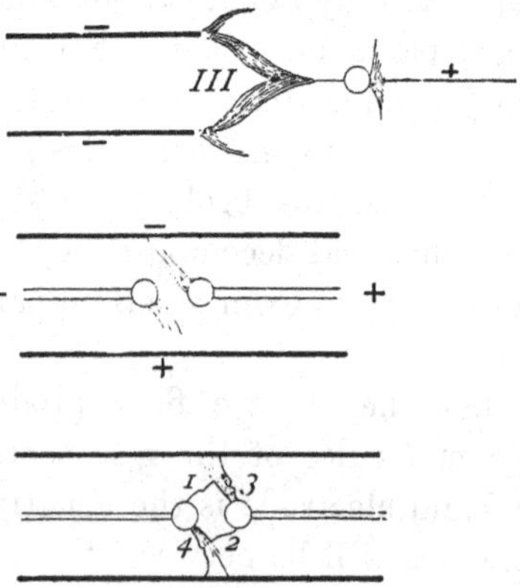

Fig. 87.

206.] Jaumann (l. c.) gives some examples of brushes which are formed at places where the electromotive intensity for steady charges is not a maximum. He explains these by supposing that the variations in the density of the electricity are more rapid at some parts of the electrodes than at others, and that *ceteris paribus* the discharge takes place most readily at the places where the rate of variation of the charge is greatest. Some of these brushes are represented in Fig. 87, taken from Jaumann.

Theory of the Electric Discharge.

207.] The phenomena attending the electric discharge through gases are so beautiful and varied that they have attracted the attention of numerous observers. The attention given to these phenomena is not, however, due so much to the beauty of the experiments, as to the wide-spread conviction that there is perhaps no other branch of physics which affords us so promising an opportunity of penetrating the secret of electricity; for while the passage of this agent through a metal or an electrolyte is invisible, that through a gas is accompanied by the most brilliantly luminous effects, which in many cases are so much influenced by changes in the conditions of the discharge as to give us many opportunities of testing any view we may take of the nature of electricity, of the electric discharge, and of the relation between electricity and matter.

Though the account we have given in this chapter of the discharge through gases is very far from complete, it will probably have been sufficient to convince the student that the phenomena are very complex and very extensive. It is therefore desirable to find some working hypothesis by which they can be coordinated: the following method of regarding the discharge seems to do this to a very considerable extent.

208.] This view is, that the passage of electricity through a gas as well as through an electrolyte, and as we hold through a metal as well, is accompanied and effected by chemical changes; also that 'chemical decomposition is not to be considered merely as an accidental attendant on the electrical discharge, but as an essential feature of the discharge without which it could not occur' (*Phil. Mag.* [5], 15, p. 432, 1883). The nature of the chemical changes which accompany the discharge may be roughly described as similar to those which on Grotthus' theory of electrolysis are supposed to occur in a Grotthus chain. The way such chemical changes effect the passage of the electricity has been already described in Art. 31, when we considered the way in which a tube of electrostatic induction contracted when in a conductor. The shortening of a tube of electrostatic induction is equivalent to the passage of electricity through the conductor.

In conduction through electrolytes the signs of chemical change are so apparent both in the deposition on the electrodes of the constituents of the electrolyte and in the close connection, expressed by Faraday's Laws, between the quantity of electricity transferred through the electrolyte and the amount of chemical change produced, that no one can doubt the importance of the part played in this case by chemical decomposition in the transmission of the electric current.

209.] When electricity passes through gases, though there is (with the possible exception of Perrot's experiment, see Art. 200) no one phenomenon whose interpretation is so unequivocal as some in electrolysis, yet the consensus of evidence given by the very varied phenomena shown by the gaseous discharge seems to point strongly to the conclusion that here, as in electrolysis, the discharge is accomplished by chemical agency.

Perrot, in 1861, seems to have been the first to suggest that the discharge through gases was of an electrolytic nature. In 1882 Giese (*Wied. Ann.* 17, pp. 1, 236, 519) arrived at the same conclusion from the study of the conductivity of flames.

Before applying this view to explain in detail the laws governing the electric discharge through gases, it seems desirable to mention one or two of the phenomena in which it is most plainly suggested.

The experiments bearing most directly on this subject are those made by Perrot on the decomposition of steam by the discharge from a Ruhmkorff's coil (see Art. 200). Perrot found that when the discharge passed through steam there was an excess of oxygen given off at the positive pole and an excess of hydrogen at the negative, and that these excesses were chemically equivalent to each other and to the amount of copper deposited from a voltameter containing copper sulphate placed in series with the discharge tube. If this result should be confirmed by subsequent researches, it would be a direct and unmistakeable proof that the passage of electricity through gases, just as much as through electrolytes, is effected by chemical means. It would also show that the charge of electricity associated with an atom of an element in a gas is the same as that associated with the same atom in an electrolyte.

210.] Again, Grove (*Phil. Trans.* 1852, Part I, p. 87) made nearly forty years ago some experiments which show that the chemical action going on at the positive electrode is not the same as that at the negative. Grove made the discharge from a Ruhmkorff's coil pass between a steel needle and a silver plate, the distance between the point of the needle and the plate being about 2·5 mm.; the gas through which the discharge passed was a mixture of hydrogen and oxygen at pressures about 2 cm. of mercury. When the silver plate was positive and the needle negative a patch of oxide was formed on the plate, while if the plate were originally negative no oxidation occurred. When the silver plate had been oxidised while being used as a positive electrode, if the current were reversed so that the plate became the negative electrode, the oxide was reduced by the hydrogen and the plate became clean. When pure hydrogen was substituted for the mixture of hydrogen and oxygen no chemical action could be observed on the plate, which was however a little roughened by the discharge; if however the plate was oxidised to begin with, it rapidly deoxidised in the hydrogen, especially when it was connected with the negative pole of the coil. Reitlinger and Wächter (*Wied. Ann.* 12, p. 590, 1881) found that the oxidation was very dependent upon the quantity of water vapour present; when the gas was thoroughly dried very little oxidation took place. The effect may therefore be due to the decomposition of the water vapour into hydrogen and oxygen, an excess of oxygen going to the positive and an excess of hydrogen to the negative pole.

Ludeking (*Phil. Mag.* [5], 33, p. 521, 1892) has found that when the discharge passes through hydriodic acid gas, iodine is deposited on the positive electrode but not on the negative.

211.] Again, chemical changes take place in many gases when the electric discharge passes through them. Perhaps the best known example of this is the formation of ozone by the silent discharge through oxygen. There are however a multitude of other instances, thus ammonia, acetylene, phosphoretted hydrogen, and indeed most gases of complex chemical constitution are decomposed by the spark.

Another fact which also points to the conclusion that the discharge is accomplished by chemical means is that mentioned in Art. 38, that the halogens chlorine, bromine, and iodine, which are dissociated at high temperatures, and which at such temperatures have already undergone the chemical change which we regard as preliminary to conduction, have then lost all power of insulation and allow electricity to pass through them with ease.

Then, again, we have the very interesting result discovered by R. v. Helmholtz (*Wied. Ann.* 32, p. 1, 1887), that a gas through which electricity is passing and one in which chemical changes are known to be going on both affect a steam jet in the same way.

212.] Again, one of the most striking features of the discharge through gases is the way in which one discharge facilitates the passage of a second; the result is true whether the discharge passes between electrodes or as an endless ring, as in the experiments described in Art. 77. Closely connected with this effect is Hittorf's discovery (*Wied. Ann.* 7, p. 614, 1879) that a few galvanic cells are able to send a current through gas which is conveying the electric discharge. Schuster (*Proceedings Royal Soc.*, 42, p. 371, 1887) describes a somewhat similar effect. A large discharge tube containing air at a low pressure was divided into two partitions by a metal plate with openings round the perimeter, which served to screen off from one compartment any electrical action occurring in the other, if a vigorous discharge passed in one of these compartments, the electromotive force of about one quarter of a volt was sufficient to send a current through the air in the other.

Since such electromotive forces would not produce any discharge through air in its normal state, these experiments suggest that the chemical state of the gas has been altered by the discharge.

213.] We shall now go on to discuss more in detail the consequences of the view that dissociation of the molecules of a gas always accompanies electric discharge through gases. We notice, in the first place, that the separation of one atom from another in the molecule of a gas is very unlikely to be produced by the unaided agency of the external electric field. Let us take the case of a molecule of hydrogen as an example; we suppose that the molecule consists of two atoms, one with a positive charge, the other with an equal negative one. The most obvious assumption, which indeed is not an assumption if we accept Perrot's results, to make about the magnitude of the charges on the atoms is that each is equal in magnitude to that charge which the laws of electrolysis show to be associated with an atom of a monovalent element. We shall denote this charge by e; it is the one molecule of electricity which Maxwell speaks about in Art. 260 of the *Electricity and Magnetism*.

The electrostatic attraction between the atoms is the molecule

$$\frac{e^2}{r^2},$$

where r is the distance between them. If the other molecules of hydrogen present do not help to split up the molecule, the force tending to pull the atoms apart is

$$2Fe,$$

where F is the external electromotive intensity.

The ratio of the force tending to separate the atoms, to their electrostatic attraction, is thus $2Fr^2/e$; now at atmospheric pressure discharge will certainly take place through hydrogen if F in electrostatic units is as large as 100, while at lower pressures a very much smaller value of F will be all that is required. To be on the safe side, however, we shall suppose that $F = 10^2$; then, assuming that the electrochemical equivalent of hydrogen is 10^{-4} and that there are 10^{21} molecules per cubic centimetre at atmospheric pressure, since the mass of a cubic centimetre of hydrogen is $1/11 \times 10^3$ of a gramme, e in electromagnetic units will be $10^4/11 \times 10^{24}$, or e in electrostatic units will be about $2{\cdot}7 \times 10^{-11}$ and r is of the order 10^{-8}, hence $2Fr^2/e$, the ratio under consideration, will be about $1/1{\cdot}4 \times 10^3$; this is so small that it shows the separation of the atoms cannot be effected by the direct action of the electric field upon them when the molecule is not colliding with other molecules. If the atoms in a molecule were almost but not quite shaken apart by a collision with another molecule, the action of the electric field might be sufficient to complete the separation.

The electric field, however, by polarizing the molecules of the gas, may undoubtedly exert a much greater effect than it could produce by its direct action on a single molecule. When the gas is not polarized, the forces exerted on one molecule by its neighbours act some in one direction, others in the opposite, so that the resultant effect is very small; when, however. the medium is polarized, order is introduced into the arrangement of the molecules, and the inter-molecular forces by all tending in the same direction may produce very large effects.

214.] The arrangement of the molecules of a gas in the electric field and the tendency of the inter-molecular forces may be illustrated to some extent by the aid of a model consisting of a large number of similar small magnets suspended by long strings attached to their centres. The positive and negative atoms in the molecules of the gas are represented by the poles of the magnets, and the forces between the molecules by those between the magnets. The way the molecules tend to arrange themselves in the electric field is represented by the arrangement of the magnets in a magnetic field.

The analogy between the model and the gas, though it may serve to illustrate the forces between the molecules, is very imperfect, as the magnets are almost stationary, while the molecules are moving with great rapidity, and the collisions which occur in consequence introduce effects which are not represented in the model. The magnets, for example, would form long chains similar to those formed by iron filings when placed in the magnetic field; in the gas, however, though some of the molecules would form chains, they would be broken up into short lengths by the bombardment of other molecules. The length of these chains would depend upon the intensity of the bombardment to which they were subjected, that is upon the pressure of the gas; the greater the pressure the more intense the bombardment, and therefore the shorter the chain.

We shall call these chains of molecules Grotthus' chains, because we suppose that when the discharge passes through the gas it passes by the agency of these chains, and that the same kind of interchange of atoms goes on amongst the molecules of these chains as on Grotthus' theory of electrolysis goes on between the molecules on a Grotthus' chain in an electrolyte.

The molecules in such a chain tend to pull each other to pieces, and the force with which the last atom in the chain is attracted to the next atom will be much smaller than the force between two atoms in an isolated molecule; this atom will therefore be much more easily detached from the chain than it would from a single molecule, and thus chemical change, and therefore electric discharge, will take place much more easily than if the chains were absent.

215.] As far as the electrical effects go, it does not matter whether the effect of the electric field is merely to arrange chains which already exist scattered about in the gas, or whether it actually produces new chains; we are more concerned with the presence of such chains than with their method of production. The existence of a small number of such chains (and it only requires a most insignificant fraction of the whole number of molecules to be arranged in chains to enable the gas to convey the most intense discharge) would have important chemical results, as it would greatly increase the ability of the gas to enter into chemical combination.

$$A_1 — B_1 \qquad A_2 — B_2 \qquad A_3 — B_3 \qquad A_4 — B_4$$

Fig. 88.

216.] The way in which the electric discharge passes along such a chain of molecules is similar to the action in an ordinary Grotthus' chain. Thus, let $A_1 B_1$, $A_2 B_2$, $A_3 B_3$, &c., Fig. 88, represent consecutive molecules in such a chain, the A's being the positive atoms and the B's the negative. Let one atom, A_1, at the end of the chain be close to the positive electrode. Then when the chain breaks down the atom A_1 at the end of the chain goes to the positive electrode, B_1 the other atom in this molecule, combining with the negative atom A_2 in the next molecule, B_2 combining with A_3; the last molecule being left free and serving as a new electrode from which a new series of recombinations in a consecutive chain originates. There would thus be along the line of discharge a series of quasi-electrodes, at any of which the products of the decomposition of the gas might appear.

The whole discharge between the electrodes consists on this view in a series of non-contemporaneous discharges, these discharges travelling consecutively from one chain to the next.

The experiment described in Art. 105 shows that this discharge starts from the positive electrode and travels to the negative with a velocity comparable with that of light. The introduction of these Grotthus' chains enables us to see how the velocity of the discharge can be so great, while the velocity of the individual molecules is comparatively small. The smallness of the velocity of these molecules has been proved by spectroscopic observations;

many experiments have shown that there is no appreciable displacement in the lines of the spectrum of the gas in the discharge tube when the discharge is observed end on, while if the molecules were moving with even a very small fraction of the velocity of light, Döppler's principle shows that there would be a measurable displacement of the lines. It does not indeed require spectroscopic analysis to prove that the molecules cannot be moving with half the velocity of light; if they did it can easily be shown that the kinetic energy of the particles carrying the discharge of a condenser would have to be greater than the potential energy in the condenser before discharge.

When, however, we consider the discharge as passing along these Grotthus' chains, since the recombinations of the different molecules in the chain go on simultaneously, the electricity will pass from one end of the chain to the other in the time required for an atom in one molecule to travel to the oppositely charged atom in the next molecule in the chain. Thus the velocity of the discharge will exceed that of the individual atoms in the proportion of the length of the chain to the distance between two adjacent atoms in neighbouring molecules. This ratio may be very large, and we can understand therefore why the velocity of the electric discharge transcends so enormously that of the atoms.

217.] We thus see that the consideration of the smallness of the electromotive intensity required to produce chemical change or discharge, as well as of the enormous velocity with which the discharge travels through the gas, has led us to the conclusion that a small fraction of the molecules of the gas are held together in Grotthus' chains, while the consideration of the method by which the discharge passes along these chains indicates that the spark through the gas consists of a series of non-contemporaneous discharges, the discharge travelling along one chain, then waiting for a moment before it passes through the next, and so on. It is remarkable that many of the physicists, who have paid the greatest attention to the passage of electricity through gases, have been driven by their observations to the conclusion that the electric discharge is made up of a large number of separate discharges. The behaviour of striæ under the action of magnetic force is one of the chief reasons for coming to this conclusion. On this point Spottiswoode and Moulton (*Phil. Trans.* 1879, part 1,

p. 205) say, 'If a magnet be applied to a striated column, it will be found that the column is not simply thrown up or down as a whole, as would be the case if the discharge passed in direct lines from terminal to terminal, threading the striæ in its passage. On the contrary, each stria is subjected to a rotation or deformation of exactly the same character as would be caused if the stria marked the termination of flexible currents radiating from the bright head of the stria behind it and terminating in the hazy inner surface of the stria in question. An examination of several cases has led the authors of this paper to conclude that the currents do thus radiate from the bright head of a stria to the inner surface of the next, and that there is no direct passage from one terminal of the tube to the other.'

With regard to the way the discharge takes place, the same authors say (*Phil. Trans.* 1879, part 1, p. 201)—'If, then, we are right in supposing that the series of artificially produced hollow shells are analogous in their structures and functions to striæ, it is not difficult to deduce, from the explanation above given, the *modus operandi* of an ordinary striated discharge. The passage of each of the intermittent pulses from the bright surface of a stria towards the hollow surface of the next may well be supposed, by its inductive action, to drive from the next stria a similar pulse, which in its turn drives one from the next stria, and so on. . . . The passage of the discharge is due in both cases to an action consisting of an independent discharge from one stria to the next, and the idea of this action can perhaps be best illustrated by that of a line of boys crossing a brook on stepping stones, each boy stepping on the stone which the boy in front of him has left.'

Goldstein (*Phil. Mag.* [5] 10, p. 183, 1880) expresses much the same opinion. He says: 'By numerous comparisons, and taking account of all apparently essential phenomena, I have been led to the following view:—

'The kathode-light, each bundle of secondary negative light, as well as each layer of positive light, represent each a separate current by itself, which begins at the part of each structure turned towards the kathode, and ends at the end of the negative rays or of the stratified structure, without the current flowing in one structure propagating itself into the next, without the electricity which flows through one also traversing the rest in order.

'I suspect, then, that as many new points of departure of the discharge are present in a length of gas between two electrodes as this shows of secondary negative bundles or layers—that as according to experiments repeatedly mentioned all the properties and actions of the discharge at the kathode are found again at the secondary negative light and with each layer of positive light, the intimate action is the same with these as it is with those.'

218.] Thus, if we regard a stria as a bundle of Grotthus' chains in parallel rendered visible, the bright parts of the stria corresponding to the ends of the chain, the dull parts to the middle, the conclusion of the physicists just quoted are almost identical with those we arrived at by the consideration of the chains. We therefore regard the stratification of the discharge as evidence of the existence of these chains, and suppose that a stria is in fact a bundle of Grotthus' chains.

219.] As far as phenomena connected with the electric discharge are concerned, the Grotthus' chain is the unit rather than the molecule ; now the length of this chain is equal to the length of a stria, which is very much greater than the diameter of a molecule, than the average distance between two molecules, or even than the mean free path of a molecule : thus the structure of a gas, as far as phenomena connected with the electric discharge are concerned, is on a very much coarser scale than its structure with reference to such properties as gaseous diffusion where the fundamental length is that of the mean free path of the molecules.

220.] Peace's discovery that the density—which we shall call the critical density—at which the 'electric strength' of the gas is a minimum depends upon the distance between the electrodes, proves that the gas, when in an electric field sufficiently intense to produce discharge, possesses a structure whose length scale is comparable with the distance between the electrodes when these are near enough together to influence the critical density. As this distance is very much greater than any of the lengths recognized in the ordinary Kinetic Theory of Gases, the gas when under the influence of the electric field must have a structure very much coarser than that recognized by that theory. In our view this structure consists in the formation of Grotthus' chains.

221.] The striations are only clearly marked within somewhat narrow limits of pressure. But it is in accordance with the conclusion which all who have studied the spark have arrived at—that there is complete continuity between the bright well-defined spark which occurs at high pressures and the diffused glow which represents the discharge at high exhaustions—to suppose that they always exist in the spark discharge, but that at high pressures they are so close together that the bright and dark parts cease to be separable by the eye.

The view we have taken of the action of the Grotthus' chains in propagating the electric discharge, and the connection between these chains and the striations, does not require that every discharge should be visibly striated; on the contrary, since the striations will only be visible when there is great regularity in the disposition of these chains, we should expect that it would only be under somewhat exceptional circumstances that the conditions would be regular enough to give rise to visible striations.

222.] We shall now proceed to consider more in detail the application of the preceding ideas to the phenomena of the electric discharge. The first case we shall consider is the calculation of the potential difference required to produce discharge under various conditions.

It is perhaps advisable to begin with the caution that in comparing the potential differences required to *produce* discharge through a given gas we must be alive to the fact that the condition of the gas is altered for a time by the passage of the discharge. Thus, when the discharges follow each other so rapidly that the interval between two discharges is not sufficiently long to allow the gas to return to its original condition before the second discharge passes, this discharge is in reality passing through a gas whose nature is a function of the electrical conditions. Thus, though this gas may be called hydrogen or oxygen, it is by no means identical with the gas which was called by the same name before the discharge passed through it. When the discharges follow each other with great rapidity the supply of dissociated molecules left by preceding discharges may be so large that the discharge ceases to be disruptive, and is analogous to that through a very hot gas whose molecules are dissociated by the heat.

The measurements of the potential differences required to send the first spark through a gas are thus more definite in their interpretation than measurements of potential gradients along the path of a nearly continuous discharge.

The striations on the preceding view of the discharge may, since they are equivalent to a bundle of Grotthus' chains, be regarded as forming a series of little electrolytic cells, the beginning and the end of a stria corresponding to the electrodes of the cell. Let F be the electromotive intensity of the field, λ the length of a stria, then when unit of electricity passes through the stria the work done on it by the electric field is $F\lambda$. The passage of the electricity through the stria is accompanied just as in the case of the electrolytic cell, by definite chemical changes, such as the decomposition of a certain number of molecules of the gas; thus if w is the increase in the potential energy of the gas due to the changes which occur when unit of electricity passes through the stria, then neglecting the heat produced by the current we have by the Conservation of Energy

$$F\lambda = w,$$

or the difference in potential between the beginning and end of a stria is equal to w. If the chemical and other changes which take place in the consecutive striæ are the same, the potential difference due to each will be the same also. There is however one stria which is under different conditions from the others, viz. that next the negative electrode, i. e. the negative dark space. For in the body of the gas, the ions set free at an extremity of the stria, are set free in close proximity to the ions of opposite sign at the extremity of an adjacent stria. In the stria next the electrode the ions at one end are set free against a metallic surface. The experiments described in the account we have already given of the discharge show that the chemical changes which take place at the cathode are abnormal; one reason for this no doubt is the presence of the metal, which makes many chemical changes possible which could not take place if there were nothing but gas present. This stria is thus under exceptional circumstances and may differ in size and fall of potential from the other striæ. Hittorf's experiments, Art. 140, show that the fall of potential at the cathode is abnormally great. If we call this potential fall K and consider the case of discharge between two parallel metal plates; the discharge on this view, starting from the positive electrode, goes consecutively across a number n of similar striæ, one of which reaches up to the positive electrode, the fall of potential across each of these is w; the discharge finally crosses the stria in contact with the negative electrode in which the fall of potential is K; thus V, the total fall of potential as the discharge goes from the positive to the negative electrode, is given by the equation

$$V = K + nw. \tag{1}$$

If l is the distance between the plates, λ_0 the length of the stria next the cathode, λ the length of the other stria, then

$$n = \frac{l - \lambda_0}{\lambda}.$$

Substituting this value for n in (1) we get

$$V = \left(K - \frac{w\lambda_0}{\lambda}\right) + \frac{l}{\lambda} w,$$

which may be written

$$V = K' + al. \tag{2}$$

According to this equation the curve representing the relation between potential difference and spark length for constant pressure is a straight line which does not pass through the origin. The curves we have given from the papers by Paschen and Peace show that this is very approximately true. The curves show that for air K' would at atmospheric pressure be about 600 volts from Paschen's experiments and about 400 volts from Peace's.

If R is the electromotive intensity required to produce a spark of length l between two parallel infinite plates, then since $R = V/l$

$$R = \frac{K'}{l} + a. \tag{3}$$

Since K' is positive, the electromotive intensity required to produce discharge increases as the length of the spark diminishes; in other words, the electric strength of a thin layer of gas is greater than that of a thick layer. The electric strength will sensibly increase as soon as K'/l become appreciable in comparison with a, this will occur as soon as l ceases to be a very large multiple of the length of a stria. Thus the thickness of the layer when the 'electric strength' begins to vary appreciably is comparable with the length of a stria at the pressure at which the discharge takes place; this length is very large when compared with molecular distances or with the mean free path of the molecules of the gas; hence we see why the change in the 'electric strength' of a gas takes place when the spark length is very large in comparison with lengths usually recognized in the Kinetic Theory of Gases.

According to formula (3), the curve representing the relation between electromotive intensity and spark length is a rectangular hyperbola; this is confirmed by the curves given by Dr. Liebig for air, carbonic acid, oxygen and coal gas (see Fig. 19), and by those given by Mr. Peace for air.

223.] The preceding formulæ are not applicable when the distance between the electrodes is less than λ_0 the length of the stria next the cathode. But if the discharge passes through the gas and is not carried by metal dust torn from the electrodes we can easily see that the electric strength must increase as the distance between the electrodes diminishes. For as we have seen, the molecules which are active in carrying the discharge are not torn in pieces by the direct action of the electric field but by the attraction of the neighbouring molecules in the Grotthus' chain. Now when we push the electrodes so near together that the distance between them is less than the normal length of the chain, we take away some of the molecules from the chain and so make it more difficult for the molecules which remain to split up any particular molecule into atoms, so that in order to effect this splitting up we must increase the number of chains in the field, in other words, we must increase the electromotive intensity.

Peace's curves, Fig. 27, showing the relation between the potential difference and spark length are exceedingly flat in the neighbourhood of the critical spark length. This shows that the potential difference required to produce discharge increases very slowly at first as the spark length is shortened to less than the length of a Grotthus' chain.

We now proceed to consider the relation between the spark potential and the pressure. As we have already remarked, the length of a Grotthus' chain depends upon the density of the gas; the denser the gas the shorter the chain: this is illustrated by the way in which the striæ lengthen out when the pressure is reduced. The experiments which have been made on the connection between the length of a stria and the density of the gas are not sufficiently decisive to enable us to formulate the exact law connecting these two quantities, we shall assume however that it is expressed by the equation

$$\lambda = \beta\rho^{-k},$$

where λ is the length of a stria, ρ the density of the gas, and β, k positive constant.

Equation (1) involves K the fall of potential at the cathode and w the fall along a stria as well as λ. Warburg's experiments (Art. 160) show that the cathode fall K is almost independent of the pressure, and although no observations have been made on the influence of a change in the pressure on the value of w, it is not likely that w any more than K depends to any great extent upon the pressure. If we substitute the preceding value of λ in equation (2) we get

$$V = K' + \frac{l}{\beta}\rho^k w.$$

Both Paschen's and Peace's experiments show that when the spark length is great enough to include several striæ the curve representing the relation between the spark potential and density for a constant spark length, though very nearly straight, is slightly convex to the axis along which the densities are measured. This shows that k is slightly, but only slightly, greater than unity.

224.] It is interesting to trace the changes which take place in the conditions of discharge between two electrodes at a fixed distance apart as the pressure of the gas gradually diminishes.

When the pressure is great the striæ are very close together, so that if the distance between the electrodes is a millimetre or more, a large number of striæ will be crowded in between them. As the pressure diminishes the striæ widen out, and fewer and fewer of them can find room to squeeze in between the electrodes, and as the number of striæ between the electrodes diminishes, the potential required to produce a spark diminishes also, each stria that is squeezed out corresponding to a definite diminution in the spark potential. This diminution in potential will go on until the striæ have all been eliminated with the exception of one. There can now be no further reduction in the number of striæ as the pressure diminishes, and the Grotthus' chain which is left, and which is required to split up the molecules to allow the discharge to take place, gets curtailed as the pressure falls by a larger and larger fraction of its natural length, and therefore has greater and greater difficulty in effecting the decomposition of the molecules, so that the electric strength of the gas will now increase as the pressure diminishes. There will thus be a density at which the electric strength of the gas is a minimum, and that density will be the one at which the length of the stria next the cathode is equal or nearly equal to the distance between the electrodes. Thus the length of a stria at the minimum strength will have to be very much less when the electrodes are very near together than when they are far apart, and since the stria-length is less the density at which the 'electric strength' is a minimum will be very much greater when the electrodes are near together than when they are far apart. This is most strikingly exemplified in Mr. Peace's experiments, for when the distance between the electrodes was reduced from 1/5 to 1/100 of a millimetre the critical pressure was raised from 30 to 250 mm. of mercury. The mean free path of a molecule of air at a pressure of 30 mm. is about 1/400 of a millimetre.

225.] The existence of a critical pressure, or pressure at which the electric strength is a minimum, when the discharge passes between electrodes can thus be explained if we recognize the formation of Grotthus' chains in the gas, and the theory leads to the conclusion which, as we have seen, is in accordance with the facts, that the critical pressure depends on the spark length.

226.] We have seen that when the distance between the electrodes is less than the length of the stria next the negative electrode, the intensity of the field required to produce discharge will increase as the distance between the electrodes diminishes. Peace's observations show that this increase is so rapid that the potential difference between the electrode when the spark passes increases when the spark length is diminished, or in other words, that the electromotive intensity increases more rapidly than the reciprocal of the length of a Grotthus' chain. This will explain the remarkable results observed by Hittorf (Art. 170) and Lehmann (Art. 170) when the electrodes were placed very near together in a gas at a somewhat low pressure. In such cases it was found that the discharge instead of passing in the straight line between the electrodes took a very roundabout course. To explain this, suppose that in the experiment shown in Fig. 68

the electrodes are nearer together than the length of the chain next the electrode, i.e. the negative dark space; then if the discharge passed along the shortest path between the plates, the potential difference required would, by Peace's experiments, considerably exceed K, the normal cathode potential fall; if however the discharge passed as in the figure along a line of force, whose length is greater than the negative dark space, the potential difference required would be K plus that due to any small positive column which may exist in the discharge. The latter part of the potential difference is small compared with K, so that the potential difference required to produce discharge along this path will only be a little in excess of K, while that required to produce discharge along the shortest path would, by Peace's experiments, be considerably greater than K, the discharge will therefore pass as in the figure in preference to taking the shortest path.

227.] Since a term in the expression (1) for the potential difference required to produce a spark of given length is inversely proportional to the length of a stria, anything which diminishes the length of a stria will tend to increase this potential difference. Now the length of a stria is influenced by the size of the discharge tube as soon as the length becomes comparable with the diameter of the tube; the narrower the tube the shorter are the striæ. Hence we should expect to find that it would require a greater potential difference to produce at a given pressure a spark through a narrow tube than through a wide one. This is confirmed by the experiments made by De la Rue and Hugo Müller, described in Art. 169.

228.] We do not at present know enough about the laws which govern the passage of electricity from a gas to a solid, or from a solid to a gas, to enable us to account for the difference between the appearances presented by the discharge at the cathode and anode of a vacuum tube; it may, however, be well to consider one or two points which must doubtless influence the behaviour of the discharge at the two electrodes.

We have seen (Art. 108) that the positive column in the electric discharge starts from the positive electrodes, and that with the exception of the negative rays, no part of the discharge seems to begin at the cathode; we have also seen that the potential differences in the neighbourhood of the cathode are much greater than those near the anode. These results might at first sight seem inconsistent with the experiments we have described (Art. 40) on the electrical effect on metal surfaces of ultra-violet light and incandescence. In these experiments we saw that under such influences negative electricity escaped with great ease from a metallic electrode, while, on the other hand, positive electricity had great difficulty in doing so. In the ordinary discharge through gases it seems, on the contrary, to be the positive electricity which escapes with ease, while the negative only escapes with great difficulty. We must remember, however, that the vehicle conveying the electricity may not be the same in the two cases. When ultra-violet light is incident on a metal plate, there seems to be nothing in the phenomena inconsistent with the hypothesis that the negative electrification is carried away by the vapour or dust of the metal. In the case of vacuum tubes, however, the electricity is doubtless conveyed for the most part by the gas and not by the metal. In order to get the electricity from the gas into the metal, or from the metal into the gas, something equivalent to chemical combination must take place between the metal and the gas. Some experiments have been made on this point by Stanton (*Proc. Roy. Soc.* 47, p. 559, 1890), who found that a hot copper or iron rod connected to earth only discharged the electricity from a positively electrified conductor in its neighbourhood when chemical action was visibly going on over the surface of the rod, e.g. when it was being oxidised in an atmosphere of oxygen. When it was covered with a film of oxide it did not discharge the adjacent conductor; if when coated with oxide it was placed in an atmosphere of hydrogen it discharged the electricity as long as it was being deoxidised, but as soon as the deoxidation was complete the leakage of the electricity stopped. On the other hand, when the conductor was negatively electrified, it leaked even when no apparent chemical action was taking place. I have myself observed (*Proc. Roy. Soc.* 49, p. 97, 1891) that the facility with which electricity passed from a gas to a metal was much increased when chemical action took place. If this is the case, the question as to the relative ease with which the electricity escapes from the two electrodes through a vacuum tube, depends upon whether a positively or negatively electrified surface more readily enters into chemical combination with the adjacent gas, while the sign of the electrification of a metal surface under the influence of ultra-violet light may, on the other hand, depend upon whether the 'Volta-potential' (see Art. 44) for the metal in its solid state is less or greater than for the dust or vapour of the metal.

229.] In framing any theory of the difference between the positive and negative electrodes, we must remember that at the electrodes we have either two different substances or the same substance in two different states in contact, and it is in accordance with what we know of the electrical effects produced by the contact of different substances that the gas in the immediate neighbourhood of the electrodes should be polarized, that is, that the molecular tubes of induction in the gas should tend to point in a definite direction relatively to the outward drawn normals to the electrode: let us suppose that the polarization is such that the negative ends of the tubes are the nearest to the electrode: we may regard the molecules of the gas as being under the influence of a couple tending to twist them into this position. If now this electrode is the cathode, then before these molecules are available for carrying the discharge, they must be twisted right round against the action of an opposing couple, so that to produce discharge at this electrode the electric field must be strong enough to twist the molecules out of their original alignment into the opposite one, it must therefore be stronger than in the body of the gas where the opposing couple does not exist: a polarization of this kind would therefore make the cathode potential gradient greater than that in the body of the gas.

CHAPTER III.

Conjugate Functions.

230.] The methods given by Maxwell for solving problems in Electrostatics by means of Conjugate Functions are somewhat indirect, since there is no rule given for determining the proper transformation for any particular problem. Success in using these methods depends chiefly upon good fortune in guessing the suitable transformation. The use of a general theorem in Transformations given by Schwarz (*Ueber einige Abbildungsaufgaben*, Crelle 70, pp. 105–120, 1869), and Christoffel (*Sul problema delle temperature stazionarie*, Annali di Matematica, I. p. 89, 1867), enables us to find by a direct process the proper transformations for electrostatical problems in two dimensions when the lines over which the potential is given are straight. We shall now proceed to the discussion of this method which has been applied to Electrical problems by Kirchhoff (*Zur Theorie des Condensators*, Gesammelte Abhandlungen, p. 101), and by Potier (Appendix to the French translation of Maxwell's *Electricity and Magnetism*); it has also been applied to Hydrodynamical problems by Michell (*On the Theory of Free Stream Lines*, Phil. Trans. 1890, A. p. 389), and Love (*Theory of Discontinuous Fluid Motions in two dimensions*, Proc. Camb. Phil. Soc. 7, p. 175, 1891).

231.] The theorem of Schwarz and Christoffel is that any polygon bounded by straight lines in a plane, which we shall call the z plane, where $z = x + \iota y$, x and y being the Cartesian coordinates of a point in this plane, can be transformed into the axis of ξ in a plane which we shall call the t plane, where $t = \xi + \iota\eta$, ξ and η being the Cartesian coordinates of a point in this plane; and that points inside the polygon in the z plane transform into points on one side of the axis of ξ. The transformation which effects this is represented by the equation

$$\frac{dz}{dt}=C(t-t_1)^{\frac{a_1}{\pi}-1}(t-t_2)^{\frac{a_2}{\pi}-1}\dots(t-t_r)^{\frac{a_r}{\pi}-1}\dots(t-t_n)^{\frac{a_n}{\pi}-1}, \quad (1)$$

where $a_1, a_2, \dots a_n$ are the internal angles of the polygon in the z plane; $t_1, t_2, \dots t_n$ are real quantities and are the coordinates of points on the axis of ξ corresponding to the angular points of the polygon in the z plane.

To prove this proposition, we remark that the argument of dz/dt, that is the value of θ when dz/dt is expressed in the form $R\epsilon^{\iota\theta}$ where R is real, remains unchanged as long as z remains real and does not pass through any one of the values $t_1, t_2, \dots t_n$; in other words, the part of the real axis of t between the points t_r and t_{r+1} corresponds to a straight line in the plane of z.

We must now investigate what happens when t passes through one of the points such as t_r on the axis of ξ. With centre t_r describe a small semi-circle BDC on the positive side of the axis of ξ, and consider the change in dz/dt as t passes round BDC from B to C.

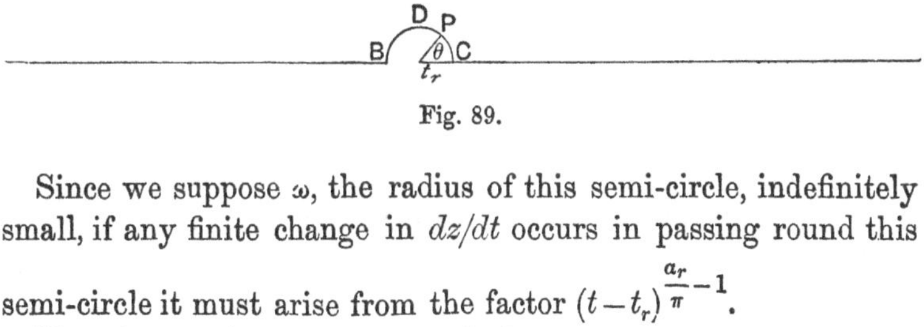

Fig. 89.

Since we suppose ω, the radius of this semi-circle, indefinitely small, if any finite change in dz/dt occurs in passing round this semi-circle it must arise from the factor $(t-t_r)^{\frac{a_r}{\pi}-1}$.

Now for a point on the semi-circle BDC

$$t-t_r=\omega\epsilon^{\iota\theta},$$

$$(t-t_r)^{\frac{a_r}{\pi}-1}=\omega^{\frac{a_r}{\pi}-1}\epsilon^{\iota\left(\frac{a_r}{\pi}-1\right)\theta},$$

hence, since θ decreases from π to zero as the point travels round the semi-circle, the argument of $(t-t_r)^{\frac{a_r}{\pi}-1}$, and therefore of dz/dt, is increased by $\pi-a_r$, that is the line corresponding to the portion $t_r\, t_{r+1}$ of the axis of ξ makes with the line corresponding to the portion $t_{r-1}t_r$ the angle $\pi-a_r$; in other words, the internal angle of the polygon in the z plane at the point corresponding to t_r is α_r.

If we imagine a point to travel along the axis of ξ in the plane of t from $t = -\infty$ to $t = +\infty$ and then back again from $+\infty$ to $-\infty$ along a semi-circle of infinite radius with its centre at the origin of coordinates in the t plane, then, as long as the point is on the axis of ξ, the corresponding point in the plane z is on one of the sides of the polygon. To find the path in z corresponding to the semi-circle in t we put

$$t = R\epsilon^{\iota\theta},$$

where R is very great and is subsequently made infinite: equation (1) then becomes

$$\frac{dz}{dt} = CR^{\frac{\alpha_1+\alpha_2+\dots\alpha_n}{\pi}-n}\epsilon^{\iota\left\{\frac{\alpha_1+\alpha_2+\dots\alpha_n}{\pi}-n\right\}\theta}, \tag{2}$$

since R is infinite compared with any of the quantities $t_1, t_2, \dots t_n$.

Since along the semi-circle

$$dt = \iota R\epsilon^{\iota\theta}d\theta,$$

equation (2) becomes

$$dz = \iota CR^{\frac{1+\alpha_2+\dots\alpha_n}{\pi}-(n-1)}\epsilon^{\iota\left\{\frac{\alpha_1+\alpha_2+\dots\alpha_n}{\pi}-(n-1)\right\}\theta}d\theta,$$

or

$$z = CR^{\frac{\alpha_1+\alpha_2+\dots\alpha_n}{\pi}-(n-1)}\frac{\epsilon^{\iota\left\{\frac{\alpha_1+\alpha_2+\alpha_n}{\pi}-(n-1)\right\}\theta}}{\frac{\alpha_1+\alpha_2+\dots\alpha_n}{\pi}-(n-1)}.$$

Thus the path in the z plane corresponding to the semi-circle in the plane of z is a portion of a circle subtending an angle $\alpha_1+\alpha_2+\dots\alpha_n-(n-1)\pi$ at the origin, and whose radius is zero or infinite according as

$$\frac{\alpha_1+\alpha_2+\dots\alpha_n}{\pi}-(n-1)$$

is ~~positive or negative~~.

If this quantity is zero, then equation (2) becomes

$$\frac{dz}{dt} = \frac{C}{R\epsilon^{\iota\theta}} = \frac{C}{t},$$

hence

$$\begin{aligned} z &= C\log t + A \\ &= C\log R + \iota C\theta + A, \end{aligned}$$

where A is the constant of integration.

Thus as the point in the t plane moves round the semi-circle the point in the z plane will travel over a length $C\pi$ of a straight line parallel to the axis of y at an infinite distance from the origin.

232.] Since by equation (1) the value of dz/dt cannot vanish or become infinite for values of t inside the area bounded by the axis of ξ and the infinite semi-circle, this area can be conformably transformed to the area bounded by the polygon in the z plane.

233.] When we wish to transform any given polygon in the z plane into the axis of ξ in the t plane we have the values of $\alpha_1, \alpha_2, \dots \alpha_n$ given. As regards the values of $t_1, t_2, \dots t_n$ some may be arbitrarily assumed while others will have to be determined from the dimensions of the polygon. Whatever the values of $t_1, t_2, \dots t_n$, the transformation (1) will transform the axis of ξ into a polygon whose internal angles have the required values. In order that this polygon should be similar to the given one we require $n-3$ conditions to be satisfied; hence as regards the n quantities $t_1, t_2, \dots t_n$, the values of 3 of them may be arbitrarily assumed, while the remaining $n-3$ must be determined from the dimensions of the polygon in the z plane.

234.] The method of applying the transformation theorem to the solution of two dimensional problems in Electrostatics in which the boundaries of the conductors are planes, is to take the polygon whose sides are the boundaries of the conductors, which we shall speak of as the polygon in the z plane, and transform it by the Schwarzian transformation into the real axis in a new plane, which we shall call the t plane. If ψ represents the potential function, ϕ the stream function, and $w = \phi + \iota\psi$, the condition that ψ is constant over the conductors may be represented by a diagram in the w plane consisting of lines parallel to the real axis in this plane: we must transform these lines by the Schwarzian transformation into the real axis in the t plane. Thus corresponding to a point on the real axis in the t plane we have a point in the boundary of a conductor in the z plane and a point along a line of constant potential in the w plane, and we make this potential correspond to the potential of the conductor in the electrostatical problem whose solution we require.

In this way we find

$$x + \iota y = f(t),$$
$$\phi + \iota\psi = F(t),$$

where f and F are known functions; eliminating t between these equations we get

$$\phi + \iota\psi = \chi(x + \iota y),$$

which gives us the solution of our problem.

235.] We shall now proceed to consider the application of this method to some special problems. The first case we shall consider is the one discussed by Maxwell in Art. 202 of the *Electricity and Magnetism*, in which a plate bounded by a straight edge and at potential V is placed above and parallel to an infinite plate at zero potential. The diagrams in the z and w planes are given in Figs. 90 and 91 respectively.

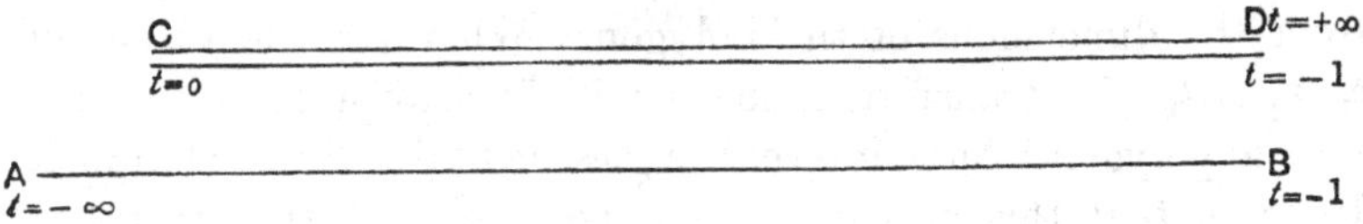

Fig. 90.

Fig 91.

The boundary of the z diagram consists of the infinite straight line AB, the two sides of the line CD, and an arc of a circle stretching from $x = -\infty$ on the line AB to $x = +\infty$ on the line CD. We may assume arbitrarily the values of t corresponding to three corners of the diagram, we shall thus assume $t = -\infty$ at the point $x = -\infty$ on the line AB, $t = -1$ at the point $x = +\infty$ on the same line, and $t = 0$ at C. The internal angles of the polygon are zero at B and 2π at C; hence by equation (1), Art. 231, the Schwarzian transformation of the diagram in the z plane to the real axis of the t plane is

$$\frac{dz}{dt} = C\frac{t}{t+1}. \tag{3}$$

The diagram in the w plane consists of two parallel straight lines; the internal angle at G, the point corresponding to $t = -1$, is zero; hence the Schwarzian transformation to the real axis of t is

$$\frac{dw}{dt} = B\frac{1}{t+1}. \tag{4}$$

From (3) we have

$$z = x + \iota y = C\{t - \log(t+1) + \iota\pi\}, \tag{5}$$

where the constant has been chosen so as to make $y = 0$ from $t = -\infty$ to -1. When t passes through the value -1, the value of y increases by $C\pi$, so that if h is the distance between the plates

$$h = C\pi,$$

hence we have

$$x + \iota y = \frac{h}{\pi}\{t - \log(t+1) + \iota\pi\}. \tag{6}$$

From (4) we have

$$w = \phi + \iota\psi = B\{\log(t+1) - \iota\pi\};$$

where the constant of integration has been chosen so as to make $\psi = 0$ from $t = -\infty$ to $t = -1$. As t passes through the value -1, ψ diminishes by $B\pi$. Hence, as the infinite plate is at zero potential and the semi-infinite one at potential V, we have

$$V = -B\pi,$$

or

$$\phi + \iota\psi = -\frac{V}{\pi}\{\log(t+1) - \iota\pi\}. \tag{7}$$

Eliminating t from equations (6) and (7), we get

$$x + \iota y = \frac{h}{V}\{\phi + \iota\psi - \frac{V}{\pi}(1 + \epsilon^{-(\phi+\iota\psi)\frac{\pi}{V}})\},$$

which is the transformation given in Maxwell's *Electricity and Magnetism*, Art. 202.

For many purposes, however, it is desirable to retain t in the expressions for the coordinates x and y and for the potential and current functions ψ and ϕ.

Thus to find the quantity of electricity on a portion of the underneath side of the semi-infinite plate, we notice that on this side of the plate t ranges from -1 to 0, and that at a distance from the edge of the plate which is a large multiple of h, t is approximately -1. In this case we have by (6), if x be the distance from the edge of the plate corresponding to t,

$$x = \frac{h}{\pi}\{t - \log(1+t)\},$$

or since $t = -1$ approximately

$$\log(t+1) = -\left\{\frac{\pi x}{h} + 1\right\}.$$

The surface density σ of the electricity on a conductor is equal to

$$-\frac{1}{4\pi}\frac{d\psi}{d\nu},$$

where $d\nu$ is an element of the outward drawn normal to the conductor. When, as in the present case, the conductors are parallel to the axis of x, $d\nu = \pm dy$, the $+$ or $-$ sign being taken according as the outward drawn normal is the positive or negative direction of y; i.e. the positive sign is to be taken at the upper surface of the plates, the negative sign at the lower. We thus have

$$\sigma = \mp\frac{1}{4\pi}\frac{d\psi}{dy} = \mp\frac{1}{4\pi}\frac{d\phi}{dx}.$$

Since $$\sigma = -\frac{1}{4\pi}\frac{d\psi}{d\nu}$$

and $$\frac{d\psi}{d\nu} = \frac{d\phi}{ds},$$

where ds is an element of the section of the conductor

$$\sigma = -\frac{1}{4\pi}\frac{d\phi}{ds}$$

$$= -\frac{1}{4\pi}\frac{d\phi}{dt}\frac{dt}{ds}.$$

The quantity of electricity on a strip of unit depth (the depth being measured at right angles to the plane of x, y) is equal to

$$\int\sigma ds = -\frac{1}{4\pi}\int\frac{d\phi}{dt}\frac{dt}{ds}ds$$

$$= -\frac{1}{4\pi}\{\phi(t_2) - \phi(t_1)\},$$

where t_1, t_2 are the values of t at the beginning and end of the strip, t_2 being algebraically greater than t_1.

The quantity of electricity on the strip of breadth x is equal to

$$\frac{1}{4\pi}\{\phi_t - \phi_0\},$$

and this by equation (7) is equal to

$$-\frac{1}{4\pi}\frac{V}{\pi}\log(t+1)$$

$$=\frac{V}{4\pi h}\left\{x+\frac{h}{\pi}\right\}.$$

Thus the quantity of electricity on the lower side of the plate is the same as if the density were uniform and equal to that on an infinite plate, the breadth of the strip being increased by h/π. This, however, only represents the electricity on the lower side of the plate, there is also a considerable quantity of electricity on the top of the plate. To find an expression for the quantity of electricity on a strip of breadth x, we notice that on the top of the plate t ranges from zero to infinity, and that when x is a large multiple of h, t is very large; in this case the solution of the equation

$$x=\frac{h}{\pi}\{t-\log(1+t)\}$$

is approximately

$$t=\pi\frac{x}{h}+\log\left\{1+\frac{\pi x}{h}\right\},$$

and the quantity of electricity on a strip of breadth x is $\frac{1}{4\pi}\{\phi_0-\phi_t\}$, and thus by equation (7) is equal to

$$\frac{V}{4\pi^2}\log(t+1)$$

$$=\frac{V}{4\pi^2}\log\left\{1+\frac{\pi x}{h}+\log\left(1+\frac{\pi x}{h}\right)\right\}.$$

Thus the quantity of electricity on an infinitely long strip is infinite, though its ratio to the quantity of electricity on the lower side of the strip is infinitely small.

The surface density $\pm d\phi/4\pi dx$ of the distribution of electricity on the semi-infinite plate is by equations (6) and (7) equal to

$$\mp\frac{V}{4\pi h}\frac{1}{t}.$$

On the underneath side of the plate t is very nearly equal to -1 when the distance from the edge of the plate is a large multiple of h, so that in this case the density soon reaches a constant value. On the upper side of the plate, however, when x is a large multiple of h, t is approximately equal to

$$\frac{\pi x}{h},$$

so that the density varies inversely as the distance from the edge of the plate.

The capacity of a breadth x of the upper plate, i.e. the ratio of the charge on both surfaces to V, is

$$\frac{x}{4\pi h}\left[1+\frac{h}{\pi x}+\frac{h}{\pi x}\log\left\{1+\frac{\pi x}{h}+\log\left(1+\frac{\pi x}{h}\right)\right\}\right].$$

We see by the principle of images that the distribution of electricity on the upper plate is the same as would ensue if, instead of the infinite plate at zero potential, we had another semi-infinite parallel plate at potential $-V$, at a distance $2h$ below the upper plate, and therefore that in this case the capacity of a breadth x, when x/h is large, of either plate is app oximately

$$\frac{x}{8\pi h}\left[1+\frac{h}{\pi x}+\frac{h}{\pi x}\log\left\{1+\frac{\pi x}{h}+\log\left(1+\frac{\pi x}{h}\right)\right\}\right].$$

236.] The next case we shall consider is the one discussed by Maxwell in Art. 195, in which a semi-infinite conducting plane is placed midway between two parallel infinite conducting planes, maintained at zero potential; we shall suppose that the potential of the semi-infinite plane is V. The diagrams in the z and w planes are given in Figs. 92 and 93 respectively.

Fig. 92.

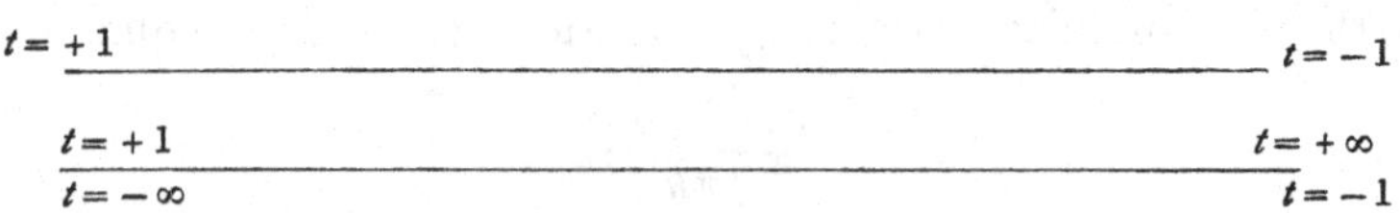

Fig. 93.

The boundary of the z diagram consists of the infinite line AB, the two sides of the semi-infinite line CD, and the infinite line EF. We shall assume $t = 0$ at C, $t = -\infty$ at the point $x = -\infty$ on the line AB, $t = -1$ at the point $x = +\infty$ on the same line, then by symmetry $t = +1$ at the point $x = +\infty$ on the line EF, and $t = +\infty$ at the point $x = -\infty$ on the same line. The internal angles of the polygon are zero at B and E, and 2π at C, hence by equation (1) the Schwarzian transformation of the diagram in the z plane to the real axis in the t plane is

$$\frac{dz}{dt} = \frac{Ct}{(t+1)(t-1)}. \tag{8}$$

The diagram in the w plane consists of three parallel lines, or rather one line and the two sides of another; in Fig. 93 the upper side of the lower line corresponds to the conductor EF, the lower side to the conductor AB. The internal angles occur at the points corresponding to $t = -1$ and to $t = +1$ and are both zero; hence the transformation which turns the diagram in the w plane to the real axis in the t plane is

$$\frac{dw}{dt} = \frac{B}{(t+1)(t-1)}. \tag{9}$$

From (8) we have

$$z = x + \iota y = \tfrac{1}{2} C \{ \log \{t^2 - 1\} - \iota\pi \}, \tag{10}$$

where the constant of integration has been determined so as to make $x = 0$, $y = 0$ at C. When t passes through the values ± 1 the value of y increases by $-\frac{1}{2}C\pi$, hence if h is the distance of the semi-infinite plane from either of the two infinite ones we have

$$-\tfrac{1}{2}C\pi = h,$$

or

$$x + \iota y = \frac{h}{\pi}\{\iota\pi - \log(t^2 - 1)\}. \tag{11}$$

From equation (9) we have

$$w = \phi + \iota\psi = \frac{V}{\pi}\log\frac{t-1}{t+1}. \tag{12}$$

From this equation we get

$$t^2 - 1 = \frac{4}{\left(\epsilon^{\frac{1}{2}\frac{\pi}{V}(\phi+\iota\psi)} - \epsilon^{-\frac{1}{2}\frac{\pi}{V}(\phi+\iota\psi)}\right)^2}.$$

Substituting this value of t^2-1 in (11), we get

$$x+\iota y = \frac{h}{\pi}\left[\iota\pi - 2\log 2 + 2\log\left\{\epsilon^{\frac{1}{2}\frac{\pi}{V}(\phi+\iota\psi)} - \epsilon^{-\frac{1}{2}\frac{\pi}{V}(\phi+\iota\psi)}\right\}\right]$$

$$= \frac{h}{\pi}\left[\iota\pi - 2\log 2 + \log\left\{\epsilon^{\frac{\pi\phi}{V}} + \epsilon^{-\frac{\pi\phi}{V}} - 2\cos\frac{\pi\psi}{V}\right\}\right.$$

$$\left. + 2\iota\tan^{-1}\left\{\frac{\left(\epsilon^{\frac{1}{2}\frac{\pi\phi}{V}} + \epsilon^{-\frac{1}{2}\frac{\pi\phi}{V}}\right)}{\epsilon^{\frac{1}{2}\frac{\pi\phi}{V}} - \epsilon^{-\frac{1}{2}\frac{\pi\phi}{V}}}\tan\frac{\pi\psi}{2V}\right\}\right],$$

which is equivalent to the result given in Maxwell, Art. 195.

The quantity of electricity on a portion whose length is CP and breadth unity of the lower side of the plane CD is

$$\frac{1}{4\pi}\{\phi_P - \phi_C\}.$$

Now $\phi_C = 0$, and when CP is large compared with h, t is very nearly equal to -1, hence if CP $= x$ we have in this case from (11)

$$x = -\frac{h}{\pi}\{\log 2 + \log(t+1)\},$$

and from (12)

$$\phi_P = \frac{V}{\pi}\{\log 2 - \log(t+1)\},$$

hence

$$\phi_P = \frac{V}{\pi}\left\{2\log 2 + \frac{\pi x}{h}\right\},$$

and the quantity of electricity on the strip is

$$\frac{V}{4\pi h}x\left\{1 + \frac{2h}{\pi x}\log 2\right\}.$$

That is, it is the same as if the distribution were uniform and the same as for two infinite plates with the breadth of the strip increased by $\frac{2h}{\pi}\log 2$.

237.] To find the correction for the thickness of the semi-infinite plate, we shall solve by the Schwarzian method the problem of a semi-infinite plate of finite thickness and rectangular section placed midway between two infinite plates. The two infinite plates are at zero potential, the semi-infinite one at potential V.

The diagram in the z plane is represented in Fig. 94. The boundary consists of the infinite line AB, the semi-infinite line CD, the finite line CE, the semi-infinite line EF and the infinite line GH. We shall assume $t = -\infty$ at the point on the line AB where x is equal to $-\infty$, $t = -1$ at the point on the

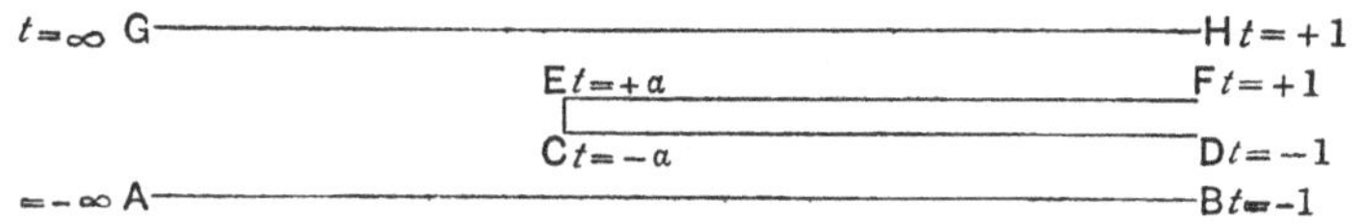

Fig. 94.

same line where $x = +\infty$: $t = -a$ at C $(a < 1)$, $t = +a$ at E, $t = +1$ at the point on the line GH where $x = +\infty$ and $t = +\infty$ at the point on the same line where $x = -\infty$. The internal angles of the polygon are

$$0 \text{ when } t = \pm 1, \quad \frac{3\pi}{2} \text{ when } t = \pm a,$$

hence the transformation which transforms the boundary of the z diagram into the real axis of the t plane is

$$\begin{aligned}\frac{dz}{dt} &= \frac{C(t+a)^{\frac{1}{2}}(t-a)^{\frac{1}{2}}}{(t+1)(t-1)} \\ &= \frac{C(t^2-a^2)^{\frac{1}{2}}}{t^2-1} \\ &= \frac{C}{\{t^2-a^2\}^{\frac{1}{2}}} + \tfrac{1}{2}C(1-a^2)\frac{1}{\{t^2-a^2\}^{\frac{1}{2}}}\left\{\frac{1}{t-1} - \frac{1}{t+1}\right\}. \qquad (13)\end{aligned}$$

The first term on the right-hand side is integrable, and the second and third become integrable by the substitutions $u = 1/t - 1$ and $u = 1/t + 1$ respectively. Integrating (13) we find

$$\begin{aligned} z = {} & C\log\{t + \sqrt{t^2-a^2}\} - C\log\sqrt{-a^2} \\ & + \tfrac{1}{2}(1-a^2)^{\frac{1}{2}}C\log\left[\frac{(t-a^2-\sqrt{1-a^2}\sqrt{t^2-a^2})}{(t+a^2+\sqrt{1-a^2}\sqrt{t^2-a^2})}\frac{(t+1)}{(t-1)}\right], \qquad (14)\end{aligned}$$

where the constant has been chosen so as to make both x and y vanish when $t = 0$.

If $2h$ is the thickness of the semi-infinite plate and $2H$ the distance between the infinite plates, then when t passes through the value unity y increases by $H-h$. When t is nearly unity we may put

$$t = 1 + R\iota^{\iota\theta},$$

where R is small, and θ changes from π to zero as t passes through unity. When t is approximately 1, equation (13) becomes

$$\frac{dz}{dt} = \tfrac{1}{2}C(1-a^2)^{\frac{1}{2}}\frac{1}{t-1},$$

hence the increase in z as t passes through 1 is

$$\tfrac{1}{2}C(1-a^2)^{\frac{1}{2}}\left[\log R + \iota\theta\right]_\pi^0$$

$$= -\frac{\iota\pi}{2}C(1-a^2)^{\frac{1}{2}},$$

but since the increase in z when t passes through this value is $\iota(H-h)$, we have

$$H-h = -C\frac{\pi}{2}(1-a^2)^{\frac{1}{2}}.$$

When t changes from $+\infty$ to $-\infty$, z diminishes by $\iota 2H$; but when t is very large, equation (13) becomes

$$\frac{dz}{dt} = \frac{C}{t},$$

$$z = C\log t.$$

Now $$t = R\epsilon^{\iota\theta},$$

where R is infinite, and θ changes from 0 to π as t changes from $+\infty$ to $-\infty$; but as t changes from plus to minus infinity, z increases by

$$C\left[\log R + \iota\theta\right]_0^\pi$$

$$= \iota C\pi,$$

and since the *diminution* in z is $\iota 2H$, we have

$$H = -C\frac{\pi}{2}.$$

Thus $$h = H\{1-\sqrt{1-a^2}\},$$

or $$a = \sqrt{\frac{h(2H-h)}{H^2}}.$$

The diagram in the w plane is the same as in Art. 236, hence we have

$$\phi + \iota\psi = \frac{V}{\pi}\log\frac{t-1}{t+1}. \tag{15}$$

The quantity of electricity on the portion of the semi-infinite plate between O, the point midway between C and E, and P a point on the upper surface of the boundary, is

$$\frac{1}{4\pi}\{\phi_O - \phi_P\}.$$

Now at O, $t = 0$, hence $\phi_O = 0$, and if EP is large compared with H, t at P is approximately equal to 1. In this case we find from (14), writing $EP = x$,

$$x = C\log\left\{\frac{1+\sqrt{1-a^2}}{a}\right\} + \tfrac{1}{2}C\{1-a^2\}^{\frac{1}{2}}\log\frac{a^2}{2(1-a^2)}$$
$$+ \tfrac{1}{2}C\{1-a^2\}^{\frac{1}{2}}\log(t-1).$$

Substituting for C and a their values in terms of H and h we get

$$-\log(t-1) = \frac{\pi}{H-h}\left\{x + \frac{H}{\pi}\log\frac{2H-h}{h}\right.$$
$$\left. + \frac{H-h}{\pi}\log\frac{h(2H-h)}{2(H-h)^2}\right\}. \tag{16}$$

But from equation (15)

$$\phi_P = \frac{V}{\pi}\{\log(t-1) - \log 2\},$$

since t at P is approximately equal to 1. Hence the quantity of electricity on the strip OP is

$$\frac{V}{4\pi\{H-h\}}\left\{x + \frac{H}{\pi}\log\frac{2H-h}{h} + \frac{H-h}{\pi}\log\frac{h(2H-h)}{(H-h)^2}\right\}.$$

Thus the breadth of the strip, which must be added to allow for the concentration of the electricity near the boundary, is

$$\frac{H}{\pi}\log\frac{2H-h}{h} + \frac{H-h}{\pi}\log\frac{h(2H-h)}{(H-h)^2}.$$

If h is very small this reduces to

$$\frac{2H}{\pi}\log 2,$$

which was the result obtained in Art. 236.

The density of the electricity at the point x on the top of the semi-infinite plate is $-\frac{1}{4\pi}\frac{d\phi}{dx}$, now

$$\frac{d\phi}{dx} = \frac{d\phi}{dt}\frac{dt}{dx}$$
$$= \frac{2V}{\pi(t+1)(t-1)}\frac{(t+1)(t-1)}{C(t^2-a^2)^{\frac{1}{2}}}$$
$$= \frac{V}{\pi C}\frac{2}{(t^2-a^2)^{\frac{1}{2}}}$$
$$= -\frac{V}{H}\frac{1}{(t^2-a^2)^{\frac{1}{2}}}.$$

Hence the density of the electricity on the plate is

$$\frac{V}{4\pi H}\frac{1}{(t^2-a^2)^{\frac{1}{2}}}.$$

This is infinite at the edges C and E. When EP is a large multiple of H, $t = 1$ approximately, and the density is

$$\frac{V}{4\pi H}\frac{1}{\{1-a^2\}^{\frac{1}{2}}},$$

or since
$$(1-a^2)^{\frac{1}{2}} = \frac{H-h}{H},$$

the density is uniform and equal to

$$\frac{1}{4\pi}\frac{V}{H-h}.$$

238.] Condensers are sometimes made by placing one cube inside another; in order to find the capacity of a condenser of this kind we shall investigate the distribution of electricity on a system of conductors such as that represented in Fig. 95, where ABC is maintained at zero potential and FED at potential V.

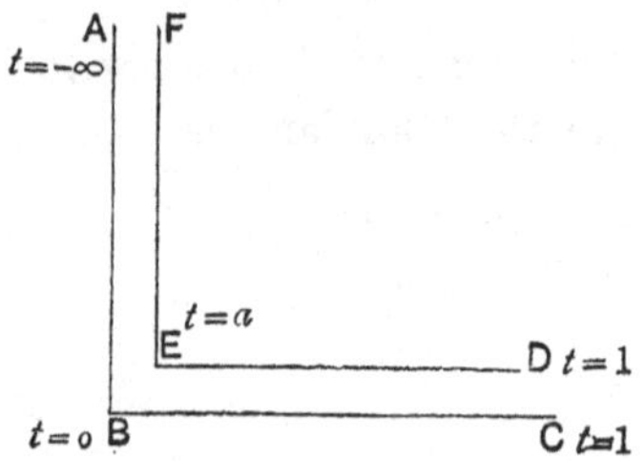

Fig. 95.

The diagram in the z plane is bounded by the lines AB, BC, DE, EF; we shall assume that $t = -\infty$ at the point on the line AB where $y = +\infty$, $t = 0$ at B, $t = 1$ at the point on BC where

$x = +\infty$, and $t = a$ at E, where a is a quantity greater than unity which has to be determined by the geometry of the system. The internal angles of the polygon in the z plane are $\pi/2$ at B, zero at C, $3\pi/2$ at E. The transformation which turns the boundary of the z polygon into the real axis in the t plane is by equation (1) expressed by the equation

$$\frac{dz}{dt} = \frac{C(a-t)^{\frac{1}{2}}}{t^{\frac{1}{2}}(1-t)}. \qquad (17)$$

The diagram in the w plane consists of the real axis and a line parallel to it. The internal angle of the polygon is at $t = 1$ and is equal to zero, hence the transformation which turns this diagram into the real axis of t is

$$\frac{dw}{dt} = \frac{B}{1-t},$$

or

$$\phi + \iota\psi = \iota V - \frac{V}{\pi}\log(1-t),$$

since V is the increment in ψ when t passes through the value 1.

To integrate (17) put

$$t = a\frac{u^2}{1+u^2}.$$

We have then

$$\frac{dz}{du} = \frac{2Ca}{(1+u^2)\{1-(a-1)u^2\}}$$

$$= 2C\left\{\frac{1}{1+u^2} + \frac{a-1}{1-(a-1)u^2}\right\}.$$

Hence

$$z = 2C\tan^{-1}u + \sqrt{a-1}\,C\log\left(\frac{1+\sqrt{a-1}\,u}{1-\sqrt{a-1}\,u}\right)$$

$$= 2C\sin^{-1}\sqrt{\frac{t}{a}} + \sqrt{a-1}\,C\log\left\{\frac{\sqrt{a-t}+\sqrt{a-1}\sqrt{t}}{\sqrt{a-t}-\sqrt{a-1}\sqrt{t}}\right\}; \qquad (18)$$

where the constants have been chosen so as to make x and y vanish when $t = 0$.

When $t = a$, we have

$$x + \iota y = C\pi + \sqrt{a-1}\,C\iota\pi.$$

Hence if h and k are the coordinates of E referred to the axes BC, AB, we have

$$h = C\pi,$$
$$k = C\sqrt{a-1}\,\pi.$$

We can also deduce these equations from equation (17) by the process used to determine the constants in Art. 237.

We may write (18) in the form

$$x + \iota y = \frac{2h}{\pi}\sin^{-1}\sqrt{\frac{t}{a}} + \frac{k}{\pi}\log\left\{\frac{(\sqrt{a-t}+\sqrt{a-1}\sqrt{t})^2}{a(1-t)}\right\}. \quad (19)$$

The quantity of electricity on the strip BP, where P is a point on BC, is equal to

$$-\frac{1}{4\pi}\{\phi_P - \phi_B\}$$

$$= \frac{1}{4\pi}\frac{V}{\pi}\log(1-t_p).$$

Now if BP is large compared with k, the value of t at P is approximately unity; from (19) we get the more accurate value

$$-\log(1-t) = \frac{\pi}{k}x - \frac{2h}{k}\sin^{-1}\sqrt{\frac{1}{a}} - 2\log\left\{2\sqrt{\frac{a-1}{a}}\right\},$$

$$= \pi\frac{x}{k} - \frac{2h}{k}\tan^{-1}\frac{h}{k} - 2\log\frac{2k}{\sqrt{h^2+k^2}}.$$

Hence the quantity of electricity on the strip is

$$-\frac{V}{4\pi k}\left\{x - \frac{2h}{\pi}\tan^{-1}\frac{h}{k} + \frac{2k}{\pi}\log\frac{\sqrt{h^2+k^2}}{2k}\right\}.$$

Hence the quantity is the same as if the electricity were distributed with the uniform density $-V/4\pi k$ over a strip whose breadth was less than BP by

$$\frac{2h}{\pi}\tan^{-1}\frac{h}{k} - \frac{2k}{\pi}\log\frac{\sqrt{h^2+k^2}}{2k}.$$

In the important case when $h = k$, this becomes

$$\frac{h}{2} + \frac{h}{\pi}\log 2.$$

The surface density of the electricity at any point on BC or ED is

$$\mp \frac{V}{4\pi^2 C}\sqrt{\frac{t}{a-t}},$$

the $-$ or $+$ sign being taken according as the point is on BC or ED. This expression vanishes at B and is infinite at E.

At P, a point on BC at some distance from B, t is approximately unity, so that the surface density is

$$-\frac{V}{4\pi^2 C\sqrt{a-1}}$$

$$= -\frac{V}{4\pi k}.$$

This result is of course obvious, but it may be regarded as affording a verification of the preceding solution.

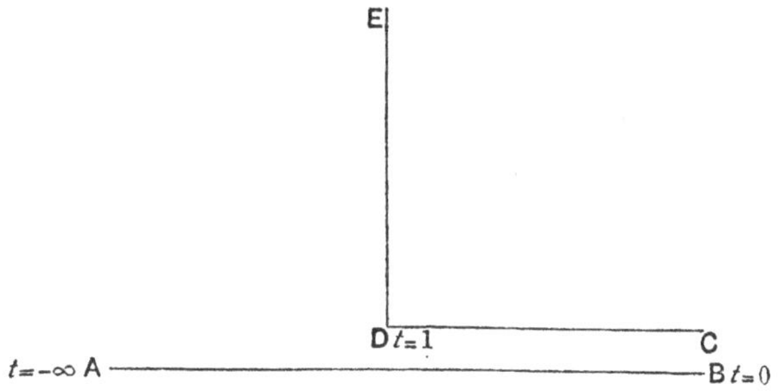

Fig. 96.

239.] Another case of some interest is that represented in Fig. 96, where we have an infinite plane AB at potential V in presence of a conductor at zero potential bounded by two semi-infinite planes CD, DE at right angles to each other. The diagram in the z plane is bounded by the lines AB, CD, DE and a quadrant of a circle whose radius is infinite. We shall assume $t = -\infty$ at the point on the line AB where $x = -\infty$, $t = 0$ at the point on the same line where $x = +\infty$, $t = 1$ at D. The internal angles of the polygon in the z plane are zero at B and $3\pi/2$ at D. The transformation which turns the boundary of the z polygon into the real axis in the t plane is therefore, by equation (1),

$$\frac{dz}{dt} = \frac{C(1-t)^{\frac{1}{2}}}{t}. \tag{20}$$

The diagram in the w plane consists of two straight lines parallel to the real axis, the internal angle being zero at the point $t = 0$; hence we have

$$w = \phi + \iota\psi = \frac{V}{\pi}\log t,$$

since the plane AB is at potential V and CDE at potential zero.

Integrating equation (20), we find when $t > 0 <, 1$

$$z = x + \iota y = C\left(2\sqrt{1-t} - \log\frac{1+\sqrt{1-t}}{1-\sqrt{1-t}}\right), \tag{21}$$

where no constant of integration is needed if the origin of coordinates is taken at D where $t = +1$. If h is the distance between CD and AB, then z increases by ιh when t changes sign, hence we have by equation (20), by the process similar to that by which we deduced the constant in Art. (237),

$$h = -C\pi;$$

so that (21) becomes, $0 < t < 1$,

$$x + \iota y = \frac{h}{\pi}\left\{\log\frac{1+\sqrt{1-t}}{1-\sqrt{1-t}} - 2\sqrt{1-t}\right\}. \tag{22}$$

The quantity of electricity on a strip DP where P is a point on DC is

$$\frac{V}{4\pi^2}\log t_P,$$

if t_P is the value of t at P. If DP is large compared with h, t_P will be very nearly zero; the value of $\log t_P$ is then readily got by writing (22) in the form

$$x + \iota y = \frac{h}{\pi}\{2\log(1+\sqrt{1-t}) - \log t - 2\sqrt{1-t}\}.$$

So that if $x =$ DP, we have approximately,

$$-\log t_P = \frac{\pi}{h}\left\{x - \frac{2h}{\pi}\log 2 + \frac{2h}{\pi}\right\},$$

$$= \frac{\pi}{h}\left\{x + \frac{2h}{\pi}(1 - \log 2)\right\}.$$

Thus the quantity of electricity on DP is

$$-\frac{V}{4\pi h}\left\{x+\frac{2h}{\pi}(1-\log 2)\right\}.$$

We can prove in a similar way that if Q is a point on DE the charge on DQ is equal to

$$\frac{V}{2\pi^2}\log\left(\frac{\pi\,\mathrm{DQ}}{2h}\right).$$

240.] If the angle CDE, instead of being equal to $\pi/2$, were equal to π/n, the transformation of the diagram in the z plane to the real axis of t could be effected by the relation

$$\frac{dz}{dt}=\frac{C(t-1)^{\frac{n-1}{n}}}{t}.$$

241.] We shall now proceed to discuss a problem which enables us to estimate the effect produced by the slit between the guard-ring and the plate of a condenser on the capacity of the condenser.

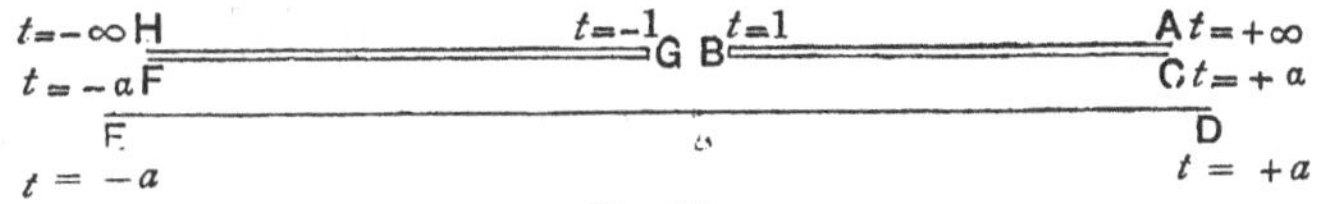

Fig. 97.

Fig. 98.

When the plate and the guard-ring are of finite thickness the integration of the differential equation between z and t involves the use of Elliptic Functions. In the two limiting cases when the thickness of the plate is infinitely small or infinitely great, the necessary integrations can however be effected by simpler means.

We shall begin with the case where the thickness of the plate is very small, and consider the distribution of electricity on two semi-infinite plates separated by a finite interval $2k$ and placed parallel to an infinite plane at the distance h from it.

We shall suppose that the two semi-infinite plates are at the same potential V, and that the infinite plate is at potential zero. The diagrams in the z and w planes are represented in Figs. 97 and 98.

The diagram in the z plane is bounded by the infinite straight line ED, the two sides AB and BC of the semi-infinite line on the right, the two sides FG, GH of the semi-infinite line on the left, and a semi-circle of infinite radius. A point traversing the straight portion of the boundary might start from A and travel to B on the upper side of the line on the right, then from B to C along the under side, from D to E along the infinite straight line, from F to G on the under side of the line on the left and from G to H on the upper side of this line. We shall suppose that $t = +\infty$ at A, $t = +1$ at B, $t = +a\,(a < 1)$ at C, $t = -a$ at F, $t = -1$ at G, $t = -\infty$ at H. The internal angles of the polygon in the z plane are 2π at B, zero at C, zero at F, and 2π at G; hence the transformation which turns the diagram in the z plane into the real axis of t is expressed by the relation

$$\frac{dz}{dt} = C\frac{t^2-1}{t^2-a^2}. \qquad (23)$$

The diagram in the w plane consists of two straight lines parallel to the real axis and the potential changes by V when t passes through the values $\pm a$: hence we easily find

$$\phi + \iota\psi = \frac{V}{\pi}\log\frac{t+a}{t-a} + \iota V. \qquad (24)$$

We have from equation (23)

$$z = C\left\{t - \frac{(1-a^2)}{2a}\log\frac{t-a}{t+a} + \frac{(1-a^2)}{2a}\iota\pi\right\}, \qquad (25)$$

where the constant of integration has been chosen so as to make $x = 0$, $y = 0$ when $t = 0$. The axis of x is ED, the axis of y the line at right angles to this passing through the middle of GB.

If $2k$ is the width of the gap and h the vertical distance between the plates, $x = k$, $y = h$, when $t = 1$, hence we have by (25)

$$k = C\left\{1 - \frac{(1-a^2)}{2a}\log\frac{1-a}{1+a}\right\},$$

$$h = C\frac{(1-a^2)}{2a}\pi.$$

Hence a is determined by the equation

$$k = \frac{h}{\pi}\left\{\frac{2a}{1-a^2} + \log\frac{1+a}{1-a}\right\}. \qquad (26)$$

The quantity of electricity on the lower side of the semi-infinite plate between B and P is, since t increases from P to B,

$$\frac{1}{4\pi}\{\phi_P - \phi_B\},$$

or by (24)

$$\frac{V}{4\pi^2}\left\{\log\frac{t_P+a}{t_P-a} - \log\frac{1+a}{1-a}\right\}.$$

But by (25) if BP $= x-k$, we have

$$x-k = C\left[t_P - 1 - \frac{1-a^2}{2a}\left\{\log\frac{t_P-a}{t_P+a} - \log\frac{1-a}{1+a}\right\}\right].$$

Hence if Q is the quantity of electricity on the lower side of the plate between B and P,

$$x-k = C(t_P - 1) + \frac{4\pi h}{V}.Q,$$

$$Q = \frac{V}{4\pi h}\{x-k+C(1-t_P)\},$$

or since $t_P = a$ approximately, if P is a considerable distance from B, we have

$$Q = \frac{V}{4\pi h}\{x-k+C(1-a)\}. \tag{27}$$

The quantity of electricity Q_1 on the upper side of the plate, from A to B, is equal to

$$\frac{1}{4\pi}(\phi_B - \phi_A),$$

or since $t = +\infty$ at A, and therefore ϕ_A vanishes, we have

$$Q_1 = -\frac{V}{4\pi^2}\log\frac{1-a}{1+a}. \tag{28}$$

We can by equation (26) easily express a in terms of k/h, when this ratio is either very small or very large. We shall begin by considering the first case, which is the one that most frequently occurs in practice.

We see from (26) that when k/h is very small, a is very small and is approximately equal to

$$\frac{\pi}{4}\frac{k}{h}.$$

The corresponding value of C is $\frac{1}{2}k$, hence, neglecting $(k/h)^3$,

$$Q = \frac{V}{4\pi h}\left\{x - \tfrac{1}{2}k - \frac{\pi}{8}\frac{k^2}{h}\right\},$$

$$Q_1 = \frac{V}{4\pi^2}2a$$

$$= \frac{V}{4\pi h}\frac{k}{2}.$$

Hence $Q+Q_1$, the whole quantity of electricity between A and P, is approximately equal to

$$\frac{V}{4\pi h}\left\{x - \frac{\pi}{8}\frac{k^2}{h}\right\}.$$

Hence the quantity of electricity on the plate of the condenser is to the present degree of approximation the same as if the electricity were uniformly distributed over the plate with the density it would have if the slit were absent, provided that the area of the plate is increased by that of a strip whose width is

$$k - \frac{\pi}{8}\frac{k^2}{h};$$

thus the breadth of the additional strip is very approximately half that of the slit.

We pass on now to the case when h/k is very small. We see from equation (26) that in this case a is very nearly equal to unity, the approximate values of a and C being given by the equations

$$1 - a = \frac{h}{\pi k},$$

$$C = k.$$

Hence by equations (27) and (28) we have

$$Q = \frac{V}{4\pi h}\left\{x - k + \frac{h}{\pi}\right\},$$

$$Q_1 = \frac{V}{4\pi^2}\log\frac{2\pi k}{h}.$$

So that the total charge $Q+Q_1$ on AP is equal to

$$\frac{V}{4\pi h}\left[x - k + \frac{h}{\pi}\left\{1 + \log\frac{2\pi k}{h}\right\}\right],$$

and thus the width of the additional strip is

$$\frac{h}{\pi}\left\{1+\log\frac{2\pi k}{h}\right\}.$$

242.] We have hitherto supposed that the potentials of the plates ABC and FGH are the same; we can however easily modify the investigation so as to give the solution of the case when ABC is maintained at the potential V_1 and FGH at the potential V_2. The relation between z and t will not be affected by this change, but the relation between w and t will now be represented by the equation

$$\phi+\iota\psi=\frac{V_2}{\pi}\log(t+a)-\frac{V_1}{\pi}\log(t-a)+\iota V_1.$$

The quantity of electricity between B and P, a point on the lower side of the plate, is

$$\frac{1}{4\pi}\{\phi_P-\phi_B\}.$$

Now if BP is large, t at P is approximately equal to a, and

$$\phi_P=\frac{V_2}{\pi}\log 2a-\frac{V_1}{\pi}\log(t-a);$$

but by equation (25) we have when t is nearly equal to a,

$$-\log(t-a)=\frac{\pi}{h}(x-Ca)-\log 2a,$$

hence $$\phi_P=\frac{V_2-V_1}{\pi}\log 2a+\frac{V_1}{h}(x-Ca).$$

When h/k is large a is small and approximately equal to $\pi k/4h$, and this equation becomes

$$\phi_P=\frac{V_2-V_1}{\pi}\log\frac{\pi k}{2h}+\frac{V_1}{h}x.$$

Since $t=1$ at B and a is small, we see that ϕ_B is approximately equal to $a(V_1+V_2)/\pi$ or $(V_1+V_2)k/4h$, hence the quantity of electricity between B and P is approximately equal to

$$\frac{V_1}{4\pi h}x-\frac{V_2-V_1}{4\pi^2}\log\frac{2h}{\pi k}-\frac{V_1+V_2}{4\pi h}\frac{k}{4}.$$

The charge Q_1 on the upper side of the plate ABC between a point P′ vertically above P and B is, since t increases from B to P′, equal to

$$\frac{1}{4\pi}\{\phi_B - \phi_{P'}\}.$$

Now
$$\phi_{P'} = \frac{V_2}{\pi}\log(t_{P'}+a) - \frac{V_1}{\pi}\log(t_{P'}-a),$$

which, since $t_{P'}$ is large, may be written as

$$\phi_{P'} = \frac{V_2 - V_1}{\pi}\log t_{P'}.$$

When BP′ is large, $t_{P'}$ is large also, and by equation (25) is approximately equal to x/C, that is to $2x/k$, thus

$$\phi_{P'} = \frac{V_2 - V_1}{\pi}\log\frac{2x}{k},$$

$$\phi_B = (V_1 + V_2)\frac{k}{4h};$$

and therefore Q_1, the charge on the upper part of the plate, is given by the equation

$$Q_1 = \frac{(V_1+V_2)}{4\pi}\frac{k}{4h} - \frac{(V_2 - V_1)}{4\pi^2}\log\frac{2x}{k};$$

thus $Q+Q_1$, the sum of the charges on the upper and lower portions, is given by the equation

$$Q+Q_1 = \frac{V_1}{4\pi h}x - \frac{(V_2-V_1)}{4\pi^2}\left\{\log\frac{2h}{\pi k} + \log\frac{2x}{k}\right\}.$$

243.] We shall now proceed to discuss the other extreme case of the guard-ring, that in which the depth of the slit is infinite. We shall begin with the case when the guard-ring and the condenser plate are at the same potential. The diagram in the w plane is the same as that in Art. 239, while the diagram in the z plane is represented in Fig. 99. The boundary of this diagram consists of the semi-infinite lines AB, BC at right angles to each other, the infinite line DE parallel to BC, the semi-infinite line FG which is in the same straight line as BC, and the semi-infinite line GH at right angles to FG. We shall suppose $t = +\infty$ at A, $t = +1$ at B, $t = +a$ $(a < 1)$ at C, $t = -a$

at F, $t=-1$ at G, and $t=-\infty$ at H. The internal angles of the polygon in the z plane are $3\pi/2$ at B and G and zero at C and F.

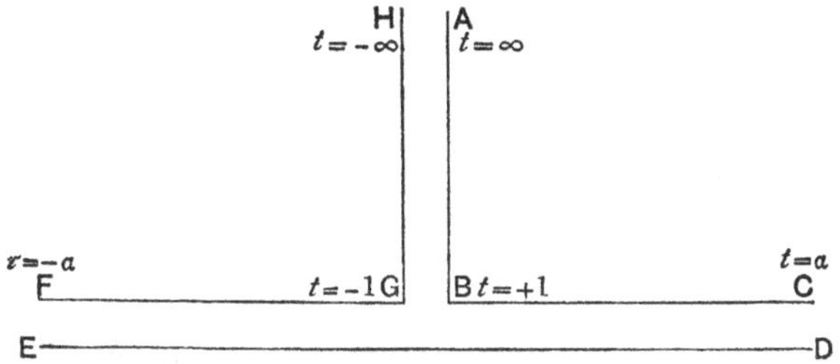

Fig. 99.

Thus the transformation which turns the boundary in the z plane into the real axis of the t plane is expressed by the equation

$$\frac{dz}{dt}=C\frac{(t^2-1)^{\frac{1}{2}}}{t^2-a^2}.$$

If we are dealing with the portion of the boundary for which t is less than unity, it is more convenient to write this equation as

$$\frac{dz}{dt}=C\iota\frac{(1-t^2)^{\frac{1}{2}}}{t^2-a^2}$$

$$=C\iota\left[\frac{1-a^2}{2a}\frac{1}{(1-t^2)^{\frac{1}{2}}}\left\{\frac{1}{t-a}-\frac{1}{t+a}\right\}-\frac{1}{(1-t^2)^{\frac{1}{2}}}\right].$$

Integrating, we find

$$z=-C\iota\left[\frac{\sqrt{1-a^2}}{2a}\log\frac{(1-at+\sqrt{1-a^2}\sqrt{1-t^2})(t+a)}{(1+at+\sqrt{1-a^2}\sqrt{1-t^2})(t-a)}+\sin^{-1}t\right]$$
$$+C\pi\frac{\sqrt{1-a^2}}{2a},$$

where the constant of integration has been chosen so as to make $x=0$, $y=0$ when $t=0$; ED is the axis of x, and the axis of y is midway between AB and GH. Writing D for $-C\iota$, the preceding equation takes the form

$$z=D\sin^{-1}t+D\frac{\sqrt{1-a^2}}{2a}\log\frac{(1-at+\sqrt{1-a^2}\sqrt{1-t^2})(t+a)}{(1+at+\sqrt{1-a^2}\sqrt{1-t^2})(t-a)}$$
$$+D\iota\pi\frac{\sqrt{1-a^2}}{2a}. \qquad (29)$$

Now if $2k$ is the width of the slit, and h the distance of the plate of the condenser from the infinite plate, $x = k$, $y = h$ when $t = 1$, hence from (29)

$$k = D\frac{\pi}{2},$$

$$h = D\frac{\pi}{2}\frac{\sqrt{1-a^2}}{a},$$

or

$$a^2 = \frac{k^2}{h^2+k^2}.$$

The relation between w and t is the same as in Art. 239, and we have

$$w = \phi + \iota\psi = \frac{V}{\pi}\log(t+a) - \frac{V}{\pi}\log(t-a) + \iota V.$$

The quantity of electricity Q on the plate of the condenser between A and P, a point on BC at some considerable distance from B, is

$$\frac{1}{4\pi}\{\phi_P - \phi_A\};$$

since t is infinite at the point corresponding to A, we see that ϕ_A is zero, hence

$$Q = \frac{1}{4\pi}\phi_P$$

$$= \frac{V}{4\pi^2}\log\frac{t_P+a}{t_P-a}.$$

Now the point P corresponds to a point in the t plane where t is very nearly equal to a; hence we have approximately by (29)

$$\log\frac{t_P+a}{t_P-a} = \frac{\pi}{h}\left(x - D\sin^{-1}a - \frac{h}{\pi}\log(1-a^2)\right)$$

$$= \frac{\pi}{h}\left(x - \frac{2k}{\pi}\sin^{-1}\frac{k}{\sqrt{h^2+k^2}} - \frac{h}{\pi}\log\frac{h^2}{h^2+k^2}\right).$$

Thus $$Q = \frac{V}{4\pi h}\left\{x - \frac{2k}{\pi}\sin^{-1}\frac{k}{\sqrt{h^2+k^2}} - \frac{h}{\pi}\log\frac{h^2}{h^2+k^2}\right\}.$$

In the case which occurs most frequently in practice, that in which k is small compared with h, we have, neglecting $(k/h)^2$,

$$Q = \frac{V}{4\pi h}x;$$

that is, the quantity of electricity on the plate is the same as if the distribution were uniform and the width of the plate were increased by half the breadth of the slit.

The quantity of electricity on the face AB of the slit is equal to

$$\frac{V}{4\pi^2}\log\frac{1+a}{1-a},$$

or, substituting the value for a previously found,

$$\frac{V}{4\pi^2}\log\left\{\frac{1+\dfrac{k}{\sqrt{h^2+k^2}}}{1-\dfrac{k}{\sqrt{h^2+k^2}}}\right\},$$

and this when k/h is small is equal to

$$\frac{V}{4\pi h}\frac{2k}{\pi}.$$

Thus $2/\pi$ of the increase in the charge on ABC, over the value it would have if the surface density were uniformly $V/4\pi h$ on BC, is on the side AB of the slit, and $(\pi-2)/\pi$ is on the face of the plate of the condenser.

244.] A slight modification of the preceding solution will enable us to find the distribution of electricity on the conductors when ABC and FGH are no longer at the same potential. If V_1 is the potential of ABC, V_2 that of FGH, then the relation between z and t will remain the same as before, while the relation between w and t will now be expressed by the equation

$$w=\phi+\iota\psi=\frac{V_2}{\pi}\log(t+a)-\frac{V_1}{\pi}\log(t-a)+\iota V_1,$$

$$\text{or}\quad \phi+\iota\psi=\frac{V_2-V_1}{\pi}\log(t+a)+\frac{V_1}{\pi}\log(t+a)-\frac{V_1}{\pi}\log(t-a)+\iota V_1.$$

Hence the quantity of electricity on QBP where Q is a point on AB at some distance from B will exceed the quantity that would be found from the results of the preceding Article by

$$\frac{V_2-V_1}{4\pi^2}\log\frac{t_P+a}{t_Q+a}.$$

Since P is a point on BC at some distance from B, t_P is approximately equal to a, and since a is small and t_Q large we may replace t_Q+a by t_Q; making these substitutions the preceding expression becomes

$$\frac{V_2-V_1}{4\pi^2}\log\frac{2a}{t_Q}. \tag{30}$$

When t is large, the relation between z and t, which is given by the equation

$$\frac{dz}{dt}=C\frac{(t^2-1)^{\frac{1}{2}}}{t^2-a^2},$$

is by integrating this equation found to be

$$x-k+\iota(y-h)$$
$$=C\log(t+\sqrt{t^2-1})+\frac{\sqrt{1-a^2}}{2a}C\left\{\sin^{-1}\frac{1-at}{t-a}-\sin^{-1}\frac{1+at}{t+a}\right\},$$

or substituting for C (the ιD of the preceding Article) its value $\iota 2k/\pi$, we have

$$x-k+\iota(y-h)$$
$$=\iota\frac{2k}{\pi}\log(t+\sqrt{t^2-1})+\iota\frac{h}{\pi}\left\{\sin^{-1}\frac{1-at}{t-a}-\sin^{-1}\frac{1+at}{t+a}\right\}.$$

Hence, when t is large we have approximately

$$\log 2t=\frac{\pi}{2k}(y-h).$$

Substituting this value for $\log t_Q$ in the expression (30), we find that the correction to be applied on account of the difference of potential between ABC and FGH to the expression given by Art. 243 for the quantity of electricity on QBP is

$$-\frac{(V_2-V_1)}{4\pi^2}\left\{\log\frac{\sqrt{h^2+k^2}}{4k}+\frac{\pi}{2k}(y-h)\right\},$$

where $y-h=$ BQ.

245.] The indirect method given by Maxwell, *Electrostatics*, Chap. XII, in which we begin by assuming an arbitrary relation between z and w of the form

$$x+\iota y=F(\phi+\iota\psi),$$

and then proceed to find the problems in electrostatics which can be solved by this relation, leads to some interesting results when elliptic functions are employed. Thus, let us assume

$$x+\iota y=b\,\mathrm{sn}(\phi+\iota\psi), \tag{31}$$

and suppose that ϕ is the potential and ψ the stream function.

Let k be the modulus of the elliptic functions, $2K$ and $2\iota K'$ the real and imaginary periods. Let us trace the equipotential surface for which $\phi = K$; we have

$$x + \iota y = b \operatorname{sn}(K + \iota\psi)$$
$$= \frac{b}{\operatorname{dn}(\psi, k')}, \qquad (32)$$

where $\operatorname{dn}(\psi, k')$ denotes that the modulus of the elliptic function is k', that is $\sqrt{1-k^2}$, and not k. From equation (32) we see that $y = 0$, and

$$x = \frac{b}{\operatorname{dn}(\psi, k')}.$$

Now $\operatorname{dn}(\psi, k')$ is always positive, its greatest value is unity when $\psi = 0$, or an even multiple of K', its least value is k when ψ is an odd multiple of K', thus the equation

$$x + \iota y = \frac{b}{\operatorname{dn}(\psi, k')}$$

represents the portion of the axis of x between $x = b$ and $x = b/k$.

If we put $\phi = -K$, we have

$$x + \iota y = b \operatorname{sn}(-K + \iota\psi),$$
$$= -\frac{b}{\operatorname{dn}(\psi, k')};$$

hence the equipotential surface, $-K$, consists of the portion of the axis of x between $x = -b$ and $x = -b/k$.

Thus the transformation (31) solves the case of two infinite plane strips AB, CD, Fig. 100, of finite and equal widths, $b(1-k)/k$, in one plane placed so that their sides are parallel to each other.

Fig. 100.

In the above investigation the potential difference is $2K$. The quantity of electricity on the top of the strip CD is equal to the difference in the values of ψ at C and D divided by 4π. Now the difference in the values of ψ at C and D is K', hence the quantity of electricity on the top of the strip is

$$\frac{1}{4\pi} K'.$$

There is an equal quantity of electricity on the bottom of the strip, so that the total charge on CD is

$$\frac{1}{4\pi} 2K'.$$

The difference of potential between the strips is $2K$, hence the capacity of the strip per unit length measured parallel to z is

$$\frac{1}{4\pi}\frac{K'}{K}.$$

The modulus k of the elliptic functions is the ratio of BC to AD, that is the ratio of the shortest to the longest distance between points in the lines AB and CD. The values of K and K' for given values of k are tabulated in Legendre's *Traité des Fonctions Elliptiques*: so that with these tables the capacity of two strips of any width can be readily found.

When k is small, that is when the breadth of either of the strips is large compared with the distance between them, K and K' are given approximately by the following equations,

$$K = \frac{\pi}{2},$$

$$K' = \log(4/k) = \log(4\,AD/BC).$$

Hence in this case the capacity is approximately,

$$\frac{1}{2\pi^2}\log(4\,AD/BC).$$

Returning to the general case, if σ is the surface density of the electricity at the point P on one of the strips AB, we have

$$\sigma = \frac{1}{4\pi}\frac{d\psi}{dx};$$

and since

$$x = -\frac{b}{\text{dn}(\psi, k')},$$

$$-\frac{dx}{d\psi} = bk'^2 \text{sn}(\psi, k')\,\text{cn}(\psi, k')/\text{dn}^2(\psi, k')$$

$$= \frac{1}{b}\{x^2 - b^2\}^{\frac{1}{2}}\{b^2 - k^2x^2\}^{\frac{1}{2}}$$

$$= \frac{k}{b}\sqrt{CP.DP.AP.BP};$$

hence

$$\sigma = -\frac{b}{4\pi k}\frac{1}{\sqrt{CP.DP.AP.BP}}.$$

The solution of the case of two strips at equal and opposite potentials, includes that of a strip at potential K in front of an infinite plane at potential zero. The solution of this case can be deduced directly from the transformation

$$x + \iota y = b \operatorname{dn}(\phi + \iota\psi),$$

if ψ be taken as the potential and ϕ as the stream function.

246.] *Capacity of a Pile of Plates*, Fig. 101. If we put

$$\epsilon^{\frac{x+\iota y}{b}} = \operatorname{sn}(\phi + \iota\psi), \tag{33}$$

then when $\phi = K$

$$\epsilon^{\frac{x+\iota y}{b}} = \operatorname{sn}(K + \iota\psi) = \frac{1}{\operatorname{dn}(\psi, k')}. \tag{34}$$

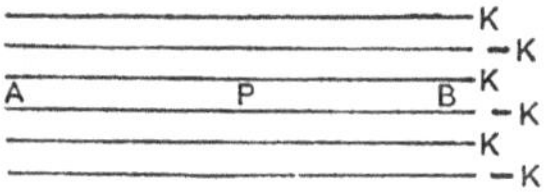

Fig. 101.

Thus, since $\operatorname{dn}(\psi, k')$ is always real and positive,

$$y = 0, \; y = 2\pi b, \; y = 4\pi b, \; \&c.,$$

while x varies between the values x_1, x_2, where

$$\left.\begin{aligned} \epsilon^{\frac{x_1}{b}} &= 1, \\ \epsilon^{\frac{x_2}{b}} &= \frac{1}{k}. \end{aligned}\right\} \tag{35}$$

When $\phi = -K$,

$$\epsilon^{\frac{x+\iota y}{b}} = \operatorname{sn}(-K + \iota\psi) = -\frac{1}{\operatorname{dn}(\psi, k')},$$

hence, since $\operatorname{dn}(\psi, k')$ is always real and positive,

$$y = \pi b, \; y = 3\pi b, \; y = 5\pi b, \; \&c.,$$

while x varies between the same values as before. Thus, if in equation (33) we take ϕ to be the potential and ψ the stream function, the equation will give the electrical distribution over a pile of parallel strips of finite width, $x_2 - x_1$, the distance between the consecutive strips being πb, alternate strips being at the same potential. The potential of one set of plates is K, that of the other $-K$.

The quantity of electricity on one side of one of the strips per unit length parallel to z is, as in Art. 245, equal to $K'/4\pi$, and since the charge on either side is the same, the total charge on the strips is $K'/2\pi$. The potential difference is $2K$, hence the capacity of one of the strips per unit length is equal to

$$\frac{K'}{4\pi K}.$$

We see from equation (35) that

$$k = \epsilon^{-\frac{(x_2 - x_1)}{b}};$$

but $x_2 - x_1 = d$, the breadth of one of the strip, hence

$$k = \epsilon^{-\frac{d}{b}}.$$

Having found k from this equation, we can by Legendre's Tables find the values of K and K', and hence the capacity of the strips. When the breadth of the strips is large compared with the distance between them, d/b is large, hence k is small; in this case we have approximately

$$K = \frac{\pi}{2},$$

$$K' = \log(4/k) = \log(4\epsilon^{\frac{d}{b}})$$
$$= 2\log 2 + \frac{d}{b},$$

so that the capacity of one strip is

$$\frac{1}{2\pi^2}\left\{2\log 2 + \frac{d}{b}\right\}.$$

Returning to the general case, the surface density of the electricity at a point P on the positive side of one of the strips AB is equal to

$$\frac{1}{4\pi}\frac{d\psi}{dx}.$$

But by equation (34)

$$\frac{1}{b}\epsilon^{\frac{x}{b}}\frac{dx}{d\psi} = k'^2 \operatorname{sn}(\psi, k')\operatorname{cn}(\psi, k')/\operatorname{dn}^2(\psi, k').$$

Substituting the values of

$$\operatorname{sn}(\psi, k'), \quad \operatorname{cn}(\psi, k'), \quad \operatorname{dn}(\psi, k')$$

in terms of $\epsilon^{\frac{x}{b}}$, we get

$$\frac{d\psi}{dx} = \frac{1}{b} \frac{\epsilon^{\frac{(x_2-x_1)}{2b}}}{\left\{\left(\epsilon^{\frac{(x-x_1)}{b}} - \epsilon^{-\frac{(x-x_1)}{b}}\right)\left(\epsilon^{\frac{(x_2-x)}{b}} - \epsilon^{-\frac{(x_2-x)}{b}}\right)\right\}^{\frac{1}{2}}}.$$

Hence the surface density is equal to

$$\frac{1}{4\pi b} \frac{\epsilon^{\frac{AB}{2b}}}{\left\{\left(\epsilon^{\frac{AP}{b}} - \epsilon^{-\frac{AP}{b}}\right)\left(\epsilon^{\frac{BP}{b}} - \epsilon^{-\frac{BP}{b}}\right)\right\}^{\frac{1}{2}}}.$$

The distribution of electricity on any one of the plates is evidently the same as if the plate were placed midway between two infinite parallel plates at potential zero, the distance between the two infinite plates being $2\pi b$.

247.] *Capacity of a system of $2n$ plates arranged radially and making equal angles with each other, the alternate plates being at the same potential, the extremities of the plates lying on two coaxial right circular cylinders.* Let us put

$$\left(\frac{x+\iota y}{b}\right)^n = \mathrm{sn}\,(\phi + \iota\psi),$$

or, transforming to polar coordinates r and θ,

$$\left(\frac{r}{b}\right)^n \epsilon^{\iota n\theta} = \mathrm{sn}\,(\phi + \iota\psi).$$

Then, as before, we see that when $\phi = K$, $n\theta = 0$ or 2π, or 4π, and so on, and when $\phi = -K$, $n\theta = \pi$ or 3π, or 5π, &c.; hence this transformation solves the case of $2n$ plates arranged radially, making angles π/n with each other, one set of n plates being at the potential K, the other set at the potential $-K$. When $\phi = K$, we have

$$\left(\frac{r}{b}\right)^n = \frac{1}{\mathrm{dn}\,(\psi, k')}.$$

Hence if r_1 and r_2 are the smallest and greatest distances of the edges of a plate from the line to which all the plates converge, we have

$$\left(\frac{r_1}{b}\right)^n = 1,$$

$$\left(\frac{r_2}{b}\right)^n = \frac{1}{k'},$$

or

$$k = \left(\frac{r_1}{r_2}\right)^n.$$

The total charge on both sides of one of the plates is, as before, $K'/2\pi$, and since the potential difference is $2K$ the capacity of the plate is $K'/4\pi K$. When r_1 is small compared with r_2, k is small, and we have then approximately

$$K = \frac{\pi}{2},$$

$$K' = \log(4/k) = \log 4 + n \log(r_2/r_1).$$

Thus the capacity of a plate is in this case approximately

$$\frac{1}{2\pi^2}\{\log 4 + n \log(r_2/r_1)\}.$$

Returning to the general case, the surface density of the electricity on one side of a plate is equal to

$$\frac{1}{4\pi}\frac{d\psi}{dr};$$

but since
$$\left(\frac{r}{b}\right)^n = \frac{1}{\operatorname{dn}(\psi, k')},$$

$$\frac{n}{b}\left(\frac{r}{b}\right)^{n-1}\frac{dr}{d\psi} = k'^2 \operatorname{sn}(\psi, k') \operatorname{cn}(\psi, k')/\operatorname{dn}^2(\psi, k').$$

Substituting for the elliptic functions their values in terms of r, we find when $\phi = K$

$$\frac{d\psi}{dr} = \frac{nb^n r^{n-1}}{k\{(r^{2n} - r_1^{2n})(r_2^{2n} - r^{2n})\}^{\frac{1}{2}}}.$$

Thus the surface density is equal to

$$\frac{1}{4\pi}\frac{nr_2^n r^{n-1}}{\{(r^{2n} - r_1^{2n})(r_2^{2n} - r^{2n})\}^{\frac{1}{2}}}.$$

When $n = 1$, this case coincides with that discussed in Art. 245.

248.] Let us next put

$$x + \iota y = b \operatorname{cn}(\phi + \iota\psi),$$

and take ψ for the potential, and ϕ for the stream function. Then when $\psi = 0$, we have

$$x + \iota y = b \operatorname{cn} \phi,$$

hence $y = 0$, and x can have any value between $\pm b$: thus the equipotential surface for which ψ is zero is the portion of the axis of x between $x=-b$, and $x=+b$. When $\psi=K'$,

$$x+\iota y = b\,\mathrm{cn}\,(\phi+\iota K')$$
$$= -\frac{b\iota\,\mathrm{dn}\,\phi}{k\,\mathrm{sn}\,\phi};$$

hence $x=0$, and y ranges from $+bk'/k$ to $+\infty$ and from $-bk'/k$ to $-\infty$. Hence the section of the equipotential surface for which $\psi=K'$ is the portion of the axis of y included between these limits. Thus the section of the conductors over which the distribution of electricity is given by this transformation is similar to that represented in Fig. 102, where the axis of x is *vertical*.

To find the quantity of electricity on AB we notice that $\phi=0$ at A and is equal to $2K$ at B, hence the quantity of electricity on one side of AB is equal to $K/2\pi$, thus the total charge on AB is K/π. The difference of potential between AB and CD or EF is K', so that the capacity of AB is equal to

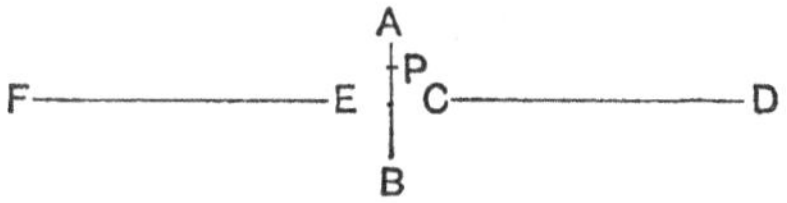

Fig. 102.

$$\frac{1}{\pi}\frac{K}{K'}.$$

The modulus k of the elliptic functions is given by the equation

$$\frac{k'}{k} = \frac{\{1-k^2\}^{\frac{1}{2}}}{k} = \frac{\mathbf{EC}}{\mathbf{AB}}.$$

If AB is very large compared with EC then k is very nearly unity, and in this case we have

$$K = \log(4/k') = \log(4\mathbf{AB}/\mathbf{EC}),$$

$$K' = \frac{\pi}{2};$$

so that the capacity of AB is

$$\frac{2}{\pi^2}\log(4\,\mathbf{AB}/\mathbf{EC}).$$

The surface density of the electricity at a point P on either side of AB is (without any limitation as to the value of k) equal to

$$\frac{1}{4\pi}\frac{d\phi}{dx},$$

and since $x = b \operatorname{cn} \phi$,

$$\frac{dx}{d\phi} = -b \operatorname{sn}\phi \operatorname{dn}\phi$$

$$= -\frac{k}{b}(b^2-x^2)^{\frac{1}{2}}\left\{\frac{k'^2}{k^2}b^2+x^2\right\}^{\frac{1}{2}}$$

$$= -\frac{k}{b}\,\text{CP}\sqrt{\text{AP.BP}};$$

hence the surface density is equal to

$$-\frac{b}{4\pi k}\frac{1}{\text{CP}\sqrt{\text{AP. BP}}}.$$

249.] We pass on now to consider the transformation

$$\epsilon^{\frac{x+\iota y}{b}} = \operatorname{cn}(\phi+\iota\psi),$$

where ϕ is taken as the potential and ψ as the stream function.

Over the equipotential surface for which $\phi = 0$, we have

$$\epsilon^{\frac{x+\iota y}{b}} = \operatorname{cn}(\iota\psi)$$

$$= \frac{1}{\operatorname{cn}(\psi, k')}.$$

Hence $y = 0,\ \pm\pi b,\ \pm 2\pi b, \ldots;$

while x ranges from 0 to infinity.

For the equipotential surface for which $\phi = K$, we have

$$\epsilon^{\frac{x+\iota y}{b}} = \operatorname{cn}(K+\iota\psi)$$

$$= -\iota k'\frac{\operatorname{sn}(\psi, k')}{\operatorname{dn}(\psi, k')}.$$

Hence $y = \pm\frac{1}{2}\pi b,\ \pm\frac{3}{2}\pi b,\ \pm\frac{5}{2}\pi b \ldots,$

while x ranges from minus infinity to a value x_1 given by the equation

$$\epsilon^{\frac{x_1}{b}} = \frac{k'}{k}.$$

Thus this transformation gives the distribution of electricity on a pile of semi-infinite parallel plates at equal intervals πb apart, maintained at potential zero when in presence of another pile of semi-infinite parallel plates at the same distance apart maintained at potential K, the planes of the second set of plates being midway between those of the first. The second set of plates project a distance x_1 into the first set, x_1 being given by the equation $\epsilon^{x_1/b} = k'/k$. If the edges of the second set of plates are outside the first set, then x_1 is negative and numerically equal to the distance between the planes containing the ends of the two sets of plates. The system of conductors is represented in Fig. 103.

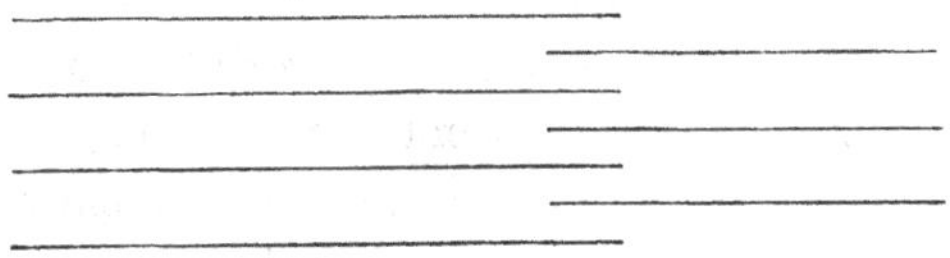

Fig. 103.

The quantity of electricity on the two sides of one of the plates is $K'/2\pi$, hence the capacity of such a plate is

$$\frac{K'}{2\pi K}.$$

If the ends of the two sets of plates are in the same plane, then $x_1 = 0$, and therefore $k' = k$, so that $K' = K$; hence the capacity of each plate is in this case $1/(2\pi)$.

When the plates do not penetrate and are separated by a distance which is large compared with the distance between two parallel plates, x_1 is negative and large compared with b, hence k' is small, and therefore k nearly equal to unity; in this case

$$K' = \frac{\pi}{2},$$

$$K = \log(4/k'),$$

$$= \log 4 + \frac{x'}{b},$$

where $x' = -x_1$.

Thus the capacity of a plate in this case is approximately equal to

$$\frac{b}{4(b \log 4 + x')}.$$

The surface density at a point on one of the first set of plates at a distance x from the edge is easily shewn by the methods previously used to be equal, whatever be the value of k, to

$$-\frac{1}{4\pi kb}\frac{\epsilon^{\frac{x}{b}}}{\sqrt{(\epsilon^{\frac{2x}{b}}-1)(\epsilon^{\frac{2x}{b}}+\epsilon^{\frac{2x_1}{b}})}}.$$

250.] The transformation

$$\left(\frac{x+\iota y}{b}\right)^n = \text{cn}\,(\phi+\iota\psi),$$

with ϕ as the potential and ψ as the stream function, gives the solution of the case represented in Fig. 104; where the $2n$ outer planes at potential zero are supposed to extend to infinity, the $2n$ inner planes at potential K bisect the angles between the outer planes, and OA$=b$.

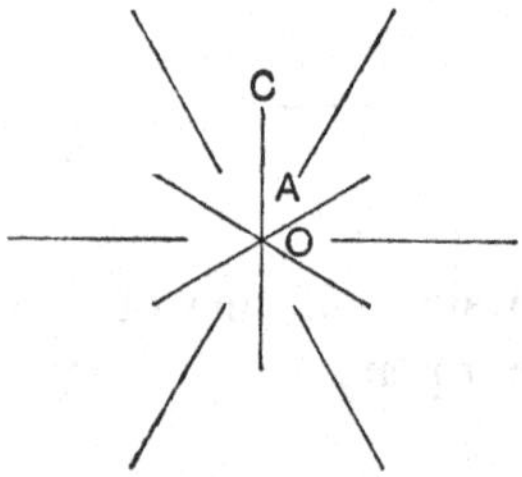

Fig. 104.

We can easily prove that in this case the quantity of electricity on the outer plates is equal to nK'/π, so that the capacity of the system is equal to

$$\frac{n}{\pi}\frac{K'}{K},$$

when the modulus of the elliptic functions is determined by the relation

$$\left(\frac{\text{OC}}{\text{OA}}\right)^n = \frac{k'}{k}.$$

251.] The transformation

$$x+\iota y = b\,\text{dn}\,(\phi+\iota\psi),$$

where ϕ is the potential and ψ the stream function, gives the solution of the case represented in Fig. 105, in which a finite

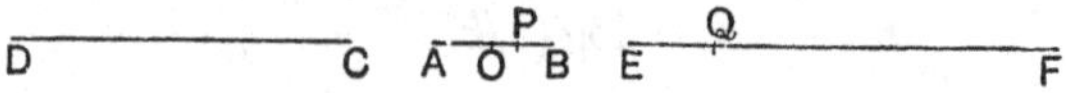

Fig. 105.

plate is placed in the space between two semi-infinite plates. For when $\phi = 0$, we have

$$\begin{aligned} x+\iota y &= b\,\text{dn}\,\iota\psi \\ &= b\frac{\text{dn}\,(\psi, k')}{\text{cn}\,(\psi, k')}; \qquad (36)\end{aligned}$$

hence $y = 0$, and x ranges from $+b$ to $+\infty$ and from $-b$ to $-\infty$, thus giving the portions EF, CD of the figure.

When $\phi = K$, we have

$$x + \iota y = b\,\mathrm{dn}\,(K + \iota\psi)$$
$$= bk' \frac{\mathrm{cn}\,(\psi, k')}{\mathrm{dn}\,(\psi, k')}; \qquad (37)$$

hence $y = 0$, and x ranges between $\pm bk'$, thus giving the portion AB of the figure.

The quantity of electricity on the two sides of the plate AB is equal to K'/π, hence the capacity of this plate is equal to

$$\frac{1}{\pi}\frac{K'}{K},$$

where the modulus k of the elliptic functions is given by the equation

$$k' = \{1 - k^2\}^{\frac{1}{2}} = \mathrm{OA}/\mathrm{OC}.$$

When AC is small compared with AB, k' is nearly equal to unity, and k is therefore small, in this case we have approximately

$$K = \frac{\pi}{2},$$
$$K' = \log(4/k)$$
$$= \log 4 + \tfrac{1}{2}\log\frac{\mathrm{OC}^2}{\mathrm{AC.BC}};$$

so that in this case the capacity of the plate AB is equal to

$$\frac{1}{\pi^2}\left\{\log\frac{\mathrm{OC}^2}{\mathrm{AC.BC}} + 2\log 4\right\}.$$

Returning to the general case, the surface density of the electricity on one side of the plate AB at a point P is equal to

$$\frac{1}{4\pi}\frac{d\psi}{dx}.$$

Using equation (37) we find that this is equal to

$$\frac{b}{4\pi}\frac{1}{\{(b^2 - x^2)(b^2k'^2 - x^2)\}^{\frac{1}{2}}},$$

which may be written in the form

$$\frac{b}{4\pi}\frac{1}{\{\mathrm{AP.BP.CP.EP}\}^{\frac{1}{2}}}.$$

The surface density at a point Q on EF may be shown in a similar way, using (36), to be equal to

$$-\frac{b}{4\pi}\frac{1}{\{(x^2-b^2)(x^2-b^2k'^2)\}^{\frac{1}{2}}},$$

which is equal to

$$-\frac{b}{4\pi}\frac{1}{\{\text{AQ}.\text{BQ}.\text{CQ}.\text{EQ}\}^{\frac{1}{2}}}.$$

252.] If we put

$$\epsilon^{\frac{x+\iota y}{b}} = \text{dn}\,(\phi+\iota\psi),$$

and take as before ϕ for the potential and ψ for the stream function, then since, when $\phi = 0$,

$$\epsilon^{\frac{x+\iota y}{b}} = \text{dn}\,(\iota\psi)$$
$$= \frac{\text{dn}\,(\psi, k')}{\text{cn}\,(\psi, k')},$$

we have $y = 0$, $y = \pm\pi b$, $y = \pm 2\pi b\ldots$, while x ranges from 0 to $+\infty$. Thus the equipotential surfaces for which ϕ vanishes are a pile of parallel semi-infinite plates stretching from the axis of y to infinity along the positive direction of x, the distance between two adjacent plates being πb.

When $\phi = K$, we have

$$\epsilon^{\frac{x+\iota y}{b}} = \text{dn}\,(K+\iota\psi)$$
$$= k'\frac{\text{cn}\,(\psi, k')}{\text{dn}\,(\psi, k')};$$

thus $y = 0$, $y = \pm\pi b$, $y = \pm 2\pi b\ldots$, while x ranges from $-\infty$ to $-x_1$, where x_1 is given by the equation

$$\epsilon^{-\frac{x_1}{b}} = k'. \qquad (38)$$

Thus the equipotential surfaces for which $\phi = K$ are a pile of parallel semi-infinite plates stretching from $-\infty$ to a distance x_1 from the previous set of plates. The distance between adjacent plates in this set is again πb, and the planes of the plates in this set are the continuations of those of the plates in the set at potential zero. This system of conductors is represented in Fig. 106.

The quantity of electricity on both sides of one of the plates at potential zero is $-K'/2\pi$, hence the capacity of such a plate is

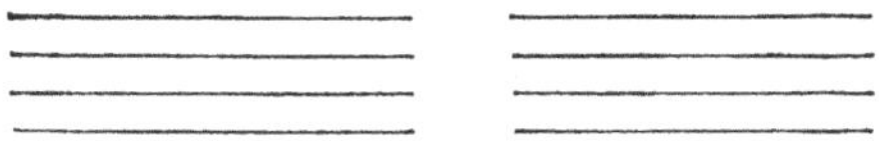

Fig. 106.

$$\frac{1}{2\pi}\frac{K'}{K},$$

the modulus of the elliptic functions being given by equation (38).

When the distance between the edges of the two sets of plates is large compared with the distance between two adjacent parallel plates, then x_1 is large compared with b, so that k' is small; in this case we have approximately

$$K' = \frac{\pi}{2},$$

$$\begin{aligned} K &= \log(4/k') \\ &= \log 4 + \frac{x_1}{b}; \end{aligned}$$

hence the capacity of a plate is equal to

$$\frac{b}{4\,(x_1 + b\log 4)}.$$

The surface density of the electricity at a point P on one of the planes at potential zero is in the general case easily proved to be equal to

$$-\frac{1}{4\pi b}\frac{\epsilon^{\frac{x}{b}}}{\left\{(\epsilon^{\frac{2x}{b}}-1)(\epsilon^{\frac{2x}{b}}-\epsilon^{-\frac{2x_1}{b}})\right\}^{\frac{1}{2}}}.$$

253.] The transformation

$$\left(\frac{x+\iota y}{b}\right)^n = \operatorname{dn}(\phi + \iota\psi),$$

where ϕ is the potential and ψ the stream function and n a positive integer, gives the solution of the case shown in Fig. 107, when the potential of the outer radial plates is zero and that of the inner K. The $2n$ outer plates make equal angles with each other and extend to infinity.

The quantity of electricity on both sides of one of the outer plates is $-K'/2\pi$; since there are $2n$ of these plates the capacity of the system is

$$\frac{n}{\pi}\frac{K'}{K},$$

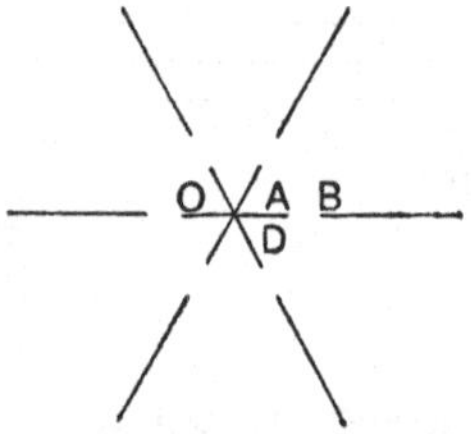

Fig. 107.

the modulus of the Elliptic Functions being given by the equation

$$k' = \{1-k^2\}^{\frac{1}{2}} = \left(\frac{OA}{OB}\right)^n.$$

254.] We have only considered those applications of elliptic function to electrostatics where the expression for the capacity of the electrical system proves to be such that it can be readily calculated in any special case by the aid of Legendre's Tables. There are many other transformations which are of great interest analytically, though the want of tables of the special functions involved makes them of less interest for experimental purposes than those we have considered. Thus, for example, the transformation

$$x + \iota y = Z(\phi + \iota\psi),$$

where Z is the function introduced by Jacobi and defined by the equation

$$Z(u) = \int^u \mathrm{dn}^2 u\, du - \frac{E}{K},$$

if ψ is the potential and ϕ the stream function, gives the distribution of electricity in the important case of a condenser formed by two parallel and equal plates of finite breadth.

CHAPTER IV.

ELECTRICAL WAVES AND OSCILLATIONS.

255.] THE properties of electrical systems in which the distribution of electricity varies periodically and with sufficient rapidity to call into play the effects of electric inertia, are so interesting and important that they have attracted a very large amount of attention ever since the principles which govern them were set forth by Maxwell in his *Electricity and Magnetism*. We shall in this Chapter consider the theory of such vibrating electrical systems, while the following Chapter will contain an account of some remarkable experiments by which the properties of such systems have been exhibited in a very striking way.

256.] We shall begin by writing down the general equations which we shall require in discussing the transmission of electric disturbances through a field in which both insulators and conductors are present.

Let F, G, H be the components of the vector potential parallel to the axes of x, y, z respectively, P, Q, R the components of the electromotive intensity, and a, b, c those of the magnetic induction in the same directions, let ϕ be the electrostatic potential, σ the specific resistance of the conductor, μ and μ' the magnetic permeabilites of the conductor and dielectric respectively, and K and K' the specific inductive capacities of the conductor and dielectric respectively, then we have

$$\left.\begin{aligned} P &= -\frac{dF}{dt} - \frac{d\phi}{dx}, \\ Q &= -\frac{dG}{dt} - \frac{d\phi}{dy}, \\ R &= -\frac{dH}{dt} - \frac{d\phi}{dz}. \end{aligned}\right\} \qquad (1)$$

We have also
$$a = \frac{dH}{dy} - \frac{dG}{dz},$$

hence
$$\frac{da}{dt} = \frac{d}{dy}\frac{dH}{dt} - \frac{d}{dz}\frac{dG}{dt},$$

so that
$$\frac{da}{dt} = \frac{dQ}{dz} - \frac{dR}{dy};$$

similarly
$$\left.\begin{aligned}\frac{db}{dt} &= \frac{dR}{dx} - \frac{dP}{dz},\\[1mm] \frac{dc}{dt} &= \frac{dP}{dy} - \frac{dQ}{dx}.\end{aligned}\right\} \qquad (2)$$

If a, β, γ are the components of the magnetic force, u, v, w those of the total current, then (Maxwell's *Electricity and Magnetism*, Art. 607)

$$\left.\begin{aligned}4\pi u &= \frac{d\gamma}{dy} - \frac{d\beta}{dz},\\[1mm] 4\pi v &= \frac{da}{dz} - \frac{d\gamma}{dx},\\[1mm] 4\pi w &= \frac{d\beta}{dx} - \frac{da}{dy}.\end{aligned}\right\} \qquad (3)$$

In the metal the total current is the sum of the conduction and polarization currents; the conduction current parallel to x is P/σ, the polarization current $\frac{K}{4\pi}\frac{dP}{dt}$, or if P varies as $\epsilon^{\iota pt}$, the polarization current is $\frac{K}{4\pi}\iota p.P$. Thus the ratio of the conduction to the polarization current is $\frac{4\pi}{K\sigma\iota p}$, and since σ in electromagnetic measure is of the order 10^4 for the commoner metals and K in the same measure of the order 10^{-21}, we see that unless the vibrations are comparable in rapidity with those of light we may neglect the polarization current in the metal in comparison with the conduction current. Thus in the conductor we have

$$\frac{4\pi}{\sigma}P = \frac{d\gamma}{dy} - \frac{d\beta}{dz} = \frac{1}{\mu}\left(\frac{dc}{dy} - \frac{db}{dz}\right),$$

and therefore by (2) we have, assuming

$$\frac{dP}{dx} + \frac{dQ}{dy} + \frac{dR}{dz} = 0,$$

similarly

$$\left.\begin{aligned} \nabla^2 P &= \frac{4\pi\mu}{\sigma}\frac{dP}{dt};\\ \nabla^2 Q &= \frac{4\pi\mu}{\sigma}\frac{dQ}{dt},\\ \nabla^2 R &= \frac{4\pi\mu}{\sigma}\frac{dR}{dt}. \end{aligned}\right\} \qquad (4)$$

It follows from equation (2) that a, b, c satisfy equations of the same form.

In the dielectric there is only the polarization current, the component of which parallel to x is $\frac{K'}{4\pi}\frac{dP}{dt}$; hence in the dielectric we have

$$K'\frac{dP}{dt} = \frac{d\gamma}{dy} - \frac{d\beta}{dz} = \frac{1}{\mu'}\left(\frac{dc}{dy} - \frac{db}{dz}\right),$$

and therefore by (2)

similarly

$$\left.\begin{aligned} \nabla^2 P &= \mu' K'\frac{d^2P}{dt^2};\\ \nabla^2 Q &= \mu' K'\frac{d^2Q}{dt^2},\\ \nabla^2 R &= \mu' K'\frac{d^2R}{dt^2}. \end{aligned}\right\} \qquad (5)$$

We shall suppose that the effects are periodic and of frequency $p/2\pi$, so that the components of the electromotive intensity, as well as of the magnetic induction, will all vary as $\epsilon^{\iota pt}$ and will not explicitly involve the time in any other way. We shall also suppose that the electric waves are travelling parallel to the axis of z, so that the variables before enumerated will contain $\epsilon^{\iota mz}$ as a factor, m being a quantity which it is one of the objects of our investigation to determine. With these assumptions we see that d/dt may be replaced by ιp, and d/dz by ιm.

Alternating Electric Currents in Two Dimensions.

257.] The cases relating to alternating currents which are of the greatest practical importance are those in which the currents flow along metallic wires. As the analysis, however, in these cases is somewhat complicated, we shall begin by considering the two dimensional problem, as this, though of comparatively small practical importance, enables us by the aid of simple analysis to illustrate some important properties possessed by alternating currents.

The case we shall first consider is that of an infinite conducting plate bounded by the planes $x = h$, $x = -h$, immersed in a dielectric. We shall suppose that plane waves of electromotive intensity are advancing through the dielectric, and that these waves impinge on the plate. We shall suppose also that the waves fall on both sides of the plate and are symmetrical with respect to it. These waves when they strike against the plate will be reflected from it, so that there will on either side of the plate be systems of direct and reflected waves.

Let P and R denote the components of the electromotive intensity parallel to the axes of x and z respectively, the component parallel to the axis of y vanishing since the case is one in two dimensions. Then in the dielectric the part of R due to the direct wave will be of the form

$$B\epsilon^{\iota(mz+lx+pt)},$$

while the part due to the reflected wave will be of the form

$$C\epsilon^{\iota(mz-lx+pt)}.$$

Thus in the dielectric on one side of the plate

$$R = B\epsilon^{\iota(mz+lx+pt)} + C\epsilon^{\iota(mz-lx+pt)}. \tag{1}$$

If V is the velocity with which electromagnetic disturbances are propagated through the dielectric, we have by equation (5), Art. 256, since $\mu'K' = 1/V^2$,

$$\frac{d^2R}{dx^2} + \frac{d^2R}{dz^2} = \frac{1}{V^2}\frac{d^2R}{dt^2},$$

hence

$$l^2 + m^2 = \frac{p^2}{V^2}.$$

If λ is the wave length of the incident wave, θ the angle between the normal to the wave front and the axis of x, we have, since

$$p = \frac{2\pi}{\lambda}V,$$

$$l = \frac{2\pi}{\lambda}\cos\theta, \quad m = \frac{2\pi}{\lambda}\sin\theta.$$

Since Q vanishes, we have

$$\frac{dP}{dx} + \frac{dR}{dz} = 0.$$

Substituting the value of R from equation (1), we find

$$P = -\frac{m}{l}\{B\epsilon^{\iota(mz+lx+pt)} - C\epsilon^{\iota(mz-lx+pt)}\}. \qquad (2)$$

The resultant electromotive intensity in the incident wave is

$$\frac{B}{\cos\theta}\epsilon^{\iota(mz+lx+pt)},$$

in the reflected wave

$$\frac{C}{\cos\theta}\epsilon^{\iota(mz-lx+pt)}.$$

Let us now consider the electromotive intensity in the conducting plate; in this region we have, by (4), Art. 256, if μ is the magnetic permeability and σ the specific resistance of the plate,

$$\frac{d^2R}{dx^2} + \frac{d^2R}{dz^2} = \frac{4\pi\mu}{\sigma}\frac{dR}{dt},$$

or, since R varies as $\epsilon^{\iota(mz+pt)}$,

$$\frac{d^2R}{dx^2} = n^2R,$$

where

$$n^2 = m^2 + \frac{4\pi\mu\iota p}{\sigma}.$$

The solution of this, since the electromotive intensity is symmetrical with respect to the plane $x = 0$, is of the form

$$R = A(\epsilon^{nx} + \epsilon^{-nx})\epsilon^{\iota(mz+pt)}, \qquad (3)$$

and since

$$\frac{dP}{dx} + \frac{dR}{dz} = 0,$$

$$P = -\frac{\iota m}{n}A(\epsilon^{nx} - \epsilon^{-nx})\epsilon^{\iota(mz+pt)}. \qquad (4)$$

If a, b, c are the components of magnetic induction, then

$$\frac{da}{dt} = \frac{dQ}{dz} - \frac{dR}{dy},$$

$$\frac{db}{dt} = \frac{dR}{dx} - \frac{dP}{dz},$$

$$\frac{dc}{dt} = \frac{dP}{dy} - \frac{dQ}{dx};$$

hence $a = 0$, $c = 0$, and

$$b = \frac{n^2 - m^2}{\iota pn}A(\epsilon^{nx} - \epsilon^{-nx})\epsilon^{\iota(mz+pt)} \text{ in the plate,} \qquad (5)$$

$$b = \frac{l^2 + m^2}{lp}(B\epsilon^{\iota lx} - C\epsilon^{-\iota lx})\epsilon^{\iota(mz+pt)} \text{ in the dielectric.} \qquad (6)$$

We can get the expression for the magnetic force in the dielectric very simply by the method given in Art. 9. In the incident wave the resultant electromotive intensity is

$$\frac{B}{\cos\theta}\epsilon^{\iota(mz+lx+pt)},$$

hence the polarization is

$$\frac{K'}{4\pi}\frac{B}{\cos\theta}\epsilon^{\iota(mz+lx+pt)},$$

where K' is the specific inductive capacity of the dielectric. The Faraday tubes are moving with the velocity V, hence by equations (4), Art. (9), the magnetic force due to their motion is

$$VK'\frac{B}{\cos\theta}\epsilon^{\iota(mz+lx+pt)}.$$

The magnetic induction corresponding to this magnetic force is equal, since $\mu'K'$ equals $1/V^2$, to

$$\frac{B}{V\cos\theta}\epsilon^{\iota(mz+lx+pt)},$$

which is the first term on the right in equation (6). We may show in a similar way that the magnetic force due to the motion of the Faraday tubes in the reflected wave is equal to the second term on the right in equation (6).

We must now consider the conditions which hold at the junction of the plate and the dielectric. These may be expressed in many different ways: they are, however, when the conductors are at rest, equivalent to the conditions that the tangential electromotive intensity, and the tangential magnetic force, are continuous. Thus when $x = h$ we must have both R and b/μ continuous. The first of these conditions gives

$$A(\epsilon^{nh}+\epsilon^{-nh}) = B\epsilon^{\iota lh}+C\epsilon^{-\iota lh}, \tag{7}$$

the second

$$\frac{n^2-m^2}{\mu n}A(\epsilon^{nh}-\epsilon^{-nh}) = \frac{\iota(l^2+m^2)}{\mu' l}(B\epsilon^{\iota lh}-C\epsilon^{-\iota lh}). \tag{8}$$

Since $$n^2-m^2 = \frac{4\pi\mu\iota p}{\sigma},$$

and $$\frac{l^2+m^2}{l} = \frac{2\pi}{\lambda\cos\theta},$$

and for all known dielectrics μ' may without sensible error be put equal to unity, equation (8) may be written

$$\frac{2p}{\sigma n} A\,(\epsilon^{nh} - \epsilon^{-nh}) = \frac{1}{\lambda \cos\theta}(B\epsilon^{\iota lh} - C\epsilon^{-\iota lh}). \tag{9}$$

From (7) and (9) we get

$$A\,\{\epsilon^{nh} + \epsilon^{-nh} + \frac{2p\lambda\cos\theta}{\sigma n}(\epsilon^{nh} - \epsilon^{-nh})\} = 2B\epsilon^{\iota lh}, \tag{10}$$

$$A\,\{\epsilon^{nh} + \epsilon^{-nh} - \frac{2p\lambda\cos\theta}{\sigma n}(\epsilon^{nh} - \epsilon^{-nh})\} = 2C\epsilon^{-\iota lh}. \tag{11}$$

It will be convenient to express A, B, C in terms of the total current through the plate. If w is the intensity of the current parallel to z in the plate, $w = R/\sigma$, hence by (3)

$$w = \frac{A}{\sigma}(\epsilon^{nx} + \epsilon^{-nx})\,\epsilon^{\iota(mz+pt)}.$$

If $I_0\epsilon^{\iota(mz+pt)}$ is the total current passing through unit width measured parallel to y, of the plate,

$$I_0\epsilon^{\iota(mz+pt)} = \int_{-h}^{+h} w\,dx\,;$$

hence

$$I_0 = \frac{2A}{\sigma n}(\epsilon^{nh} - \epsilon^{-nh}), \tag{12}$$

so that

$$w = \tfrac{1}{2} n I_0 \frac{(\epsilon^{nx} + \epsilon^{-nx})}{(\epsilon^{nh} - \epsilon^{-nh})}\,\epsilon^{\iota(mz+pt)}. \tag{13}$$

Let us now suppose that the frequency of the vibrations is so small that nh is a small quantity, this will be the case if $h\sqrt{4\pi\mu p/\sigma}$ is small. When nh and therefore nx is small, equation (13) becomes approximately

$$w = \frac{I_0}{2h}\,\epsilon^{\iota(mz+pt)}\,;$$

thus the current in the plate is distributed uniformly across it. When nh is small, equations (12), (10) and (11) become approximately

$$I_0 = \frac{4Ah}{\sigma},$$

$$A\,(1 + 4\pi V h\,\sigma^{-1}\cos\theta) = B\epsilon^{\iota lh},$$

$$A\,(1 - 4\pi V h\,\sigma^{-1}\cos\theta) = C\epsilon^{-\iota lh}.$$

Thus corresponding to the current $I_0 \cos(pt+mz)$ in the plate, we find

$$\left.\begin{aligned} R &= \frac{\sigma I_0}{2h}\cos(pt+mz), \\ P &= \frac{\sigma I_0 mx}{2h}\sin(pt+mz), \\ b &= 4\pi\mu I_0 \frac{x}{2h}\cos(pt+mz) \end{aligned}\right\} \text{in the plate.}$$

Thus, since mx is exceedingly small, we see that the maximum electromotive intensity parallel to the boundary of the plate is exceedingly large compared with the maximum at right angles to it.

In the dielectric we have

$$R = \frac{\sigma I_0}{2h}\cos(pt+mz)\cos l(x-h) - 2\pi I_0 V\cos\theta\sin(pt+mz)\sin l(x-h),$$

$$P = \frac{1}{2h}\sigma I_0 \tan\theta\sin(pt+mz)\sin l(x-h) - 2\pi I_0 V\sin\theta\cos(pt+mz)\cos l(x-h),$$

$$b = -\frac{\sigma I_0}{2Vh\cos\theta}\sin(pt+mz)\sin l(x-h) + 2\pi I_0\cos(pt+mz)\cos l(x-h).$$

Thus at the surface of the plate where $x = h$

$$R = \frac{\sigma I_0}{2h}\cos(pt+mz),$$

$$P = -2\pi I_0 V\sin\theta\cos(pt+mz),$$

$$b = 2\pi I_0\cos(pt+mz).$$

Thus at the surface of the plate $P/R = -4\pi Vh\sigma^{-1}\sin\theta$. If the plate is metallic this quantity is exceedingly large unless the plate is excessively thin or θ very small, so that in the dielectric the resultant electromotive intensity at the surface of the plate is along the normal, this is in striking contrast to the effect inside the plate where P/R is very small. The Faraday tubes in the dielectric close to the plate are thus at right angles to the plate, while in the plate they are parallel to it; hence by Art. 10 the electric momentum in the dielectric close to the plate is parallel to the axis of z, or parallel to the plate, while in the plate itself it is parallel to the axis of x, or in the direction of motion from the outside of the plate to the inside. If $4\pi Vh\cos\theta = \sigma$, then

$C = 0$; in this case there is no reflected wave; the wave reflected from one side of the plate is annulled by the direct wave coming through the plate from the other side. It is worthy of remark that the only one of the quantities we have considered whose value either in the interior of the plate or near to the plate in the dielectric depends sensibly upon θ, the direction of motion of the incident wave, is the normal electromotive intensity in the dielectric and in the plate.

258.] We shall now proceed to discuss the case when nh is large. We shall begin by considering the distribution of current in the plate. We have by (13)

$$w = \tfrac{1}{2} I_0 n \frac{(\epsilon^{nx} + \epsilon^{-nx})}{(\epsilon^{nh} + \epsilon^{-nh})} \epsilon^{\iota(mz+pt)},$$

and since nh is large this equation may be written as

$$w = \tfrac{1}{2} I_0 n \epsilon^{-n(h-x)} \epsilon^{\iota(mz+pt)}. \tag{14}$$

Now
$$n^2 = m^2 + \frac{4\pi\mu\iota p}{\sigma}$$
$$= \frac{p^2}{V^2}\sin^2\theta + \frac{4\pi\mu\iota p}{\sigma}.$$

Now p^2/V^2 is very small compared with $4\pi\mu p/\sigma$ if the plate conducts as well as a metal, unless the vibrations are quicker than those of light. When the current makes a million vibrations per second $(p^2/V^2)/(4\pi\mu p/\sigma)$ is approximately $5 \times 10^{-16}(\sigma/\mu)$, and is thus excessively small unless the resistance is enormously greater than that of acidulated water; we may therefore without appreciable error write

$$n^2 = \frac{4\pi\mu\iota p}{\sigma},$$

and
$$n = \sqrt{2\pi\mu p/\sigma}\,(1+\iota) = n_1(1+\iota) \text{ say, where}$$
$$n_1 = \sqrt{2\pi\mu p/\sigma}.$$

Substituting this value for n, and taking the real part of (14), we have

$$w = \sqrt{\pi\mu p/\sigma}\, I_0 \epsilon^{-n_1(h-x)} \cos\left\{mz - n_1(h-x) + pt + \frac{\pi}{4}\right\}.$$

The presence of the factor $\epsilon^{-n_1(h-x)}$ in this expression shows that the current diminishes in geometrical progression as $h-x$ increases in arithmetical progression, and that it will practically vanish as soon as $n_1(h-x)$ is a small multiple of unity. We thus get the very interesting result that an alternating current does not distribute itself uniformly over the cross-section of the conductor through which it is flowing but concentrates itself towards the outside of the conductor. When the vibrations are very rapid the currents are practically confined to a thin skin on the outside of the conductor. The thickness of this skin will diminish as n_1 increases; we shall take $1/n_1$ as the measure of its thickness.

This inequality in the distribution of alternating currents is explicitly stated in Art. 690 of Maxwell's *Electricity and Magnetism*, but its importance was not recognised until it was brought into prominence and its consequences developed by the investigations of Mr. Heaviside and Lord Rayleigh, and the experiments of Professor Hughes.

The amount of this concentration is very remarkable in the magnetic metals, even for comparatively slow rates of alternation of the current. Let us for example take the case of a current making 100 vibrations per second, and suppose that the plate is made of soft iron for which we may put $\mu = 10^3$, $\sigma = 10^4$. In this case $p = 2\pi \times 10^2$, and n_1 or $\{2\pi\mu p/\sigma\}^{\frac{1}{2}}$ is approximately equal to 20; thus at a depth of half a millimetre from the surface of such a plate, the maximum intensity of the current will only be $1/\epsilon$ or ·368 of its value at the surface. At the depth of 1 millimetre it will only be ·135, at 2 millimetres ·018, and at 3 millimetres ·0025, or the 1/400 part of its value at the exterior. Thus in such a plate, with the assigned rate of alternations, the currents will practically cease at the depth of about 2 mm. and will be reduced to about 1/7 of their value at the depth of one millimetre. Thus in this case the currents, and therefore the magnetic force, are confined to a layer not more than 3 millimetres thick.

The thickness of the 'skin' for copper is about 13 times that for soft iron.

The preceding results apply to currents making 100 vibrations per second; when we are dealing with such alternating currents as are produced by the discharge of a Leyden Jar, where there may be millions of alternations per second, the thickness of the 'skin' in soft iron is often less than the hundredth part of a millimetre.

Returning to the determination of P, R, and b for this case, we find from equations (3), (4), and (12) in the plate

$$R = \{\pi\mu p\sigma\}^{\frac{1}{2}} I_0 \epsilon^{-n_1(h-x)} \cos\left\{mz + n_1(x-h) + pt + \frac{\pi}{4}\right\},$$

$$P = \tfrac{1}{2} m\sigma I_0 \epsilon^{-n_1(h-x)} \cos\left\{mz + n_1(x-h) + pt - \frac{\pi}{2}\right\},$$

$$b = 2\pi\mu I_0 \epsilon^{-n_1(h-x)} \cos\{mz + n_1(x-h) + pt\}.$$

Thus we see that in this case, as well as when nh was small, P/R is in general very small, so that the resultant electromotive intensity is nearly parallel to the surface of the plate.

In the dielectric we have by equations (10), (11), and (12) when nh is large;

$$R = \{\pi\mu p\sigma\}^{\frac{1}{2}} I_0 \cos\left\{mz + pt + \frac{\pi}{4}\right\} \cos l(x-h)$$
$$- 2\pi V \cos\theta I_0 \sin(mz+pt) \sin l(x-h),$$

$$P = \{\pi\mu p\sigma\}^{\frac{1}{2}} \tan\theta I_0 \sin\left\{mz + pt + \frac{\pi}{4}\right\} \sin l(x-h)$$
$$- 2\pi V \sin\theta I_0 \cos(mz+pt) \cos l(x-h),$$

$$b = 2\pi I_0 \cos(mz+pt) \cos l(x-h)$$
$$- \frac{1}{V}\{\pi\mu p\sigma\}^{\frac{1}{2}} \sec\theta I_0 \sin\left(mz + pt + \frac{\pi}{4}\right) \sin l(x-h).$$

At the surface of the plate these become

$$R = \{\pi\mu p\sigma\}^{\frac{1}{2}} I_0 \cos\left(mz + pt + \frac{\pi}{4}\right),$$

$$P = -2\pi V \sin\theta I_0 \cos(mz+pt),$$

$$b = 2\pi I_0 \cos(mz+pt),$$

and we see, as before, that in general P/R is very large, so that the electromotive intensity near the plate in the dielectric is approximately at right angles to it.

Thus, as in the case of the slower vibrations, the momentum is tangential in the dielectric and normal in the plate.

If we compare the expressions for the components of the electromotive intensity in the dielectric given above with those given in the preceding article, we see that, except close to the plate, they are very approximately the same.

Periodic Currents in Cylindrical Conductors, and the Rate of Propagation of Electric Disturbances along them.

259.] We shall now proceed to consider the case which is most easily realized in practice, that in which electrical disturbances are propagated along long straight cylindrical wires, such for example as telegraph wires or sub-marine cables.

A peculiar feature of electrical problems in which infinitely long straight cylinders play a part, is the effect produced by the presence of other conductors, even though these are such a long way off that it might have appeared at first sight that their influence could have been neglected. This is exemplified by the well-known formula for the capacity of two coaxial cylinders. If $\mathbf{a}$ and $\mathbf{b}$ are the radii of the two cylinders the capacity per unit length is proportional to $1/\log(\mathbf{b}/\mathbf{a})$. Thus, even though the cylinders were so far apart that the radius of the outer cylinder was 100 times that of the inner, yet if the distance were further increased until the outer radius was 10,000 times the inner, the capacity of the condenser would be halved, though similar changes in the distances between concentric spheres would hardly have affected their capacity to an appreciable extent. For this reason we shall, though it involves rather more complex analysis, suppose that our cylinder is surrounded by conductors, and the results we shall obtain will enable us to determine when the effects due to the conductors can legitimately be neglected.

260.] The case we shall investigate is that of a cylindrical metallic wire surrounded by a dielectric, while beyond the dielectric we have another conductor; the dielectric is bounded by concentric cylinders whose inner and outer radii are $\mathbf{a}$ and $\mathbf{b}$ respectively. If $\mathbf{b}/\mathbf{a}$ is a very large quantity, we have a case approximating to an aerial telegraph wire, while when $\mathbf{b}/\mathbf{a}$ is not large the case becomes that of a sub-marine cable.

In the dielectric between the conductors there are convergent and divergent waves of Faraday tubes, the incidence of which on the conductors produces the currents through them.

261.] We shall take the axis of the cylinders as the axis of z, and suppose that the electric field is symmetrical round this axis; then if the components of the electric intensity and magnetic induction vary as $\epsilon^{\iota(mz+pt)}$, the differential equations by which these quantities are determined are of the form

$$\frac{d^2f}{dr^2}+\frac{1}{r}\frac{df}{dr}-n^2f=0,$$

where r denotes the distance of a point from the axis of z. The complete solution of this equation is expressed by

$$f=AJ_0(\iota nr)+BK_0(\iota nr)^*. \qquad (1)$$

Here $J_0(x)$ represents Bessel's function of zero order, and

$$K_0(x)=(C+\log 2-\log x)J_0(x)+2\{J_2(x)-\tfrac{1}{2}J_4(x)+\tfrac{1}{3}J_6(x)\ldots\}; \qquad (2)$$

where C is Gauss' constant, and is equal to ·5772157.......

When the real part of ιn is finite, $J_0(\iota nr)$ is infinite when r is infinite (Heine, *Kugelfunctionen*, vol. i. p. 248), so that in any region where r may become infinite we must have $A=0$ in equation (1). Again, $K_0(\iota nr)$ becomes infinite when r vanishes, so that in any region in which r can vanish $B=0$.

We shall find the following approximate equations very useful in our subsequent work.

When x is small

$$J_0(\iota x)=1, \qquad J_0'(\iota x)=-\tfrac{1}{2}\iota x.$$

$$K_0(\iota x)=\log\frac{2\gamma}{\iota x}, \qquad K_0'(\iota x)=-\frac{1}{\iota x};$$

where $\log\gamma=·5772157\ldots,$

and $J_0'(\iota x)$ is written for $\dfrac{dJ_0(\iota x)}{d(\iota x)}$.

When x is very large

$$J_0(\iota x)=\frac{\epsilon^x}{\sqrt{2\pi x}}; \qquad J_0'(\iota x)=-\frac{\iota\epsilon^x}{\sqrt{2\pi x}}.$$

$$K_0(\iota x)=\epsilon^{-x}\sqrt{\frac{\pi}{2x}}; \qquad K_0'(\iota x)=\iota\epsilon^{-x}\sqrt{\frac{\pi}{2x}}.$$

(See Heine, *Kugelfunctionen*, vol. i. p. 248).

262.] We shall now proceed to apply these results to the investigation of the propagation of electric disturbances along the

* Heine, *Kugelfunctionen*, vol. i. p. 189.

wire. The axis of the wire is taken as the axis of z; P, Q, R are the components of the electromotive intensity parallel to the axes of x, y, z respectively; a, b, c are the components of the magnetic induction parallel to these axes: μ, σ are respectively the magnetic permeability and specific resistance of the wire, μ', σ' the values of the same quantities for the external conductor, K is the specific inductive capacity of the dielectric between the wire and the outer conductor. We shall suppose that the magnetic permeability of the dielectric is unity, and that V is the velocity of propagation of electromagnetic action through this dielectric. We shall begin by considering the equations which hold in the dielectric: it is from this region that the Faraday tubes come which produce the currents in the conductor. We shall assume, as before, that the components of the electromotive intensity vary as $\epsilon^{\iota(mz+pt)}$.

The differential equation satisfied by R, the z component of the electromotive intensity in the dielectric, is (Art. 256)

$$\frac{d^2R}{dx^2}+\frac{d^2R}{dy^2}+\frac{d^2R}{dz^2}=\frac{1}{V^2}\frac{d^2R}{dt^2},$$

or, since R varies as $\epsilon^{\iota(mz+pt)}$,

$$\frac{d^2R}{dx^2}+\frac{d^2R}{dy^2}-k^2R=0, \tag{3}$$

where $$k^2=m^2-\frac{p^2}{V^2}.$$

If we introduce cylindrical coordinates r, θ, z, this equation may be written

$$\frac{d^2R}{dr^2}+\frac{1}{r}\frac{dR}{dr}+\frac{1}{r^2}\frac{d^2R}{d\theta^2}-k^2R=0;$$

but since the electric field is symmetrical about the axis of z, R is independent of θ, hence this equation becomes

$$\frac{d^2R}{dr^2}+\frac{1}{r}\frac{dR}{dr}-k^2R=0,$$

the solution of which by Art. 261 is, C and D being constants,

$$R=\{CJ_0(\iota kr)+DK_0(\iota kr)\}\,\epsilon^{\iota(mz+pt)}. \tag{4}$$

Both the J and K functions have to be included, as r can neither vanish nor become infinite in the dielectric. This equation indicates the presence of converging and diverging waves of Faraday tubes in the dielectric. If the currents in the wire are in planes through the axis of z, and if S is the component of the electromotive intensity along r, then

$$P = S\frac{x}{r}, \quad Q = S\frac{y}{r};$$

hence, since S is a function of r, z, and t, and not of θ, we may write

$$P = \frac{d\chi}{dx}, \quad Q = \frac{d\chi}{dy}; \tag{5}$$

where χ is a function we proceed to determine. Since P and Q satisfy equations of the form

$$\left.\begin{aligned} &\frac{d^2P}{dx^2} + \frac{d^2P}{dy^2} - k^2P = 0, \\ \text{we have} \quad &\frac{d^2\chi}{dx^2} + \frac{d^2\chi}{dy^2} - k^2\chi = 0. \\ \text{But} \quad &\frac{dP}{dx} + \frac{dQ}{dy} + \frac{dR}{dz} = 0, \end{aligned}\right\} \tag{6}$$

so that by equations (5) and (6)

$$k^2\chi + \frac{dR}{dz} = 0.$$

We thus have the following expressions for P, Q, R,

$$\left.\begin{aligned} P &= -\frac{\iota m}{k^2}\frac{d}{dx}\{CJ_0(\iota kr) + DK_0(\iota kr)\}\,\epsilon^{\iota(mz+pt)}, \\ Q &= -\frac{\iota m}{k^2}\frac{d}{dy}\{CJ_0(\iota kr) + DK_0(\iota kr)\}\,\epsilon^{\iota(mz+pt)}, \\ R &= \{CJ_0(\iota kr) + DK_0(\iota kr)\}\,\epsilon^{\iota(mz+pt)}. \end{aligned}\right\} \tag{7}$$

To find a, b, c, the components of the magnetic induction, we have

$$\frac{da}{dt} = \frac{dQ}{dz} - \frac{dR}{dy},$$

$$\frac{db}{dt} = \frac{dR}{dx} - \frac{dP}{dz},$$

$$\frac{dc}{dt} = \frac{dP}{dy} - \frac{dQ}{dx}.$$

From these equations we find

$$\left.\begin{aligned} a &= \frac{(m^2-k^2)}{\iota pk^2}\frac{d}{dy}\{CJ_0(\iota kr)+DK_0(\iota kr)\}\,\epsilon^{\iota(mz+pt)}, \\ b &= -\frac{(m^2-k^2)}{\iota pk^2}\frac{d}{dx}\{CJ_0(\iota kr)+DK_0(\iota kr)\}\,\epsilon^{\iota(mz+pt)}, \\ c &= 0; \end{aligned}\right\} \quad (8)$$

thus the resultant magnetic induction is equal to

$$\frac{m^2-k^2}{\iota pk^2}\frac{d}{dr}\{CJ_0(\iota kr)+DK_0(\iota kr)\}\,\epsilon^{\iota(mz+pt)},$$

and the lines of magnetic force are circles with their centres along the axis of z and their planes at right angles to it.

We now proceed to consider the wire. The differential equation satisfied by R in the wire is

$$\frac{d^2R}{dx^2}+\frac{d^2R}{dy^2}+\frac{d^2R}{dz^2}=\frac{4\pi\mu}{\sigma}\frac{dR}{dt}.$$

Transforming this equation to cylindrical coordinates it becomes, since R is independent of θ,

$$\frac{d^2R}{dr^2}+\frac{1}{r}\frac{dR}{dr}-n^2R=0,$$

where, as usual, $$n^2=m^2+\frac{4\pi\mu\iota p}{\sigma}.$$

Since r can vanish in the wire, the solution of this equation is

$$R=AJ_0(\iota nr)\,\epsilon^{\iota(mz+pt)},$$

where A is a constant.

We can deduce the expressions for P and Q from R in the same way as for the dielectric, and we find

$$\left.\begin{aligned} P &= -\frac{\iota m}{n^2}A\frac{d}{dx}J_0(\iota nr)\,\epsilon^{\iota(mz+pt)}, \\ Q &= -\frac{\iota m}{n^2}A\frac{d}{dy}J_0(\iota nr)\,\epsilon^{\iota(mz+pt)}, \\ R &= AJ_0(\iota nr)\,\epsilon^{\iota(mz+pt)}; \end{aligned}\right\} \quad (9)$$

and also

$$\left.\begin{aligned} a &= \frac{m^2-n^2}{\iota pn^2}A\frac{d}{dy}J_0(\iota nr)\,\epsilon^{\iota(mz+pt)}, \\ b &= -\frac{m^2-n^2}{\iota pn^2}A\frac{d}{dx}J_0(\iota nr)\,\epsilon^{\iota(mz+pt)}, \\ c &= 0. \end{aligned}\right\} \quad (10)$$

The resultant magnetic induction is at right angles to r and z and equal to

$$\frac{m^2-n^2}{\iota pn^2} A \frac{d}{dr} J_0(\iota nr)\,\epsilon^{\iota(mz+pt)}.$$

In the outer conductor the differential equations are of the same form, but their solution will be expressed by the K functions and not by the J's, since r can be infinite in the outer conductor. We find if

$$n'^2 = m^2 + \frac{4\pi\mu'\iota p}{\sigma'},$$

that in the outer conductor, E being a constant,

$$\left.\begin{aligned} P &= -\frac{\iota m}{n'^2} E \frac{d}{dx} K_0(\iota n'r)\,\epsilon^{\iota(mz+pt)}, \\ Q &= -\frac{\iota m}{n'^2} E \frac{d}{dy} K_0(\iota n'r)\,\epsilon^{\iota(mz+pt)}, \\ R &= EK_0(\iota n'r)\,\epsilon^{\iota(mz+pt)}; \end{aligned}\right\} \qquad (11)$$

$$\left.\begin{aligned} a &= \frac{m^2-n'^2}{\iota pn'^2} E \frac{d}{dy} K_0(\iota n'r)\,\epsilon^{\iota(mz+pt)}, \\ b &= -\frac{m^2-n'^2}{\iota pn'^2} E \frac{d}{dx} K_0(\iota n'r)\,\epsilon^{\iota(mz+pt)}, \\ c &= 0. \end{aligned}\right\} \qquad (12)$$

The resultant magnetic induction is equal to

$$\frac{m^2-n'^2}{\iota pn'^2} E \frac{d}{dr} K_0(\iota n'r)\,\epsilon^{\iota(mz+pt)}.$$

The boundary conditions at the surfaces of separation of the dielectric and the metals are (1) that the electromotive intensity parallel to the surface of separation is continuous, (2) that the magnetic *force* parallel to the surface is also continuous. Hence if $\mathbf{a}$, $\mathbf{b}$ are respectively the inner and outer radii of the layer of dielectric, condition (1) gives

$$\left.\begin{aligned} AJ_0(\iota n\mathbf{a}) &= CJ_0(\iota k\mathbf{a}) + DK_0(\iota k\mathbf{a}), \\ EK_0(\iota n'\mathbf{b}) &= CJ_0(\iota k\mathbf{b}) + DK_0(\iota k\mathbf{b}). \end{aligned}\right\} \qquad (13)$$

Condition (2) gives, writing $J_0'(x)$ for $dJ_0(x)/dx$, and $K_0'(x)$ for $dK_0(x)/dx$, and substituting for m^2-n^2, m^2-k^2, $m^2-n'^2$ the values $-4\pi\mu\iota p/\sigma$, p^2/V^2, $-4\pi\mu'\iota p/\sigma'$ respectively,

$$\left.\begin{aligned} \frac{4\pi\iota}{\sigma n} AJ_0'(\iota n\mathbf{a}) &= -\frac{p}{V^2 k}\{CJ_0'(\iota k\mathbf{a}) + DK_0'(\iota k\mathbf{a})\}, \\ \frac{4\pi\iota}{\sigma' n'} EK_0'(\iota n'\mathbf{b}) &= -\frac{p}{V^2 k}\{CJ_0'(\iota k\mathbf{b}) + DK_0'(\iota k\mathbf{b})\}. \end{aligned}\right\} \qquad (14)$$

Eliminating A and E from equations (13) and (14), we get

$$C\left(\frac{4\pi\iota}{\sigma n}J_0'(\iota n\mathbf{a})J_0(\iota k\mathbf{a})+\frac{p}{V^2k}J_0(\iota n\mathbf{a})J_0'(\iota k\mathbf{a})\right)$$

$$+D\left(\frac{4\pi\iota}{\sigma n}J_0'(\iota n\mathbf{a})K_0(\iota k\mathbf{a})+\frac{p}{V^2k}J_0(\iota n\mathbf{a})K_0'(\iota k\mathbf{a})\right)=0,$$

$$C\left(\frac{4\pi\iota}{\sigma'n'}K_0'(\iota n'\mathbf{b})J_0(\iota k\mathbf{b})+\frac{p}{V^2k}K_0(\iota n'\mathbf{b})J_0'(\iota k\mathbf{b})\right)$$

$$+D\left(\frac{4\pi\iota}{\sigma'n'}K_0'(\iota n'\mathbf{b})K_0(\iota k\mathbf{b})+\frac{p}{V^2k}K_0(\iota n'\mathbf{b})K_0'(\iota k\mathbf{b})\right)=0.$$

Eliminating C and D from these equations, we get

$$\left(\frac{4\pi\iota}{\sigma n}J_0'(\iota n\mathbf{a})J_0(\iota k\mathbf{a})+\frac{p}{V^2k}J_0(\iota n\mathbf{a})J_0'(\iota k\mathbf{a})\right)\times$$

$$\left(\frac{4\pi\iota}{\sigma'n'}K_0'(\iota n'\mathbf{b})K_0(\iota k\mathbf{b})+\frac{p}{V^2k}K_0(\iota n'\mathbf{b})K_0'(\iota k\mathbf{b})\right)$$

$$=\left(\frac{4\pi\iota}{\sigma n}J_0'(\iota n\mathbf{a})K_0(\iota k\mathbf{a})+\frac{p}{V^2k}J_0(\iota n\mathbf{a})K_0'(\iota k\mathbf{a})\right)\times$$

$$\left(\frac{4\pi\iota}{\sigma'n'}K_0'(\iota n'\mathbf{b})J_0(\iota k\mathbf{b})+\frac{p}{V^2k}K_0(\iota n'\mathbf{b})J_0'(\iota k\mathbf{b})\right). \quad (15)$$

This equation gives the relation between the wave length $2\pi/m$ along the wire and the frequency $p/2\pi$ of the vibration. To simplify this equation, we notice that $k\mathbf{a}$, $k\mathbf{b}$ are both very small quantities, for, as we shall subsequently find, k, when the electrical waves are very long, is inversely proportional to the wave length, while when the waves are short k is small compared with the reciprocal of the wave length; we may therefore assume that when the waves transmitted along the cable are long compared with its radii, $k\mathbf{a}$ and $k\mathbf{b}$ are very small. But in this case we have approximately,

$$J_0(\iota k\mathbf{a})=1,\quad J_0(\iota k\mathbf{b})=1,$$

$$J_0'(\iota k\mathbf{a})=-\tfrac{1}{2}\iota k\mathbf{a},\quad J_0'(\iota k\mathbf{b})=-\tfrac{1}{2}\iota k\mathbf{b};$$

$$K_0(\iota k\mathbf{a})=\log\frac{2\gamma}{\iota k\mathbf{a}},\quad K_0(\iota k\mathbf{b})=\log\frac{2\gamma}{\iota k\mathbf{b}},$$

$$K_0'(\iota k\mathbf{a})=-\frac{1}{\iota k\mathbf{a}},\quad K_0'(\iota k\mathbf{b})=-\frac{1}{\iota k\mathbf{b}}.$$

Making these substitutions, equation (15) reduces to

$$k^2 = -\frac{p}{4\pi V^2}\Big[\sigma n\Big(\frac{1}{\mathbf{a}} + \tfrac{1}{2}k^2\mathbf{a}\log\frac{2\gamma}{\iota k\mathbf{b}}\Big)\frac{J_0(\iota n\mathbf{a})}{J_0'(\iota n\mathbf{a})}$$
$$-\sigma' n'\Big(\frac{1}{\mathbf{b}} + \tfrac{1}{2}k^2\mathbf{b}\log\frac{2\gamma}{\iota k\mathbf{a}}\Big)\frac{K_0(\iota n'\mathbf{b})}{K_0'(\iota n'\mathbf{b})}$$
$$-\frac{p}{8\pi V^2}\frac{\sigma n}{\mathbf{a}}\frac{\sigma' n'}{\mathbf{b}}(\mathbf{b}^2-\mathbf{a}^2)\frac{J_0(\iota n\mathbf{a})K_0(\iota n'\mathbf{b})}{J_0'(\iota n\mathbf{a})K_0'(\iota n'\mathbf{b})}\Big]\frac{1}{\log(\mathbf{b}/\mathbf{a})}.$$

Now since both $k\mathbf{a}$ and $k\mathbf{b}$ are very small,

$$k^2\mathbf{a}^2\log\frac{2\gamma}{\iota k\mathbf{b}}, \quad k^2\mathbf{b}^2\log\frac{2\gamma}{\iota k\mathbf{a}}$$

will be exceedingly small quantities unless $\mathbf{a}$ is so much smaller than $\mathbf{b}$ that $\log(2\gamma/\iota k\mathbf{a})$ is comparable with $1/k^2\mathbf{b}^2$. This would require such a disproportion between $\mathbf{b}$ and $\mathbf{a}$ as to be scarcely realizable in practice on a planet of the size of the earth; we may therefore write the preceding equation in the form

$$k^2 = -\frac{\iota p^2}{V^2}\Big[\frac{\mu}{n\mathbf{a}}\frac{J_0(\iota n\mathbf{a})}{J_0'(\iota n\mathbf{a})} - \frac{\mu'}{n'\mathbf{b}}\frac{K_0(\iota n'\mathbf{b})}{K_0'(\iota n'\mathbf{b})}$$
$$-\tfrac{1}{2}\iota\frac{p^2}{V^2}(\mathbf{b}^2-\mathbf{a}^2)\frac{\mu}{n\mathbf{a}}\frac{\mu'}{n'\mathbf{b}}\frac{J_0(\iota n\mathbf{a})}{J_0'(\iota n\mathbf{a})}\frac{K_0(\iota n'\mathbf{b})}{K_0'(\iota n'\mathbf{b})}\Big]\frac{1}{\log(\mathbf{b}/\mathbf{a})}\cdots, \quad (16)$$

where we have put $n^2 = 4\pi\mu\iota p/\sigma$, $n'^2 = 4\pi\mu'\iota p/\sigma'$. We showed in Art. 258 that we were justified in doing this when the electrical vibrations are not so rapid as to be comparable in frequency with those of light.

We see from (16) that k^2 is given by an equation of the form

$$k^2 = -\frac{\iota p^2}{V^2}\Big(\xi - \eta - \tfrac{1}{2}\frac{\iota p^2}{V^2}(\mathbf{b}^2-\mathbf{a}^2)\,\xi\eta\Big). \quad (16^*)$$

We remark that for all electrical oscillations whose wave lengths are large compared with the radii of the cable, $p^2(\mathbf{b}^2-\mathbf{a}^2)/V^2$ is an exceedingly small quantity, since it is of the order $(\mathbf{b}^2-\mathbf{a}^2)/\lambda^2$, where λ is the length of the electrical wave.

In equation (16*) we see that we can neglect the third term inside the bracket as long as both ξ and η are small compared with $2V^2/p^2(\mathbf{b}^2-\mathbf{a}^2)$.

Now
$$\xi = \frac{\mu}{n\mathbf{a}}\frac{J_0(\iota n\mathbf{a})}{J_0'(\iota n\mathbf{a})},$$

so that the large values of ξ occur when $n\mathbf{a}$ is small; and in this case, substituting the approximate values for J_0 and J_0', we see that

$$\xi = -\frac{2\mu}{\iota n^2 \mathbf{a}^2} = \frac{\sigma}{2\pi p \mathbf{a}^2} = \frac{V^2}{4\pi p^2(\mathbf{b}^2-\mathbf{a}^2)} \frac{2\sigma p}{V^2} \frac{\mathbf{b}^2-\mathbf{a}^2}{\mathbf{a}^2}.$$

Now for cables of practicable dimensions and materials conveying oscillations slower than those of light $2\sigma p(\mathbf{b}^2-\mathbf{a}^2)/V^2\mathbf{a}^2$ is an exceedingly small quantity, so that for such cases ξ is very small compared with $2V^2/p^2(\mathbf{b}^2-\mathbf{a}^2)$.

Again,
$$\eta = \frac{\mu'}{n'\mathbf{b}} \frac{K_0(\iota n'\mathbf{b})}{K_0'(\iota n'\mathbf{b})};$$

the large values of η occur when $n'\mathbf{b}$ is small. Substituting the approximate values for K_0, K_0' we find

$$\eta = -\iota\mu' \log\left(\frac{2\gamma}{\iota n'\mathbf{b}}\right).$$

This is very small compared with $\mu'/n'\mathbf{b}$, and may, as in the preceding case, be shown for all practicable cases to be very small compared with $2V^2/p^2(\mathbf{b}^2-\mathbf{a}^2)$. Hence, as both ξ and η are small compared with this quantity, we may neglect the third term inside the bracket in equation (16), which thus reduces to

$$k^2 = -\frac{\iota p^2}{V^2}\left\{\frac{\mu}{n\mathbf{a}} \frac{J_0(\iota n\mathbf{a})}{J_0'(\iota n\mathbf{a})} - \frac{\mu'}{n'\mathbf{b}} \frac{K_0(\iota n'\mathbf{b})}{K_0'(\iota n'\mathbf{b})}\right\} \frac{1}{\log(\mathbf{b}/\mathbf{a})}. \qquad (17)$$

We shall now proceed to deduce from this equation the velocity of propagation of electrical oscillations of different frequencies.

Slowly Alternating Currents.

263.] The first case we shall consider is the one where the frequency is so small that $n\mathbf{a}$ is a small quantity. In this case, since we have approximately

$$J_0(\iota n\mathbf{a})/J_0'(\iota n\mathbf{a}) = -2/\iota n\mathbf{a},$$

equation (17) becomes

$$k^2 = -\frac{\iota p^2}{V^2}\left\{\frac{2\iota\mu}{n^2\mathbf{a}^2} - \frac{\mu'}{n'\mathbf{b}} \frac{K_0(\iota n'\mathbf{b})}{K_0'(\iota n'\mathbf{b})}\right\} \frac{1}{\log(\mathbf{b}/\mathbf{a})}. \qquad (18)$$

The first term inside the bracket is very large, for it is equal to $2\iota\mu/n^2\mathbf{a}^2$ and $n\mathbf{a}$ is small; the second term in the bracket vanishes if $\mathbf{b}$ is infinite, and even if $\mathbf{b}$ is so small that $n'\mathbf{b}$ is a small quantity, we see, by substituting the values for K_0 and K_0' when the variable is small, that the ratio of the magnitude of the second term inside the bracket to that of the first is approximately equal to

$$\frac{\mu'}{2\mu} n^2 \mathbf{a}^2 \log \frac{2\gamma}{\iota n' \mathbf{b}},$$

and thus unless $n'\mathbf{b}$ is exceedingly small compared with $n\mathbf{a}$ the second term may be neglected.

Hence, since $n^2 = 4\pi\mu\iota p/\sigma$, we may write (18) in the form

$$k^2 = -\frac{p}{V^2} \frac{\iota\sigma}{2\pi\mathbf{a}^2} \frac{1}{\log(\mathbf{b}/\mathbf{a})},$$

but $k^2 = m^2 - \frac{p^2}{V^2}$, so that

$$m^2 = \frac{p^2}{V^2}\left\{1 - \frac{\iota\sigma}{2\pi p\mathbf{a}^2} \frac{1}{\log(\mathbf{b}/\mathbf{a})}\right\};$$

we have seen however that the second term in the bracket is large compared with unity, so that we have approximately

$$m^2 = -\frac{p}{V^2} \frac{\iota\sigma}{2\pi\mathbf{a}^2} \frac{1}{\log(\mathbf{b}/\mathbf{a})}.$$

If $\mathbf{R}$ is the resistance and Γ the capacity in electromagnetic measure per unit length of the wire, then since

$$\mathbf{R} = \frac{\sigma}{\pi\mathbf{a}^2}, \quad \Gamma = \frac{1}{2V^2\log(\mathbf{b}/\mathbf{a})},$$

we have $$m^2 = -\iota p \mathbf{R}\Gamma,$$

or $$m = -\{p\mathbf{R}\Gamma\}^{\frac{1}{2}}\left\{\frac{1}{\sqrt{2}} - \frac{\iota}{\sqrt{2}}\right\},$$

where the sign has been taken so as to make the real part of ιm negative. The reason for this is as follows: if $m = -a + \iota\beta$, R, the electromotive intensity parallel to the axis of the wire, will be expressed by terms of the form

$$\cos(-az + pt)\,\epsilon^{-\beta z}.$$

This represents a vibration whose phases propagated with the velocity p/a in the positive direction of z, and which dies away to $1/\epsilon$ of its original value after passing over a distance $1/\beta$; if β were negative the disturbance would increase indefinitely as it travelled along the wire. Substituting the value of a, we see that the velocity of propagation of the phases is

$$\left\{\frac{2p}{\mathbf{R}\Gamma}\right\}^{\frac{1}{2}};$$

thus the velocity of propagation is directly proportional to the square root of the frequency and inversely proportional to the square root of the product of the resistance and capacity of the wire per unit length.

The distance to which a disturbance travels before falling to $1/\epsilon$ of its original value is, on substituting the value of β, seen to be

$$\left\{\frac{2}{\mathbf{R}\Gamma p}\right\}^{\frac{1}{2}};$$

thus the distance to which a disturbance travels is inversely proportional to the square root of the product of the frequency, the resistance, and capacity per unit length.

If we take the case of a cable transmitting telephone messages of such a kind that $2\pi/p$, the period of the electrical vibrations, is 1/100 of a second, then if the copper core is 4 millimetres in diameter and the external radius of the guttapercha covering about 2·5 times that of the core, $\mathbf{R}$ is about $1{\cdot}3 \times 10^{-5}$ Ohms, or in absolute measure $1{\cdot}3 \times 10^4$. Γ is about 15×10^{-22}. Substituting these values for $\mathbf{R}$ and Γ, we find that the vibrations will travel over about 128 kilometres before falling to $1/\epsilon$ of their initial value. The velocity of propagation of the phases is about 80,000 kilometres per second. If we take an iron telegraph wire 4 mm. in diameter, $\mathbf{R}$ is about $9{\cdot}4 \times 10^4$; the capacity of such a wire placed 4 metres above the ground is stated by Hagenbach (*Wied. Ann.* 29. p. 377, 1886) to be about 10^{-22} per centimetre, hence the distance to which electrical vibrations making 100 vibrations per second would travel before falling to $1/\epsilon$ of their original value would be $\{1{\cdot}3 \times 15/9{\cdot}4\}^{\frac{1}{2}}$, or 1·43 times the distance in the preceding case: thus the messages along the aerial wire would travel about half as far again as those along the cable, the increased resistance of the iron telegraph wire being more than counterbalanced by the smaller electrostatic capacity. Since vibrations of different frequencies die away at different rates, a message such as a telephone message which is made up of vibrations whose frequencies extend over a somewhat wide range will lose its character as soon as there is any appreciable decay in the vibrations. We see from this investigation that the lower the pitch the further will the vibrations travel, so that when a piece of music is transmitted along a telephone wire the high notes suffer the most.

264.] We shall now proceed to consider the expressions when $n\mathbf{a}$ is small for the electromotive intensity and magnetic induction in the wire and dielectric in terms of the total current flowing through any cross section of the wire.

We have seen that

$$m = -\{p\mathbf{R}\Gamma\}^{\frac{1}{2}}\left\{\frac{1}{\sqrt{2}} - \frac{\iota}{\sqrt{2}}\right\},$$

hence if $$a = \{\tfrac{1}{2}p\mathbf{R}\Gamma\}^{\frac{1}{2}},$$

we may suppose that the current through the wire at z is equal

to $$I_0 \epsilon^{-az}\cos(-az+pt).$$

This is equal to $$\int_0^{\mathbf{a}} \frac{R}{\sigma} 2\pi r dr,$$

so that in this case we find by equation (9), since $J_0(\iota nr)$ can be replaced by unity as nr is small,

$$A = \frac{\sigma I_0}{\pi \mathbf{a}^2},$$

so that by (9) we have approximately

$$R = \frac{\sigma I_0}{\pi \mathbf{a}^2}\epsilon^{-az}\cos(-az+pt).$$

Thus the electromotive intensity, and therefore the current parallel to z, is uniformly distributed over the cross-section. The electromotive intensity along the radius, $\{P^2+Q^2\}^{\frac{1}{2}}$, is easily found by equation (9) to be

$$-\frac{\iota m}{2}\frac{\sigma I_0}{\pi \mathbf{a}^2} r\epsilon^{-az}\epsilon^{\iota(-az+pt)}.$$

Substituting the value of m and taking the real part, we see that it is equal to

$$\left\{\frac{p\sigma\Gamma}{\pi \mathbf{a}^2}\right\}\frac{\sigma I_0}{2\pi \mathbf{a}^2} r\epsilon^{-az}\cos\left(-az+pt+\frac{\pi}{4}\right),$$

it is thus very small compared with the intensity along the axis of the wire, so that in the wire the Faraday tubes are approximately parallel to the axis of the wire.

The magnetic induction in this case reduces approximately to

$$\frac{2\mu I_0}{\mathbf{a}^2} r\epsilon^{-az}\cos(-az+pt).$$

In the dielectric, we have by equations (7), (13), and (14), assuming that kr is small,

$$R = \frac{\sigma}{\pi a^2} I_0 \left\{1 - 2V^2\Gamma \log\frac{r}{a}\right\} \epsilon^{-az} \cos(-az+pt),$$

since from (13) and (14) $D = 2\Gamma V^2 A$.

The electromotive intensity along the radius, $\{P^2+Q^2\}^{\frac{1}{2}}$, is equal to

$$2V^2\{\pi a^2\Gamma/p\sigma\}^{\frac{1}{2}} \frac{\sigma}{\pi a^2} \frac{1}{r} I_0 \epsilon^{-az} \cos\left(-az+pt-\frac{\pi}{4}\right).$$

In this case the radial electromotive intensity is very large compared with the tangential intensity, so that in the dielectric the Faraday tubes are approximately at right angles to the wire.

The resultant magnetic induction is equal to

$$\frac{2I_0}{r} \epsilon^{-az} \cos(-az+pt).$$

265.] The interpretation of (17) is easy when na is very small, since in this case the first term inside the bracket is very large compared with the second; as na increases the discussion of the equation becomes more difficult, since the second term in the bracket is becoming comparable with the first. It will facilitate the discussion of the equation if we consider the march of the function $\iota n a J_0(\iota n a)/J_0'(\iota n a)$. Perhaps the simplest way to do this is to expand the function $xJ_0(x)/J_0'(x)$ in powers of x. Since $J_0(x)$ is a Bessel's function of zero order, we have

$$J_0''(x) + \frac{1}{x}J_0'(x) + J_0(x) = 0,$$

so that

$$\frac{xJ_0(x)}{J_0'(x)} = -1 - \frac{xJ_0''(x)}{J_0'(x)}$$

$$= -1 - x\frac{d}{dx}\log J_1(x),$$

since $J_0'(x) = -J_1(x)$, $J_1(x)$ being Bessel's function of the first order.

Let $0, x_1, x_2, x_3 \ldots$ be the positive roots of the equation

$$J_1(x) = 0,$$

then

$$J_1(x) = x\left(1-\frac{x^2}{x_1^2}\right)\left(1-\frac{x^2}{x_2^2}\right)\left(1-\frac{x^2}{x_3^2}\right)\ldots$$

so that

$$\frac{d}{dx}\log J_1(x) = \frac{1}{x} - \frac{2x}{x_1^2\left(1 - \frac{x^2}{x_1^2}\right)} - \frac{2x}{x_2^2\left(1 - \frac{x^2}{x_2^2}\right)} - \dots,$$

and therefore

$$x\frac{d}{dx}\log J_1(x) = 1 - \frac{2x^2}{x_1^2}\left(1 + \frac{x^2}{x_1^2} + \frac{x^4}{x_1^4} + \dots\right)$$
$$- \frac{2x^2}{x_2^2}\left(1 + \frac{x^2}{x_2^2} + \frac{x^4}{x_2^4} + \dots\right) + \dots$$
$$= 1 - 2x^2\left(\frac{1}{x_1^2} + \frac{1}{x_2^2} + \frac{1}{x_3^2} + \dots\right)$$
$$-2x^4\left(\frac{1}{x_1^4} + \frac{1}{x_2^4} + \frac{1}{x_3^4} + \dots\right) - \dots.$$

Thus if S_n denotes the sum of the reciprocals of the nth powers of the roots of the equation

$$J_1(x)/x = 0,$$

we have

$$\frac{xJ_0(x)}{J_0'(x)} = -2 + 2S_2x^2 + 2S_4x^4 + 2S_6x^6 + \dots.$$

Now the equation $J_1(x)/x = 0$, when expanded in powers of x is,

$$1 - \frac{x^2}{2.4} + \frac{x^4}{2.4.4.6} - \frac{x^6}{2.4.4.6.6.8} + \dots = 0.$$

Hence, if we calculate S_2, S_4, S_6 &c. by Newton's Rule, we find

$$S_2 = \frac{1}{8}, \quad S_4 = \frac{1}{12\times 16}, \quad S_6 = \frac{1}{12\times 16^2},$$

$$S_8 = \frac{1}{12\times 15\times 16^2}, \quad S_{10} = \frac{13}{9\times 15\times 16^4};$$

hence

$$\frac{xJ_0(x)}{J_0'(x)} = -2 + \frac{x^2}{4} + \frac{x^4}{96} + \frac{x^6}{1536} + \frac{x^8}{23040} + \frac{13x^{10}}{4423680} - \dots,$$

so that

$$\iota n\mathbf{a}\frac{J_0(\iota n\mathbf{a})}{J_0'(\iota n\mathbf{a})} = -2 - \frac{n^2\mathbf{a}^2}{4} + \frac{n^4\mathbf{a}^4}{96} - \frac{n^6\mathbf{a}^6}{1536} + \frac{n^8\mathbf{a}^8}{23040} - \frac{13n^{10}\mathbf{a}^{10}}{4423680},$$

and since $n^2 = 4\pi\mu\iota p/\sigma$ approximately, we have

$$\iota n \mathrm{a} \frac{J_0(\iota n \mathrm{a})}{J_0'(\iota n \mathrm{a})} = -2 - \frac{1}{96}(4\pi\mu p \mathrm{a}^2/\sigma)^2 + \frac{1}{23040}(4\pi\mu p \mathrm{a}^2/\sigma)^4 \ldots$$
$$- \iota \left\{ \frac{1}{4}(4\pi\mu p \mathrm{a}^2/\sigma) - \frac{1}{1536}(4\pi\mu p \mathrm{a}^2/\sigma)^3 + \frac{13}{4423680}(4\pi\mu p \mathrm{a}^2/\sigma)^5 \ldots \right\}. \quad (19)$$

The values of $\iota n \mathrm{a} J_0(\iota n \mathrm{a})/J_0'(\iota n \mathrm{a})$ for a few values of $4\pi\mu p \mathrm{a}^2/\sigma$ are given in the following table:—

$4\pi\mu p \mathrm{a}^2/\sigma$	$\iota n \mathrm{a} J_0(\iota n \mathrm{a})/J_0'(\iota n \mathrm{a})$
·5	$-2\{1{\cdot}001 + {\cdot}062\iota\}$
1	$-2\{1{\cdot}005 + {\cdot}125\iota\}$
1·5	$-2\{1{\cdot}012 + {\cdot}186\iota\}$
2	$-2\{1{\cdot}021 + {\cdot}25\iota\}$
2·5	$-2\{1{\cdot}032 + {\cdot}31\iota\}$
3	$-2\{1{\cdot}045 + 37\iota\}$

From this table we see that even when $4\pi\mu p \mathrm{a}^2/\sigma$ is as large as unity, we may still as an approximation put

$$\iota n \mathrm{a} J_0(\iota n \mathrm{a})/J_0'(\iota n \mathrm{a})$$

equal to -2, and k^2 will continue to be given by (18).

266.] We must consider now the relative values of the terms inside the bracket in (18) when $n\mathrm{a}$ is comparable with unity. In the case of aerial telegraph wires it is conceivable that there may be cases in which though $n\mathrm{a}$ is not large $n'\mathrm{b}$ may be so; but when this is the case we have by Art. 261

$$K_0(\iota n' \mathrm{b}) = -\iota K_0'(\iota n' b),$$

so that since $n'\mathrm{b}$ is very large the second term inside the bracket in equation (18) will be small compared with the first, hence we have

$$k^2 = -\frac{p}{V^2} \frac{\iota\sigma}{2\pi \mathrm{a}^2} \frac{1}{\log(\mathrm{b}/\mathrm{a})},$$

which is the same value as in Art. 263.

In all telegraph cables where the external conductor is water, and in all but very elevated telegraph wires where the external conductor is wet earth, the value of σ' will so greatly exceed that of σ that unless b is more than a thousand times as great as a, $n'\mathsf{b}$ will be very small if the value of $n\mathsf{a}$ is comparable with unity. In this case however by Art. 261,

$$\frac{K_0(\iota n'\mathsf{b})}{K_0'(\iota n'\mathsf{b})} = -\iota n'\mathsf{b}\log\frac{2\gamma}{\iota n'\mathsf{b}},$$

so that equation (18) becomes

$$k^2 = \frac{p^2}{V^2}\left\{\frac{2\mu}{n^2\mathsf{a}^2} + \mu'\log\frac{2\gamma}{\iota n'\mathsf{b}}\right\}\frac{1}{\log(\mathsf{b}/\mathsf{a})}.$$

Since $n'\mathsf{b}$ is very small while $n\mathsf{a}$ is comparable with unity, the second term inside the brackets will be very large compared with the first, hence this equation may be written

$$k^2 = \frac{p^2}{V^2}\log\frac{2\gamma}{\iota n'\mathsf{b}}\frac{\mu'}{\log(\mathsf{b}/\mathsf{a})}; \tag{20}$$

or $$m^2 = \frac{p^2}{V^2}\left\{1 + \log\frac{2\gamma}{\iota n'\mathsf{b}}\frac{\mu'}{\log(\mathsf{b}/\mathsf{a})}\right\}.$$

Thus approximately

$$m^2 = \frac{p^2}{V^2}\log\frac{2\gamma}{\iota n'\mathsf{b}}\frac{\mu'}{\log(\mathsf{b}/\mathsf{a})},$$

and since $$n'^2 = 4\pi\mu'\iota p/\sigma',$$

$$m^2 = \tfrac{1}{2}\frac{p^2}{V^2}\left\{\log\frac{\sigma'\gamma^2}{\mu'\pi\mathsf{b}^2 p} + \frac{\iota\pi}{2}\right\}\frac{\mu'}{\log(\mathsf{b}/\mathsf{a})},$$

hence we have approximately

$$m = \frac{1}{\sqrt{2}}\frac{p}{V}\left\{\frac{\mu'\log\dfrac{\sigma'\gamma^2}{\mu'\pi\mathsf{b}^2 p}}{\log(\mathsf{b}/\mathsf{a})}\right\}^{\frac{1}{2}}\left\{1 + \iota\frac{\pi}{4}\frac{1}{\log(\sigma'\gamma^2/\mu'\pi\mathsf{b}^2 p)}\right\}, \tag{21}$$

where the plus sign has been taken so as to make the real part of ιm negative. This equation corresponds to a vibration whose phases are propagated with the velocity

$$V\left\{\frac{\log(\mathsf{b}^2/\mathsf{a}^2)}{\mu'\log(\sigma'\gamma^2/\mu'\pi\mathsf{b}^2 p)}\right\}^{\frac{1}{2}},$$

and which fades away to $1/\epsilon$ of its original value after passing over a distance

$$\frac{4}{\pi}\frac{V}{p\mu'^{\frac{1}{2}}}\{\log(\mathsf{b}^2/\mathsf{a}^2)\times\log(\sigma'\gamma^2/\mu'\pi\mathsf{b}^2 p)\}^{\frac{1}{2}}.$$

This case presents many striking peculiarities. In the first place we see that to our order of approximation both the velocity of propagation of the phases and the rate of decay of the vibrations are independent of the resistance of the wire. These quantities depend somewhat on the resistance of the external conductor, but only to a comparatively small extent even on that, as σ' only enters their expressions as a logarithm. The velocity of propagation of the phases only varies slowly with the frequency, as p only occurs in its expression as a logarithm. The rate of decay, i. e. the real part of ιm, is proportional to the frequency and thus varies more rapidly with this quantity than when $n\mathrm{a}$ is small, as in that case the rate of decay is proportional to the square root of the frequency (Art. 263). We see from the preceding investigation that for sending periodic disturbances along a cable, the frequency being such as to make $n'\mathrm{b}$ a very small quantity, we do not gain any appreciable advantage by making the core of a good conductor like copper rather than of an inferior one like iron *unless* the conditions are such as to make $n\mathrm{a}$ small compared with unity. We see too that the distance to which the disturbance travels before it falls to $1/\epsilon$ of its original value increases with the resistance of the external conductor. We shall show in a subsequent article that the heat produced per second in the external conductor is very large compared with that produced in the same time in the wire, thus the dissipation of energy is controlled by the external conductor and not by the wire.

The preceding results will continue true as long as $n'\mathrm{b}$ is small, even though the frequency of the electrical vibrations gets so large that $n\mathrm{a}/\mu$ is a very large quantity; for when $n\mathrm{a}$ is large we have by Art. 261,

$$J'_0(\iota n\mathrm{a}) = -\iota J_0(\iota n\mathrm{a}),$$

so that equation (16) becomes

$$k^2 = \frac{p^2}{V^2}\left\{\frac{\mu}{n\mathrm{a}} + \mu' \log\frac{2\gamma}{\iota n'\mathrm{b}}\right\}\frac{1}{\log(\mathrm{b}/\mathrm{a})}.$$

Since $n\mathrm{a}/\mu$ is large and $n'\mathrm{b}$ small the second term inside the bracket is large compared with the first, so that we get the same value of k^2 as that given by equation (20).

267.] The next case we have to consider is that in which both

$n\mathbf{a}$ and $n'\mathbf{b}$ are very large; when this is the case we know by Art. 261 that

$$J'_0(\iota n\mathbf{a}) = -\iota J_0(\iota n\mathbf{a}), \quad K'_0(\iota n'\mathbf{b}) = \iota K_0(\iota n'\mathbf{b}).$$

Making these substitutions, equation (17) becomes

$$k^2 = \frac{p^2}{V^2}\left\{\frac{\mu}{n\mathbf{a}} + \frac{\mu'}{n'\mathbf{b}}\right\}\frac{1}{\log(\mathbf{b}/\mathbf{a})}, \tag{22}$$

or
$$m^2 = \frac{p^2}{V^2}\left\{1 + \left(\frac{\mu}{n\mathbf{a}} + \frac{\mu'}{n'\mathbf{b}}\right)\frac{1}{\log(\mathbf{b}/\mathbf{a})}\right\}$$

$$= \frac{p^2}{V^2}\left\{1 - \frac{\iota}{\sqrt{8\pi p}}\left(\sqrt{\frac{\mu\sigma}{\mathbf{a}^2}} + \sqrt{\frac{\mu'\sigma'}{\mathbf{b}^2}}\right)\frac{1}{\log(\mathbf{b}/\mathbf{a})}\right\}$$

approximately. Since the second term inside the bracket is small compared with unity, extracting the square root we have,

$$m = -\frac{p}{V}\left\{1 - \frac{\iota}{\sqrt{32\pi p}}\left(\sqrt{\frac{\mu\sigma}{\mathbf{a}^2}} + \sqrt{\frac{\mu'\sigma'}{\mathbf{b}^2}}\right)\frac{1}{\log(\mathbf{b}/\mathbf{a})}\right\}. \tag{23}$$

This represents a vibration travelling approximately with the velocity V and dying away to $1/\epsilon$ of its initial value after traversing a distance

$$4V\sqrt{\frac{2\pi}{p}}\left\{\sqrt{\frac{\mu\sigma}{\mathbf{a}^2}} + \sqrt{\frac{\mu'\sigma'}{\mathbf{b}^2}}\right\}^{-1}\log(\mathbf{b}/\mathbf{a}).$$

Since the imaginary part of m is small compared with the real part, the vibration will travel over many wave lengths before its amplitude is appreciably reduced. From the expression for the rate of decay in this case we see that when the wire is surrounded by a very much worse conductor than itself, as is practically always the case with cables, the distance to which these very rapid oscillations will travel will be governed mainly by the outside conductor, and will be almost independent of the resistance and permeability of the wire; no appreciable advantage therefore would in this case be derived by using a well-conducting but expensive material like copper for the wire. In aerial wires the decay will be governed by the conductivity of the earth rather than by that of the wire, unless the height of the wire above the ground, which we may take to be comparable with $\mathbf{b}$, is so great that $\mu'\sigma'/\mathbf{b}^2$ is not large compared with $\mu\sigma/\mathbf{a}^2$.

Experiments which confirm the very important conclusion that these rapid oscillations travel with the velocity V, that is with the velocity of light through the dielectric, will be described in the next chapter.

268.] As rapidly alternating currents are now very extensively employed, it will be useful to determine the components of the electromotive intensity both in the wire and in the dielectric in terms of the total current passing through the wire. Let this current at the point z and time t be represented by the real part of $I_0 \epsilon^{\iota(mz+pt)}$. The line integral of the magnetic force taken round any circuit is equal to 4π times the current through that circuit. Now by equation (10) the magnetic force at the surface of the wire is

$$\frac{4\pi\iota}{\sigma n} A J'_0(\iota n\mathrm{a})\,\epsilon^{\iota(mz+pt)}.$$

Since the line integral of this round the surface of the wire is equal to $4\pi I_0 \epsilon^{\iota(mz+pt)}$, we have

$$A = -\frac{\iota\sigma n}{2\pi\mathrm{a}} \frac{I_0}{J'_0(\iota n\mathrm{a})}.$$

Substituting this value for A in equation (9), we find that in the wire

$$R = -\frac{\iota\sigma n}{2\pi\mathrm{a}} \frac{I_0}{J'_0(\iota n\mathrm{a})} J_0(\iota nr)\,\epsilon^{\iota(mz+pt)}; \qquad (24)$$

where the real part of the expression on the right-hand side is to be taken. When $n\mathrm{a}$ and nr are very large, we have by Art. 261

$$J'_0(\iota n\mathrm{a}) = -\iota\frac{\epsilon^{n\mathrm{a}}}{\sqrt{2\pi n\mathrm{a}}}, \qquad J_0(\iota nr) = \frac{\epsilon^{nr}}{\sqrt{2\pi nr}};$$

substituting these values in (24), we find

$$R = \left\{\frac{\mu p\sigma}{\pi\mathrm{a}r}\right\}^{\frac{1}{2}} I_0 \epsilon^{-\{2\pi\mu p/\sigma\}^{\frac{1}{2}}(\mathrm{a}-r)} \cos(\psi), \qquad (25)$$

where $\qquad \psi = mz + pt - (2\pi\mu p/\sigma)^{\frac{1}{2}}(\mathrm{a}-r) + \frac{\pi}{4}.$

Similarly, we find by equation (9) that the radial electromotive intensity $(P^2+Q^2)^{\frac{1}{2}}$ is given by the equation

$$\{P^2+Q^2\}^{\frac{1}{2}} = -\frac{p}{V}\frac{\sigma I_0}{2\pi\sqrt{\mathrm{a}r}}\,\epsilon^{-\{2\pi\mu p/\sigma\}^{\frac{1}{2}}(\mathrm{a}-r)} \sin\left(\psi-\frac{\pi}{4}\right). \qquad (26)$$

The resultant magnetic force is by equation (10) equal to

$$\frac{2}{\sqrt{\mathrm{a}r}} I_0 \epsilon^{-\{2\pi\mu p/\sigma\}^{\frac{1}{2}}(\mathrm{a}-r)} \cos\left(\psi-\frac{\pi}{4}\right).$$

Since all these expressions contain the factor $\epsilon^{-(2\pi\mu p/\sigma)^{\frac{1}{2}}(\mathbf{a}-r)}$, we see that the magnitudes of the electromotive intensity and of the magnetic force must, since $n\mathbf{a}$—and therefore $(2\pi\mu p/\sigma)^{\frac{1}{2}}\mathbf{a}$—is by hypothesis very large, diminish very rapidly as the distance from the surface of the wire increases. The maximum values of these quantities at the distance $(\sigma/2\pi\mu p)^{\frac{1}{2}}$ from the surface are only $1/\epsilon$ of their values at the boundary, and they diminish in geometrical progression as the distance from the surface increases in arithmetical progression. Thus the currents and magnetic forces are, as in Art. 258, practically confined to a skin on the outside of the wire. We have taken $(\sigma/2\pi\mu p)^{\frac{1}{2}}$ as the measure of the thickness of this 'skin.' For currents making 100 vibrations per second, the skin for soft iron having a magnetic permeability of 1000 is about half a millimetre thick, for copper it is about thirteen times as great. For currents making a million vibrations per second, such as can be produced by discharging Leyden jars, the thickness of the skin for soft iron —since we know that this substance retains its magnetic properties even in these very rapidly alternating magnetic fields (J. J. Thomson, *Phil. Mag.* Nov. 1891, p. 460)—is about 1/200 of a millimetre, for copper it is about 1/15 of a millimetre. In these cases there is enormous concentration of the current, and since the currents produced by the discharge of a Leyden jar, though they only last for a short time, are very intense whilst they last, the condition of the outer layers of the wires whilst the discharge is passing through them is very interesting, as they are conveying currents of enormously greater density than would be sufficient to melt them if the currents were permanent instead of transient.

This concentration of the current, or 'throttling' as it is sometimes called, produces a great increase in the apparent resistance of the wire, since it reduces so largely the area which is available for the passage of the current. If in equation (25) we put $r = \mathbf{a}$, we get maximum value of $R = (\mu p\sigma/\pi\mathbf{a}^2)^{\frac{1}{2}} \times$ (maximum value of the current through the wire), thus we may look upon $(\mu p\sigma/\pi\mathbf{a}^2)^{\frac{1}{2}}$ as the apparent resistance per unit length of the wire to these alternating currents. This resistance increases indefinitely with the rate of alternation of the current; we see too that it is inversely proportional to the circumference of the wire instead of to the area as for steady currents. This is what we should expect, since the currents are concentrated in the region of the circumference. The resistance of the solid wire to these alternating currents is the same as that to steady currents of a tube of the same material, the outside of the tube coinciding with the outside of the wire, and the thickness of the tube being $1/\sqrt{2}$ times the thickness of the skin.

We see by comparing equations (25) and (26) that the electromotive intensity parallel to the axis of the wire is very large compared with the radial electromotive intensity in the wire, so that in the wire the Faraday tubes are approximately parallel to its axis.

269.] Let us now consider the expressions for the electromotive intensities and magnetic force in the dielectric; we find by equations (8) and (24), assuming $k\mathrm{a}$ and $k\mathrm{b}$ small, $n\mathrm{a}$, $n'\mathrm{b}$ large,

$$D = 2\iota V^2 k^2 I_0/p.$$

Hence, using (22), we have in the dielectric when kr is small,

$$R = \left\{\left(\frac{\mu p \sigma}{\pi \mathrm{a}^2}\right)^{\frac{1}{2}} - \left[\left(\frac{\mu p \sigma}{\pi \mathrm{a}^2}\right)^{\frac{1}{2}} + \left(\frac{\mu' p \sigma'}{\pi \mathrm{b}^2}\right)^{\frac{1}{2}}\right] \frac{\log r/\mathrm{a}}{\log \mathrm{b}/\mathrm{a}}\right\} I_0 \cos\phi,$$

where
$$\phi = mz + pt + \frac{\pi}{4},$$

while the radial electromotive intensity is

$$\frac{2VI_0}{r}\cos(mz+pt),$$

and the resultant magnetic force

$$\frac{2I_0}{r}\cos(mz+pt).$$

We see that the maximum value of the radial electromotive intensity is very great compared with that of the tangential, so that in the dielectric the Faraday tubes are approximately radial. The momentum due to these tubes is, by Art. 12, at right angles both to the tubes and the magnetic force, so that in the dielectric it is parallel to the axis of the wire, while in the wire itself it is radial. Thus for these rapidly alternating currents the momentum in the dielectric follows the wire. The radial polarization in the dielectric is $K/4\pi$ times the radial electromotive intensity, and since

$$K = 1/V^2,$$

it is equal to
$$\frac{I_0}{2\pi V r}\cos(mz+pt).$$

If the Faraday tubes in the dielectric are moving with velocity V at right angles to their length, i.e. parallel to the wire, the magnetic force due to these moving tubes is, by Art. 9, at right angles both to the direction of motion, i.e. to the axis of the wire, and to the direction of the tubes, i.e. to the radius, and the magnitude of the magnetic force being, by (4), Art 9, $4\pi V$ times the polarization, is

$$\frac{2I_0}{r}\cos(mz+pt),$$

which is the expression we have already found. Hence we may regard the magnetic force in the field as due to the motion through it of the radial Faraday tubes, these moving parallel to the wire with the velocity with which electromagnetic disturbances are propagated through the dielectric.

In the outer conductor when $n'r$ is large

$$R = -I_0\left\{\frac{\mu' p\sigma'}{\pi \mathbf{b} r}\right\}^{\frac{1}{2}}\epsilon^{-(2\pi\mu' p/\sigma')^{\frac{1}{2}}(r-\mathbf{b})}\cos\phi',$$

where
$$\phi' = mz+pt-(2\pi\mu' p/\sigma')^{\frac{1}{2}}+\frac{\pi}{4}.$$

The radial electromotive intensity is

$$\frac{1}{2\pi}\frac{p}{V}\frac{\sigma' I_0}{\sqrt{\mathbf{b} r}}\epsilon^{-(2\pi\mu' p/\sigma')^{\frac{1}{2}}(r-\mathbf{b})}\cos\left(\phi'+\frac{\pi}{4}\right).$$

The resultant magnetic force is perpendicular to r and equal to

$$\frac{2I_0}{\sqrt{\mathbf{b} r}}\epsilon^{-(2\pi\mu' p/\sigma')^{\frac{1}{2}}(r-\mathbf{b})}\cos\left(\phi'-\frac{\pi}{4}\right).$$

We see from these equations that unless $p\sigma'$ is comparable with V^2 the tangential electromotive intensity will be large compared with the radial.

Transmission of Arbitrary Disturbances along Wires.

270.] Since vibrations with different periods travel at different rates, we cannot without further investigation determine the rate at which an arbitrary disturbance communicated to a limited portion of the wire will travel along it. In order to deduce an expression which would represent completely the way in which an arbitrary disturbance is propagated, we should have to make use of the general relation between m and p given by equation (18). This relation is however too complicated to allow of the necessary integrations being effected. The complication arises from the vibrations whose frequencies are so great that $2\pi\mu p a^2/\sigma$ is no longer a small quantity; such vibrations however die away more rapidly than the slower ones, so that when the distance from the origin of disturbance is considerable the latter are the only vibrations whose effects are felt. For such vibrations, we have by Art. 263

$$\iota p = -\frac{m^2}{\mathrm{R}\Gamma}.$$

Hence a term in the expression for R of the form

$$F(a)\epsilon^{-\frac{m^2}{\mathrm{R}\Gamma}t}\cos m(z-a),$$

where a is any constant and $F(a)$ denotes an arbitrary function of a, will satisfy the electrical conditions. By Fourier's theorem, however,

$$\frac{1}{2\pi}\int_{-\infty}^{+\infty}\int_{-\infty}^{+\infty} F(a)\epsilon^{-\frac{m^2}{\mathrm{R}\Gamma}t}\cos m(z-a)\,dm\,da, \tag{27}$$

is equal to $F(z)$ when $t=0$. Hence this integral, since it satisfies the equations of the electric field, will be the expression for the disturbance on the wire at z at the time t of the disturbance, which is equal to $F(z)$ when $t=0$. When the disturbance is originally confined to a space close to the origin, $F(a)$ vanishes unless a is very small; the expression (27) becomes in this case

$$\mathbf{F}\int_{-\infty}^{+\infty}\epsilon^{-\frac{m^2}{\mathrm{R}\Gamma}t}\cos mz\,dm, \tag{28}$$

where $$\mathbf{F}=\int F(a)\,da.$$

Since $$\int_{-\infty}^{-\infty}\epsilon^{-a^2x^2}\cos 2bx\,dx = \frac{\sqrt{\pi}}{a}\epsilon^{-\frac{b^2}{a^2}},$$

we see by (28) that the disturbance at time t and place z will be equal to

$$\mathbf{F}\{\pi\mathrm{R}\Gamma/t\}^{\frac{1}{2}}\epsilon^{-z^2\mathrm{R}\Gamma/4t}. \tag{29}$$

Thus at a given point on the wire the disturbance will vary as

$$\frac{1}{\sqrt{t}}\epsilon^{-\frac{c}{t}},$$

where c is a constant. The rise and fall of the disturbance with the time is represented in Fig. 108, where the ordinates represent

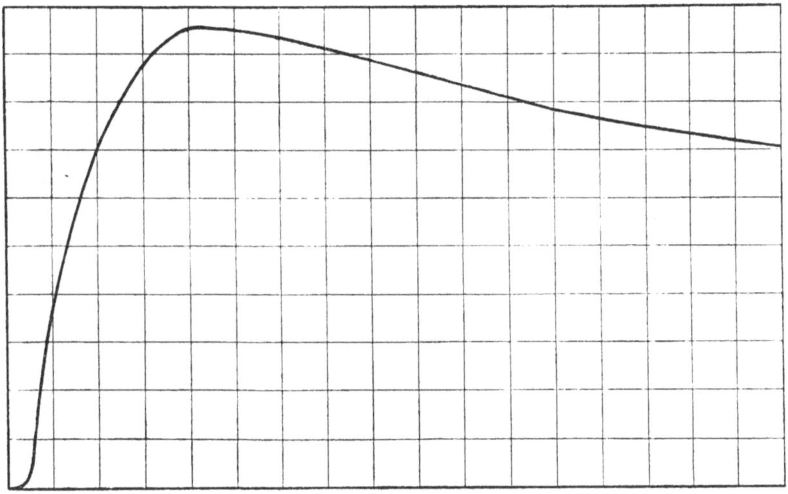

Fig. 108.

the intensity of the disturbance and the abscissae the time. It will be noticed that the disturbance remains very small until t approaches $c/4$, when it begins to increase with great rapidity, reaching its maximum value when $t = 2c$; when t is greater than this the disturbance diminishes, but fades away from its maximum value much more slowly than it approached it.

Since the disturbance rises suddenly to its maximum value we may with propriety call T, the time which elapses before this value is attained at a given point, the time taken by the disturbance to travel to that point. We see from (29) that

$$T = \tfrac{1}{2}z^2\,\mathbf{R}\Gamma. \tag{30}$$

Thus the time taken by the disturbance to travel a distance z is proportional to z^2, it is also proportional to the product of the resistance and capacity per unit length.

By dividing z by T we get the so-called 'velocity of the current along the wire;' this by (30) is equal to

$$\frac{2}{z\mathbf{R}\Gamma}. \tag{31}$$

The velocity thus varies inversely as the length of the cable, and for short lengths it may be very great. The preceding formula would in fact, unless z were greater than $2/V\mathbf{R}\Gamma$, indicate a velocity of propagation greater than V. This however is impossible, and the error arises from our using the equation $\iota p = -m^2/\mathbf{R}\Gamma$ instead of the accurate equation (18). By our approximate equation vibrations of infinite frequency travel with infinite velocity, in reality we have seen (Art. 267) that they travel with the velocity V. These very rapid vibrations however die away very quickly, and when we get to a distance equal to a small multiple of $2/V\mathbf{R}\Gamma$ they will practically have disappeared, and at such distances we may trust the expressions (31).

A considerable number of experiments have been made on the time required to transmit messages on both aerial and submarine cables; the results of some of these, made on aerial telegraph iron wires 4 mm. in diameter, are given in the accompanying table taken from a paper by Hagenbach (*Wied. Ann.* 29. p. 377):—

Observer.	Length of line in kilometres.	Time taken for message to travel (T.)	$10^{20}T/$ (square of length of line in centimetres).
Fizeau and Gounelle .	314	·003085	313
Walker . . .	885	·02943	376
Mitchel . . .	977	·02128	223
Gould and Walker .	1681	·07255	257
Guillemin . . .	1004	·028	278
Plantamour and Hirsch	132·6	·00895	5090
Löwy and Stephan .	863	·024	322
Albrecht . . .	1230	·059	390
Hagenbach . . .	284·8	·00176	217

Hagenbach proved by making experiments with lines of different lengths that the time taken by a message to travel along a line was proportional to the square of the length of the line.

If we apply the formula

$$T = \tfrac{1}{2}z^2 R\Gamma$$

to Hagenbach's experiment in the above table, where

$$z = 284{\cdot}8 \times 10^5,$$
$$R = 9{\cdot}4 \times 10^4,$$

and (by estimation) $\Gamma = 10^{-22}$,

we find $T = {\cdot}0038$, whereas Hagenbach found ·0017. The agreement is not good, but we must remember that with delicate receiving instruments it will be possible to detect the disturbance before it reaches its maximum value, so that we should expect the observed time to be less than that at which the effect is a maximum. In Hagenbach's experiment the line was about 4 times the length which, according to the formula, would have made the disturbance travel with the velocity of light, so that it would seem to have been long enough to warrant the application of a formula which assumes that the shorter waves would have become so reduced in amplitude that their effects might be neglected.

When the wire is of length l, we know by Fourier's Theorem that any initial disturbance R may be represented by the equation

$$R = \left(A_1 \sin \frac{\pi z}{l} + B_1 \cos \frac{\pi z}{l}\right) + \left(A_2 \sin \frac{2\pi z}{l} + B_2 \cos \frac{2\pi z}{l}\right) + \left(A_3 \sin \frac{3\pi z}{l} + B_3 \cos \frac{3\pi z}{l}\right) + \ldots.$$

Since $\iota p = -m^2/\mathrm{R}\Gamma$,

the value of R after a time t has elapsed will be represented by the equation

$$R = \left(A_1 \sin \frac{\pi z}{l} + B_1 \cos \frac{\pi z}{l}\right) \epsilon^{-\frac{\pi^2}{l^2}\frac{t}{\mathrm{R}\Gamma}} + \left(A_2 \sin \frac{2\pi z}{l} + B_2 \cos \frac{2\pi z}{l}\right) \epsilon^{-\frac{4\pi^2}{l^2}\frac{t}{\mathrm{R}\Gamma}} + \ldots$$
$$\ldots \left(A_s \sin \frac{s\pi z}{l} + B_s \cos \frac{s\pi z}{l}\right) \epsilon^{-\frac{s^2\pi^2}{l^2}\frac{t}{\mathrm{R}\Gamma}} + \ldots.$$

For a full discussion of the transmission of signals along cables the reader is referred to a series of papers by Lord Kelvin at the beginning of Vol. II of his Collected Papers.

Relation between the External Electromotive Intensity
and the Current.

271.] We have hitherto only considered the total electromotive intensity and have not regarded it as made up of two parts, one due to external causes and the other due to the induction of the alternating currents in the conductors and dielectric. For some purposes, however, it is convenient to separate the electromotive intensity into these two parts, and to find the relation between the currents and the external electromotive intensity acting on the system.

We may conveniently regard the external electromotive intensity as arising from an electrostatic potential ϕ satisfying the equation $\nabla^2 \phi = 0$. We suppose that, as in the preceding investigation, all the variables contain the factor $\epsilon^{\iota(mz+pt)}$. Since ϕ varies as $\epsilon^{\iota mz}$, the equation $\nabla^2 \phi = 0$ is equivalent to

$$\frac{d^2\phi}{dr^2} + \frac{1}{r}\frac{d\phi}{dr} - m^2\phi = 0.$$

The solution of this is, in the wire

$$\phi = L J_0(\iota mr)\, \epsilon^{\iota(mz+pt)},$$

in the dielectric

$$\phi = \{M J_0(\iota mr) + N K_0(\iota mr)\}\, \epsilon^{\iota(mz+pt)},$$

in the outer conductor

$$\phi = S K_0(\iota mr)\, \epsilon^{\iota(mz+pt)}.$$

If, as before, **a** and **b** are the radii of the internal and external boundaries of the dielectric, we have, since ϕ is continuous,

$$L J_0(\iota m\mathbf{a}) = M J_0(\iota m\mathbf{a}) + N K_0(\iota m\mathbf{a}),$$
$$S K_0(\iota m\mathbf{b}) = M J_0(\iota m\mathbf{b}) + N K_0(\iota m\mathbf{b}).$$

The excess of the normal electromotive intensity due to the electrostatic potential in the dielectric over that in the wire is equal to

$$\iota m \{L J_0'(\iota m\mathbf{a}) - (M J_0'(\iota m\mathbf{a}) + N K_0'(\iota m\mathbf{a})\}\, \epsilon^{\iota(mz+pt)};$$

substituting the value for $L - M$ in terms of N from the preceding equation, this becomes

$$\iota m \frac{N}{J_0(\iota m\mathbf{a})} \{J_0'(\iota m\mathbf{a}) K_0(\iota m\mathbf{a}) - J_0(\iota m\mathbf{a}) K_0'(\iota m\mathbf{a})\}.$$

Now $$J_0'(\iota m\mathbf{a})K_0(\iota m\mathbf{a})-J_0(\iota m\mathbf{a})K_0'(\iota m\mathbf{a})=\frac{1}{\iota m\mathbf{a}},$$

for let $$u=J_0'(x)K_0(x)-J_0(x)K_0'(x),$$

then $$\frac{du}{dx}=J_0''(x)K_0(x)-J_0(x)K_0''(x),$$

but $$J_0''(x)+\frac{1}{x}J_0'(x)-J_0(x)=0,$$

$$K_0''(x)+\frac{1}{x}K_0'(x)-K_0(x)=0;$$

substituting the values of $J_0''(x)$, $K_0''(x)$ from these equations, we find

$$\frac{du}{dx}=-\frac{1}{x}\{J_0'(x)K_0(x)-J_0(x)K_0'(x)\}$$

$$=-\frac{u}{x},$$

hence $$u=\frac{C}{x},$$

where C is a constant. Substituting from Art. 261 the values for $J_0(x)$, $J_0'(x)$, $K_0(x)$, $K_0'(x)$ when x is very small, we find that C is equal to unity.

Thus when $r=\mathbf{a}$, the normal electromotive intensity due to the electrostatic potential in the dielectric exceeds that in the wire by

$$\frac{N}{\mathbf{a}J_0(\iota m\mathbf{a})}\epsilon^{\iota(mz+pt)}.$$

Similarly we may show that when $r=\mathbf{b}$ the normal electromotive intensity in the dielectric exceeds that in the outer conductor by

$$-\frac{M}{\mathbf{b}K_0(\iota m\mathbf{b})}\epsilon^{\iota(mz+pt)}.$$

Now the electromotive intensities arising from the induction of the currents are continuous, so that the discontinuity in the total normal intensity must be equal to the discontinuity in the components arising from the electrostatic potential. By equations (7), (9), (14) the total normal intensity in the dielectric at the surface of separation exceeds that in the wire by

$$A\frac{m}{n}J_0'(\iota n\mathbf{a})\left\{\frac{n^2-m^2}{\mu(k^2-m^2)}-1\right\}\epsilon^{\iota(mz+pt)};$$

hence we have

$$A\frac{m}{n}J_0'(\iota n\mathbf{a})\left\{\frac{n^2-m^2}{\mu(k^2-m^2)}-1\right\}=\frac{N}{\mathbf{a}J_0(\iota m\mathbf{a})}. \qquad (32)$$

Similarly

$$E\frac{m}{n'}K_0'(\iota n'\mathbf{b})\left\{\frac{n'^2-m^2}{\mu'(k^2-m^2)}-1\right\}=-\frac{M}{\mathbf{b}K_0(\iota m\mathbf{b})}. \qquad (33)$$

By equations (10) and (12)

$$2\pi\mathbf{a}\frac{(n^2-m^2)}{\mu p n}AJ_0'(\iota n\mathbf{a})\epsilon^{\iota(mz+pt)},$$

$$2\pi\mathbf{b}\frac{(n'^2-m^2)}{\mu' p n'}EK_0'(\iota n'\mathbf{b})\epsilon^{\iota(mz+pt)}$$

are respectively the line integrals of the magnetic force round the circumference of the wire and the inner circumference of the outer conductor, hence they are respectively 4π times the current through the wire, and 4π times the current through the wire plus that through the dielectric. Unless however the radius of the outer conductor is enormously greater than that of the wire, the current through the wire is infinite in comparison with that through the dielectric: for the electromotive intensity R is of the same order in the wire and in the dielectric; the current density in the wire is R/σ, that in the dielectric $(K/4\pi)\,dR/dt$, or $K\iota pR/4\pi$, or $\iota pR/4\pi V^2$. Now for metals σ is of the order 10^4; and since V^2 is 9×10^{20}, we see that even if there are a million alternations per second the intensity of the current in the wire to that in the dielectric is roughly as 2×10^{11} is to unity; thus, unless the area through which the polarization currents flow exceeds that through which the conduction currents flow in a ratio which is impracticable in actual experiments, we may neglect the polarization currents in comparison with the conduction ones, so that

$$\frac{\mathbf{a}(n^2-m^2)}{\mu n}AJ_0'(\iota n\mathbf{a})=\frac{\mathbf{b}(n'^2-m^2)}{\mu' n'}EK_0'(\iota n'\mathbf{b}). \qquad (34)$$

Returning to equations (32) and (33), we notice that

$$(m^2-n^2)/\mu(k^2-m^2),$$

which is equal to

$$4\pi\iota V^2/\sigma p,$$

is very large when σp is small compared with V^2. Now σ for metals is of the order 10^4, and V^2 is equal to 9×10^{20}; so that unless p is of the order 10^{16} at least, that is unless the vibrations are as rapid as those of light, $(m^2-n^2)/\mu(m^2-k^2)$ is exceedingly large. Even when the conductivity is no better than that of sea-water, where σ may be taken to be of the order 10^{10}, this quantity will be very large unless there are more than a thousand million vibrations per second. Hence in equations (32) and (33) we may neglect the second terms inside the brackets on the left-hand sides, and write

$$\left.\begin{aligned} A\frac{m}{n}\frac{(n^2-m^2)}{\mu(k^2-m^2)}J_0'(\iota n\mathbf{a}) &= \frac{N}{\mathbf{a}J_0(\iota m\mathbf{a})}, \\ E\frac{m}{n'}\frac{(n'^2-m^2)}{\mu'(k^2-m^2)}K_0'(\iota n'\mathbf{b}) &= -\frac{M}{\mathbf{b}K_0(\iota m\mathbf{b})}; \end{aligned}\right\} \tag{35}$$

hence by (34), we have

$$\frac{N}{J_0(\iota m\mathbf{a})} = -\frac{M}{K_0(\iota m\mathbf{b})}. \tag{36}$$

Let $\mathbf{E}$ be the external electromotive intensity parallel to the axis of the wire at its surface, then

$$\mathbf{E} = -\iota m\{MJ_0(\iota m\mathbf{a}) + NK_0(\iota m\mathbf{a})\}\epsilon^{\iota(mz+pt)},$$

or by equation (36)

$$= -\iota mN\{K_0(\iota m\mathbf{a}) - K_0(\iota m\mathbf{b})\}\epsilon^{\iota(mz+pt)}.$$

Since both $m\mathbf{a}$ and $m\mathbf{b}$ are very small, we have approximately by Art. 261,

$$K_0(\iota m\mathbf{a}) = \log\frac{2\gamma}{\iota m\mathbf{a}}, \quad K_0(\iota m\mathbf{b}) = \log\frac{2\gamma}{\iota m\mathbf{b}},$$

hence we have

$$\mathbf{E} = -\iota mN\log(\mathbf{b}/\mathbf{a})\epsilon^{\iota(mz+pt)},$$

or by equation (35), since $J_0(\iota m\mathbf{a}) = 1$,

$$\mathbf{E} = -\frac{\iota m^2\mathbf{a}}{n}\frac{n^2-m^2}{\mu(k^2-m^2)}J_0'(\iota n\mathbf{a})\log(\mathbf{b}/\mathbf{a})A\epsilon^{\iota(mz+pt)}.$$

But by Art. 263 we have, if $I_0\epsilon^{\iota(mz+pt)}$ is the total current through the wire,

$$I_0 = \frac{2\pi\mathbf{a}\iota}{\sigma n}AJ_0'(\iota n\mathbf{a}),$$

hence, since

$$n^2 - m^2 = 4\pi\mu\iota p/\sigma,$$
$$m^2 - k^2 = p^2/V^2,$$
$$\mathbf{E} = 2\iota p \frac{m^2}{\frac{p^2}{V^2}} \log(\mathrm{b}/\mathrm{a}) . I_0 \epsilon^{\iota(mz+pt)}.$$

But by equation (18)

$$m^2 = \frac{p^2}{V^2}\left\{1 - \frac{1}{4\pi p}\left(\frac{n\sigma}{\mathrm{a}}\frac{J_0(\iota n\mathrm{a})}{J_0'(\iota n\mathrm{a})} - \frac{n'\sigma'}{\mathrm{b}}\frac{K_0(\iota n'\mathrm{b})}{K_0'(\iota n'\mathrm{b})}\right)\frac{1}{\log \mathrm{b}/\mathrm{a}}\right\}.$$

hence

$$\mathbf{E} = 2\iota p\left\{\log\frac{\mathrm{b}}{\mathrm{a}} - \frac{1}{4\pi p}\left(\frac{n\sigma}{\mathrm{a}}\frac{J_0(\iota n\mathrm{a})}{J_0'(\iota n\mathrm{a})} - \frac{n'\sigma'}{\mathrm{b}}\frac{K_0(\iota n'\mathrm{b})}{K_0'(\iota n'\mathrm{b})}\right)\right\} I_0\epsilon^{\iota(mz+pt)}. \quad (37)$$

272.] Now, as in Art. 263, when both $n\mathrm{a}$ and $n'\mathrm{b}$ are small, the last term inside the bracket will be small compared with the others; so that we may write equation (37) in the form

$$\mathbf{E} = 2\iota p\left\{\log\frac{\mathrm{b}}{\mathrm{a}} - \frac{1}{4\pi p}\frac{n\sigma}{\mathrm{a}}\frac{J_0(\iota n\mathrm{a})}{J_0'(\iota n\mathrm{a})}\right\} I,$$

where I is the total current through the wire and is equal to

$$I_0\epsilon^{\iota(mz+pt)}.$$

From the expressions for $\iota n\mathrm{a}J_0(\iota n\mathrm{a})/J_0'(\iota n\mathrm{a})$ given in Art. 265, we see that we may write this equation

$$\mathbf{E} = 2\iota p\left\{\log\frac{\mathrm{b}}{\mathrm{a}} + \frac{\sigma}{2\pi\iota p\mathrm{a}^2}\left[1 + \frac{1}{12\times 16}(4\pi\mu p\mathrm{a}^2/\sigma)^2 - \frac{1}{12\times 15\times 16^2}(4\pi\mu p\mathrm{a}^2/\sigma)^4 + \ldots + \iota\left(\frac{1}{8}(4\pi\mu p\mathrm{a}^2/\sigma) - \frac{1}{12\times 16^2}(4\pi\mu p\mathrm{a}^2/\sigma)^3 + \frac{13}{9\times 15\times 16^4}(4\pi\mu p\mathrm{a}^2/\sigma)^5 + \ldots\right)\right]\right\} I,$$

or

$$\mathbf{E} = \iota p\left\{2\log\frac{\mathrm{b}}{\mathrm{a}} + \frac{1}{2}\mu - \frac{1}{48}\frac{\pi^2\mu^3p^2\mathrm{a}^4}{\sigma^2} + \frac{13}{8640}\frac{\pi^4\mu^5p^4\mathrm{a}^8}{\sigma^4}\ldots\right\} I$$
$$+ \frac{\sigma}{\pi\mathrm{a}^2}\left\{1 + \frac{1}{12}\frac{\pi^2p^2\mu^2\mathrm{a}^4}{\sigma^2} - \frac{1}{180}\frac{\pi^4p^4\mu^4\mathrm{a}^8}{\sigma^4}\ldots\right\} I. \quad (38)$$

We may write this as

$$\mathbf{E} = \mathbf{P}\iota p I + \mathbf{Q} I, \quad (39)$$

or since $$\iota p I = \frac{dI}{dt},$$

as $$\mathbf{E} = \mathbf{P}\frac{dI}{dt} + \mathbf{Q}I.$$

If L is the coefficient of self-induction and R the resistance of a circuit through which a current I is flowing, we have

$$\text{external electromotive force} = L\frac{dI}{dt} + RI.$$

By the analogy of this equation with (39) we may call **P** the self-induction and **Q** the resistance of the cable per unit length for these alternating currents. **Q** has been called the 'impedance' of unit length of the circuit by Mr. Heaviside, and this term is preferable to resistance as it enables the latter to be used exclusively for steady currents.

By comparing (39) with (38), we see that

$$\left.\begin{aligned} \mathbf{P} &= 2\log\frac{\mathrm{b}}{\mathrm{a}} + \frac{1}{2}\mu - \frac{1}{48}(\mu^3 p^2\pi^2\mathrm{a}^4/\sigma^2) + \frac{13}{8640}(\mu^5 p^4\pi^4\mathrm{a}^8/\sigma^4) - \dots, \\ \mathbf{Q} &= \frac{\sigma}{\pi\mathrm{a}^2}\left\{1 + \frac{1}{12}(\mu^2 p^2\pi^2\mathrm{a}^4/\sigma^2) - \frac{1}{180}(\mu^4 p^4\pi^4\mathrm{a}^8/\sigma^4) + \dots\right\}. \end{aligned}\right\} \quad (40)$$

These results are the same as those given in equation (18), Art. 690, of Maxwell's *Electricity and Magnetism*, with the exception that μ is put equal to unity in that equation and in it A is written instead of $2\log(\mathrm{b}/\mathrm{a})$.

We see from these equations that as the rate of alternation increases, the impedance increases while the self-induction diminishes; both these effects are due to the influence of the rate of alternation on the distribution of the current. As the rate of alternation increases the current gets more and more concentrated towards the surface of the wire; the effective area of the wire is thus diminished and the resistance therefore increased. On the other hand, the concentration of the current on the surface of the wire increases the average distance between the portions of the currents in the wire, and diminishes that between the currents in the wire and those flowing in the opposite direction in the outer conductor; both these effects diminish the self-induction of the system of currents.

The expression for **Q** does not to our degree of approximation involve b at all, while b only enters into the first term of the expression for P, which is independent of the frequency; thus, as long as na is very small, the presence of the outer conductor does not affect the impedance, nor the way in which the self-induction varies with the frequency. When $p = 0$ the self-induction per unit length is $2 \log(\mathbf{b}/\mathbf{a}) + \frac{1}{2}\mu$. Since μ for soft iron may be as great as 2000, the self-induction per unit length of straight iron wires will be enormously greater than that of wires made of the non-metallic metals.

273.] We shall now pass on to the case when na is large and n'b small, so that $n\sigma J_0(\iota n\mathbf{a})/p\mathbf{a}J_0'(\iota n\mathbf{a})$ is small compared with $n'\sigma' K_0(\iota n'\mathbf{b})/p\mathbf{b}K_0'(\iota n'\mathbf{b})$. These conditions are compatible if the specific resistance of the outer conductor is very much greater than that of the wire. In this case equation (37) becomes

$$\mathbf{E} = 2\iota p\left\{\log\frac{\mathbf{b}}{\mathbf{a}} + \frac{1}{4\pi p}\frac{n'\sigma'}{\mathbf{b}}\frac{K_0(\iota n'\mathbf{b})}{K_0'(\iota n'\mathbf{b})}\right\}I.$$

Since n'b is small, we have approximately

$$K_0(\iota n'\mathbf{b}) = \log(2\gamma/\iota n'\mathbf{b}), \quad K_0'(\iota n'\mathbf{b}) = -1/\iota n'\mathbf{b};$$

hence $$\mathbf{E} = 2\iota p\left\{\log \mathbf{b}/\mathbf{a} + \mu' \log(\gamma/\sqrt{\pi\mu'\mathbf{b}^2 p/\sigma'}) - \iota 3\mu'\frac{\pi}{4}\right\}.$$

Thus the coefficient of self-induction in this case is

$$2\log(\mathbf{b}/\mathbf{a}) + 2\mu'\log(\gamma/\sqrt{\pi\mu' p\mathbf{b}^2/\sigma'}),$$

and the impedance $$\tfrac{3}{2}\pi p\mu'.$$

It is worthy of remark that to our order of approximation neither the impedance nor the self-induction depends upon the resistance of the wire. This is only what we should expect for the self-induction, for since na is large the currents will all be on the surface of the wire; the configuration of the currents has thus reached a limit beyond which it is not affected by the resistance of the wire. It should be noticed that the conditions na large and n'b small make the impedance $\frac{3}{2}\pi p\mu'$ large compared with the resistance $\sigma/\pi\mathbf{a}^2$ for steady currents.

Very Rapid Currents.

274.] We must now consider the case where the frequency is so great that $n\mathbf{a}$ and $n'\mathbf{b}$ are very large; in this case, by Art. 261,

$$J_0'(\iota n\mathbf{a}) = -\iota J_0(\iota n\mathbf{a}), \quad K_0'(\iota n'\mathbf{b}) = \iota K_0(\iota n'\mathbf{b}),$$

so that equation (37) becomes

$$\mathbf{E} = 2\iota p\left\{\log\frac{\mathbf{b}}{\mathbf{a}} + \left[\left(\frac{\sigma\mu}{4\pi p\mathbf{a}^2}\right)^{\frac{1}{2}} + \left(\frac{\sigma'\mu'}{4\pi p\mathbf{b}^2}\right)^{\frac{1}{2}}\right]\left(\frac{1}{\sqrt{2}} - \frac{\iota}{\sqrt{2}}\right)\right\}I; \quad (41)$$

we see from this equation that the self-induction $\mathbf{P}$ is given by the equation

$$\mathbf{P} = 2\log(\mathbf{b}/\mathbf{a}) + (\sigma\mu/2\pi p\mathbf{a}^2)^{\frac{1}{2}} + (\sigma'\mu'/2\pi p\mathbf{b}^2)^{\frac{1}{2}}, \quad (42)$$

and the impedance $\mathbf{Q}$ by

$$\mathbf{Q} = (\sigma\mu p/2\pi\mathbf{a}^2)^{\frac{1}{2}} + (\sigma'\mu' p/2\pi\mathbf{b}^2)^{\frac{1}{2}}. \quad (43)$$

In a cable the conductivity of the outer conductor is very much less than that of the core, so that $\sigma'/\mathbf{b}^2$ will be large compared with $\sigma/\mathbf{a}^2$; thus the self-induction and impedance of a cable are both practically independent of the resistance of the wire and depend mainly upon that of the outer conductor. The limiting value of the self-induction when the frequency is indefinitely increased is $2\log(\mathbf{b}/\mathbf{a})$; as this does not involve μ it is the same for iron as for copper wires. The difference between the self-induction per unit length of the cable for infinitely slow and infinitely rapid vibrations is by equations (40) and (42) equal to $\mu/2$. The impedance of the circuit increases indefinitely with the frequency of the alternations.

If we trace the changes in the values of the self-induction and impedance as the frequency p increases, we see from Arts. 272, 273, 274 that when this is so small that $n\mathbf{a}$ is a small quantity the self-induction decreases and the impedance increases by an amount proportional to the square of the frequency. When the frequency increases so that $n\mathbf{a}$ is considerable while $n'\mathbf{b}$ is small, the self-induction varies very slowly with the frequency while the impedance is directly proportional to it. When the frequency is so great that both $n\mathbf{a}$ and $n'\mathbf{b}$ are large the self-induction approaches the limit $2\log(\mathbf{b}/\mathbf{a})$, while the impedance is proportional to the square root of the frequency.

Flat Conductors.

275.] In many experiments flat strips of metal in parallel planes are used instead of wires, with the view of diminishing the self-induction; these are generally arranged so that the direct and return currents flow along adjacent and parallel strips. When the frequency of the vibrations is very large, the positive and negative currents endeavour to get as near together as possible, they will thus flow on the surfaces of the strips which are nearest each other. If the distance between the planes of the strips is small in comparison with their breadth we may consider them as a limiting case of the cable, when the specific resistance of the wire is the same as that of the outer conductor, and when the values of $\mathbf{a}$ and $\mathbf{b}$ are indefinitely great, their difference however remaining finite and equal to the distance between the strips. If I' is the current flowing across unit width of the strip, then, since with our previous notation I is the current flowing over the circumference of the cable,

$$I' = I/2\pi\mathbf{a}.$$

Since $\mathbf{b} = \mathbf{a} + d$, where d is very small compared with $\mathbf{a}$,

$$\log\frac{\mathbf{b}}{\mathbf{a}} = \frac{d}{\mathbf{a}} \text{ approximately.}$$

Making these substitutions, equation (41) becomes

$$\mathbf{E} = 2\iota p\left\{d + \sqrt{\frac{\mu\sigma}{2\pi p}}(1-\iota)\right\} 2\pi I'.$$

Thus, in this case the self-induction per unit length is

$$4\pi\left\{d + \sqrt{\frac{\mu\sigma}{2\pi p}}\right\},$$

and the impedance

$$4\pi\sqrt{\frac{\mu\sigma p}{2\pi}}.$$

276.] Though, as we have just seen, it is possible to regard the case of two parallel metal slabs as a particular case of the cable, yet inasmuch as the geometry of the particular case is much simpler than that of the cable, the case is one where points of theory are most conveniently discussed; it is therefore advisable to treat it independently. We shall suppose that we have two slabs of the same metal, the adjacent faces of the slabs being parallel and separated by the distance $2h$; we shall take the plane parallel to these faces and midway between them as the plane of yz, the axis of x being normal to the faces. We shall suppose that all the variable quantities vary as $\epsilon^{\iota(mz+pt)}$ and are independent of y. The slabs are supposed also to extend to infinity in directions parallel to y and z and to be infinitely thick.

Let σ be the specific resistance of the slabs, V the velocity of propagation of electrodynamic action through the dielectric which separates them. Then, using the same notation as before, since all the quantities are independent of y, the differential equations satisfied by the components of the electromotive intensity are by Art. 262

$$\frac{d^2 R}{dx^2} = k^2 R \text{ in the dielectric,}$$

and

$$\frac{d^2 R}{dx^2} = n^2 R$$

in either of the slabs.

Thus, in the dielectric we may put

$$R = (A\epsilon^{kx} + B\epsilon^{-kx})\,\epsilon^{\iota(mz+pt)},$$

$$P = -\frac{\iota m}{k}(A\epsilon^{kx} - B\epsilon^{-kx})\,\epsilon^{\iota(mz+pt)},$$

in the slab for which x is positive

$$R = C\epsilon^{-nx}\,\epsilon^{\iota(mz+pt)},$$

$$P = \frac{\iota m}{n} C\epsilon^{-nx}\,\epsilon^{\iota(mz+pt)},$$

and in that for which x is negative

$$R = D\epsilon^{nx}\,\epsilon^{\iota(mz+pt)},$$

$$P = -\frac{\iota m}{n} D\epsilon^{nx}\,\epsilon^{\iota(mz+pt)},$$

the real part of n being taken positive in both cases.

Since R is continuous when $x = \pm h$, we have

$$\left.\begin{aligned} A\epsilon^{kh} + B\epsilon^{-kh} &= C\epsilon^{-nh}, \\ A\epsilon^{-kh} + B\epsilon^{kh} &= D\epsilon^{-nh}. \end{aligned}\right\} \qquad (44)$$

Since the magnetic force parallel to the surface is continuous, we have, if μ is the magnetic permeability of the slab,

$$\frac{k^2-m^2}{k}(A\epsilon^{kh}-B\epsilon^{-kh})=\frac{m^2-n^2}{\mu n}C\epsilon^{-nh},$$

$$\frac{k^2-m^2}{k}(A\epsilon^{-kh}-B\epsilon^{kh})=-\frac{m^2-n^2}{\mu n}D\epsilon^{-nh}.$$

Eliminating C and D by the aid of equations (44), we have

$$\left.\begin{aligned} A\left(\frac{k^2-m^2}{k}-\frac{m^2-n^2}{\mu n}\right)\epsilon^{kh} &= B\left(\frac{k^2-m^2}{k}+\frac{m^2-n^2}{\mu n}\right)\epsilon^{-kh}, \\ A\left(\frac{k^2-m^2}{k}+\frac{m^2-n^2}{\mu n}\right)\epsilon^{-kh} &= B\left(\frac{k^2-m^2}{k}-\frac{m^2-n^2}{\mu n}\right)\epsilon^{kh}. \end{aligned}\right\} \qquad (45)$$

From these equations we get

$$A^2=B^2.$$

The solution $A=B$ corresponds to the current flowing in the same direction in the two slabs, the other solution corresponds to the case when the current flows in one direction in one slab and in the opposite direction in the other; it is this case we shall proceed to investigate. Putting $A=-B$, equation (45) becomes

$$\frac{k^2-m^2}{k}(\epsilon^{kh}+\epsilon^{-kh})+\frac{n^2-m^2}{\mu n}(\epsilon^{kh}-\epsilon^{-kh})=0; \qquad (46)$$

but

$$k^2-m^2=-p^2/V^2,$$

$$n^2-m^2=4\pi\mu\iota p/\sigma,$$

and kh is very small, thus (46) becomes approximately

$$\frac{p^2}{V^2k}=\frac{4\pi\iota p}{\sigma n}kh,$$

or

$$k^2=-\frac{p^2}{V^2}\frac{\iota n\sigma}{4\pi hp},$$

so that

$$m^2=\frac{p^2}{V^2}\left\{1-\frac{\iota n\sigma}{4\pi hp}\right\}. \qquad (47)$$

As we have remarked before, $4\pi\mu p/\sigma$ is in the case of metals very large compared with m^2, so that we have approximately

$$n^2=4\pi\mu\iota p/\sigma,$$

and therefore approximately by equation (47)

$$m^2 = \frac{p^2}{V^2}\left\{1 + \left(\frac{\sigma\mu}{8\pi h^2 p}\right)^{\frac{1}{2}}(1-\iota)\right\}. \tag{48}$$

Thus, if $m = \xi + \iota\eta$, we have

$$\xi^2 - \eta^2 = \frac{p^2}{V^2}\left\{1 + \left(\frac{\sigma\mu}{8\pi h^2 p}\right)^{\frac{1}{2}}\right\},$$

$$2\xi\eta = -\frac{p^2}{V^2}\left(\frac{\sigma\mu}{8\pi h^2 p}\right)^{\frac{1}{2}}.$$

But if ω is the velocity with which the phases are propagated along the slab, $\omega^2 = p^2/\xi^2$, so that we have

$$\frac{1}{\omega^2} = \frac{\eta^2}{p^2} + \frac{1}{V^2}\left\{1 + \left(\frac{\sigma\mu}{8\pi h^2 p}\right)^{\frac{1}{2}}\right\},$$

thus $1/\omega^2$ is never less than $1/V^2$, or ω is never greater than V, so that the velocity of propagation of the phases along the slab can never exceed the rate at which electrodynamic action travels through the dielectric.

If the frequency is so high that $\sigma\mu/8\pi h^2 p$ is small, then we have by equation (48)

$$m = -\frac{p}{V}\left\{1 - \frac{\iota}{2}\left(\frac{\sigma\mu}{8\pi h^2 p}\right)^{\frac{1}{2}}\right\} \text{ approximately.}$$

This equation represents a disturbance propagated with the velocity V, whose amplitude fades away to $1/\epsilon$ of its original value after traversing a distance

$$2Vh\left\{\frac{8\pi p}{\sigma\mu}\right\}^{\frac{1}{2}}.$$

If the frequency is so low that $\sigma\mu/8\pi h^2 p$ is large, then we have approximately by equation (48)

$$m = -\frac{p}{V}\left\{\frac{\sigma\mu}{4\pi h^2 p}\right\}^{\frac{1}{4}}\left(\cos\frac{\pi}{8} - \iota\sin\frac{\pi}{8}\right),$$

or

$$m = -\frac{p}{V}\left\{\frac{\sigma\mu}{4\pi h^2 p}\right\}^{\frac{1}{4}}(\cdot 92 - \iota\cdot 38).$$

This corresponds to a vibration propagated with the velocity

$$1\cdot 08\, V\{4\pi h^2 p/\sigma\mu\}^{\frac{1}{4}},$$

and fading away to $1/\epsilon$ of its original amplitude after traversing a distance

$$2\cdot 6\, V\{4\pi h^2/\mu\sigma p^3\}^{\frac{1}{4}}.$$

If the total current through a slab per unit width is represented by the real part of $I_0\epsilon^{\iota(mz+pt)}$, then, when the frequency is so great that $\sigma\mu/8\pi h^2 p$ is a small quantity and therefore the real part of m large compared with the imaginary part, we have since

$$I_0\epsilon^{\iota(mz+pt)} = \int_h^\infty \frac{R}{\sigma}dx = \frac{C}{\sigma n}\epsilon^{-nh}\epsilon^{\iota(mz+pt)},$$

$$C = \sigma n\epsilon^{nh} I_0;$$

hence by (44) $$A = -B = \frac{\sigma n I_0}{2kh}.$$

We have therefore in the dielectric

$$R = \sigma I_0\sqrt{2}n'(x/h)\cos\left(mz+pt+\frac{\pi}{4}\right),$$

$$P = 4\pi m I_0(V^2/p)\cos(mz+pt),$$

$$b = -4\pi I_0\cos(mz+pt),$$

where $$n' = \{2\pi\mu p/\sigma\}^{\frac{1}{2}}.$$

In the metal slab we have on the side where x is positive,

$$R = \sigma I_0\sqrt{2}n'\epsilon^{-n'(x-h)}\cos\left(mz+pt-n'(x-h)+\frac{\pi}{4}\right),$$

$$P = -\sigma I_0 m\epsilon^{-n'(x-h)}\sin(mz+pt-n'(x-h)),$$

$$b = -4\pi\mu I_0\epsilon^{-n'(x-h)}\cos(mz+pt-n'(x-h)).$$

We see from these equations that P/R is very large in the dielectric and very small in the metal slab, thus the Faraday tubes are at right angles to the conductor in the dielectric and parallel to it in the metal slab.

Mechanical Force between the Slabs.

277.] This may be regarded as consisting of two parts, (1) an attractive force, due to the attraction of the positive electricity of one slab on the negative of the other, (2) a repulsive force, due to the repulsion between the positive currents in one slab and the negative in the other. To calculate the first force we notice that since $V^2/p\sigma$ is very large, the value of P in the conductor is very small compared with the value in the dielectric, and may without appreciable error be neglected; hence if e is the surface density of the electricity on the slab and K the specific inductive capacity of the dielectric,

$$4\pi e = -K\,4\pi m\,(V^2/p)\,I_0 \cos(mz+pt).$$

The force on the slab per unit area is equal to $Pe/2$; substituting the values of P and e this becomes

$$2\pi K m^2 (V^4/p^2)\,I_0^{\,2}\cos^2(mz+pt).$$

The force due to the repulsion between the currents in the slabs per unit volume is equal to the product of the magnetic induction b into w, the intensity of the current parallel to z. Since

$$4\pi\mu w = -\frac{db}{dx},$$

the force per unit volume is equal to

$$-\frac{1}{8\pi\mu}\frac{db^2}{dx},$$

hence the repulsive force per unit area of the surface of the slab

$$= -\int_h^\infty \frac{1}{8\pi\mu}\frac{db^2}{dx}\,dx$$

$$= \frac{1}{8\pi\mu}(b^2)_{x=h} = 2\pi\mu I_0^{\,2}\cos^2(mz+pt).$$

When the alternations are so rapid that the vibrations travel with the velocity of light

$$V^2 m^2 = p^2,$$

and since $K = 1/V^2$, the attraction between the slabs is equal to

$$2\pi I_0^{\,2}\cos^2(mz+pt),$$

while the repulsion is

$$2\pi\mu I_0^{\,2}\cos^2(mz+pt),$$

hence the resultant repulsion is equal to

$$2\pi(\mu-1)\,I_0^{\,2}\cos^2(mz+pt).$$

If the slabs are non-magnetic $\mu = 1$, so that for these very rapid vibrations the electrostatic attraction just counterbalances the electromagnetic repulsion. Mr. Boys (*Phil. Mag.* [5], 31, p. 44, 1891) found that the mechanical forces between two conductors carrying very rapidly alternating currents was too small to be detected, even by the marvellously sensitive methods for measuring small forces which he has perfected, and which would have enabled him to detect forces comparable in magnitude with those due to the electrostatic charges or to the repulsion between the currents.

Propagation of Longitudinal Waves of Magnetic Induction along Wires.

278.] In the preceding investigations the current has been along the wire and the lines of magnetic force have formed a series of co-axial circles, the axis of these circles being that of the wire. Another case, however, of considerable practical importance is when these relations of the magnetic force and current are interchanged, the current flowing in circles round the axis of the wire while the magnetic force is mainly along it. This condition might be realized by surrounding a portion of the wire by a short co-axial solenoid, then if alternating currents are sent through this solenoid periodic magnetic forces parallel to the wire will be started. We shall in this article investigate the laws which govern the transmission of such forces along the wire. The problem has important applications to the construction of transformers ; in some of these the primary coil is wound round one part of a closed magnetic circuit, the secondary round another. This arrangement will not be efficient if there is any considerable leakage of the lines of magnetic force between the primary and the secondary. We should infer from general considerations that the magnetic leakage would increase with the rate of alternation of the current through the primary. For let us suppose that an alternating current passes through an insulated ring imbedded in a cylinder of soft iron surrounded by air, the straight axis of the ring coinciding with the axis of the cylinder. The variations in the intensity of the current through this ring will induce other currents in the iron in its neighbourhood; the magnetic action of these currents will, on the whole, cause the component of the magnetic force along the axis of the cylinder to be less and the radial component greater than if the current through the ring were steady; in which case there are no currents in the iron. Thus the effect of the changes in the intensity of the current through the primary will be to squeeze as it were the lines of magnetic force out of the iron and make them complete their circuit through the air. Thus when the field is changing quickly, the lines of magnetic force, instead of taking a long path through the medium of high permeability, will take a short path, even though the greater part of it is through a medium of low permeability such as air. The case is quite analogous to the difference between the path of a steady current and that of a rapidly alternating one. A steady current flows along the path of least resistance, a rapidly alternating one along the path with least self-induction. Thus, for example, if we have two wires in parallel, one very long but made of such highly conducting material that the total resistance is small, the other wire short but of such a nature that the resistance is large, then when the current is steady by far the greater part of it will travel along the long wire; if however the current is a rapidly alternating one, the greater part of it will travel along the short wire because the self-induction is smaller than for the long wire, and for these rapidly alternating currents the resistance is a secondary consideration.

In the magnetic problem the iron corresponds to the good conductor, the air to the bad one. When the field is steady the lines of force prefer to take a long path through the iron rather than a short one through the air; they will thus tend to keep within the iron; when however the magnetic field is a very rapidly alternating one, the paths of the lines of force will tend to be as short as possible, whatever the material through which they pass. The lines of force will thus in this case leave the iron and complete their circuit through the air.

We shall consider the case of a right circular soft iron cylinder where the lines of magnetic force are in planes through the axis taken as that of z, the corresponding system of currents flowing round circles whose axis is that of the cylinder. The cylinder is surrounded by a dielectric which extends to infinity. Let a, b, c be the components of the magnetic induction parallel to the axes of x, y, z respectively; then, since the component of the magnetic induction in the xy plane is at right angles to the axis of the cylinder, we may put

$$a = \frac{d\chi}{dx}, \quad b = \frac{d\chi}{dy}.$$

Let us suppose that a, b, c all vary as $\epsilon^{\iota(mz+pt)}$.

Now in the iron cylinder a, b, c all satisfy differential equations of the form

$$\frac{d^2c}{dx^2} + \frac{d^2c}{dy^2} = n^2c,$$

where

$$n^2 = m^2 + 4\pi\mu\iota p/\sigma,$$

μ being the magnetic permeability and σ the specific resistance of the cylinder.

In the dielectric outside the cylinder the differential equation satisfied by the components of the magnetic induction is of the form

$$\frac{d^2c}{dx^2} + \frac{d^2c}{dy^2} = k^2c,$$

where

$$k^2 = m^2 - \frac{p^2}{V^2},$$

and V is the velocity with which electromagnetic disturbances are propagated through the dielectric.

We have also

$$\frac{da}{dx} + \frac{db}{dy} + \frac{dc}{dz} = 0.$$

The solution of these equations is easily seen to be, in the iron cylinder,

$$c = AJ_0(\iota nr)\epsilon^{\iota(mz+pt)},$$

$$a = -\frac{\iota m}{n^2}A\frac{d}{dx}J_0(\iota nr)\epsilon^{\iota(mz+pt)},$$

$$b = -\frac{\iota m}{n^2}A\frac{d}{dy}J_0(\iota nr)\epsilon^{\iota(mz+pt)};$$

while in the dielectric, since r can become infinite,

$$c = CK_0(\iota kr)\epsilon^{\iota(mz+pt)},$$

$$a = -\frac{\iota m}{k^2}C\frac{d}{dx}K_0(\iota kr)\epsilon^{\iota(mz+pt)},$$

$$b = -\frac{\iota m}{k^2}C\frac{d}{dy}K_0(\iota kr)\epsilon^{\iota(mz+pt)}.$$

Let $\mathbf{a}$ be the radius of the cylinder, then when $r = \mathbf{a}$ the tangential magnetic force in the cylinder is equal to that in the dielectric, hence

$$\frac{A}{\mu}J_0(\iota n\mathbf{a}) = CK_0(\iota k\mathbf{a});$$

since the radial magnetic induction is continuous, we have

$$\frac{m}{n} A J_0'(\iota n \mathrm{a}) = \frac{m}{k} C K_0'(\iota k \mathrm{a}).$$

Eliminating A and C from these equations, we get

$$\frac{\iota n \mathrm{a}}{\mu} \frac{J_0(\iota n \mathrm{a})}{J_0'(\iota n \mathrm{a})} = \iota k \mathrm{a} \frac{K_0(\iota k \mathrm{a})}{K_0'(\iota k \mathrm{a})}, \tag{49}$$

an equation which will enable us to find m when p is known.

Let us begin with the case when the frequency of the alternations is small enough to allow of the currents being nearly uniformly distributed over the cross-section of the cylinder. In this case we have approximately

$$J_0(\iota n \mathrm{a}) = 1, \quad J_0'(\iota n \mathrm{a}) = -\tfrac{1}{2} \iota n \mathrm{a},$$

so that equation (49) becomes

$$-\frac{2}{\mu} = \iota k \mathrm{a} \frac{K_0(\iota k \mathrm{a})}{K_0'(\iota k \mathrm{a})}. \tag{50}$$

Since for soft iron $2/\mu$ is a small quantity, the right-hand side of this equation and therefore $k\mathrm{a}$ must be small; but in this case we have approximately

$$K_0(\iota k \mathrm{a}) = \log(2\gamma/\iota k \mathrm{a}),$$

$$K_0'(\iota k \mathrm{a}) = -\frac{1}{\iota k \mathrm{a}},$$

so that (50) becomes

$$-\frac{2}{\mu} = k^2 \mathrm{a}^2 \log(2\gamma/\iota k \mathrm{a}). \tag{51}$$

To solve this equation consider the solution of

$$x \log x = -y,$$

when y is small. If $x = -y/\log y$, then

$$x \log x = -y \left\{1 + \frac{\log\log(1/y)}{\log(1/y)}\right\},$$

but when y is small $\log\log(1/y)$ is small compared with $\log(1/y)$, so that an approximate solution of the equation is

$$x = -y/\log y.$$

If we apply this result to equation (51), we find that the approximate solution of that equation is

$$k^2 = -\frac{4}{a^2\mu}\frac{1}{\log(\mu\gamma^2)}.$$

Now $$k^2 = m^2 - \frac{p^2}{V^2},$$

and since the value we have just found for k is in any practicable case very large compared with p^2/V^2, we see that $k^2 = m^2$ approximately, so that

$$m = \frac{\iota 2}{a}\left\{\frac{1}{\mu}\frac{1}{\log(\mu\gamma^2)}\right\}^{\frac{1}{2}}.$$

Thus since in the expression for c there is the factor

$$\epsilon^{\iota m z} \text{ or } \epsilon^{-\frac{2}{a}z\left\{\frac{1}{\mu\log(\mu\gamma^2)}\right\}^{\frac{1}{2}}},$$

we see that the magnetic force will die away to $1/\epsilon$ of its value at a distance

$$\tfrac{1}{2}a\{\mu\log(\mu\gamma^2)\}^{\frac{1}{2}}$$

from its origin.

279]. In the last case the current was uniformly distributed over the cross-section. We can investigate the effect of the concentration of the current at the boundary of the cylinder by supposing that $n a$ is large compared with unity though small compared with μ. In this case, since approximately

$$J_0'(\iota n a) = -\iota J_0(\iota n a),$$

equation (49) becomes

$$-\frac{n a}{\mu} = \iota k a\frac{K_0(\iota k a)}{K_0'(\iota k a)}.$$

Since the left-hand side of this equation is small, $\iota k a$ is also small, so that by Art. 261 we may write this equation as

$$-\frac{n a}{\mu} = k^2 a^2 \log(2\gamma/\iota k a). \tag{52}$$

This equation gives a value for k^2 which is very large compared with p^2/V^2, so that approximately $m = k$. We also see that k or m is small compared with n, we may therefore put

$$n = \{4\pi\mu\iota p/\sigma\}^{\frac{1}{2}}.$$

Thus equation (52) becomes

$$k^2 \mathrm{a}^2 \log \frac{\iota k \mathrm{a}}{2\gamma} = \left\{\frac{4\pi p \mathrm{a}^2}{\mu\sigma}\right\}^{\frac{1}{2}} \epsilon^{\frac{\iota\pi}{4}},$$

or putting $\iota k\mathrm{a}/2\gamma = q$,

$$q^2 \log q^2 = -\frac{1}{2\gamma^2}\left\{\frac{4\pi p \mathrm{a}^2}{\mu\sigma}\right\}^{\frac{1}{2}} \epsilon^{\frac{\iota\pi}{4}}.$$

To solve this equation put $q^2 = w\epsilon^{\iota\psi}$; equating real and imaginary parts, we get

$$w \log w \cos\psi - w\psi \sin\psi = -\frac{1}{\gamma^2}\left\{\frac{\pi p \mathrm{a}^2}{2\mu\sigma}\right\}^{\frac{1}{2}},$$

$$w \log w \sin\psi + w\psi \cos\psi = -\frac{1}{\gamma^2}\left\{\frac{\pi p \mathrm{a}^2}{2\mu\sigma}\right\}^{\frac{1}{2}}.$$

Since w is very small, the terms in $\log w$ are much the most important; an approximate solution of these equations is, therefore, since the solution of $x \log x = -y$, is $x = -y/\log y$,

$$w = -\frac{\dfrac{1}{\gamma^2}\left\{\dfrac{\pi p \mathrm{a}^2}{\mu\sigma}\right\}^{\frac{1}{2}}}{\log \dfrac{1}{\gamma^2}\left\{\dfrac{\pi p \mathrm{a}^2}{\mu\sigma}\right\}^{\frac{1}{2}}},$$

$$\psi = \frac{\pi}{4}.$$

Hence, since $k = m$ and $k^2\mathrm{a}^2 = -4\gamma^2 w \epsilon^{\frac{\iota\pi}{4}}$, we find

$$m\mathrm{a} = 2\gamma\sqrt{-w}\left\{\cos\frac{\pi}{8} + \iota \sin\frac{\pi}{8}\right\}$$

$$= 2\gamma w^{\frac{1}{2}}\left(\cos\frac{5\pi}{8} + \iota \sin\frac{5\pi}{8}\right).$$

Thus, since in the expression for c there is the factor $\epsilon^{\iota m z}$, we see that c will fade away to $1/\epsilon$ of its initial value at a distance from the origin equal to

$$\frac{\mathrm{a}}{2\gamma w^{\frac{1}{2}}} \operatorname{cosec} \frac{5\pi}{8},$$

or substituting the value of w just found,

$$\frac{\mathrm{a}}{2} \operatorname{cosec} \frac{5\pi}{8} \left\{\frac{\mu\sigma}{\pi p \mathrm{a}^2}\right\} \left\{\log \gamma^2 \left(\frac{\mu\sigma}{\pi p \mathrm{a}^2}\right)^{\frac{1}{2}}\right\}^{\frac{1}{2}}.$$

This distance is much shorter than the corresponding one

when the current was uniformly distributed over the cross-section of the wire, and the important factor varies as $\mu^{\frac{1}{4}}$ instead of $\mu^{\frac{1}{2}}$. Thus the leakage of the lines of magnetic force out of the iron cylinder is much greater when the alternations are rapid than when they are slow. This is in accordance with the conclusion we came to from general reasoning at the beginning of Art. 278.

The result of this investigation points strongly to the advisability of very fine lamination of the core of a transformer, so as to get a uniform distribution of magnetic force over the iron and thus avoid magnetic leakage. There are many other advantages gained by fine lamination, of which one, more important than the effect we are considering, is the diminution in the quantity of heat dissipated by eddy currents. We shall proceed to consider in the next article the dissipation of energy by the currents in the wire.

Dissipation of Energy by the Heat produced by Alternating Currents.

280.] A great deal of light is thrown on the laws which govern the decay of currents in conductors by the consideration of the circumstances which affect the amount of heat produced in unit time by these currents. As we have obtained the expressions for these currents we could determine their heating effect by direct integration; we shall however proceed by a different method for the sake of introducing a very important theorem due to Professor Poynting, and given by him in his paper 'On the Transfer of Energy in the Electromagnetic Field,' *Phil. Trans.* 1884, Part II, p. 343. The theorem is that

$$\frac{K}{4\pi}\iiint\left(P\frac{dP}{dt}+Q\frac{dQ}{dt}+R\frac{dR}{dt}\right)dx\,dy\,dz$$

$$+\frac{\mu}{4\pi}\iiint\left(a\frac{da}{dt}+\beta\frac{d\beta}{dt}+\gamma\frac{d\gamma}{dt}\right)dx\,dy\,dz$$

$$+\iiint(X\dot{x}+Y\dot{y}+Z\dot{z})\,dx\,dy\,dz+\iiint(Pp+Qq+Rr)\,dx\,dy\,dz$$

$$=\frac{1}{4\pi}\iint\{l(R'\beta-Q'\gamma)+m(P'\gamma-R'a)+n(Q'a-P'\beta)\}\,dS,$$

where the volume integrals on the left-hand side are taken throughout the volume contained by the closed surface S, of which dS is an element and l, m, n the direction cosines of the normal drawn outwards.

P, Q, R are the components of the electromotive intensity.

α, β, γ those of the magnetic force.

X, Y, Z those of the mechanical force acting on the body in consequence of the passage of currents through it.

$\dot{x}$, $\dot{y}$, $\dot{z}$ the components of the velocity of a point in the body.

p, q, r the components of the conduction currents.

P', Q', R' the parts of the components of the electromotive intensity which do not depend upon the motion of the body.

K the specific inductive capacity and μ the magnetic permeability.

The following proof of this theorem is taken almost verbatim from Professor Poynting's paper. Let u, v, w be the components of the total current, which is the sum of the polarization and conduction currents; we have, since the components of the former are respectively

$$\frac{K}{4\pi}\frac{dP}{dt}, \quad \frac{K}{4\pi}\frac{dQ}{dt}, \quad \frac{K}{4\pi}\frac{dR}{dt},$$

$$\frac{K}{4\pi}\frac{dP}{dt} = u - p,$$

$$\frac{K}{4\pi}\frac{dQ}{dt} = v - q,$$

$$\frac{K}{4\pi}\frac{dR}{dt} = w - r.$$

Hence

$$\frac{K}{4\pi}\iiint\left(P\frac{dP}{dt} + Q\frac{dQ}{dt} + R\frac{dR}{dt}\right)dx\,dy\,dz$$

$$= \iiint\{P(u-p) + Q(v-q) + R(w-r)\}\,dx\,dy\,dz$$

$$= \iiint(Pu + Qv + Rw)\,dx\,dy\,dz$$

$$- \iiint(Pp + Qq + Rr)\,dx\,dy\,dz. \qquad (53)$$

Now (Maxwell's *Electricity and Magnetism*, Vol. II, Art. 598),

$$P = c\dot{y} - b\dot{z} - \frac{dF}{dt} - \frac{d\psi}{dx} = c\dot{y} - b\dot{z} + P',$$

$$Q = a\dot{z} - c\dot{x} - \frac{dG}{dt} - \frac{d\psi}{dy} = a\dot{z} - c\dot{x} + Q',$$

$$R = b\dot{x} - a\dot{y} - \frac{dH}{dt} - \frac{d\psi}{dz} = b\dot{x} - a\dot{y} + R',$$

where P', Q', R' are the parts of P, Q, R which do not contain the velocities.

Thus

$$\begin{aligned} Pu + Qv + Rw &= (c\dot{y} - b\dot{z})\,u + (a\dot{z} - c\dot{x})\,v + (b\dot{x} - a\dot{y})\,w + P'u + Q'v + R'w, \\ &= -\{(vc - wb)\,\dot{x} + (wa - uc)\,\dot{y} + (ub - va)\,\dot{z}\} + P'u + Q'v + R'w, \\ &= -\{X\dot{x} + Y\dot{y} + Z\dot{z}\} + P'u + Q'v + R'w; \end{aligned}$$

where X, Y, Z are the components of the mechanical force per unit volume (Maxwell, Vol. II, Art. 603).

Substituting this value for $Pu + Qv + Rw$ in (53) and transposing, we obtain

$$\frac{K}{4\pi}\iiint\left(P\frac{dP}{dt} + Q\frac{dQ}{dt} + R\frac{dR}{dt}\right)dx\,dy\,dz$$

$$+ \iiint(X\dot{x} + Y\dot{y} + Z\dot{z})\,dx\,dy\,dz + \iiint(Pp + Qq + Rr)\,dx\,dy\,dz$$

$$= \iiint(P'u + Q'v + R'w)\,dx\,dy\,dz. \qquad (54)$$

Now

$$4\pi u = \frac{d\gamma}{dy} - \frac{d\beta}{dz},$$

$$4\pi v = \frac{da}{dz} - \frac{d\gamma}{dx},$$

$$4\pi w = \frac{d\beta}{dx} - \frac{da}{dy}.$$

Substituting these values for u, v, w in the right-hand side of equation (54), that side of the equation becomes

$$\frac{1}{4\pi}\iiint\left\{P'\left(\frac{d\gamma}{dy} - \frac{d\beta}{dz}\right) + Q'\left(\frac{da}{dz} - \frac{d\gamma}{dx}\right) \;\wedge\; \left(\frac{d\beta}{dx} - \frac{da}{dy}\right)\right\}dx\,dy\,dz$$

$$= \frac{1}{4\pi}\iiint\left[\left\{R'\frac{d\beta}{dx} - Q'\frac{d\gamma}{dx}\right\} + \left\{P'\frac{d\gamma}{dy} - R'\frac{da}{dy}\right\} + \left\{Q'\frac{da}{dz} - P'\frac{d\beta}{dz}\right\}\right]dx\,dy\,dz.$$

Integrating by parts, we find that the expression is equal to

$$\frac{1}{4\pi}\iint (R'\beta - Q'\gamma)\,dydz + \frac{1}{4\pi}\iint (P'\gamma - R'a)\,dxdz$$
$$+\frac{1}{4\pi}\iint (Q'a - P'\beta)\,dxdy$$
$$-\frac{1}{4\pi}\iiint \left(\beta\frac{dR'}{dx} - \gamma\frac{dQ'}{dx} + \gamma\frac{dP'}{dy} - a\frac{dR'}{dy} + a\frac{dQ'}{dz} - \beta\frac{dP'}{dz}\right)dxdydz,$$

the double integrals being taken over the closed surface. This expression may be written as

$$\frac{1}{4\pi}\iint \{l(R'\beta - Q'\gamma) + m(P'\gamma - R'a) + n(Q'a - P'\beta)\}\,dS$$
$$-\frac{1}{4\pi}\iiint \Big\{a\left(\frac{dQ'}{dz} - \frac{dR'}{dy}\right) + \beta\left(\frac{dR'}{dx} - \frac{dP'}{dz}\right)$$
$$+\gamma\left(\frac{dP'}{dy} - \frac{dQ'}{dx}\right)\Big\}\,dxdydz,$$

where dS is an element of the surface and l, m, n are the direction cosines of the normal to the surface drawn outwards.

But
$$\frac{dQ'}{dz} - \frac{dR'}{dy} = \frac{d}{dt}\left(\frac{dH}{dy} - \frac{dG}{dz}\right)$$
$$= \frac{da}{dt} = \mu\frac{da}{dt}.$$

Similarly
$$\frac{dR'}{dx} - \frac{dP'}{dz} = \frac{db}{dt} = \mu\frac{d\beta}{dt},$$
$$\frac{dP'}{dy} - \frac{dQ'}{dx} = \frac{dc}{dt} = \mu\frac{d\gamma}{dt}.$$

Hence we see that the right-hand side of (54) is equal to

$$\frac{1}{4\pi}\iint \{l(R'\beta - Q'\gamma) + m(P'\gamma - R'a) + n(Q'a - P'\beta)\}\,dS$$
$$-\frac{\mu}{4\pi}\iiint \left(a\frac{da}{dt} + \beta\frac{d\beta}{dt} + \gamma\frac{d\gamma}{dt}\right)dxdydz.$$

Transposing the last term to the other side of the equation, we get

$$\frac{K}{4\pi}\iiint\left(P\frac{dP}{dt}+Q\frac{dQ}{dt}+R\frac{dR}{dt}\right)dx\,dy\,dz$$

$$+\frac{\mu}{4\pi}\iiint\left(a\frac{da}{dt}+\beta\frac{d\beta}{dt}+\gamma\frac{d\gamma}{dt}\right)dx\,dy\,dz$$

$$+\iiint(X\dot{x}+Y\dot{y}+Z\dot{z})\,dx\,dy\,dz+\iiint(Pp+Qq+Rr)\,dz\,dy\,dz$$

$$=\frac{1}{4\pi}\iint\{l(R'\beta-Q'\gamma)+m(P'\gamma-R'a)+n(Q'a-P'\beta)\}\,dS, \quad (55)$$

which is the theorem we set out to prove.

Now the electrostatic energy inside the closed surface is (Maxwell, Art. 631)

$$\frac{1}{2}\iiint(Pf+Qg+Rh)\,dx\,dy\,dz,$$

or since $\quad f=\dfrac{K}{4\pi}P, \quad g=\dfrac{K}{4\pi}Q, \quad h=\dfrac{K}{4\pi}R,$

$$\frac{K}{8\pi}\iiint(P^2+Q^2+R^2)\,dx\,dy\,dz.$$

The electromagnetic energy inside the same surface is (Maxwell, Art. 635)

$$\frac{1}{8\pi}\iiint(aa+b\beta+c\gamma)\,dx\,dy\,dz,$$

or

$$\frac{\mu}{8\pi}\iiint(a^2+\beta^2+\gamma^2)\,dx\,dy\,dz.$$

Thus the first two integrals on the left-hand side of equation (55) express the gain per second in electric and magnetic energy. The third integral expresses the work done per second by the mechanical forces. The fourth integral expresses the energy transformed per second in the conductor into heat, chemical energy, and so on. Thus the left-hand side expresses the total gain in energy per second within the closed surface, and equation (55) expresses that this gain in energy may be regarded as coming across the bounding surface, the amount crossing that surface per second being expressed by the right-hand side of that equation.

Thus we may regard the change in the energy inside the closed surface as due to the transference of energy across that surface; the energy moving at right angles both to **H**, the resultant magnetic force, and to **E**, the resultant of P', Q', R'. The amount of energy which in unit time crosses unit area at right angles to the direction of the energy flow is $\mathbf{HE}\sin\theta/4\pi$, where θ is the angle between **H** and **E**. The direction of the energy flow is related to those of **H** and **E** in such a way that the rotation of a positive screw from **E** to **H** would be accompanied by a translation in the direction of the flow of energy.

Equation (55) justifies us in asserting that we shall arrive at correct results as to the changes in the distribution of energy in the field if we regard the energy as flowing in accordance with the laws just enunciated: it does not however justify us in asserting that the flow of energy at any point *must* be that given by these laws, for we can find an indefinite number of quantities u_s, v_s, w_s of the dimensions of flow of energy which satisfy the condition

$$\iint (lu_s + mv_s + nw_s)\,dS = 0,$$

where the integration is extended over any closed surface. Hence, we see that if the components of the flow of energy were

$$\begin{aligned} &R'\beta - Q'\gamma + \Sigma u_s \text{ instead of } R'\beta - Q'\gamma,\\ &P'\gamma - R'\alpha + \Sigma v_s \text{ instead of } P'\gamma - R'\alpha,\\ &Q'\alpha - P'\beta + \Sigma w_s \text{ instead of } Q'\alpha - P'\beta, \end{aligned}$$

the changes in the distribution of energy would still be those which actually take place.

Though Professor Poynting's investigation does not give a unique solution of the problem of finding the flow of energy at any point in the electromagnetic field, it is yet of great value, as the solution which it does give is simple and one that readily enables us to form a consistent and vivid representation of the changes in the distribution of energy which are going on in any actual case that we may have under consideration. Several applications of this theorem are given by Professor Poynting in the paper already quoted, to which we refer the reader. We shall now proceed to apply it to the determination of the rate of heat production in wires at rest traversed by alternating currents.

281.] Since the currents are periodic, P^2, Q^2, R^2, α^2, β^2, γ^2 will be of the form

$$A + B\cos(2pt+\theta),$$

where A and B do not involve the time; hence the first two integrals on the left-hand side of equation (55) will be multiplied by factors which, as far as they involve t, will be of the form $\sin(2pt+\theta)$; hence, if we consider the mean value of these terms over a time involving a great many oscillations of the currents, they may be neglected: the gain or loss of energy represented by these terms is periodic, and at the end of a period the energy is the same as at the beginning. The third term on the left-hand side vanishes in our case because the wires are at rest, and since $\dot{x}$, $\dot{y}$, $\dot{z}$ vanish P', Q', R' become identical with P, Q, R.

Thus when the effects are periodic we see that equation (55) leads to the result that the mean value with respect to the time of

$$\iiint (Pp + Qq + Rr)\, dx dy dz$$

is equal to that of

$$\frac{1}{4\pi}\iint \{l(R\beta - Q\gamma) + m(P\gamma - R\alpha) + n(Q\alpha - P\beta)\}\, dS.$$

The first of these expressions is, however, the mean rate of heat production, and in the case of a wire whose electrical state is symmetrical with respect to its axis, the value of the quantity under the sign of integration is the same at each point of the circumference of a circle whose plane is at right angles to the axis of the wire; hence in this case we have the result:

The mean rate of heat production per unit length of the wire is equal to the mean value of

$$\tfrac{1}{2}\mathbf{a}\,(\text{tangential electromotive intensity}) \times (\text{tangential magnetic force}), \qquad (56)$$

a, as before, being the radius of the wire.

282.] Let us apply this result to find the rate of heat production in the wire and in the outer conductor of a cable when the current is parallel to the axis of the wire. By the methods of Art. 268, we see that if the total current through the wire at the point z is equal to the real part of

$$I_0 \epsilon^{\iota(mz+pt)},$$

or if $m = -a + \iota\beta$, to

$$I_0 \epsilon^{-\beta z} \cos(-az + pt),$$

then, Art. 268, equation (24), the electromotive intensity R in the wire parallel to the axis of z is equal to the real part of

$$-\frac{\iota\sigma n}{2\pi\mathrm{a}} \frac{J_0(\iota nr)}{J_0'(\iota n\mathrm{a})} I_0 \epsilon^{-\beta z} \epsilon^{\iota(-az+pt)}. \tag{57}$$

If we neglect the polarization currents in the dielectric in comparison with the conduction currents through the wire, then the line integral of the magnetic force round the inner surface of the outer conductor must equal $4\pi I_0 \epsilon^{\iota(mz+pt)}$; using this principle we see that E in equation (11), Art. 262, equals $-\iota n'\sigma' I_0/2\pi\mathrm{b} K_0'(\iota n'\mathrm{b})$, and hence the electromotive intensity parallel to z in the outer conductor is equal to the real part of

$$-\frac{\iota\sigma' n'}{2\pi\mathrm{b}} \frac{K_0(\iota n'r)}{K_0'(\iota n'\mathrm{b})} I_0 \epsilon^{-\beta z} \epsilon^{\iota(-az+pt)}, \tag{58}$$

the notation being the same as in Art. 262.

The tangential magnetic force at the surface of the wire is (Art. 262)

$$\frac{2I_0}{\mathrm{a}} \epsilon^{-\beta z} \cos(-az + pt), \tag{59}$$

while that at the surface of the outer conductor is, if we neglect the polarization currents in the dielectric in comparison with the conduction currents through the wire,

$$\frac{2I_0}{\mathrm{b}} \epsilon^{-\beta z} \cos(-az + pt). \tag{60}$$

Let us now consider the case when the rate of alternation of the current is so slow that both $n\mathrm{a}$ and $n'\mathrm{b}$ are small quantities. When $n\mathrm{a}$ is small $J_0'(\iota n\mathrm{a}) = -\iota n\mathrm{a}/2$, while $J_0(\iota n\mathrm{a}) = 1$ approximately; hence, putting $r = \mathrm{a}$ in (57), we find that the tangential electromotive intensity is

$$\frac{\sigma}{\pi\mathrm{a}^2} I_0 \epsilon^{-\beta z} \cos(-az + pt).$$

Hence by (56) and (59) the rate of heat production in the wire is equal to the mean value of

$$\frac{\sigma}{\pi\mathrm{a}^2} I_0^2 \epsilon^{-2\beta z} \cos^2(-az + pt),$$

that is to $$\frac{\sigma}{2\pi \mathbf{a}^2} I_0^2 \epsilon^{-2\beta z}.$$

Let us now consider the rate of heat production in the outer conductor; since $n'\mathbf{b}$ is very small, we have approximately

$$K_0(\iota n'\mathbf{b}) = \log(2\gamma/\iota n'\mathbf{b}), \quad K_0'(\iota n'\mathbf{b}) = -1/\iota n'\mathbf{b}.$$

Making these substitutions in (58), we see that the tangential electromotive intensity at the surface of the outer conductor is equal to the real part of

$$-\frac{\sigma' n'^2}{2\pi} \log(2\gamma/\iota n'\mathbf{b}) I_0 \epsilon^{-\beta z} \epsilon^{\iota(-az+pt)},$$

and since $n'^2 = 4\pi\mu'\iota p/\sigma'$, the real part of this expression is

$$2\mu' p \log(\gamma\sqrt{\sigma'/\pi\mu' p\mathbf{b}^2}) I_0 \epsilon^{-\beta z} \sin(-az+pt)$$
$$-\tfrac{3}{2}\pi\mu' p I_0 \epsilon^{-\beta z} \cos(-az+pt).$$

Hence by (56) and (60) the rate of heat production in the outer conductor is equal to

$$\tfrac{3}{4}\pi\mu' p I_0^2 \epsilon^{-2\beta z},$$

since the mean value with respect to the time of

$$\sin(-az+pt)\cos(-az+pt)$$

is zero. Thus, when $n'\mathbf{b}$ is small, the rate of heat production in the outer conductor is independent both of the radius and specific resistance of that conductor. The ratio of the heat produced in unit time in the wire to that produced in the outer conductor is thus $2\sigma/3\pi^2\mathbf{a}^2\mu' p$, which is very large since we have assumed that $n^2\mathbf{a}^2$, i.e. $4\pi\mu p\mathbf{a}^2/\sigma$, is a small quantity; in this case, therefore, by far the larger proportion of the heat is produced in the wire. This explains the result found in Art. 263 that the rate of decay of the vibrations is nearly independent of the resistance of the outer conductor and depends almost wholly upon that of the wire.

283.] When the frequency is so great that $n\mathbf{a}$ is large though $n'\mathbf{b}$ is still small, then $J_0(\iota n\mathbf{a}) = \iota J_0'(\iota n\mathbf{a})$, so that by (57) the tangential electromotive intensity at the surface of the wire is equal to the real part of

$$\frac{\sigma}{2\pi\mathbf{a}} \{4\pi\mu\iota p/\sigma\}^{\frac{1}{2}} I_0 \epsilon^{-\beta z} \epsilon^{\iota(-az+pt)},$$

which is equal to

$$\frac{\sigma}{2\pi \mathbf{a}}\{2\pi\mu p/\sigma\}^{\frac{1}{2}} I_0 \epsilon^{-\beta z}\{\cos(-az+pt)-\sin(-az+pt)\}.$$

Hence by (56) and (59) the mean rate of heat production in the wire is equal to

$$\frac{\sigma}{4\pi \mathbf{a}}(2\pi\mu p/\sigma)^{\frac{1}{2}} I_0{}^2 \epsilon^{-2\beta z}.$$

Since $n'\mathbf{b}$ is supposed to be small the rate of heat production in the outer conductor is as before

$$\frac{3\pi}{4}\mu' p I_0{}^2 \epsilon^{-2\beta z},$$

hence the ratio of the amount of heat produced in unit time in the wire to that produced in the outer conductor is

$$\frac{\mu}{\mu'}\left\{\frac{2\sigma}{9\pi^3 p\mu \mathbf{a}^2}\right\}^{\frac{1}{2}}.$$

Thus, since $n^2\mathbf{a}^2$ and so $4\pi p\mu\mathbf{a}^2/\sigma$ is very large by hypothesis, we see that unless μ/μ' is very large this ratio will be very small; in other words the greater part of the heat is produced in the outer conductor; this is in accordance with the result obtained in Art. 266, which showed that the rate of decay of the vibrations was independent of the resistance of the wire.

284.] When the frequency is so high that both $n\mathbf{a}$ and $n'\mathbf{b}$ are large, then the expression for the heat produced in the wire is that just found. To find the heat produced in the outer conductor we have, when $n'\mathbf{b}$ is very large,

$$K_0(\iota n'\mathbf{b}) = -\iota K_0{}'(\iota n'\mathbf{b});$$

hence by (58) the tangential electromotive intensity in the outer conductor is equal to the real part of

$$-\frac{\sigma'}{2\pi\mathbf{b}}(4\pi\mu'\iota p/\sigma')^{\frac{1}{2}} I_0 \epsilon^{-\beta z}\epsilon^{\iota(-az+pt)},$$

which is equal to

$$-\frac{\sigma'}{2\pi\mathbf{b}}(2\pi\mu' p/\sigma')^{\frac{1}{2}} I_0 \epsilon^{-\beta z}\{\cos(-az+pt)-\sin(-az+pt)\}.$$

Hence by (56) and (60) the mean rate of heat production in the outer conductor is

$$\frac{\sigma'}{4\pi\mathbf{b}}\{2\pi\mu' p/\sigma'\}^{\frac{1}{2}} I_0^2 \epsilon^{-2\beta z}.$$

Thus the ratio of the heat produced in unit time in the wire to that produced in the same time in the outer conductor is

$$\left\{\frac{\mu\sigma}{\mathrm{a}^2}\right\}^{\frac{1}{2}} \Big/ \left\{\frac{\mu'\sigma'}{\mathrm{b}^2}\right\}^{\frac{1}{2}},$$

so that if, as is generally the case in cables, σ' is very much greater than σ, by far the larger part of the heat will be produced in the outer conductor.

Heat produced by Foucault Currents in a Transformer.

285.] We shall now proceed to consider the case discussed in Art. 278, where the lines of magnetic force are in planes through the axis of the wire, the currents flowing in circles in planes at right angles to this axis. This case is one which is of great practical importance, as the conditions approximate to those which obtain in the soft iron cylindrical core of an induction coil or a transformer; in this case the windings of the primary coil are in planes at right angles to the axis of the iron cylinder, while the lines of magnetic force due to the primary coil are in planes passing through this axis. When a variable current is passing through the primary coil, currents are induced which heat the core and the heat thus produced is wasted as far as the production of useful work is concerned; it is thus a matter of importance to investigate the laws which govern its development, so that the apparatus may be designed in such a way as to reduce this waste to a minimum. We shall suppose that the magnetic force parallel to the axis at the surface of the wire is represented by the real part of

$$H\epsilon^{\iota(mz+pt)},$$

or if $m = \alpha + \iota\beta$, by

$$H\epsilon^{-\beta z}\cos(\alpha z + pt).$$

The magnetic force at the surface of the cylinder is the most convenient quantity in which to express the rate of heat production, for it is due entirely to the external field and is not, when the field is uniform, affected by the currents in the wire itself.

Using the notation of Art. 278 we see by the results of that article that in the wire

$$c = AJ_0(\iota nr)\epsilon^{\iota(mz+pt)}.$$

The tangential electromotive intensity Θ is given by the equation

$$\frac{dc}{dt} = -\frac{1}{r}\frac{d}{dr}(r\Theta);$$

hence
$$\Theta = \frac{p}{n} A J_0'(\iota n r)\,\epsilon^{\iota(mz+pt)};$$

but since at the surface of the wire, c is equal to the real part of

$$\mu H \epsilon^{-\beta z}\epsilon^{\iota(az+pt)},$$

we see that at the surface $\Theta =$ real part of

$$\frac{p}{n}\frac{J_0'(\iota n \mathrm{a})}{J_0(\iota n \mathrm{a})}\mu H \epsilon^{-\beta z}\epsilon^{\iota(az+pt)}. \qquad (61)$$

Let us first take the case when the radius of the wire is so small that $n\mathrm{a}$ is small; in this case we have, since approximately

$$J_0(x) = 1 - \frac{x^2}{2^2} + \frac{x^4}{2^2 4^2},$$

$$J_0'(x) = -\tfrac{1}{2}x\left\{1 - \frac{x^2}{8}\right\},$$

and
$$n^2 = 4\pi\mu\iota p/\sigma,$$

$\Theta =$ real part of

$$-\tfrac{1}{2}\iota \mathrm{a} p\left\{1 - \frac{\pi\mu\iota p \mathrm{a}^2}{2\sigma}\right\}\mu H \epsilon^{-\beta z}\epsilon^{\iota(az+pt)}$$

$$= \tfrac{1}{2}\mu \mathrm{a} p H \epsilon^{-\beta z}\sin(az+pt) - \frac{1}{4\sigma}\pi\mu^2 p^2 \mathrm{a}^3 H \epsilon^{-\beta z}\cos(az+pt).$$

But by equation (56) the rate of heat production in the wire per unit length is equal to the mean value

$$-\tfrac{1}{2}\mathrm{a}\Theta H \epsilon^{-\beta z}\cos(az+pt);$$

where the minus sign has been taken because (Art. 280) ΘH is proportional to the rate of flow of energy in the direction of translation of a right-handed screw twisting from Θ to H; in this case this direction is radially *outwards*.

Thus the rate of heat production in the wire is

$$\frac{1}{16\sigma}\pi\mu^2 p^2 \mathrm{a}^4 H^2 \epsilon^{-2\beta z},$$

and is thus proportional to the conductivity, so that good conductors will in this case absorb more energy than bad ones.

Let us now apply this result to find the energy absorbed in the core of a transformer or induction coil. We shall suppose that the core consists of iron wire of circular section, the wires being insulated from each other by the coating of rust with which they are covered. We shall consider the case when the magnetic force due to the primary coil is uniform both along the axis of the coil and over its cross-section. When the external magnetic force is uniform along z, the axis of a wire, the currents induced in the wire by the variation of the magnetic force flow in circles whose planes are at right angles to z, and the intensities of the currents are independent of the value of z. Under these circumstances the currents in the wire do not give rise to any magnetic force outside it. The magnetic force outside the wires will thus be due entirely to the primary coil, and as this magnetic force is uniform over the cross-section it will be the same for each of the wires, so that we can apply the preceding investigation to the wires separately. In order to use the whole of the iron, the magnetic force must be approximately uniformly distributed over the cross-section of the wires; for this to be the case $n\mathbf{a}$ must be small, as we have seen that when $n\mathbf{a}$ is large the magnetic force is confined to a thin skin round each wire. For soft iron, for which we may put $\mu = 10^3$, $\sigma = 10^4$, the condition that $n\mathbf{a}$ is small implies that when the primary current makes one hundred alternations per second, the radius of the wire should not be more than half a millimetre. If now the total cross-section of the iron is kept constant so as to keep the magnetic induction through the core constant, we have, if $\mathbf{N}$ is the number of wires, $\mathbf{A}$ the total cross-section of the iron,

$$\mathbf{N}\pi\mathbf{a}^2 = \mathbf{A}.$$

The heat produced in all the wires per unit length of core in one second is, if H is the maximum magnetic force due to the coil,

$$\frac{\mathbf{N}}{16\sigma}\pi\mu^2 p^2 \mathbf{a}^4 H^2,$$

or

$$\frac{\mathbf{A}^2}{16\pi\sigma\mathbf{N}}\mu^2 p^2 H^2,$$

and is thus *inversely proportional to the number of wires.* We may therefore diminish the waste of energy due to the heat produced by the induced currents in the wires by increasing the number of wires in the core. We thus arrive at the practical rule that to diminish the waste of work by eddy currents the core should be made up of as fine wire as possible. In many transformers the iron core is built up of thin plates instead of wires; when this is the case the advantage of a fine sub-division of the core is even more striking than for wires, for we can easily prove that the work wasted by eddy currents is inversely proportional to the square of the number of plates (see J. J. Thomson, *Electrician*, 28, p. 599, 1892).

If γ is the current flowing through the primary coil and N the number of turns of this coil per centimetre, then

$$H \cos pt = 4\pi N\gamma,$$

and $$\tfrac{1}{2}H^2 = 16\pi^2 N^2 (\text{mean value of } \gamma^2),$$

thus in the case of a cylindrical core of radius $\mathbf{a}$ the heat produced in one second in a length l of the core will be

$$2\pi^3 \mu^2 p^2 \mathbf{a}^4 N^2 l (\text{mean value of } \gamma^2)/\sigma.$$

If $\mathbf{Q}$ is the *impedance* of a circuit (Art. 272) the heat produced in unit time is equal to

$$\mathbf{Q}\ (\text{mean value of } \gamma^2);$$

thus the core will increase the impedance of the primary coil by

$$2\pi^3 \mu^2 p^2 \mathbf{a}^4 N^2 l/\sigma.$$

286.] Let us now consider the case when $n\mathbf{a}$ is large; here we have

$$J_0'(\iota n\mathbf{a}) = -\iota J_0(\iota n\mathbf{a}),$$

and since $$n^2 = 4\pi\mu\iota p/\sigma,$$

we see by (61), putting α and β equal to zero, that

$$\Theta = \text{real part of}$$

$$-\sqrt{\frac{p\mu\sigma}{4\pi}}\,\epsilon^{\frac{\iota\pi}{4}} H\epsilon^{\iota pt}$$

$$= -\sqrt{\frac{p\mu\sigma}{8\pi}}\, H\{\cos pt - \sin pt\}.$$

But by equation (56) the rate of heat production per unit length is equal to the mean value of

$$-\tfrac{1}{2}\mathbf{a}\Theta H \cos pt,$$

and is thus equal to $$\frac{1}{8}\sqrt{\frac{p\mu\sigma}{\pi\,2}}\,\mathbf{a}H^2.$$

We can show, as before, that this corresponds to an increase in the impedance of the primary circuit equal to

$$4\pi^2 l N^2 \{p\mu\sigma/2\pi\}^{\frac{1}{2}} \mathbf{a}.$$

In this case the heat produced is proportional to the square root of the specific resistance of the core, so the worse the conductivity of the core the greater the amount of heat produced by eddy currents, whereas in the case when $n\mathbf{a}$ was small, the greater the conductivity of the core the greater was the loss due to heating.

When $n\mathbf{a}$ is large, the heat produced varies as the circumference of the core instead of, as in the previous case, as the square of the area; it also varies much more slowly with the frequency and magnetic permeability. This is due to the fact that when $n\mathbf{a}$ is large the currents are not uniformly distributed over the core but confined to a thin layer on the outside, the thickness of this layer diminishing as the magnetic permeability or the frequency increases; thus, though an increase in μ or p may be accompanied by an increase in the intensity of the currents, it will also be attended by a diminution in the area over which the currents are spread, and thus the effect on the heat produced of the increase in p or μ will not be so great as in the previous case when $n\mathbf{a}$ was small, and when no limitation in the area over which the current was spread accompanied an increase in the frequency or magnetic permeability.

If we compare the absorption of energy when $n\mathbf{a}$ is large by cores of iron and copper of the same size subject to alternating currents of the same frequency, we find—since for iron μ may be taken as 10^3 and σ as 10^4, while for copper $\mu = 1$, $\sigma = 1600$,—that the absorption of energy by the iron core is between 70 and 80 times that by the copper. The greater absorption by the iron can be very easily shown by an experiment of the kind figured in Art. 85, in which two coils are placed in the circuit connecting the outer coatings of two Leyden Jars; in one of these coils an exhausted bulb is placed, while the core in which the heat produced is to be measured is placed in the other. When the oscillating current produced by the discharge of the jars passes through the coils a brilliant discharge passes through the exhausted bulb in A, if the coil B is empty or if it contains a copper cylinder; if however an iron cylinder of the same size replaces the copper one, the discharge in the bulb is at once extinguished, showing that the iron cylinder has absorbed a great deal more energy than the copper one. This experiment also shows that iron retains its magnetic properties even when the forces to which it is exposed are reversed, as in this experiment, millions of times in a second.

287.] Another remarkable result is that though a cylinder or tube of a non-magnetic metal does not stop the discharge in the bulb in **A**, yet if a piece of glass tubing of the same size is coated with thin tinfoil or Dutch metal, or if it has a film of silver deposited upon it, it will check the discharge very decidedly. We are thus led to the somewhat unexpected result that a thin layer of metal when exposed to very rapidly alternating currents may absorb more energy than a thick layer. The following investigation affords the explanation of this, and shows that there is a certain thickness for which the heat produced is a maximum. This result can easily be verified by the arrangement just described, for if an excessively thin film of silver is deposited on a beaker very little effect is produced on the discharge in the bulb placed in **A**, but if successive layers of very thin tinfoil are wrapped round the beaker over the silver film the brightness of the discharge in **A** at first rapidly diminishes, it however soon increases again, and when a few layers of tinfoil have been wrapped round the beaker the discharge becomes almost as bright as if the beaker were away.

To investigate the theory of this effect we shall calculate the energy absorbed by a metal tube of circular cross-section, when placed inside a primary coil whose windings are in planes at right angles to the axis of the tube; this coil is supposed to be long, and uniformly wound, so that the distribution of magnetic force and current is the same in all planes at right angles to its axis. We shall use the same notation as before; the only symbols which it is necessary to define again are **a** and **b**, which are respectively the internal and external radius of the tube, and V the velocity with which electromagnetic action is propagated through the dielectric inside the tube. The magnetic force outside the tube is represented by the real part of $H\epsilon^{\iota pt}$, and this force is due entirely to the currents in the primary coil.

Then γ, the magnetic force parallel to the axis of the tube, may be (Art. 262) expressed by the following equations,

$$\gamma = AJ_0(\iota kr)\epsilon^{\iota pt} \text{ in the dielectric inside the tube,}$$

$$\gamma = \{BJ_0(\iota nr) + CK_0(\iota nr)\}\epsilon^{\iota pt} \text{ in the tube itself.}$$

Here $k^2 = -p^2/V^2$, $n^2 = 4\pi\mu\iota p/\sigma$, thus they represent the quantities represented by the same symbols in previous investigations, if in these we put $m = 0$.

Let $\mathbf{I}$ denote the tangential current at right angle to r and the axis of the cylinder, then

$$4\pi\mathbf{I} = -\frac{d\gamma}{dr},$$

if Θ is the tangential electromotive intensity in the same direction, then in the dielectric

$$\mathbf{I} = \frac{K}{4\pi}\frac{d\Theta}{dt}$$

$$= \frac{K}{4\pi}\iota p\Theta,$$

so that

$$\Theta = -\frac{V^2}{\iota p}\frac{d\gamma}{dr},$$

since $1/K = V^2$.

In the tube

$$\Theta = \sigma\mathbf{I}$$

$$= -\frac{\sigma}{4\pi}\frac{d\gamma}{dr}.$$

Since γ is continuous, we have

$$AJ_0(\iota k\mathbf{a}) = BJ_0(\iota n\mathbf{a}) + CK_0(\iota n\mathbf{a}),$$

and since Θ is continuous, we have

$$\frac{V^2k}{p}AJ_0'(\iota k\mathbf{a}) = \frac{\sigma\iota n}{4\pi}\{BJ_0'(\iota n\mathbf{a}) + CK_0'(\iota n\mathbf{a})\}.$$

Since $\iota k = p/V$, $\iota k\mathbf{a}$ will be very small, hence we may put

$$J_0(\iota k\mathbf{a}) = 1, \quad J_0'(\iota k\mathbf{a}) = -\tfrac{1}{2}\iota k\mathbf{a}.$$

Making these substitutions and remembering that

$$J_0(\iota n\mathbf{a})K_0'(\iota n\mathbf{a}) - J_0'(\iota n\mathbf{a})K_0(\iota n\mathbf{a}) = -\frac{1}{\iota n\mathbf{a}},$$

we find

$$B = -A\left\{K_0'(\iota n\mathbf{a}) + \frac{\iota n\mathbf{a}}{2\mu}K_0(\iota n\mathbf{a})\right\}\iota n\mathbf{a},$$

$$C = A\left\{J_0'(\iota n\mathbf{a}) + \frac{\iota n\mathbf{a}}{2\mu}J_0(\iota n\mathbf{a})\right\}\iota n\mathbf{a}.$$

To determine A we have the condition that when

$$r = \mathbf{b}, \quad \gamma = H\epsilon^{\iota pt},$$

hence
$$H\epsilon^{\iota pt} = \{BJ_0(\iota n\mathbf{b}) + CK_0(\iota n\mathbf{b})\}\,\epsilon^{\iota pt}.$$

In order to find the heat produced in the tube we require the value of Θ when $r = \mathbf{b}$; but here

$$\Theta = -\frac{\sigma\iota n}{4\pi}\{BJ_0'(\iota n\mathbf{b}) + CK_0'(\iota n\mathbf{b})\}\epsilon^{\iota pt}.$$

Eliminating B and C from these equations, we find

$$-\Theta = \text{real part of } H\epsilon^{\iota pt}\frac{\sigma\iota n}{4\pi}\times$$

$$\frac{\left\{J_0'(\iota n\mathbf{a})K_0'(\iota n\mathbf{b}) - J_0'(\iota n\mathbf{b})K_0'(\iota n\mathbf{a}) + \dfrac{\iota n\mathbf{a}}{2\mu}[J_0(\iota n\mathbf{a})K_0'(\iota n\mathbf{b}) - J_0'(\iota n\mathbf{b})K_0(\iota n\mathbf{a})]\right\}}{J_0'(\iota n\mathbf{a})K_0(\iota n\mathbf{b}) - J_0(\iota n\mathbf{b})K_0'(\iota n\mathbf{a}) + \dfrac{\iota n\mathbf{a}}{2\mu}[J_0(\iota n\mathbf{a})K_0(\iota n\mathbf{b}) - J_0(\iota n\mathbf{b})K_0(\iota n\mathbf{a})]}.$$

The effect we are considering is one which is observed when the rate of alternation of the current is very high, so that both $n\mathbf{a}$ and $n\mathbf{b}$ are very large; but when this is the case

$$J_0(\iota n\mathbf{a}) = \frac{\epsilon^{n\mathbf{a}}}{\sqrt{2\pi n\mathbf{a}}}, \qquad J_0'(\iota n\mathbf{a}) = -\frac{\iota\epsilon^{n\mathbf{a}}}{\sqrt{2\pi n\mathbf{a}}},$$

$$K_0(\iota n\mathbf{a}) = \epsilon^{-n\mathbf{a}}\sqrt{\frac{\pi}{2n\mathbf{a}}}, \qquad K_0'(\iota n\mathbf{a}) = \iota\epsilon^{-n\mathbf{a}}\sqrt{\frac{\pi}{2n\mathbf{a}}},$$

$$J_0(\iota n\mathbf{b}) = \frac{\epsilon^{n\mathbf{b}}}{\sqrt{2\pi n\mathbf{b}}}, \qquad J_0'(\iota n\mathbf{b}) = -\frac{\iota\epsilon^{n\mathbf{b}}}{\sqrt{2\pi n\mathbf{b}}},$$

$$K_0(\iota n\mathbf{b}) = \epsilon^{-n\mathbf{b}}\sqrt{\frac{\pi}{2n\mathbf{b}}}, \qquad K_0'(\iota n\mathbf{b}) = \iota\epsilon^{-n\mathbf{b}}\sqrt{\frac{\pi}{2n\mathbf{b}}};$$

making these substitutions and writing h for $\mathbf{b}-\mathbf{a}$, we find

$-\Theta =$ real part of

$$\frac{\sigma n}{4\pi}\,\frac{\epsilon^{nh} - \epsilon^{-nh} + \dfrac{n\mathbf{a}}{2\mu}(\epsilon^{nh} + \epsilon^{-nh})}{\epsilon^{nh} + \epsilon^{-nh} + \dfrac{n\mathbf{a}}{2\mu}(\epsilon^{nh} - \epsilon^{-nh})}\,H\epsilon^{\iota pt}. \qquad (62)$$

Now since $n\mathbf{a}$ is very large, $n\mathbf{a}/\mu$ is also very large for the non-magnetic metals, and even for the magnetic metals if the frequency of the currents in the primary is exceedingly large;

but when this is the case, then, unless h is so small that $n^2 \mathbf{a} h/\mu$ is no longer large, we may write equation (62) as

$$-\Theta = \text{real part of}$$

$$\frac{\sigma n}{4\pi} \frac{\epsilon^{nh} + \epsilon^{-nh}}{\epsilon^{nh} - \epsilon^{-nh}} H \epsilon^{\iota pt}. \tag{63}$$

Since $n = \{4\pi\mu\iota p/\sigma\}^{\frac{1}{2}}$, we may write $n = n_1(1+\iota)$, where $n_1 = \{2\pi\mu p/\sigma\}^{\frac{1}{2}}$, and equation (63) becomes

$$\Theta = -\frac{\sigma n_1}{4\pi} \frac{\epsilon^{2n_1 h} - \epsilon^{-2n_1 h} + 2\sin 2n_1 h}{\epsilon^{2n_1 h} + \epsilon^{-2n_1 h} - 2\cos 2n_1 h} H \cos pt$$

$$+ \frac{\sigma n_1}{4\pi} \frac{\epsilon^{2n_1 h} - \epsilon^{-2n_1 h} - 2\sin 2n_1 h}{\epsilon^{2n_1 h} + \epsilon^{-2n_1 h} - 2\cos 2n_1 h} H \sin pt. \tag{64}$$

In calculating the part of the energy flowing into the tube which is converted into heat, we need only consider the part which flows across the outer surface of the tube, because the energy flowing across the inner surface is equal to that which flows into the dielectric inside the tube, and since there is no dissipation of energy in this region the average of the flow of energy across the inner surface of the tube must vanish. Hence the amount of heat produced in unit time in the tube is by equation (36) equal to the mean value of

$$-\tfrac{1}{2} \mathbf{b} \Theta H \cos pt,$$

where the $-$ sign has been taken because the translatory motion of a right-handed screw twisting from Θ to H is radially outwards; this by (64) is equal to

$$\frac{\sigma n_1 \mathbf{b}}{16\pi} \frac{(\epsilon^{2n_1 h} - \epsilon^{-2n_1 h} + 2\sin 2n_1 h)}{\epsilon^{2n_1 h} + \epsilon^{-2n_1 h} - 2\cos 2n_1 h} H^2;$$

when $n_1 h$ is very large this is equal to

$$\frac{\sigma n_1 \mathbf{b}}{16\pi} H^2,$$

which is (Art. 286), as it ought to be, the same as for a solid cylinder of radius $\mathbf{b}$.

When h is small and $n^2 \mathbf{a} h/\mu$ not large we must take into account terms which we have neglected in arriving at the preceding expression.

In this case, we find from (62) that

$$\Theta = -\frac{(\pi p^2 \mathbf{a}^2 h/\sigma) H \cos pt}{1 + 4\pi^2 p^2 \mathbf{a}^2 h^2/\sigma^2} + \tfrac{1}{2}\frac{p \mathbf{a} H \sin pt}{1 + 4\pi^2 p^2 \mathbf{a}^2 h^2/\sigma^2}, \qquad (65)$$

so that the rate of heat production is

$$\tfrac{1}{4}\frac{(\pi p^2 \mathbf{a}^2 \mathbf{b} h/\sigma) H^2}{1 + 4\pi^2 p^2 \mathbf{a}^2 h^2/\sigma^2}.$$

Thus it vanishes when $h = 0$, and is a maximum when

$$h = \frac{\sigma}{2\pi \mathbf{a} p};$$

the rate of heat production is then

$$\tfrac{1}{16} p \mathbf{b} \mathbf{a} H^2,$$

and bears to the rate when the tube is solid the ratio

$$\frac{\pi p \mathbf{a}}{n_1 \sigma} : 1,$$

which is equal to $n_1 \mathbf{a}/2\mu$.

Since $n_1 \mathbf{a}/\mu$ is very large the heat produced in a tube of this thickness is very much greater than that produced in a solid cylinder.

Let us take the case of a tin tube whose internal radius is 3 cm. surrounded by a primary coil conveying a current making a hundred thousand vibrations per second, then since in this case

$$\sigma = 1 \cdot 3 \times 10^4, \quad \mathbf{a} = 3, \quad p = 2\pi \times 10^5, \quad \mu = 1,$$

the thickness which gives the maximum heat production is about 1/90 of a millimetre, and the heat produced is about 26 times as much as would be produced in a solid tin cylinder of the same radius as the tube.

We see from equation (65) that the amplitude of Θ diminishes as the thickness of the plate increases, but that when the plate is indefinitely thin the phases of the tangential electromotive intensity and of the tangential magnetic force differ by a quarter-period; the product of these quantities will thus be proportional to $\sin 2pt$, and as the mean value of this vanishes there is no energy converted into heat in the tube. As the thickness of the tube increases the amplitude of Θ diminishes, but the phase of Θ gets more nearly into unison with that of H. We may regard Θ as made up of two oscillations, one being in the same phase as H while the phase of the other differs from that of H by a quarter-period. The amplitude of the second component diminishes as the thickness of the tube increases, while that of the first reaches a maximum when $h = \sigma/2\pi \mathbf{a} p$.

In the investigation of the heat produced when h is small, $n\mathbf{a}/\mu$ has been assumed large. We can however easily show that unless this is the case the heat produced in a thin tube will not exceed that produced in a solid cylinder.

Vibrations of Electrical Systems.

288.] If the distribution of electricity on a system in electrical equilibrium is suddenly disturbed, the electricity will redistribute itself so as to tend to go back to the distribution it had when in electrical equilibrium; to effect this redistribution electric currents will be started. The currents possess kinetic energy which is obtained at the expense of the potential energy of the original distribution of electricity; this kinetic energy will go on increasing until the distribution of electricity is the same as it was in the state from which it was displaced. As this state is one of equilibrium its potential energy is a minimum. The kinetic energy which the system has acquired will carry it through this state, and the system will go on losing kinetic and reacquiring potential energy until the kinetic energy has all disappeared. The system will then retrace its steps, and if there is no dissipation of energy will again regain the distribution of electricity from which it started. The distribution of electricity on the system will thus oscillate backwards and forwards; we shall in the following articles endeavour to calculate the time taken by such oscillations for some of the simpler electrical systems.

Electrical Oscillations when Two Equal Spheres are connected by a Wire.*

289.] The first case we shall consider is that of two equal spheres, or any two bodies possessing equal electric capacities, connected by a straight wire. This case can be solved at once by means of the analysis given at the beginning of this chapter.

Let us take the point on the wire midway between the spheres as the origin of coordinates, and the axis of the wire as

* See J. J. Thomson, *Proc. Lond. Math. Soc.* 19, p. 542, 1888.

the axis of z. We shall suppose that the electrostatic potential has equal and opposite values at points on the wire equidistant from the origin and on opposite sides of it. Then using the same notation as in Art. 271, we may put

$$\phi = L\,(\epsilon^{\iota mz} - \epsilon^{-\iota mz})\,J_0\,(\iota mr)\,\epsilon^{\iota pt}, \text{ in the wire,}$$
$$= L\,(\epsilon^{\iota mz} - \epsilon^{-\iota mz})\,\epsilon^{\iota pt}$$

approximately, since mr will be very small. Thus **E**, the external electromotive intensity parallel to the wire, is equal to

$$-\iota m L\,(\epsilon^{\iota mz} + \epsilon^{-\iota mz})\,\epsilon^{\iota pt}.$$

If $2l$ is the length of the wire, then the potential of the sphere at the end $z = l$, will be

$$2\,\iota L \sin ml\,\epsilon^{\iota pt}.$$

If C is the capacity of the sphere at one end of the wire, the quantity of electricity on the sphere is

$$2\,\iota CL \sin ml\,\epsilon^{\iota pt},$$

and this increases at the rate

$$-2\,CpL \sin ml\,\epsilon^{\iota pt}.$$

Now the increase in the charge of the sphere must equal the current flowing through the wire at the point $z = l$, hence if I denotes this current, we have

$$I = -2\,CpL \sin ml\,\epsilon^{\iota pt},$$

but by equation (39) of Art. 272 we have

$$\mathbf{E} = (\iota p\mathbf{P} + \mathbf{Q})\,I,$$

whence substituting the values for **E** and I when $z = l$, we get

$$-2\,\iota mL \cos ml\,\epsilon^{\iota pt} = -(\iota p\mathbf{P} + \mathbf{Q})\,2\,CpL \sin ml\,\epsilon^{\iota pt},$$

or $$m \cot ml = -\iota p\,(\iota p\mathbf{P} + \mathbf{Q})\,C. \qquad (66)$$

290.] Let us first consider the case when the wave length of electrical vibrations is very much longer than the wire; here ml is very small, so that equation (66) becomes

$$\frac{1}{l} = -\iota p\,(\iota p\mathbf{P} + \mathbf{Q})\,C. \qquad (67)$$

The values of **P** and **Q**, the self-induction and impedance of the wire, are given in equation (40) of Art. 272; they depend upon the frequency of the electrical vibrations. When this is so slow that na is a small quantity, a being the radius of the wire, then approximately

$$\mathbf{P} = \frac{L}{2l},$$

$$\mathbf{Q} = \frac{R}{2l},$$

where L is the coefficient of self-induction and R the resistance of the whole wire for steady currents.

Substituting these values in (67), we get

$$(\iota p)^2 L + \iota p R + \frac{2}{C} = 0,$$

or

$$\iota p = -\frac{R}{2L} \pm \iota \sqrt{\frac{2}{CL} - \frac{R^2}{4L^2}}. \tag{68}$$

Since the various quantities which fix the state of the electric field contain $\epsilon^{\iota p t}$ as a factor, we see that when $8L > CR^2$ these quantities will be proportional to

$$\epsilon^{-\frac{R}{2L}t} \cos\left\{\left(\frac{2}{CL} - \frac{R^2}{4L^2}\right)^{\frac{1}{2}} t + a\right\},$$

where a is a constant.

This represents an oscillation whose period is

$$2\pi \Big/ \left\{\frac{2}{CL} - \frac{R^2}{4L^2}\right\}^{\frac{1}{2}},$$

and whose amplitude dies away to $1/\epsilon$ of its original value after the time $2L/R$.

Thus, if $2/CL$ is greater than $R^2/4L^2$, that is if R^2 is less than $8L/C$, the charges on the spheres will undergo oscillations like those performed by a pendulum in a resisting medium.

Suppose, for example, that the electrical connection between the spheres is broken, and let one sphere A be charged with positive, the other sphere B with an equal quantity of negative electricity; if now the electrical connection between the spheres is restored, the positive charge on A and the negative on B will diminish until after a time both spheres are free from electrification. They will not however remain in this state, for negative electricity will begin to appear on A, positive on B, and these charges will increase in amount until (neglecting the resistance of the circuit connecting the spheres) the charges on A and B appear to be interchanged, there being now on A the same quantity of negative electricity as there was initially on B, while the charge on B is the same as that originally on A. When the negative charge on A has reached this value it begins to decrease, and after a time both spheres are again free from electrification. After this positive electricity begins to reappear on A, and increases until the charge on A is the same as it was to begin with; this positive charge then decreases, vanishes, and is replaced by a negative one as before. The system thus behaves as if the charges vibrated backwards and forwards between the spheres. The changes which take place in the electrical charges on the spheres are of course accompanied by currents in the wire, these currents flowing sometimes in one direction, sometimes in the opposite.

When the circuit has a finite resistance the amplitude of the oscillations gradually diminishes, while if the resistance is greater than $(8\, L/C)^{\frac{1}{2}}$ there will not be any vibrations at all, but the charges will subside to zero without ever changing sign; in this case the current in the connecting wire is always in one direction.

291.] If we assume that the wave length of the electrical vibrations is so great that the current may be regarded as uniform all along the wire, and that the vibrations are so slow that the current is uniformly distributed across the wire, the discharge of a condenser can easily be investigated by the following method, which is due to Lord Kelvin (*Phil. Mag.* [4], 5, p. 393, 1853). Let Q be the quantity of electricity on one of the plates of a condenser whose capacity is C' and whose plates, like those of a Leyden Jar, are supposed to be close together; also let R be the resistance and L the coefficient of self-induction for steady currents of the wire connecting the plates. The electromotive force tending to increase Q is $-Q/C'$; of this $R\,dQ/dt$ is required to overcome the resistance and $L\,d^2Q/dt^2$ to overcome the inertia of the circuit; hence we have

$$L\frac{d^2Q}{dt^2} + R\frac{dQ}{dt} + \frac{Q}{C'} = 0. \tag{69}$$

The solution of this equation is, if

$$\frac{1}{C'L} > \frac{R^2}{4L^2},$$

$$Q = A\epsilon^{-\frac{R}{2L}t}\cos\left\{\left(\frac{1}{C'L} - \frac{R^2}{4L^2}\right)^{\frac{1}{2}}t + \beta\right\},$$

where A and β are arbitrary constants.

In this case we have an oscillatory discharge whose frequency is equal to

$$\left(\frac{1}{C'L}-\frac{R^2}{4L^2}\right)^{\frac{1}{2}}.$$

When

$$\frac{1}{C'L}<\frac{R^2}{4L^2},$$

the solution of equation (69) is

$$Q=\epsilon^{-\frac{R}{2L}t}\left\{A\epsilon^{\left(\frac{R^2}{4L^2}-\frac{1}{C'L}\right)^{\frac{1}{2}}t}+B\epsilon^{-\left(\frac{R^2}{4L^2}-\frac{1}{C'L}\right)^{\frac{1}{2}}t}\right\},$$

where A and B are arbitrary constants. In this case the discharge is not oscillatory.

To compare the results of this investigation with those of the previous one, we must remember that the capacities which occur in the two investigations are measured in somewhat different ways. The capacity C in the first investigation is the ratio of the charge on the condenser to ϕ its potential; in the second investigation C' is the ratio of the charge to 2ϕ, the difference between the potentials of the plates, so that to compare the results we must put $C' = C/2$; if we do this the results given by the two investigations are identical.

292.] The existence of electrical vibrations seems to have been first suspected by Dr. Joseph Henry in 1842 from some experiments he made on the magnetization of needles placed in a coil in circuit with a wire which connected the inside to the outside coating of a Leyden Jar. He says (*Scientific Writings* of Joseph Henry, Vol. I, p. 201, Washington, 1886): 'This anomaly which has remained so long unexplained, and which at first sight appears at variance with all our theoretical ideas of the connection of electricity and magnetism, was after considerable study satisfactorily referred by the author to an action of the discharge of the Leyden jar which had never before been recognised. The discharge, whatever may be its nature, is not correctly represented (employing for simplicity the theory of Franklin) by the simple transfer of an imponderable fluid from one side of the jar to the other, the phenomenon requires us to admit the existence of a principal discharge in one direction, and then several reflex actions backward and forward, each more feeble than the preceding, until the equilibrium is obtained. All the facts are shown to be in accordance with this hypothesis, and a ready explanation is afforded by it of a number of phenomena which are to be found in the older works on electricity but which have until this time remained unexplained.'

In 1853, Lord Kelvin published (*Phil. Mag.* [4], 5, p. 393, 1853) the results we have just given in Art. 291, thus proving by the laws of electrical action that electrical vibrations must be produced when a Leyden Jar is short circuited by a wire of not too great resistance.

From 1857 to 1862, Feddersen (*Pogg. Ann.* 103, p. 69, 1858; 108, p. 497, 1859; 112, p. 452, 1861; 113, p. 437, 1861; 116, p. 132, 1862) published accounts of some beautiful experiments by which he demonstrated the oscillatory character of the jar discharge. His method consisted in putting an air break in the wire circuit joining the two coatings of the jar. When the current through this wire is near its maximum intensity a spark passes across the circuit, but when the current is near its minimum value the electromotive force is not sufficient to spark across the air break, which at these periods therefore is not luminous. Thus the image of the air space formed by reflection from a rotating mirror will be drawn out into a series of bright and dark spaces, the interval between two dark spaces depending of course on the speed of the mirror and the frequency of the electrical vibrations. Feddersen observed this appearance of the image of the air space, and he proved that the oscillatory character of the discharge was destroyed by putting a large resistance in circuit with the air space, by showing that in this case the image of the air space was a broad band of light gradually fading away in intensity instead of a series of bright and dark spaces. This experiment, which is a very beautiful one, can be repeated without difficulty. To excite the vibrations the coatings of the jar should be connected to the terminals of an induction coil or an electric machine. It is advisable to use a large jar with its coatings connected by as long a wire as possible. By connecting the coatings of the jar by a circuit with very large self-induction, Dr. Oliver Lodge (*Modern Views of Electricity*, p. 377) has produced such slow electrical vibrations that the sounds generated by the successive discharges form a musical note.

293.] In the course of the investigation in Art. 290 we have made two assumptions, (1) that ml is small, (2) that na is also small, which implies that the currents are uniformly distributed across the section of the discharging circuit. This condition is however very rarely fulfilled, as the electrical oscillations which are produced by the discharge of a condenser are in general so rapid that the currents in the discharging circuit fly to the outside of the wire instead of distributing themselves uniformly across it; when the currents do this, however, the resistance of the circuit depends on the frequency of the electrical vibrations, and the investigation of Art. 290 has to be modified. Before proceeding to the discussion of this case we shall write down the conditions which must hold when the preceding investigation is applicable.

In the first place, ml is to be small; now by Art. 263 we have when $n\mathbf{a}$ is small,

$m^2 = -\iota p$ (resistance of unit length of the wire) $\times$
(capacity of unit length of wire),

hence
$$m^2 l^2 = -\tfrac{1}{2}\iota p R l \Gamma,$$

where, as before, R is the resistance of the whole of the discharging circuit, while Γ is the capacity of unit length of the wire.

But by equation (68) when the discharge is oscillatory, we have

$$\iota p = -\frac{R}{2L} \pm \iota \left\{\frac{2}{LC} - \frac{R^2}{4L^2}\right\}^{\frac{1}{2}};$$

thus the modulus of ιp is equal to

$$\left\{\frac{2}{LC}\right\}^{\frac{1}{2}},$$

hence, when ml is small,

$$\frac{R\Gamma l}{\sqrt{CL}}$$

must be small.

The other condition is that $n\mathbf{a}$ is small, which since

$$n^2 = m^2 + \frac{4\pi\mu\iota p}{\sigma},$$

and ml is also small, is equivalent to the condition that

$$4\pi\mu\iota p\mathbf{a}^2/\sigma$$

should be small. Since the modulus of ιp is equal to $\{2/LC\}^{\frac{1}{2}}$, we see that if $n^2\mathbf{a}^2$ is small,

$$4\pi\mu\mathbf{a}^2\{2/LC\}^{\frac{1}{2}}/\sigma$$

must be small. The capacity C which occurs in this expression is measured in electromagnetic units, its value in such measure is only $1/V^2$ (where 'V' is the ratio of the units and $V^2 = 9 \times 10^{20}$) of its value in electrostatic measure. Thus the expression which has to be small to ensure the condition we are considering, contains the large factor 3×10^{10}, so that to fulfil this condition the capacity and self-induction of the circuit must be very large when the discharging circuit consists of metal wire of customary dimensions. Thus, to take an example, suppose two spheres each one metre in radius are connected by a copper wire 1 millimetre in diameter. In this case

$$C = 1/9 \times 10^{18},\ \sigma = 1600,\ \mathbf{a} = \cdot 05,\ \mu = 1,$$

substituting these values we find that to ensure $n\mathbf{a}$ being small, the self-induction of the circuit must be comparable with the enormously large value 10^{11}, which is comparable with the self-induction of a coil with 10,000 turns of wire, the coil being about half a metre in diameter.

The result of this example is sufficient to show that it is only when the self-induction of the circuit or the capacity of the condenser is exceptionally large that a theory based on the assumption that $n\mathbf{a}$ is a small quantity is applicable, it is therefore important to consider the case where $n\mathbf{a}$ is large and the currents in the discharging circuit are on the surface of the wire.

294.] The theory of this case is given in Art. 274, and we see from equations (42) and (43) of that Article that when the frequency of the vibrations is so great that $n\mathbf{a}$ and $n'\mathbf{b}$ (using the notation of Art. 274, and supposing that the wire connecting the spheres is a cable whose external radius is $\mathbf{b}$) are large quantities, equation (66) of Art. 289 becomes

$$m \cot ml = -2\iota p\,\{\iota p \log \mathbf{b}/\mathbf{a} + (\iota p)^{\frac{1}{2}}(\mu\sigma/4\pi\mathbf{a}^2)^{\frac{1}{2}} + (\iota p)^{\frac{1}{2}}(\mu'\sigma'/4\pi\mathbf{b}^2)^{\frac{1}{2}}\}C.$$

Retaining the condition that ml is small, which will be the case when the wave length of the electrical vibrations is very much greater than the length of the discharging circuit, this equation becomes

$$\frac{1}{Cl} = -\iota p\,[\iota p\, 2\log(\mathbf{b}/\mathbf{a}) + (\iota p)^{\frac{1}{2}}\, 2\,\{(\mu\sigma/4\pi\mathbf{a}^2)^{\frac{1}{2}} + (\mu'/\sigma'4\pi\mathbf{b}^2)^{\frac{1}{2}}\}],$$

which we shall write as

$$\frac{2}{C} = -\iota p \{\iota p L' + 2(\iota p)^{\frac{1}{2}} S\}, \tag{70}$$

where L' is the coefficient of self-induction of the discharging circuit for infinitely rapid alternating currents, and S is written for

$$\{(\sigma\mu/4\pi \mathbf{a}^2)^{\frac{1}{2}} + (\sigma'\mu'/4\pi \mathbf{b}^2)^{\frac{1}{2}}\}\, 2l.$$

By Art. 274, $$L' = L - \mu l,$$

where L is the self-induction of the circuit for steady currents.

If we write x for ιp, equation (70) becomes

$$x^2 L' + 2x^{\frac{3}{2}} S + \frac{2}{C} = 0,$$

hence

$$\left(x^2 L' + \frac{2}{C}\right)^2 = 4x^3 S^2,$$

or

$$L'^2 x^4 - 4S^2 x^3 + 4\frac{L'}{C} x^2 + \frac{4}{C^2} = 0, \tag{71}$$

a biquadratic equation to determine x.

If electrical oscillations take place the roots of this equation must be imaginary.

From the theory of the biquadratic equation (Burnside and Panton, *Theory of Equations*, § 68)

$$ax^4 + 4bx^3 + 6cx^2 + 4dx + e = 0,$$

we know that if

$$H = ac - b^2, \qquad I = ae - 4bd + 3c^2, \qquad G = a^2 d - 3abc + 2b^3,$$
$$J = ace + 2bcd - ad^2 - eb^2 - c^3, \qquad \Delta = I^3 - 27J^2;$$

the condition that the roots of the biquadratic are all imaginary is that Δ should be positive as well as one of the two following quantities H and $a^2 I - 12H^2$.

Dividing equation (71) by L'^2, we see that for equation (71)

$$H = \frac{2}{3}\frac{1}{L'C} - \frac{S}{L'^4}, \qquad I = \frac{16}{3}\frac{1}{L'^2 C^2},$$

$$G = 2\frac{S^2}{L'^3 C}\left\{1 - \frac{CS^4}{L'^3}\right\}, \qquad J = \frac{64}{27}\frac{1}{L'^3 C^3}\left\{1 - \frac{27}{16}\frac{CS^4}{L'^3}\right\}.$$

Hence we see that $a^2 I - 12H^2$ and Δ are both positive, if

$$S^4 < 32L'^3/27C,$$

that is if

$$16l^4\{(\sigma\mu/4\pi\mathbf{a}^2)^{\frac{1}{2}} + (\sigma'\mu'/4\pi\mathbf{b}^2)^{\frac{1}{2}}\}^4 < 32L'^3/27C,$$

which is the condition that the system should execute electrical vibrations.

When the spheres are connected by a free wire and not by a cable σ'/b^2 vanishes, and the condition that the system should oscillate reduces to

$$l^2(\sigma\mu/\pi\mathrm{a}^2)^2 < 32\,L'^3/27\,l^2C.$$

The results given by Ferrari's method for solving biquadratic equations are too complicated to be of much practical value in determining the roots of equation (71), neither, since the roots are imaginary, can we apply the very convenient method known as Horner's method to determine the numerical value of these roots to any required accuracy.

295.] For the purpose of analysing the nature of the electrical oscillations it is convenient to consider separately the real and imaginary parts of ιp, the x of equation (71). The real part, supposed negative, determines the rate at which the electrical vibrations die away, while the imaginary part gives the period of these vibrations. We shall now proceed to show how equation (71) can be treated so as to admit of the real and imaginary parts of x being separately determined by Horner's method.

If we put

$$\xi = x - \frac{S^2}{L'^2},$$

equation (71) becomes

$$\xi^4 + 6H\xi^2 + 4G\xi + I - 3H^2 = 0, \qquad (72)$$

where H, G, I are the quantities whose values we have just written down. Since the coefficient of ξ^3 in this equation vanishes and since its roots are by hypothesis complex, we see that the real part of one pair of roots will be positive, that of the other pair negative: the pair of roots whose real parts are negative are those which correspond to the solution of the electrical problem. For if the real part of ξ were positive the real part of ιp would also be positive, so that such a root would correspond to an electrical vibration whose amplitude increased indefinitely with the time.

The roots of equation (72) will be of the form

$$x_1 + \iota y_1, \quad x_1 - \iota y_1, \quad -x_1 + \iota y_2, \quad -x_1 - \iota y_2.$$

We shall now proceed to show how x_1 may be uniquely determined. Since $6H$, $-4G$, $I - 3H^2$ are respectively the sums of the products of the roots of equation (72) two and two, three and three, and all together, we have

$$y_1^2+y_2^2-2x_1^2 = 6H, \tag{73}$$

$$x_1(y_1^2-y_2^2) = 2G, \tag{74}$$

$$(x_1^2+y_1^2)(x_1^2+y_2^2) = I-3H^2,$$

or $x_1^4+x_1^2(y_1^2+y_2^2)+\frac{1}{4}\{(y_1^2+y_2^2)^2-(y_1^2-y_2^2)^2 = I-3H^2.$

Eliminating $y_1^2+y_2^2$ and $y_1^2-y_2^2$ by equations (73) and (74), we get

$$4x_1^4+12Hx_1^2+(12H^2-I)-\frac{G^2}{x_1^2} = 0,$$

or putting $x_1^2 = \eta$,

$$4\eta^3+12H\eta^2+(12H^2-I)\eta-G^2 = 0. \tag{75}$$

Since the last term of this expression is negative there is at least one positive real root of this equation, and since the values given for H and I show that when Δ is positive $12H^2-I$ is essentially negative, we see by Fourier's rule that there is only one such root. But since x_1 is real the value of η will be positive, so that the root we are seeking will be the unique positive real root of equation (75), which can easily be determined by Horner's method. The value of x_1 is equal to minus the square root of this root, and knowing x_1 we can find $y_1^2/4\pi^2$, the square of the corresponding frequency uniquely from equations (73) and (74). We can in this way in any special case determine with ease the logarithmic decrement and the frequency of the vibrations.

296.] If in equation (75) we substitute the values of G, H, and I, and write

$$L'C\eta = \zeta, \qquad CS^4/L'^3 = q,$$

that equation becomes

$$\zeta^3+2\zeta^2(1-\tfrac{3}{2}q)-\zeta(3q^2-4q)-q(1-q)^2 = 0.$$

We can by successive approximations expand ζ in terms of q, and thus when CS^4/L'^3 is small approximate to the value of ζ. The first term in this expansion is

$$\zeta = (q/2)^{\frac{1}{2}},$$

or since

$$L'Cx_1^2 = \zeta,$$

$$x_1 = -\frac{S}{2^{\frac{1}{4}}C^{\frac{1}{4}}L'^{\frac{5}{4}}}.$$

The corresponding value of y_1^2 determined by equations (73) and (74) is, retaining only the lowest power of q, approximately,

$$y_1^2 = \frac{2}{L'C}\left\{1 - \frac{2^{\frac{1}{4}}SC^{\frac{1}{4}}}{L^{\frac{3}{4}}}\right\}.$$

Now $\qquad\qquad S = \{\mu\sigma/4\pi\mathbf{a}^2)^{\frac{1}{2}} + (\mu'\sigma'/4\pi\mathbf{b}^2)^{\frac{1}{2}}\}\,2l,$

and, approximately, $\qquad\qquad y_1^2 = \frac{2}{L'C},$

and $\qquad\qquad\qquad x_1 = -\frac{Sy_1^{\frac{1}{2}}}{2^{\frac{1}{2}}L},$

hence we see,

$$x_1 = -\frac{l}{L'}\{(\mu\sigma y_1/2\pi\mathbf{a}^2)^{\frac{1}{2}} + (\mu'\sigma'y_1/2\pi\mathbf{b}^2)^{\frac{1}{2}}\}.$$

But by Art. 274, the quantity enclosed in brackets is equal to Q, the impedance of unit length of the circuit when the frequency of vibration is $y_1/2\pi$; thus we have

$$x_1 = -\frac{Q}{2L'},$$

where Q is the impedance of the whole circuit.

Since $\qquad\qquad \iota p = x_1 + \iota y_1 + \frac{S^2}{L'^2}$

$$= x_1\{1 - (2q)^{\frac{1}{4}}\} + \iota y_1,$$

the real part of ιp differs from x_1 by a quantity involving q. Neglecting this term, we see that the expression for the amplitude of the vibrations contains the factor $\epsilon^{-\frac{Q}{2L'}t}$. Comparing this with the factor $\epsilon^{-\frac{R}{2L}t}$, which occurs when the oscillations are so slow that the current is uniformly distributed over the cross-section of the discharging wire, we find that to our order of approximation we may for quick vibrations use a similar formula for the decay of the amplitude to that which holds for slow vibrations, provided we use the impedance instead of the resistance, and the coefficient of self-induction for infinitely rapid vibrations instead of that for infinitely slow ones. This result is, however, only true when CS^4/L'^3 is a small quantity. Now if the external conductor is so far away that $\mu'\sigma'/\mathbf{b}^2$ is small compared with $\mu\sigma/\mathbf{a}^2$, then

$$S^4 = \{2l(\mu\sigma/4\pi\mathbf{a}^2)^{\frac{1}{2}}\}^4 = \tfrac{1}{4}l^2\mu^2R^2,$$

where R is the resistance of the whole circuit to steady currents. Substituting this value for S^4 we see that the condition that CS^4/L'^3 is a small quantity is that $Cl^2\mu^2R^2/4L'^3$ should be small. When this is the case we see that, neglecting the effect of the external conductor,

$$x_1 = -\frac{l}{L'}(\mu\sigma y_1/2\pi \mathbf{a}^2)^{\frac{1}{2}}.$$

Since x_1 is proportional to $\mu^{\frac{1}{2}}$, the rate of decay of the vibrations will be greater when the discharging wire is made of iron than when it is made of a non-magnetic metal of the same resistance. This has been observed by Trowbridge (*Phil. Mag.* [5], 32, p. 504, 1891).

297.] We have assumed in the preceding work that the length of the electrical wave is great compared with that of the wire; we have by equation (66)

$$m \cot ml = -\iota p\{\iota p\mathbf{P}+\mathbf{Q}\}C.$$

When the frequency is very high, $\iota p\mathbf{P}$ will be very large compared with $\mathbf{Q}$, hence this equation may be written as

$$m \cot ml = p^2\mathbf{P}C.$$

Now if V is the velocity of light in the dielectric, $p = Vm$, hence we have

$$\frac{\cot ml}{ml} = \frac{V^2 2\mathbf{P}lC}{2l^2}.$$

Now $2\mathbf{P}l$ is equal to L', the self-induction of the discharging circuit for infinitely rapid vibrations, and V^2C is equal to the electrostatic measure of the capacity of the sphere which we shall denote by $[C]$, hence the preceding equation may be written as

$$\frac{\cot ml}{ml} = \frac{L'[C]}{2l^2}.$$

Thus, if $L'[C]/2l^2$ is very large, ml will be very small; if, on the other hand, $L'[C]/2l^2$ is very small, $\cot ml$ will be very small, or $ml = (2j+1)\frac{\pi}{2}$ approximately, where j is an integer. Since $2\pi/m$ is the length of the electrical wave the latter will equal $4l$, $4l/3$, $4l/5$..., or the half-wave length will be an odd submultiple of the length of the discharging wire. We are limited by our investigation to the odd submultiple because we have assumed that the current in the discharging wire is symmetrical about the middle point of that wire. If we abandon this assumption we find that the half-wave length may be any submultiple of the length of the wire. The frequencies of the vibrations are thus independent of the capacity at the end of the wire provided this is small enough to make $L[C]/2l^2$ small. In this case the vibrations are determined merely by the condition that the current in the discharging wire should vanish at its extremities.

Vibrations along Wires in Multiple Arc.

298.] When the capacities of the conductors at the ends of a single wire are very small, we have seen that the gravest electrical vibration has for its wave length twice the length of the wire and that the other vibrations are harmonics of this. We shall now investigate the periods of vibration of the system when the two conductors of small capacity are connected by two or more wires in parallel. The first case we shall consider is the one represented by Fig. 109, where in the connection between the points A and F we have the loop $BCED$.

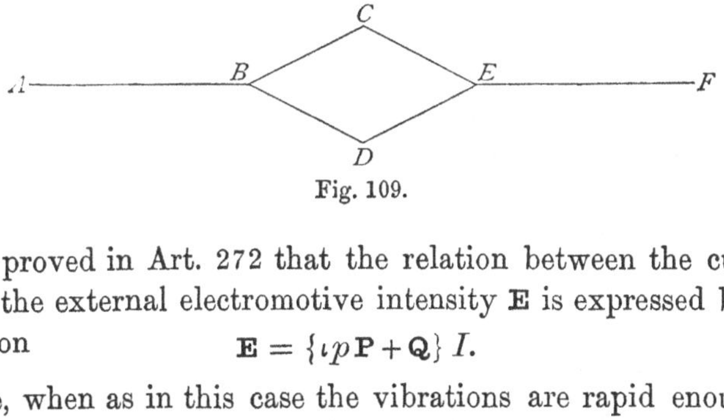

Fig. 109.

We proved in Art. 272 that the relation between the current I and the external electromotive intensity $\mathbf{E}$ is expressed by the equation

$$\mathbf{E} = \{\iota p\mathbf{P} + \mathbf{Q}\}\, I.$$

Where, when as in this case the vibrations are rapid enough to make $n\mathbf{a}$ large, the term $\iota p\mathbf{P}$ is much larger than $\mathbf{Q}$, we may therefore for our purpose write this equation as

$$\mathbf{E} = \iota p\mathbf{P}I, \qquad (76)$$

where $\mathbf{P}$ is the coefficient of self-induction of unit length of the wire for infinitely rapid vibrations.

Let the position of a point on AB be fixed by the length s_1 measured along AB from A, that of one on BCE by the length s_2 measured from B, that of one on BDE by s_3 measured also from B, and of one on EF by s_4 measured from E. Let l_1, l_2, l_3, l_4

denote the lengths AB, BCE, BDE, and EF respectively, and let P_1, P_2, P_3, P_4 denote the self-induction per unit length of these wires. Let ϕ denote the electrostatic potential, then the external electromotive intensity along a wire is $-d\phi/ds$, and as this is proportional to the current it must vanish at the ends A, F of the wire if the capacity there is, as we suppose, very small.

Hence along AB we may write, if $p/2\pi$ is the frequency,

$$\phi = a \cos m s_1 \cos pt,$$

along BCE $$\phi = (a \cos m s_2 \cos m l_1 + b \sin m s_2) \cos pt,$$

along BDE $$\phi = (a \cos m s_3 \cos m l_1 + c \sin m s_3) \cos pt,$$

and along EF $$\phi = d \cos m (s_4 - l_4) \cos pt.$$

Equating the expressions for the potential at E, we have

$$\left.\begin{aligned} a \cos m l_2 \cos m l_1 + b \sin m l_2 &= d \cos m l_4, \\ a \cos m l_3 \cos m l_1 + c \sin m l_3 &= d \cos m l_4. \end{aligned}\right\} \tag{77}$$

The current flowing along AB at B must equal the sum of the currents flowing along BCE, BDE, hence by (76) we have

$$\frac{a \sin m l_1}{P_1} = -\frac{b}{P_2} - \frac{c}{P_3}. \tag{78}$$

Again, the current along EF at E must equal the sum of the currents flowing along BCE, BDE, hence we have

$$\frac{d \sin m l_4}{P_4} = \frac{b \cos m l_2}{P_2} - \frac{a \sin m l_2 \cos m l_1}{P_2} + \frac{c \cos m l_3}{P_3} - \frac{a \sin m l_3 \cos m l_1}{P_3}. \tag{79}$$

We get from equations (77) and (78)

$$a \left\{ \frac{\sin m l_1}{P_1} - \frac{\cot m l_2 \cos m l_1}{P_2} - \frac{\cot m l_3 \cos m l_4}{P_3} \right\} = -d \cos m l_4 \left\{ \frac{\operatorname{cosec} m l_2}{P_2} + \frac{\operatorname{cosec} m l_3}{P_3} \right\}.$$

From equations (77) and (79) we get

$$d \left\{ \frac{\sin m l_4}{P_4} - \frac{\cot m l_2 \cos m l_4}{P_2} - \frac{\cot m l_3 \cos m l_4}{P_3} \right\} = -a \cos m l_1 \left\{ \frac{\operatorname{cosec} m l_2}{P_2} + \frac{\operatorname{cosec} m l_3}{P_3} \right\}.$$

Eliminating a and d from these equations, we get

$$\left\{\frac{\tan m l_1}{P_1} - \frac{\cot m l_2}{P_2} - \frac{\cot m l_3}{P_3}\right\}\left\{\frac{\tan m l_4}{P_4} - \frac{\cot m l_2}{P_2} - \frac{\cot m l_3}{P_3}\right\}$$
$$= \left\{\frac{\text{cosec}\, m l_2}{P_2} + \frac{\text{cosec}\, m l_3}{P_3}\right\}^2. \qquad (80)$$

If AB and EF are equal lengths of the same kind of wire, $l_1 = l_4$, and $P_1 = P_4$, and (80) reduces to the simple form

$$\frac{\tan m l_1}{P_1} - \frac{\cot m l_2}{P_2} - \frac{\cot m l_3}{P_3} = \pm\left\{\frac{\text{cosec}\, m l_2}{P_2} + \frac{\text{cosec}\, m l_3}{P_3}\right\};$$

taking the upper sign, we have

$$\frac{\tan m l_1}{P_1} = \frac{\cot \frac{1}{2} m l_2}{P_2} + \frac{\cot \frac{1}{2} m l_3}{P_3}, \qquad (81)$$

if we take the lower sign, we have

$$\frac{\tan m l_1}{P_1} = -\left\{\frac{\tan \frac{1}{2} m l_2}{P_2} + \frac{\tan \frac{1}{2} m l_3}{P_3}\right\}. \qquad (82)$$

Since $m = 2\pi/\lambda$, where λ is the wave length, these equations determine the wave lengths of the electrical vibrations.

If all the wires have the same radius, $P_1 = I_2 = P_3$, and equations (81) and (82) become respectively

$$\tan 2\pi \frac{l_1}{\lambda} = \cot\left(\pi \frac{l_2}{\lambda}\right) + \cot\left(\pi \frac{l_3}{\lambda}\right), \qquad (81^*)$$

and

$$\tan 2\pi \frac{l_1}{\lambda} + \tan \pi \frac{l_2}{\lambda} + \tan \pi \frac{l_3}{\lambda} = 0. \qquad (82^*)$$

From these equations we can determine the effect on the period of an alteration in the length of one of the wires. Suppose that the length of BDE is increased by δl_3, and let $\delta\lambda$ be the corresponding increase in λ, then from (81*)

$$\frac{\delta\lambda}{\lambda}\left\{l_1 \sec^2 \frac{2\pi l_1}{\lambda} + \tfrac{1}{2} l_2\, \text{cosec}^2 \frac{\pi l_2}{\lambda} + \tfrac{1}{2} l_3\, \text{cosec}^2 \frac{\pi l_3}{\lambda}\right\} = \tfrac{1}{2}\delta l_3\, \text{cosec}^2 \frac{\pi l_3}{\lambda}.$$

We see from this equation that $\delta\lambda$ and δl_3 are of the same sign, so that an increase in l_3 increases the wave length.

If we take equation (82*), we have

$$\frac{\delta\lambda}{\lambda}\left\{l_1 \sec^2 \frac{2\pi l_1}{\lambda} + \tfrac{1}{2} l_2 \sec^2 \frac{\pi l_2}{\lambda} + \tfrac{1}{2} l_3 \sec^2 \frac{\pi l_3}{\lambda}\right\} = \tfrac{1}{2}\delta l_3 \sec^2 \frac{\pi l_3}{\lambda},$$

hence, in this case also, an increase in l_3 increases λ. If l_3 is infinite the wave length is $4l_1+2l_2$ and its submultiples, as we diminish l_3 the wave length shortens, hence we see that the effect of introducing an alternative path is to shorten the wave lengths of all the vibrations. The shortening of the wave length goes on until l_3 vanishes, when the wave length of the gravest vibration is $4l_1$.

299.] The currents through the wires *BCE* and *BDE* are at *B* in the proportion of

$$\frac{\cot\frac{1}{2}ml_2}{P_2} \text{ to } \frac{\cot\frac{1}{2}ml_3}{P_3},$$

if we take the vibrations corresponding to equation (81), and in the proportion of

$$\frac{\tan\frac{1}{2}ml_2}{P_2} \text{ to } \frac{\tan\frac{1}{2}ml_3}{P_3},$$

for the vibration given by (82).

We can prove by the method of Art. 298 that if we have n wires between *B* and *F*, and if $AB = EF$,

$$\frac{\tan ml_1}{P_1} - \frac{\cot ml_2}{P_2} - \frac{\cot ml_3}{P_3} - \dots$$
$$= \pm\left\{\frac{\operatorname{cosec} ml_2}{P_2} + \frac{\operatorname{cosec} ml_3}{P_3} + \frac{\operatorname{cosec} ml_4}{P_4} + \dots\right\}.$$

It follows from this equation that if any of the wires are shortened the wave lengths of the vibrations are also shortened.

Electrical Oscillations on Cylinders.

Periods of Vibration of Electricity on the Cylindrical Cavity inside a Conductor.

300.] If on the surface of a cylindrical cavity inside a conductor an irregular distribution of electricity is produced, then on the removal of the cause producing this irregularity, currents of electricity will flow from one part of the cylinder to another to restore the electrical equilibrium, electrical vibrations will thus be started whose periods we now proceed to investigate.

Take the axis of the cylinder as the axis of z, and suppose that initially the distribution of electricity is the same on all sections at right angles to the axis of the cylinder; it will evidently remain so, and the currents which restore the electrical distribution to equilibrium will be at right angles to the axis of z.

If c is the magnetic induction parallel to z, then in the cavity filled with the dielectric c satisfies the differential equation

$$\frac{d^2 c}{dx^2} + \frac{d^2 c}{dy^2} = \frac{1}{V^2}\frac{d^2 c}{dt^2},$$

where V is the velocity of propagation of electrodynamic action through the dielectric.

In the conductor c satisfies the equation

$$\frac{d^2 c}{dx^2} + \frac{d^2 c}{dy^2} = \frac{4\pi\mu}{\sigma}\frac{dc}{dt},$$

where σ is the specific resistance and μ the magnetic permeability of the substance.

Transform these equations to polar coordinates r and θ, and suppose that c varies as $\cos s\theta\, \epsilon^{\iota pt}$; making these assumptions, the differential equation satisfied by c in the dielectric is

$$\frac{d^2 c}{dr^2} + \frac{1}{r}\frac{dc}{dr} + c\left(\frac{p^2}{V^2} - \frac{s^2}{r^2}\right) = 0,$$

the solution of which is

$$c = A \cos s\theta J_s\left(\frac{p}{V} r\right) \epsilon^{\iota pt},$$

where J_s denotes the internal Bessel's function of the sth order.

The differential equation satisfied by c in the conductor is

$$\frac{d^2 c}{dr^2} + \frac{1}{r}\frac{dc}{dr} + \left\{- \frac{4\pi\mu\iota p}{\sigma} - \frac{s^2}{r^2}\right\} c = 0.$$

Let $n^2 = 4\pi\mu\iota p/\sigma$, then the solution of this equation is

$$c = B \cos s\theta K_s(\iota n r)\, \epsilon^{\iota pt},$$

where K_s denotes the external Bessel's function of the sth order.

Since the magnetic force parallel to the surface of the cylinder is continuous, we have if a denotes the radius of the cylindrical cavity

$$A J_s\left(\frac{p}{V}\mathrm{a}\right) = \frac{B}{\mu} K_s(\iota n \mathrm{a}). \qquad (83)$$

The electromotive intensity at right angles to r is also continuous. Now the current at right angles to r and z is

$$-dc/4\pi\mu dr,$$

hence in the conductor the electromotive intensity perpendicular to r and z is $-\sigma dc/4\pi\mu dr$. In the dielectric the current is equal to the rate of increase of the electric displacement, i.e. to ιp times the electric displacement or to $\iota pK/4\pi$ times the electromotive intensity; we see that in the dielectric the electromotive intensity perpendicular to r is $-\dfrac{1}{K\iota p}\dfrac{dc}{dr}$, hence we have

$$A\frac{4\pi}{K\iota p}\frac{p}{V}J'_s\left(\frac{p}{V}\mathrm{a}\right) = \frac{B}{\mu}\iota n\sigma K'_s(\iota n\mathrm{a}). \qquad (84)$$

Eliminating A and B from (83) and (84), we get

$$\frac{4\pi}{KV}\frac{J'_s\left(\frac{p}{V}\mathrm{a}\right)}{J_s\left(\frac{p}{V}\mathrm{a}\right)} = -\sigma n\frac{K'_s(\iota n\mathrm{a})}{K_s(\iota n\mathrm{a})}. \qquad (85)$$

Now $K = \dfrac{1}{V^2}$ and $\sigma = \dfrac{4\pi\mu\iota p}{n^2}$, so that (85) may be written

$$\frac{V}{p\mathrm{a}}\frac{J'_s\left(\frac{p}{V}\mathrm{a}\right)}{J_s\left(\frac{p}{V}\mathrm{a}\right)} = \frac{\mu}{\iota n\mathrm{a}}\frac{K'_s(\iota n\mathrm{a})}{K_s(\iota n\mathrm{a})}. \qquad (86)$$

Now the wave length of the electrical vibrations will be comparable with the diameter of the cylinder, and the value of p corresponding to this will be sufficient to make $n\mathrm{a}$ exceedingly large, but when $n\mathrm{a}$ is very large we have (Heine, *Kugelfunctionen*, vol. i. p. 248)

$$K_s(\iota n\mathrm{a}) = (-\iota)^s\epsilon^{-n\mathrm{a}}\sqrt{\frac{\pi}{2n\mathrm{a}}} \text{ approximately,}$$

hence $K'_s(\iota n\mathrm{a}) = \iota K_s(\iota n\mathrm{a})$; thus the right-hand side of (86) will be exceedingly small, and an approximate solution of this equation will be

$$J'_s\left(\frac{V}{p}\mathrm{a}\right) = 0.$$

This signifies that the tangential electromotive intensity vanishes at the surface of the cylinder, or that the tubes of electrostatic induction cut its surface at right angles. The roots of the equation

$$J'_s(x) = 0,$$

for $s = 1, 2, 3$, are given in the following table taken from Lord Rayleigh's *Theory of Sound*, Vol. II, p. 266:—

$s = 1$	$s = 2$	$s = 3$
1·841	3·054	4·201
5·332	6·705	8·015
8·536	9·965	11·344
11·706		
14·864		
18·016		

Thus, when $s = 1$, the gravest period of the electrical vibrations is given by the equation

$$\frac{p}{V}\mathbf{a} = 1.841,$$

or the wave length of the vibration $2\pi V/p = \cdot 543 \times 2\pi\mathbf{a}$, and is thus more than half the circumference of the cylinder. In this case, as far as our approximations go, there is no decay of the vibrations, though if we took into account the right-hand side of (86) we should find there was a small imaginary term in the expression for p, which would indicate a gradual fading away of the vibrations. If it were not for the resistance of the conductor the oscillations would last for ever, as there is no radiation of energy away from the cylinder. The magnetic force vanishes in the conductor except just in the neighbourhood of the cavity, and the magnetic waves emitted by one portion of the walls of the cavity will be reflected from another portion, so that no energy escapes.

Metal Cylinder surrounded by a Dielectric.

301.] In this case the waves starting from one portion of the cylinder travel away through the dielectric and carry energy with them, so that the vibrations will die away independently of the resistance of the conductor.

Using the same notation as before, we have in the conducting cylinder

$$c = A \cos s\theta J_s(\iota n r)\,\epsilon^{\iota p t},$$

and in the surrounding dielectric

$$c = B \cos s\theta K_s\left(\frac{p}{V}r\right)\epsilon^{\iota p t}.$$

Since the magnetic force parallel to z is continuous, we have

$$\frac{A}{\mu} J_s(\iota n \mathrm{a}) = B K_s\left(\frac{p}{V}\mathrm{a}\right).$$

Since the electromotive intensity perpendicular to r is continuous, we have

$$\frac{A}{\mu} \iota n \sigma J'_s(\iota n \mathrm{a}) = B \frac{4\pi}{K\iota p} \frac{p}{V} K'_s\left(\frac{p}{V}\mathrm{a}\right).$$

Eliminating A and B from these equations, we get

$$\iota n \sigma \frac{J'_s(\iota n \mathrm{a})}{J_s(\iota n \mathrm{a})} = \frac{4\pi}{K\iota V} \frac{K'_s\left(\frac{p}{V}\mathrm{a}\right)}{K_s\left(\frac{p}{V}\mathrm{a}\right)},$$

or

$$\frac{1}{\iota n \mathrm{a}} \frac{J'_s(\iota n \mathrm{a})}{J_s(\iota n \mathrm{a})} = \frac{V}{\mu p \mathrm{a}} \frac{K'_s\left(\frac{p}{V}\mathrm{a}\right)}{K_s\left(\frac{p}{V}\mathrm{a}\right)}. \tag{87}$$

Now, as before, $n\mathrm{a}$ will be large, and therefore

$$J_s(\iota n \mathrm{a}) = \frac{\iota^s \epsilon^{n\mathrm{a}}}{\sqrt{2\pi n \mathrm{a}}} \text{ approximately,}$$

hence $J'_s(\iota n \mathrm{a}) = -\iota J_s(\iota n \mathrm{a})$, and the left-hand side of equation (87) is very small, so that the approximate form of (87) will be

$$K'_s\left(\frac{p}{V}\mathrm{a}\right) = 0, \tag{88}$$

which again signifies that the electromotive intensity tangential to the cylinder vanishes at its surface.

In order to calculate the approximate values of the roots of the equation $K'_s(x) = 0$, it is most convenient to use the expression for $K_s(x)$ which proceeds by powers of $1/x$. This series is expressed by the equation

$$K_s(x) = C \frac{\epsilon^{-\iota x}}{(\iota x)^{\frac{1}{2}}}\left\{1 - \frac{(1^2 - 4s^2)}{8\iota x} + \frac{(1^2-4s^2)(3^2-4s^2)}{1.2(8\iota x)^2} - \frac{(1^2-4s^2)(3^2-4s^2)(5^2-4s^2)}{1.2.3(8\iota x)^3} + \ldots\right\},$$

where C is a constant (see Lord Rayleigh, *Theory of Sound*, Vol. II, p. 271).

When $s = 1$,

$$K_1(x) = C\frac{\epsilon^{-\iota x}}{(\iota x)^{\frac{1}{2}}}\left\{1 + \frac{3}{8\iota x} - \frac{15}{2(8\iota x)^2} + \frac{105}{2(8\iota x)^3} - \ldots\right\}.$$

Thus

$$K_1'(x) = -\iota C\frac{\epsilon^{-\iota x}}{(\iota x)^{\frac{1}{2}}}\left\{1 + \frac{7}{8\iota x} + \frac{57}{128(\iota x)^2} - \frac{195}{1024(\iota x)^3}\ldots\right\}.$$

To approximate to the roots of the equation $K_1'(x) = 0$, put $\iota x = y$, and equate the first four terms inside the bracket to zero; we get

$$y^3 + \frac{7}{8}y^2 + \frac{57}{128}y - \frac{195}{1024} = 0,$$

a cubic equation to determine y. One root of this equation is real and positive, the other two are imaginary; if a is the positive root, $\beta \pm \iota\gamma$ the two imaginary roots, then we have

$$a + 2\beta = -\frac{7}{8},$$

$$2\beta a + \beta^2 + \gamma^2 = \frac{57}{128},$$

$$a(\beta^2 + \gamma^2) = \frac{195}{1024}.$$

We find by the rules for the solution of numerical equations that $a = \cdot 26$ approximately, hence

$$\beta = -\cdot 56, \qquad \gamma = \pm\cdot 64.$$

These roots are however not large enough for the approximation to be close to the accurate values.

Hence from equation (88), we see that when $s = 1$,

$$\frac{\iota p}{V}\mathbf{a} = -\cdot 56 \pm \iota\cdot 64,$$

or

$$\iota p = (-\cdot 56 \pm \iota\cdot 64)\frac{V}{\mathbf{a}}.$$

This represents a vibration whose period is $3.1\,\pi\mathbf{a}/V$, and whose amplitude fades away to $1/\epsilon$ of its original value after a time $1.8\,\mathbf{a}/V$.

The radiation of energy away from the sphere in this case is so rapid that the vibrations are practically dead beat; thus after one complete vibration the amplitude is only $\epsilon^{-1.74\pi}$, or about one two hundred and fiftieth part of its value at the beginning of the oscillation.

302.] If we consider the state of the field at a considerable distance from the cylinder and only retain in each expression the lowest power of $1/r$, we find that the magnetic induction c, the tangential and radial components Θ and R of the electric polarization in the dielectric, may be consistently represented by the following equations:

$$c = \cos\theta \frac{1}{r^{\frac{1}{2}}} \epsilon^{-.56\left(\frac{Vt-r}{a}\right)} \cos .64\left(\frac{Vt-r}{a}\right),$$

since

$$K\frac{d\Theta}{dt} = -\frac{1}{\mu}\frac{dc}{dr},$$

we have

$$\Theta = \frac{\cos\theta}{K\mu V}\frac{1}{r^{\frac{1}{2}}} \epsilon^{-.56\left(\frac{Vt-r}{a}\right)} \cos .64\left(\frac{Vt-r}{a}\right),$$

and since

$$K\frac{dR}{dt} = \frac{1}{\mu r}\frac{dc}{d\theta},$$

we have

$$R = \frac{\sin\theta}{K\mu V}\frac{a}{r^{\frac{3}{2}}} 1.34\, \epsilon^{-.56\left(\frac{Vt-r}{a}\right)}$$
$$\left\{.56\cos .64\left(\frac{Vt-r}{a}\right) - .64\sin .64\left(\frac{Vt-r}{a}\right)\right\}.$$

Thus R vanishes at all points on a series of cylinders concentric with the original one whose radii satisfy the equation

$$\cot .64\left(\frac{Vt-r}{a}\right) = 1.13,$$

the distance between the consecutive cylinders in this series is

$$1.57\,\pi a.$$

The Faraday tubes between two such cylinders form closed curves, all cutting at right angles the cylinder for which

$$\Theta = 0, \text{ or } \cos .64\left(\frac{Vt-r}{a}\right) = 0.$$

The closed Faraday tubes move away from the cylinder and are the vehicles by which the energy of the cylinder radiates into space. The axes of the Faraday tubes, i.e. the lines of electromotive intensity between two cylinders at which $R=0$, are represented in Fig. 110.

The genesis of these closed endless tubes from the unclosed ones, which originally stretched from one point to another of the cylinder, which we may suppose to have been electrified initially so that the surface density was proportional to sin θ, is shown in Fig. 111.

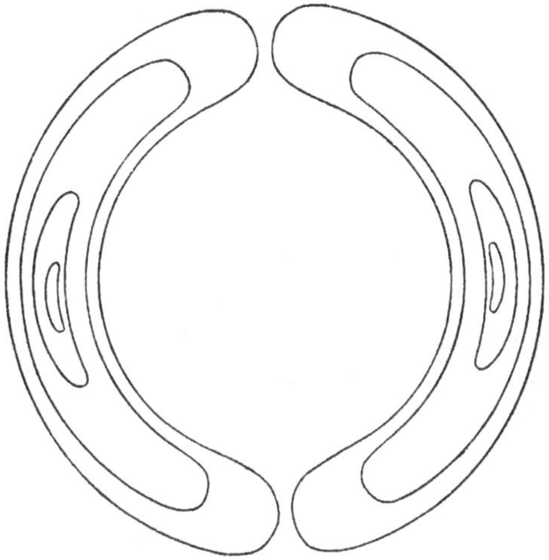

Fig. 110.

The lines represent the changes in shape in a Faraday tube which originally stretched from a positively to a negatively electrified place on the cylinder. The outer line A represents the original

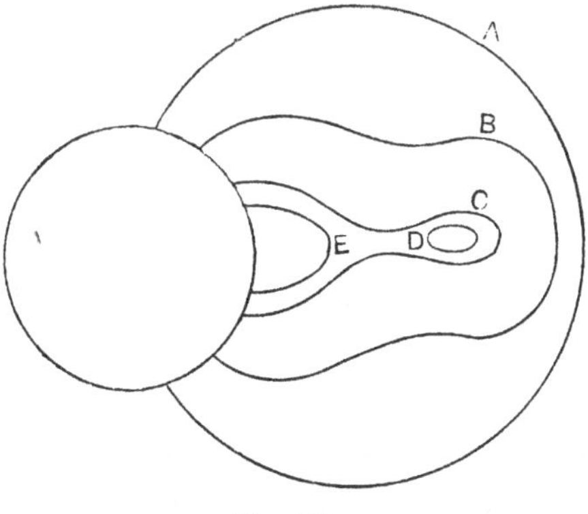

Fig. 111.

position of the tube; when the equilibrium is disturbed some of the tubes inside this one will soon run into the cylinder, and the lateral repulsion they exerted on the tube under consideration will be removed; the outside lateral pressure on this tube will now overpower the inside pressure and will produce the indentation shown in the second position B of the tube; this indentation increases until the two sides of the tube meet as in the third position C of the tube; when this takes place the tube breaks up, the outer part D travelling out into space and forming one of the closed tubes shown in Fig. 111, while the inner part E runs into the cylinder.

Decay of Magnetic Force in a Metal Cylinder.

303.] In addition to the very rapid oscillations we have just investigated there are other and slower changes which may occur in the electrical state of the cylinder. Thus, for example, a uniform magnetic field parallel to the axis of the cylinder might suddenly be removed; the alteration in the magnetic force would then induce currents in the cylinder whose magnetic action would tend to maintain the original state of the magnetic field, so that the field instead of sinking abruptly to zero would die away gradually. The rate at which the state of the system changes with the time in cases like this is exceedingly slow compared with the rate of change we have just investigated. Using the same notation as in the preceding investigation, it will be slow enough to make $p\mathrm{a}/V$ an exceedingly small quantity; when however $p\mathrm{a}/V$ is very small, $K'_s(p\mathrm{a}/V)$ is exceedingly large compared with $K_s(p\mathrm{a}/V)$, since (Heine, *Kugelfunctionen*, vol. i. p. 237) $K_s(\theta)$ is equal to

$$(-2\theta)^s \frac{d^s K_0(\theta)}{(d\theta^2)^s};$$

thus since when θ is small $K_0(\theta)$ is proportional to $\log \theta_1$, $K_s(p\mathrm{a}/V)$ is proportional to $(V/p\mathrm{a})^s$, and $K_s'(p\mathrm{a}/V)$ to $(V/p\mathrm{a})^{s+1}$; hence the right-hand side of equation (87) is exceedingly large, so that an approximate solution of that equation will be

$$J_s(\iota n\mathrm{a}) = 0.$$

We notice that this condition makes the normal electromotive intensity at the surface of the cylinder vanish, while it will be remembered that for the very rapid oscillations the tangential electromotive intensity vanished. As the normal intensity vanishes there is no electrification on the surface of the cylinder in this case.

The equation $J_s(x) = 0$ has an infinite number of roots all real, the smaller values of which from $s = 0$ to $s = 5$ are given in the following table, taken from Lord Rayleigh's *Theory of Sound*, Vol. I, p. 274.

$s = 0$	$s = 1$	$s = 2$	$s = 3$	$s = 4$	$s = 5$
2·404	3·832	5·135	6·379	7·586	8·780
5·520	7·016	8·417	9·760	11·064	12·339
8·654	10·173	11·620	13·017	14·373	15·700
11·792	13·323	14·796	16·224	17·616	18·982
14·931	16·470	17·960	19·410	20·827	22·220
18·071	19·616	21·117	22·583	24·018	25·431
21·212	22·760	24·270	25·749	27·200	28·628
24·353	25·903	27·421	28·909	30·371	31·813
37·494	29·047	30·571	32·050	33·512	34.983

This table may be supplemented by the aid of the theorem that the large roots of the equation got by equating a Bessel's function to zero form approximately an arithmetical progression whose common difference is π.

If x_q denotes a root of the equation

$$J_s(x) = 0,$$

then since p is given by the equation

$$J_s(\iota n \mathrm{a}) = 0,$$

where

$$n^2 = \frac{4\pi\mu\iota p}{\sigma},$$

we see that p_q, the corresponding value of p, is given by the equation

$$-\iota p_q = \frac{\sigma}{4\pi \mathrm{a}^2\mu} x_q^{\,2}.$$

Thus, since ιp_q is real and negative, the system simply fades away to its position of equilibrium and does not oscillate about it.

The term in c which was initially expressed by

$$A \cos s\theta J_s\left(x_q \frac{r}{\mathrm{a}}\right),$$

will after the lapse of a time t have diminished to

$$A \cos s\theta J_s\left(x_q \frac{r}{\mathrm{a}}\right) \epsilon^{-\frac{\sigma}{4\pi\mathrm{a}^2\mu} x_q^{\,2} t}.$$

If we call T the time which must elapse before the term sinks to $1/\epsilon$ of its original value, the 'time modulus' of the term, then, since

$$T = \frac{4\pi \mathrm{a}^2 \mu}{\sigma x_q^{\,2}},$$

we see that the time modulus is inversely proportional to the resistance of unit length of the cylinder and directly proportional to the magnetic permeability. Since μ/σ for iron is larger than it is for copper, the magnetic force will fade away more slowly in an iron cylinder than in a copper one.

304.] A case of great interest, which can be solved without difficulty by the preceding equations, is the one where a cylinder is placed in a uniform magnetic field which is suddenly annihilated, the lines of magnetic force being originally parallel to the axis of the cylinder. We may imagine, for example, that the cylinder is placed inside a long straight solenoid, the current through which is suddenly broken.

Since in this case everything is symmetrical about the axis of the cylinder, $s = 0$, and the values of ιp are therefore

$$-(2{\cdot}404)^2 \frac{\sigma}{4\pi \mathrm{a}^2 \mu}, \qquad -(5{\cdot}520)^2 \frac{\sigma}{4\pi \mathrm{a}^2 \mu}, \text{ \&c.}$$

Now we know from the theory of Bessel's functions that any function of r can for values of r between 0 and a be expanded in the form

$$A_1 J_0\left(x_1 \frac{r}{\mathrm{a}}\right) + A_2 J_0\left(x_2 \frac{r}{\mathrm{a}}\right) + A_3 J_0\left(x_3 \frac{r}{\mathrm{a}}\right) + \ldots,$$

where $x_1, x_2, x_3 \ldots$ are the roots of the equation

$$J_0(x) = 0.$$

Thus, initially

$$c = A_1 J_0\left(x_1 \frac{r}{\mathrm{a}}\right) + A_2 J_0\left(x_2 \frac{r}{\mathrm{a}}\right) + A_3 J_0\left(x_3 \frac{r}{\mathrm{a}}\right) + \ldots,$$

hence the value of c after a time t will be given by the equation

$$c = A_1 J_0\left(x_1 \frac{r}{\mathrm{a}}\right) \epsilon^{-\frac{\sigma}{4\pi \mathrm{a}^2 \mu} x_1^2 t} + A_2 J_0\left(x_2 \frac{r}{\mathrm{a}}\right) \epsilon^{-\frac{\sigma}{4\pi \mathrm{a}^2 \mu} x_2^2 t} + \ldots,$$

so that all we have to do is to find the coefficients A_1, A_2, $A_3 \ldots$.

We shall suppose that initially c was uniform over the section of the cylinder and equal to c_0.

Then, since

$$\int_0^{\mathrm{a}} r J_0\left(x_p \frac{r}{\mathrm{a}}\right) J_0\left(x_q \frac{r}{\mathrm{a}}\right) dr = 0$$

when p and q are different, we see that

$$c_0\int_0^{\mathrm{a}} rJ_0\left(x_q\frac{r}{\mathrm{a}}\right)dr = A_q\int_0^{\mathrm{a}} rJ_0^2\left(x_q\frac{r}{\mathrm{a}}\right)dr.$$

Now, since

$$J_0''(x)+\frac{1}{x}J_0'(x)+J_0(x) = 0,$$

$$xJ_0(x) = -\frac{d}{dx}(xJ_0'(x)),$$

hence
$$\int_0^{\mathrm{a}} rJ_0\left(x_q\frac{r}{\mathrm{a}}\right)dr = -\frac{\mathrm{a}^2}{x_q}J_0'(x_q)$$

$$= \frac{\mathrm{a}^2}{x_q}J_1(x_q).$$

Again, since

$$J_0''(x)+\frac{1}{x}J_0'(x)+J_0(x) = 0,$$

we have, multiplying by $2x^2J_0'(x)$,

$$\frac{d}{dx}\{x^2J_0'^2(x)+x^2J_0^2(x)\} = 2xJ_0^2(x),$$

hence
$$x^2\{J_0'^2(x)+J_0^2(x)\} = 2\int_0^x xJ_0^2(x)\,dx.$$

Thus, since
$$J_0(x_q) = 0,$$

$$\int_0^{\mathrm{a}} rJ_0^2\left(x_q\frac{r}{\mathrm{a}}\right)dr = \frac{1}{2}\mathrm{a}^2J_0'^2(x_q)$$

$$= \frac{1}{2}x^2J_1^2(x_q).$$

Hence, we see that

$$A_q = \frac{2c_0}{x_qJ_1(x_q)},$$

and therefore

$$c = 2c_0\Sigma\frac{1}{x_q}\frac{J_0\left(x_q\frac{r}{\mathrm{a}}\right)}{J_1(x_q)}\epsilon^{-\frac{\sigma}{4\pi\mathrm{a}^2\mu}x_q^2t}.$$

We see from this equation that immediately after the magnetic force is removed c vanishes at the surface of the cylinder; also, since the terms in the expression for c corresponding to the large roots of the equation $J_0(x) = 0$ die away more quickly than those corresponding to the smaller roots, c will ultimately be very approximately represented by the first term in the preceding expression; hence we have, since

$$J_1(2\cdot404) = \cdot519,$$

$$c = 1.6 c_0 J_0\left(2\cdot404\frac{r}{\mathrm{a}}\right)\epsilon^{-\frac{\sigma}{4\pi\mathrm{a}^2\mu}5\cdot78t}.$$

This expression is a maximum when $r = 0$ and gradually dies away to zero when $r = \mathrm{a}$, thus the lines of magnetic force fade away most quickly at the surface of the cylinder and linger longest at the centre.

The time modulus for the first term is $4\pi\mathrm{a}^2\mu/5\cdot78\sigma$. For a copper rod 1 cm. in radius for which $\sigma = 1600$, this is about 1/736 of a second; for an iron rod of the same radius for which $\mu = 1000$, $\sigma = 10^4$, it is about 2/9 of a second.

305.] The intensity of the current is $-\frac{1}{4\pi\mu}\frac{dc}{dr}$, hence at a distance r from the axis of the cylinder the intensity is

$$\frac{c_0}{2\pi\mu\mathrm{a}}\Sigma\frac{J_1\left(x_q\frac{r}{\mathrm{a}}\right)}{J_1(x_q)}\epsilon^{-\frac{\sigma}{4\pi\mathrm{a}^2\mu}x_q^2t}.$$

Since at the instant the magnetic force is destroyed, c is constant over the cross-section of the cylinder, the intensity of the current when $t = 0$ will vanish except at the surface of the cylinder, where, as the above equation shows, it is infinite. After some time has elapsed the intensity of the current will be adequately represented by the first term of the series, i.e. by

$$\frac{c_0}{2\pi\mu\mathrm{a}}\frac{J_1\left(2\cdot404\frac{r}{\mathrm{a}}\right)}{\cdot52}\epsilon^{-\frac{\sigma}{4\pi\mathrm{a}^2\mu}5\cdot78t}.$$

This vanishes at the axis of the cylinder and, as we see from tables for $J_1(x)$ (Lord Rayleigh, *Theory of Sound*, vol. I, p. 265), attains a maximum when $2\cdot404\frac{r}{\mathrm{a}} = 1\cdot841$, or at a distance from the axis about 3/4 the radius of the cylinder.

The following table, taken from the paper by Prof. Lamb on this subject (*Proc. Lond. Math. Soc.* XV, p. 143), gives the value of the total induction through the cylinder, and the electromotive force round a circuit embracing the cylinder for a series of values of t/τ, where $\tau = 4\pi\mu\mathrm{a}^2/\sigma$:—

t/τ	Total Induction	Electromotive force $/\sigma c_0$
·00	1·0000	infinite
·02	·7014	1·7332
·04	·5904	1·1430
·06	·5105	·8789
·08	·4470	·7195
·10	·3941	·6089
·20	·2178	·3168
·30	·1220	·1765
·40	·0684	·0989
·50	·0384	·0555
·60	·0215	·0311
·70	·0121	·0174
·80	·0068	·0098
·90	·0038	·0055
1·00	·0021	·0031

Rate of Decay of Currents and Magnetic Force in infinite Cylinders when the Currents are Longitudinal and the Magnetic Force Transversal.

306.] We have already considered this problem in the special case when the currents are symmetrically distributed through the cylinder in Art. 262; we shall now consider the case when the currents are not the same in all planes through the axis.

Let w be the intensity of the current parallel to the axis of the cylinder, then (Art. 256) in the cylinder w satisfies the differential equation

$$\frac{d^2 w}{dx^2} + \frac{d^2 w}{dy^2} = \frac{4\pi\mu}{\sigma}\frac{dw}{dt}.$$

If w' denotes the rate of increase in the electric displacement parallel to z in the dielectric surrounding the cylinder, then, since w' is equal to $\frac{K}{4\pi}\frac{dZ}{dt}$, where Z is the electromotive intensity parallel to z the axis of the cylinder, w' satisfies the equation

$$\frac{d^2 w'}{dx^2} + \frac{d^2 w'}{dy^2} = \frac{1}{V^2}\frac{d^2 w'}{dt^2}.$$

Let us suppose that w varies as $\cos s\theta\, \epsilon^{\iota pt}$, then transforming to cylindrical coordinates r, θ, the equation satisfied by w in the cylinder becomes

$$\frac{d^2 w}{dr^2} + \frac{1}{r}\frac{dw}{dr} + \left\{-\frac{4\pi\mu\iota p}{\sigma} - \frac{s^2}{r^2}\right\} w = 0,$$

the solution of which is

$$w = A \cos s\theta\, \epsilon^{\iota pt} J_s(\iota n r),$$

where $$n^2 = \frac{4\pi\mu\iota p}{\sigma};$$

while in the dielectric we have

$$\frac{d^2 w'}{dr^2} + \frac{1}{r}\frac{dw'}{dr} + \left(\frac{p^2}{V^2} - \frac{s^2}{r^2}\right) w' = 0,$$

the solution of which is

$$w' = B \cos s\theta \epsilon^{\iota pt} K_s\left(\frac{p}{V} r\right).$$

The electromotive intensity Z, parallel to the axis of the cylinder, is equal to σw in the cylinder and to $4\pi w'/K\iota p$ in the dielectric. At the surface of the cylinder $r = \mathrm{a}$ these must be equal, hence we have

$$\sigma A J_s(\iota n \mathrm{a}) = \frac{4\pi}{K\iota p} B K_s\left(\frac{p}{V}\mathrm{a}\right). \tag{89}$$

If Θ is the magnetic induction at right angles to r, then

$$\frac{d\Theta}{dt} = \frac{dZ}{dr},$$

or, since Θ varies as $\epsilon^{\iota pt}$,

$$\Theta = \frac{1}{\iota p}\frac{dZ}{dr}.$$

Thus, in the cylinder

$$\Theta = \frac{\sigma}{\iota p}\frac{dw}{dr}$$

$$= \frac{\sigma}{\iota p}\iota n A J_s'(\iota n \mathrm{a}), \text{ at the surface.}$$

In the dielectric

$$\Theta = \frac{1}{\iota p}\frac{4\pi}{K\iota p}\frac{p}{V} B K_s'\left(\frac{p}{V}\mathrm{a}\right), \text{ at the surface.}$$

Since the magnetic force parallel to the surface is continuous, we have

$$\left(\frac{1}{\mu}\Theta\right) \text{ in the cylinder} = \Theta \text{ in the dielectric,}$$

hence $$\frac{\sigma}{\mu}\iota n A J_s'(\iota n \mathrm{a}) = \frac{4\pi}{K\iota p}\frac{p}{V} B K_s'\left(\frac{p}{V}\mathrm{a}\right). \tag{90}$$

Eliminating A and B from (89) and (90), we have

$$\frac{\iota n \mathrm{a} J_s'(\iota n \mathrm{a})}{\mu J_s(\iota n \mathrm{a})} = \frac{p\mathrm{a}}{V}\frac{K_s'\left(\frac{p}{V}\mathrm{a}\right)}{K_s\left(\frac{p}{V}\mathrm{a}\right)}. \tag{91}$$

In this case $p\mathrm{a}/V$ is very small, so that (Art. 303) $K_s\left(\frac{p}{V}\mathrm{a}\right)$ is approximately proportional to $\left(\frac{p}{V}\mathrm{a}\right)^{-s}$, and thus

$$\frac{p}{V}\frac{K_s'\left(\frac{p}{V}\mathrm{a}\right)}{K_s\left(\frac{p}{V}\mathrm{a}\right)} = -\frac{s}{\mathrm{a}}, \text{ approximately;}$$

hence equation (91) becomes

$$\iota n \mathrm{a} J_s'(\iota n \mathrm{a}) + s\mu J_s(\iota n \mathrm{a}) = 0. \tag{92}$$

Bessel's functions, however, satisfy the relation

$$J_s'(\iota n\mathrm{a}) + \frac{s}{\iota n\mathrm{a}} J_s(\iota n\mathrm{a}) = J_{s-1}(\iota n\mathrm{a}),$$

so that (92) may be written

$$s(\mu - 1) J_s(\iota n\mathrm{a}) + \iota n\mathrm{a} J_{s-1}(\iota n\mathrm{a}) = 0.$$

For non-magnetic substances $\mu = 1$, so that this equation reduces to

$$J_{s-1}(\iota n\mathrm{a}) = 0.$$

The magnetic induction along the radius is equal to

$$-\frac{\sigma}{\iota p}\frac{1}{r}\frac{dw}{d\theta};$$

at right angles to the radius it is equal to

$$\frac{\sigma}{\iota p}\frac{dw}{dr}.$$

307.] Let us consider the case when $s = 1$. For a non-magnetic cylinder n will be given by the equation

$$J_0(\iota n\mathrm{a}) = 0;$$

thus the values of ιp will be the same as those in Art. 304, and we may put

$$w = \cos\theta\left\{A_1 J_1\left(x_1\frac{r}{\mathrm{a}}\right)\epsilon^{-\frac{\sigma}{4\pi\mathrm{a}^2}x_1^2 t} + A_2 J_1\left(x_2\frac{r}{\mathrm{a}}\right)\epsilon^{-\frac{\sigma}{4\pi\mathrm{a}^2}x_2^2 t} + \ldots\right\}, \tag{93}$$

where x_1, x_2 are the values 2·404, 5·520..., which are the roots of the equation

$$J_0(x) = 0.$$

The magnetic force along the radius is therefore

$$-\frac{4\pi \mathbf{a}^2 \sin\theta}{r}\left\{\frac{1}{x_1^2}A_1 J_1\left(x_1\frac{r}{\mathbf{a}}\right)\epsilon^{-\frac{\sigma}{4\pi\mathbf{a}^2}x_1^2 t} + \frac{1}{x_2^2}A_2 J_1\left(x_2\frac{r}{\mathbf{a}}\right)\epsilon^{-\frac{\sigma}{4\pi\mathbf{a}^2}x_2^2 t} + \ldots\right\}. \quad (94)$$

If originally the magnetic force is parallel to y and equal to H, the radial component of the magnetic force is $H\sin\theta$; hence, if we determine A_1, A_2 so that when $t = 0$ the expression (94) is equal to $H\sin\theta$, then equation (93) will give the currents generated by the annihilation of a uniform magnetic field parallel to y.

Since $$J_1(x) = -J_0'(x),$$

$$\int_0^{\mathbf{a}} r^2 J_1\left(x_p\frac{r}{\mathbf{a}}\right)dr = -\int_0^{\mathbf{a}} r^2 J_0'\left(x_p\frac{r}{\mathbf{a}}\right)dr.$$

Integrating by parts and remembering that $J_0(x_p) = 0$, we see that each of these integrals equals

$$\frac{2\mathbf{a}}{x_p}\int_0^{\mathbf{a}} r J_0\left(x_p\frac{r}{\mathbf{a}}\right)dr,$$

which is equal to

$$\frac{2\mathbf{a}^3}{x_p^2}J_1(x_p), \quad (95)$$

since $$\frac{d}{dx}\{xJ_0'(x)\} = -xJ_0(x).$$

Again, since

$$\frac{d^2 J_1(x)}{dx^2} + \frac{1}{x}\frac{dJ_1(x)}{dx} + \left(1 - \frac{1}{x^2}\right)J_1(x) = 0,$$

multiplying by

$$2x^2\frac{dJ_1}{dx},$$

we get

$$\frac{d}{dx}\left\{x^2 J_1'^2(x) + x^2\left(1 - \frac{1}{x^2}\right)J_1^2(x)\right\} = 2xJ_1^2(x).$$

Hence

$$\int_0^{\xi} xJ_1^2(x)\,dx = \frac{1}{2}\xi^2\left\{J_1'^2(\xi) + \left(1 - \frac{1}{\xi^2}\right)J_1^2(\xi)\right\}.$$

Thus, since $$\xi J_1'(\xi) + J_1(\xi) = \xi J_0(\xi),$$

we have if $$J_0(\xi) = 0,$$

$$\int_0^{\xi} xJ_1^2(x)\,dx = \frac{1}{2}\xi^2 J_1^2(\xi).$$

Hence, when x_p is a root of

$$J_0(x) = 0,$$

$$\int_0^{\mathrm{a}} r\, J_1^2\left(x_p \frac{r}{\mathrm{a}}\right) dr = \frac{1}{2}\mathrm{a}^2 J_1^2(x_p). \tag{96}$$

Now by (94)

$$H = -\frac{4\pi \mathrm{a}^2}{r}\left\{\frac{A_1}{x_1^2} J_1\left(x_1 \frac{r}{\mathrm{a}}\right) + \frac{A_2}{x_2^2} J_1\left(x_2 \frac{r}{\mathrm{a}}\right) + \dots\right\},$$

so that

$$H\int_0^{\mathrm{a}} r^2 J_1\left(x_p \frac{r}{\mathrm{a}}\right) dr = -\frac{4\pi \mathrm{a}^2}{x_p^2} A_p \int_0^{\mathrm{a}} r J_1^2\left(x_p \frac{r}{\mathrm{a}}\right) dr.$$

Hence by (95) and (96)

$$A_p = -\frac{H}{\pi \mathrm{a} J_1(x_p)}.$$

Thus by (93), the currents produced by the annihilation of a magnetic field H parallel to y are given by the equation

$$w = -\frac{H\cos\theta}{\pi \mathrm{a}} \Sigma \frac{J_1\left(x_p \frac{r}{\mathrm{a}}\right)}{J_1(x_p)} \epsilon^{-\frac{\sigma x_p^2}{4\pi \mathrm{a}^2}t}.$$

Thus the currents vanish at the axis of the cylinder; when $t = 0$ they are infinite at the surface and zero elsewhere.

When, as in the case of iron, μ is very large, the equation (92) becomes approximately $J_s(\iota n \mathrm{a}) = 0.$

The solution in this case can be worked out on the same lines as the preceding one; for the results of this investigation we refer the reader to a paper by Prof. H. Lamb (*Proc. Lond. Math. Soc.* XV, p. 270).

Electrical Oscillations on a Spherical Conductor.

308.] The equations satisfied in the electromagnetic field by the components of the magnetic induction, or of the electromotive intensity, when these quantities vary as $\epsilon^{\iota pt}$, are, denoting any one of them by F, of the form

$$\frac{d^2F}{dx^2} + \frac{d^2F}{dy^2} + \frac{d^2F}{dz^2} = -\lambda^2 F, \tag{97}$$

where in an insulator $\lambda^2 = p^2/V^2$, V being the velocity of propagation of electrodynamic action through the dielectric, and in a conductor, whose specific resistance is σ and magnetic permeability μ, $\lambda^2 = -4\pi\mu\iota p/\sigma.$

In treating problems about spheres and spherical waves it is convenient to express F as the sum of terms of the form

$$f(r)\,Y_n,$$

where $f(r)$ is a function of the distance from the centre, and Y_n a surface spherical harmonic function of the nth order. Transforming (97) to polar coordinates, we find that $f(r)$ satisfies the differential equation

$$\frac{d^2f}{dr^2} + \frac{2}{r}\frac{df}{dr} + \left(\lambda^2 - \frac{n(n+1)}{r^2}\right)f = 0.$$

We can easily verify by substitution that the solution of this equation is, writing ρ for λr,

$$f(r) = \rho^n \left\{\frac{1}{\rho}\frac{d}{d\rho}\right\}^n \left(\frac{A\epsilon^{\iota\rho} + B\epsilon^{-\iota\rho}}{\rho}\right),$$

where A and B are arbitrary constants; particular solutions of this equation are thus

$$f(r) = \rho^n \left\{\frac{1}{\rho}\frac{d}{d\rho}\right\}^n \frac{\sin\rho}{\rho}, \qquad (\alpha)$$

$$f(r) = \rho^n \left\{\frac{1}{\rho}\frac{d}{d\rho}\right\}^n \frac{\cos\rho}{\rho}, \qquad (\beta)$$

$$f(r) = \rho^n \left\{\frac{1}{\rho}\frac{d}{d\rho}\right\}^n \left\{\frac{\epsilon^{-\iota\rho}}{\rho}\right\}, \qquad (\gamma)$$

$$f(r) = \rho^n \left\{\frac{1}{\rho}\frac{d}{d\rho}\right\}^n \left\{\frac{\epsilon^{\iota\rho}}{\rho}\right\}. \qquad (\delta)$$

The first of these solutions is the only one which does not become infinite when ρ vanishes, so that it is the solution we must choose in any region where ρ can vanish; in the case of the sphere it is the function which must be used inside the sphere; we shall denote it by $S_n(\rho)$.

Outside the sphere, where ρ cannot vanish, the choice of the function must be governed by other considerations. If we are considering wave motions, then, since the solution (γ) will contain the factor $\epsilon^{\iota(pt-\rho)}$, it will correspond to a wave diverging from the sphere; the solution (δ), which contains the factor $\epsilon^{\iota(pt+\rho)}$, corresponds to waves converging on the sphere; the solutions (α) and (β) correspond to a combination of convergent and divergent waves; thus, where there is no reflection we must take (γ) if the waves are divergent, (δ) if they are convergent. In other cases we find that λ is complex and of the form $p+\iota q$; in this case (a) and (β) will be infinite at an infinite distance from the origin, while of the two solutions f and δ one will be infinite, the other zero, we must take the solution which vanishes when ρ is infinite. We shall denote (γ) by $E_n^-(\rho)$, (δ) by $E_n^+(\rho)$, and when, as we shall sometimes do, we leave the question as to which of the two we shall take unsettled until we have determined λ, we shall use the expression $E_n(\rho)$, which thus denotes one or other of (γ) and (δ).

When there is no reflection, the solution of (97) is thus expressed by

$$S_n(\rho)\, Y_n \epsilon^{\iota pt} \text{ inside the sphere,}$$

$$E_n(\rho)\, Y_n \epsilon^{\iota pt} \text{ outside the sphere.}$$

In particular when Y_n is the zonal harmonic Q_n, the solutions are

$$S_n(\rho)\, Q_n \epsilon^{\iota pt}, \quad E_n(\rho)\, Q_n \epsilon^{\iota pt}.$$

When Y_n is the first tesseral harmonic, the solutions are

$$\frac{x}{r} S_n(\rho) \frac{dQ_n}{d\mu} \epsilon^{\iota pt}, \qquad \frac{x}{r} E_n(\rho) \frac{dQ_n}{d\mu} \epsilon^{\iota pt},$$

$$\frac{y}{r} S_n(\rho) \frac{dQ_n}{d\mu} \epsilon^{\iota pt}, \qquad \frac{y}{r} E_n(\rho) \frac{dQ_n}{d\mu} \epsilon^{\iota pt},$$

where $\mu = \cos\theta$, θ being the colatitude of the intersection of the radius with the surface of the sphere.

309.] We shall now proceed to prove those properties of the functions S_n and E_n which we shall require for the subsequent investigations. The reader who desires further information about these interesting functions can derive it from the following sources :—

Stokes, 'On the Communication of Vibration from a Vibrating Body to the Surrounding Gas,' *Phil. Trans.* 1868, p. 447.

Rayleigh, 'Theory of Sound,' Vol. II, Chap. XVII.

C. Niven, 'On the Conduction of Heat in Ellipsoids of Revolution,' *Phil. Trans.* Part I, 1880, p. 117.

C. Niven, 'On the Induction of Electric Currents in Infinite Plates and Spherical Shells,' *Phil. Trans.* Part II, 1881, p. 307.

H. Lamb, 'On the Vibrations of an Elastic Sphere,' and 'On the Oscillations of a Viscous Spheroid,' *Proc. Lond. Math. Soc.*, 13, pp. 51, 189.

H. Lamb, 'On Electrical Motions in a Spherical Conductor,' *Phil. Trans.* Part II, 1883, p. 519.

V. Helmholtz, 'Wissenschaftliche Abhandlungen,' Vol. I, p. 320.
Heine, 'Kugelfunctionen,' Vol. I, p. 140.

The following propositions are for brevity expressed only for the S_n functions, since, however, their proof only depends upon the differential equations satisfied by these functions they are equally true for the functions β, γ, δ.

Since

$$S_n(\rho) = \rho^n \left\{ \frac{1}{\rho} \frac{d}{d\rho} \right\}^n \frac{\sin \rho}{\rho},$$

$$\frac{1}{\rho} \frac{d}{d\rho} \left\{ \frac{S_{n-1}}{\rho^{n-1}} \right\} = \frac{S_n}{\rho^n},$$

or

$$\rho \frac{dS_{n-1}}{d\rho} - (n-1) S_{n-1} = \rho S_n, \tag{98}$$

and therefore

$$\rho \frac{dS_n}{d\rho} - n S_n = \rho S_{n+1}. \tag{99}$$

Multiply (98) by ρ^n and differentiate with respect to ρ, and we get

$$\frac{d^2}{d\rho^2} S_{n-1} + \frac{2}{\rho} \frac{dS_{n-1}}{d\rho} - \frac{n(n-1)S_{n-1}}{\rho^2} = (n+1)\frac{S_n}{\rho} + \frac{dS_n}{d\rho}.$$

But

$$\frac{d^2 S_{n-1}}{d\rho^2} + \frac{2}{\rho} \frac{dS_{n-1}}{d\rho} + \left(1 - \frac{n(n-1)}{\rho^2}\right) S_{n-1} = 0,$$

hence

$$-\rho S_{n-1} = (n+1) S_n + \rho \frac{dS_n}{d\rho}. \tag{100}$$

From (99) and (100), we get

$$(2n+1) S_n + \rho (S_{n-1} + S_{n+1}) = 0, \tag{101}$$

and

$$(2n+1) \frac{dS_n}{d\rho} = (n+1) S_{n+1} - n S_{n-1}.$$

Again, since

$$\frac{d^2}{dr^2} S_n(\lambda r) + \frac{2}{r} \frac{d}{dr} S_n(\lambda r) + \left(\lambda^2 - \frac{n(n+1)}{r^2}\right) S_n(\lambda r) = 0,$$

and

$$\frac{d^2}{dr^2} S_n(\lambda' r) + \frac{2}{r} \frac{d}{dr} S_n(\lambda' r) + \left(\lambda'^2 - \frac{n(n+1)}{r^2}\right) S_n(\lambda' r) = 0,$$

we have

$$r^2 \left\{ S_n(\lambda' r) \frac{d^2}{dr^2} S_n(\lambda r) - S_n(\lambda r) \frac{d^2}{dr^2} S_n(\lambda' r) \right\}$$
$$+ 2r \left\{ S_n(\lambda' r) \frac{d}{dr} S_n(\lambda r) - S_n(\lambda r) \frac{d}{dr} S_n(\lambda' r) \right\}$$
$$= (\lambda'^2 - \lambda^2) r^2 S_n(\lambda r) S_n(\lambda' r),$$

and hence

$$\int_a^b r^2 S_n(\lambda r) S_n(\lambda' r)\, dr$$

$$= \frac{1}{\lambda'^2 - \lambda^2} \left\{ r^2 S_n(\lambda' r) \frac{d}{dr} S_n(\lambda r) - r^2 S_n(\lambda r) \frac{d}{dr} S_n(\lambda' r) \right\}_a^b, \quad (102)$$

so that if λ, λ' satisfy the equations

$$a^2 \left\{ S_n(\lambda' a) \frac{d}{da} S_n(\lambda a) - S_n(\lambda a) \frac{d}{da} S_n(\lambda' a) \right\} = 0,$$

$$b^2 \left\{ S_n(\lambda' b) \frac{d}{db} S_n(\lambda' b) - S_n(\lambda b) \frac{d}{db} S_n(\lambda' b) \right\} = 0,$$

then
$$\int_a^b r^2 S_n(\lambda r) S_n(\lambda' r)\, dr = 0.$$

Proceeding to the limit $\lambda' = \lambda$, we get from (102)

$$\int_a^b r^2 S_n^2(\lambda r)\, dr = -\frac{1}{2\lambda^2} \left[r^2 S_n(\lambda r) \frac{d}{dr} \left\{ r \frac{d S_n(\lambda r)}{dr} \right\} - r^3 \left\{ \frac{d S_n(\lambda r)}{dr} \right\}^2 \right]_a^b.$$

The following table of the values of the first four of the S and E functions will be found useful for the subsequent work :—

$$S_0(x) = \frac{\sin x}{x},$$

$$S_1(x) = \frac{\cos x}{x} - \frac{\sin x}{x^2},$$

$$S_2(x) = -\frac{\sin x}{x} - \frac{3\cos x}{x^2} + \frac{3\sin x}{x^3},$$

$$S_3(x) = -\frac{\cos x}{x} + \frac{6\sin x}{x^2} + \frac{15\cos x}{x^3} - \frac{15\sin x}{x^4}.$$

. . . .

$$E_0^-(x) = \frac{\epsilon^{-\iota x}}{x},$$

$$E_1^-(x) = -\frac{\epsilon^{-\iota x}}{x}\left(\iota + \frac{1}{x}\right),$$

$$E_2^-(x) = -\frac{\epsilon^{-\iota x}}{x}\left(1 - \frac{3\iota}{x} - \frac{3}{x^2}\right),$$

$$E_3^-(x) = \frac{\epsilon^{-\iota x}}{x}\left(\iota + \frac{6}{x} - \frac{15\iota}{x^2} - \frac{15}{x^3}\right).$$

. . . . .

The values of E^+ can be got from those of E^- by changing the sign of ι.

310.] We shall now proceed to the study of the oscillations of a distribution of electricity over the surface of a sphere. Let us suppose that a distribution of electricity whose surface density is proportional to a *zonal* harmonic of the nth order is produced over the surface of the sphere, and that the cause producing this distribution is suddenly removed; then, since this distribution cannot be in equilibrium unless under the influence of external forces, electric currents will start off to equalize it, and electrical vibrations will be started whose period it is the object of the following investigation to determine.

Since the currents obviously flow in planes through the axis of the zonal harmonic, which we shall take for the axis of z, there is no electromotive force round a circuit in a plane at right angles to this axis; and since the electromotive force round a circuit is equal to the rate of diminution in the number of lines of magnetic force passing through it, we see that in this case, since the motion is periodic, there can be no lines of magnetic force at right angles to such a circuit; in other words, the magnetic force parallel to the axis of z vanishes. Again, taking a small closed circuit at right angles to a radius of the sphere, we see that the electromotive force round this circuit, and therefore the magnetic force at right angles to it, vanish; hence the magnetic force has no component along the radius, and is thus at right angles to both the axis of z and the radius, so that the lines of magnetic force are a series of small circles with the axis of the harmonic for axis.

Hence, if a, b, c denote the components of magnetic induction parallel to the axes of x, y, z respectively, we may put

$$a = y\chi(r, \mu),$$
$$b = -x\chi(r, \mu),$$
$$c = 0,$$

where $\chi(r, \mu)$ denote some function of r and μ. Comparing this with the results of Art. 308, we see that inside the sphere

$$\left.\begin{aligned} a &= A\frac{y}{r}S_n(\lambda' r)\frac{dQ_n}{d\mu}\epsilon^{\iota pt}, \\ b &= -A\frac{x}{r}S_n(\lambda' r)\frac{dQ_n}{d\mu}\epsilon^{\iota pt}, \\ c &= 0, \end{aligned}\right\} \qquad (103)$$

where $\lambda'^2 = -4\pi\mu\iota p/\sigma$, and A is a constant.

Outside the sphere,

$$\left.\begin{aligned} a &= B\frac{y}{r}E_n(\lambda r)\frac{dQ_n}{d\mu}\epsilon^{\iota pt},\\ b &= -B\frac{x}{r}E_n(\lambda r)\frac{dQ_n}{d\mu}\epsilon^{\iota pt},\\ c &= 0, \end{aligned}\right\} \tag{104}$$

where $\lambda = p/V$, and B is a constant.

Since the tangential magnetic force is continuous, we have if $\mathbf{a}$ is the radius of the sphere,

$$\frac{A}{\mu}S_n(\lambda'\mathbf{a}) = BE_n(\lambda\mathbf{a}). \tag{105}$$

To get another surface condition we notice that the electromotive intensity parallel to the surface of the sphere is continuous. Now the total current through any area is equal to $1/4\pi$ times the line integral of the magnetic force round that area, hence, taking as the area under consideration an elementary one $dr\,r\sin\theta d\phi$, whose sides are respectively parallel to an element of radius and to an element of a parallel of latitude, we find, if q is the current in a meridian plane at right angles to the radius,

$$4\pi q = \frac{1}{r}\frac{d}{dr}(\gamma r),$$

where γ is the resultant magnetic force which acts tangentially to a parallel of latitude.

The electromotive intensity parallel to q is, in the conductor

$$\sigma q,$$

and in the dielectric

$$\frac{4\pi}{\iota pK}q.$$

Hence, since this is continuous, we have

$$\frac{A\sigma}{\mu}\frac{d}{d\mathbf{a}}(\mathbf{a}S_n(\lambda'\mathbf{a})) = \frac{B4\pi}{\iota pK}\frac{d}{d\mathbf{a}}\{\mathbf{a}E_n(\lambda\mathbf{a})\}. \tag{106}$$

Eliminating A and B from equations (105) and (106), we get

$$\frac{\sigma\dfrac{d}{d\mathbf{a}}\{\mathbf{a}S_n(\lambda'\mathbf{a})\}}{S_n(\lambda'\mathbf{a})} = \frac{4\pi}{\iota pK}\frac{\dfrac{d}{d\mathbf{a}}\{\mathbf{a}E_n(\lambda\mathbf{a})\}}{E_n(\lambda\mathbf{a})}. \tag{107}$$

311.] The oscillations of the surface electrification about the state of uniform distribution are extremely rapid as the wave length must be comparable with the radius of the sphere. For such rapid vibrations as these however $\lambda' \mathbf{a}$, or $\{-4\pi\mu\iota p/\sigma\}^{\frac{1}{2}}\mathbf{a}$, is very large, but when this is the case, we see from the equation

$$S_n(\rho) = \rho^n \left\{\frac{1}{\rho}\frac{d}{d\rho}\right\}^n \frac{\sin\rho}{\rho},$$

that $S_n'(\lambda'\mathbf{a})$ is approximately equal to $\pm\iota S_n(\lambda'\mathbf{a})$, so that the left-hand side of equation (107) is of the order

$$\sigma\mathbf{a}\sqrt{-\frac{4\pi\mu\iota p}{\sigma}},$$

and thus, since $1/K = V^2$,

$$\frac{\frac{d}{d\mathbf{a}}\{\mathbf{a}E_n(\lambda\mathbf{a})\}}{E_n(\lambda\mathbf{a})}$$

is of the order

$$\frac{1}{4\pi V^2}p\sigma\mathbf{a}\sqrt{-\frac{4\pi\mu\iota p}{\sigma}}, \text{ or } \sqrt{\frac{\sigma}{\mathbf{a}V}},$$

since p is comparable with $V/\mathbf{a}$.

This, when the sphere conducts as well as iron or copper, is extremely small unless $\mathbf{a}$ is less than the wave length of sodium light, while for a perfect conductor it absolutely vanishes, hence equation (107) is very approximately equivalent to

$$\frac{d}{d\mathbf{a}}\{\mathbf{a}E_n(\lambda\mathbf{a})\} = 0. \tag{108}$$

This, by the relation (101), may be written

$$E_{n+1}(\lambda\mathbf{a}) - \frac{n+1}{n}E_{n-1}(\lambda\mathbf{a}) = 0,$$

which is the form given in my paper on 'Electrical Oscillations,' *Proc. Lond. Math. Soc.* XV, p. 197.

This condition makes the tangential electromotive intensity vanish, so that the lines of electrostatic induction are always at right angles to the surface of the sphere.

312.] In order to show that the equations (103) and (104) in the preceding article correspond to a distribution of electricity over the surface of the sphere represented by a zonal harmonic Q_n of the nth order, we only need to show that the current along the radius vector varies as Q_n, for the difference between the radial currents in the sphere and in the dielectric is proportional to the rate of variation of the surface density of the electricity on the sphere, and therefore, since the surface density varies as $\epsilon^{\iota pt}$, it will be proportional to the radial current.

Consider a small area at right angles to the radius, and apply the principle that 4π times the current through this area is equal to the line integral of the magnetic force round it, we get, if P is the current along the radius and $\mu = \cos\theta$,

$$4\pi\mathsf{P} = \frac{1}{r}\frac{d}{d\mu}(\gamma\sin\theta), \tag{109}$$

where γ, as before, is the resultant magnetic force which acts along a tangent to a parallel of latitude.

By equation (103), γ is proportional to

$$\sin\theta\frac{dQ_n}{d\mu},$$

so that P is proportional to

$$\frac{d}{d\mu}\left\{\sin^2\theta\frac{dQ_n}{d\mu}\right\};$$

but

$$\frac{d}{d\mu}\left\{\sin^2\theta\frac{dQ_n}{d\mu}\right\}+n(n+1)Q_n = 0,$$

hence P, and therefore the surface density, is proportional to Q_n.

We shall now consider in more detail the case $n = 1$.

We have

$$\frac{p^2}{V^2} = \lambda^2.$$

We shall take as the solution of the equation $p^2/V^2 = \lambda^2$

$$\frac{p}{V} = \lambda,$$

and we shall take $E_n^-(\lambda r)$ as our solution, as this corresponds to a wave diverging from the sphere. Thus, equation (108) becomes

$$\frac{d}{d\mathbf{a}}\{\mathbf{a}E_1^-(\lambda\mathbf{a})\} = 0,$$

or substituting for $E_1^-(\lambda\mathbf{a})$ the value given in Art. 309,

$$\epsilon^{-\iota\lambda\mathbf{a}}\left\{\frac{1}{(\lambda\mathbf{a})^2}+\frac{\iota}{\lambda\mathbf{a}}-1\right\} = 0,$$

or

$$(\lambda\mathbf{a})^2-\iota\lambda\mathbf{a} = 1,$$

$$\lambda\mathbf{a} = \frac{\iota}{2}\pm\frac{\sqrt{3}}{2}.$$

Hence $$p = \frac{V}{\mathrm{a}}\left\{\frac{\iota}{2} + \frac{\sqrt{3}}{2}\right\},$$

taking the positive sign since the wave is divergent.

Hence, the time of vibration is $4\pi\mathrm{a}/\sqrt{3}\,V$, and the wave length $4\pi\mathrm{a}/\sqrt{3}$. The amplitude of the vibration falls to $1/\epsilon$ of its original value after a time $2\mathrm{a}/V$, that is after the time taken by light to pass across a diameter of the sphere. In the time occupied by one complete vibration the amplitude falls to $\epsilon^{-\frac{2\pi}{\sqrt{3}}}$, or about 1/35 of its original value, thus the vibrations will hardly make a complete oscillation before they become practically extinguished. This very rapid extinction of the vibrations is independent of the resistance of the conductor and is due to the emission of radiant energy by the sphere. Whenever these electrical vibrations can radiate freely they die away with immense rapidity and are practically dead beat.

If we substitute this value of λ in the expressions for the magnetic force and electromotive intensity in the dielectric, we shall find that the following values satisfy the conditions of the problem. If γ is the resultant magnetic force, acting at right angle to the meridional plane,

$$\gamma = \frac{\sin\theta\,\mathrm{a}}{r}\left\{1 - \frac{\mathrm{a}}{r} + \frac{\mathrm{a}^2}{r^2}\right\}^{\frac{1}{2}} \epsilon^{-\frac{(Vt-r)}{2\mathrm{a}}} \cos(\phi + \delta),$$

where $$\phi = \frac{\sqrt{3}}{2\mathrm{a}}(Vt - r),$$

$$\tan\delta = \frac{r-\mathrm{a}}{r+\mathrm{a}}\tan\frac{\pi}{3}.$$

If Θ is the electromotive intensity at right angles to r in the meridional plane, K the specific inductive capacity of the dielectric surrounding the sphere, then by Art. 310

$$K\Theta = \frac{\sin\theta}{Vr}\mathrm{a}\left(1 - \frac{\mathrm{a}}{r}\right)\left\{1 + \frac{\mathrm{a}}{r} + \frac{\mathrm{a}^2}{r^2}\right\}^{\frac{1}{2}} \epsilon^{-\frac{(Vt-r)}{2\mathrm{a}}} \cos(\phi + \delta'),$$

where $$\tan\delta' = \frac{\sin\frac{\pi}{3}}{\cos\frac{\pi}{3} + \frac{\mathrm{a}}{r}},$$

and V is the velocity of propagation of electromagnetic action through the dielectric. Close to the surface of the sphere $\delta = 0$, $\delta' = \pi/6$, thus γ and Θ differ in phase by $\pi/6$. At a large distance from the sphere

$$\delta = \delta',$$

so that Θ and γ are in the same phase, and we have

$$VK\,\Theta = \gamma = \frac{\sin\theta\,\mathrm{a}}{r}\,\epsilon^{-\frac{1}{2\mathrm{a}}(Vt-r)}\cos\left(\phi + \frac{\pi}{3}\right).$$

The radial electromotive intensity P is, by equation 109, given by the equation

$$KP = \frac{2\cos\theta\,\mathrm{a}^2}{Vr^2}\left\{1 - \frac{\mathrm{a}}{r} + \frac{\mathrm{a}^2}{r^2}\right\}^{\frac{1}{2}}\epsilon^{-\frac{(Vt-r)}{2\mathrm{a}}}\sin\left(\phi + \delta - \frac{\pi}{6}\right).$$

Thus at a great distance from the sphere P varies as a^2/r^2, while Θ only varies as a/r, thus the electromotive intensity is very approximately tangential. The general character of the lines of electrostatic induction is similar to that in the case of the cylinder shown in Fig. 110.

313.] The time of vibration of the electricity about the distribution represented by the second zonal harmonic is given by a cubic equation, whose imaginary roots I find to be

$$\iota\lambda\mathrm{a} = -\cdot 7 \pm 1\cdot 8\,\iota.$$

The rate of these vibrations is more than twice as fast as those about the first harmonic distribution; the rate of decay of these vibrations, though absolutely greater than in that case, is not increased in so great a ratio as the frequency, so that the system will make more vibrations before falling to a given fraction of its original value than before.

The time of vibration of the electricity about the distribution represented by the third zonal harmonic is given by a biquadratic equation whose roots are imaginary, and given by

$$\iota\lambda\mathrm{a} = -\cdot 85 \pm 2\cdot 76\,\iota,$$
$$\iota\lambda\mathrm{a} = -2\cdot 15 \pm \cdot 8\,\iota.$$

The quicker of these vibrations is more than three times faster than that about the first zonal harmonic, and there will be many more vibrations before the disturbance sinks to a given fraction of its original value. The slower vibration is of nearly the same period as that about the first harmonic, but it fades away much more rapidly than even that vibration.

The vibrations about distributions of electricity represented by the higher harmonics thus tend to get quicker as the degree of the harmonic increases, and more vibrations are made before the disturbance sinks into insignificance.

314.] We have seen in Art. 16 that a charged sphere when moving uniformly produces the same magnetic field as an element of current at its centre. If the sphere is oscillating instead of moving uniformly, we may prove (J. J. Thomson, *Phil. Mag.* [5], 28, p. 1, 1889) that if the period of its oscillations is large compared with that of a distribution of electricity over the surface of the sphere, the vibrating sphere produces the same magnetic field as an alternating current of the same period. Waves of electromotive intensity carrying energy with them travel through the dielectric, so that in this case the energy of the sphere travels into space far away from the sphere. When, however, the period of vibration of the sphere is less than that of the electricity over its surface, the electromotive intensity and the magnetic force diminish very rapidly as we recede from the sphere, the magnetic field being practically confined to the inside of the sphere, so that in this case the energy of the moving sphere remains in its immediate neighbourhood.

We may compare the behaviour of the electrified sphere with that of a string of particles of equal mass placed at equal intervals along a tightly stretched string; if one of the particles, say one of the end ones, is agitated and made to vibrate more slowly than the natural period of the system, the disturbance will travel as a wave motion along the string of particles, and the energy given to the particle at the end will be carried far away from that particle; if however the particle which is agitated is made to vibrate more quickly than the natural period of vibration of the system, the disturbance of the adjacent particles will diminish in geometrical progression, and the energy will practically be confined to within a short distance of the disturbed particle. This case possesses additional interest since it was used by Sir G. G. Stokes to explain fluorescence.

315.] To consider more closely the effect of reflection let us take the case of two concentric spherical conductors of radius a and b respectively. Then in the dielectric between the spheres, the components of magnetic induction are given by

$$a = \frac{y}{r}\left\{BE_n^+(\lambda r) + CE_n^-(\lambda r)\right\}\frac{dQ_n}{d\mu},$$

$$b = -\frac{x}{r}\left\{BE_n^+(\lambda r) + CE_n^-(\lambda r)\right\}\frac{dQ_n}{d\mu},$$

$$c = 0.$$

We may show, as in Art. 311, that if the spheres are metallic and not excessively small the electromotive intensity parallel to the surface of the spheres vanishes when $r = \mathbf{a}$ and when $r = \mathbf{b}$; thus we have

$$0 = B\frac{d}{d\mathbf{a}}\{\mathbf{a}E_n^+(\lambda\mathbf{a})\} + C\frac{d}{d\mathbf{a}}\{\mathbf{a}E_n^{-1}(\lambda\mathbf{a})\},$$

$$0 = B\frac{d}{d\mathbf{b}}\{\mathbf{b}E_n(\lambda\mathbf{b})\} + C\frac{d}{d\mathbf{b}}\{\mathbf{b}E_n^-(\lambda\mathbf{b})\}.$$

Eliminating A and B, we have

$$\frac{d}{d\mathbf{a}}\{\mathbf{a}E_n^+(\lambda\mathbf{a})\}\frac{d}{d\mathbf{b}}\{\mathbf{b}E_n^-(\lambda\mathbf{b})\} = \frac{d}{d\mathbf{a}}\{\mathbf{a}E_n^-(\lambda\mathbf{a})\}\frac{d}{d\mathbf{b}}\{\mathbf{b}E_n^+(\lambda\mathbf{b})\}.$$

When $n = 1$, this becomes

$$\tan\lambda\{\mathbf{b}-\mathbf{a}\} = \lambda\frac{\left\{\frac{1}{\mathbf{a}} - \frac{1}{\mathbf{b}}\right\}\left\{\lambda^2 + \frac{1}{\mathbf{a}\mathbf{b}}\right\}}{\left(\frac{1}{\mathbf{a}^2} - \lambda^2\right)\left(\frac{1}{\mathbf{b}^2} - \lambda^2\right) + \frac{\lambda^2}{\mathbf{a}\mathbf{b}}}. \tag{110}$$

The roots of this equation are real, so that in this case there is no decay of the vibrations apart from that arising from the resistance of the conductors.

If $\mathbf{a}$ is very small compared with $\mathbf{b}$, this equation reduces to

$$\tan\lambda\mathbf{b} = \frac{\lambda\mathbf{b}}{1-\lambda^2\mathbf{b}^2}.$$

The least root of this equation other than $\lambda = 0$, I find by the method of trial and error to be $\lambda\mathbf{b} = 2{\cdot}744$.

This case is that of the vibration of a spherical shell excited by some cause inside, here there is no radiation of the energy into space, the electrical waves keep passing backwards and forwards from one part of the surface of the sphere to another.

The wave length in this case is $2\pi\mathbf{b}/2{\cdot}744$ or $2{\cdot}29\mathbf{b}$, and is therefore less than the wave length, $4\pi\mathbf{b}/\sqrt{3}$, of the oscillations which would occur if the vibrations radiated off into space: this is an example of the general principle in the theory of vibrations that when dissipation of energy takes place either from friction, electrical resistance, or radiation, the time of vibration is increased.

In this case, since the radius of the inner sphere is made to vanish in the limit, the magnetic force inside the sphere whose radius is **b** must be expressed by that function of r which does not become infinite when r is zero, i.e. by $S_n(\lambda r)$. In the case when $n=1$, the components a, b, c of the magnetic induction are given by

$$a = \Sigma B \frac{y}{r} S_1(\lambda r)\, \epsilon^{\iota pt},$$

$$b = -\Sigma B \frac{x}{r} S_1(\lambda r)\, \epsilon^{\iota pt},$$

$$c = 0;$$

where the summation extends over all values of λ which satisfy the equation

$$\tan \lambda \mathbf{b} = \frac{\lambda \mathbf{b}}{1 - \lambda^2 \mathbf{b}^2}.$$

Let us consider the case when only the gravest vibration is excited. Let e be the surface density of the electricity, then it will be given by an equation of the form

$$e = C \cos\theta \cos pt;$$

where $p = V\lambda_1$, λ_1 being equal to $2.744/\mathbf{b}$.

By equation (109) the normal displacement current $\mathbf{P}$ is given by the equation

$$4\pi \mathbf{P} = \frac{1}{r} \frac{d}{d.\cos\theta} \{\sin\theta\, \{a^2 + b^2\}^{\frac{1}{2}}\}.$$

In this case

$$\left.\begin{aligned} a &= \frac{y}{r} B S_1(\lambda_1 r)\, \epsilon^{\iota pt}, \\ b &= -\frac{x}{r} B S_1(\lambda_1 r)\, \epsilon^{\iota pt}, \end{aligned}\right\} \qquad (111)$$

so that

$$4\pi \mathbf{P} = -\frac{2}{r} B \cos\theta\, S_1(\lambda_1 r)\, \epsilon^{\iota pt}.$$

When $r = \mathbf{b}$ the normal displacement current $= de/dt$, hence

$$-4\pi C \cos\theta\, p \sin pt = -\frac{2}{\mathbf{b}} B \cos\theta\, S_1(\lambda_1 \mathbf{b})\, \epsilon^{\iota pt}.$$

Substituting this value of $B\epsilon^{\iota pt}$ in (111), we have

$$a = \frac{y}{r} 2\pi \mathbf{b} p \sin pt\, C \frac{S_1(\lambda_1 r)}{S_1(\lambda_1 \mathbf{b})},$$

$$b = -\frac{x}{r} 2\pi \mathbf{b} p \sin pt\, C \frac{S_1(\lambda_1 r)}{S_1(\lambda_1 \mathbf{b})},$$

$$c = 0.$$

At the surface of the sphere the maximum intensity of the magnetic force is

$$2\pi \mathbf{b} p C \sin\theta,$$

or since

$$\mathbf{b} p = V\lambda_1 \mathbf{b},$$

and

$$\lambda_1 \mathbf{b} = 2{\cdot}744,$$

the maximum magnetic force is

$$2\pi \times 2{\cdot}744\, VC \sin\theta.$$

For air at atmospheric pressure VC may be as large as 25 without the electricity escaping; taking this value of VC, the maximum value of the magnetic force will be

$$431 \sin\theta;$$

this indicates a very intense magnetic field, which however would be difficult to detect on account of its very rapid rate of reversal.

Electrical Oscillations on Two Concentric Spheres of nearly equal radius.

316.] When d, the difference between the radii $\mathbf{a}$ and $\mathbf{b}$, is very small compared with $\mathbf{a}$ or $\mathbf{b}$, equation (110) becomes

$$\tan\lambda d = \frac{\lambda d(1+\lambda^2\mathbf{a}^2)}{\lambda^4\mathbf{a}^4 - \lambda^2\mathbf{a}^2 + 1}. \qquad (112)$$

There will be one root of this equation corresponding to a vibration whose wave length is comparable with $\mathbf{a}$, and other roots corresponding to wave lengths comparable with d.

When the wave length is comparable with $\mathbf{a}$, λ is comparable with $1/\mathbf{a}$, so that in this case λd is very small; when this is the case $(\tan\lambda d)/\lambda d = 1$, and equation (112) becomes approximately

$$1 = \frac{1+\lambda^2\mathbf{a}^2}{\lambda^4\mathbf{a}^4 - \lambda^2\mathbf{a}^2 + 1},$$

or

$$\lambda\mathbf{a} = \sqrt{2}.$$

The wave length $2\pi/\lambda$ is thus equal to $\pi\sqrt{2}$ times the radius of the sphere.

In this case, since the distance between the spheres is very small compared with the wave length, the tangential electromotive intensity, since it vanishes at the surface of both spheres, will remain very small throughout the space between them; the electromotive intensity will thus be very nearly radial between the spheres, and the places nearest each other on the two spheres will have opposite electrical charges. The tubes of electrostatic induction are radial, and moving at right angles to themselves traverse during a complete oscillation a distance comparable with the circumference of one of the spheres.

When the wave length is comparable with the distance between the spheres, λ is comparable with $1/d$, and $\lambda\mathbf{a}$ is therefore very large. The denominator of the right-hand side of equation (112), since it involves $(\lambda\mathbf{a})^4$, will be exceedingly large compared with the numerator, and this side of the equation will be exceedingly small, so that an approximate solution of it is

$$\tan\lambda d = 0,$$

or

$$\lambda d = n\pi,$$

where n is an integer.

The wave length $2\pi/\lambda = 2d/n$. Hence, the length of the longest wave is $2d$, and there are harmonics whose wave lengths are d, $2d/3$, $2d/4$,....

When $\lambda\mathbf{a}$ is very large, the equation on p. 373,

$$\frac{d}{dr}\{BrE_1^+(\lambda r) + CrE_1^-(\lambda r)\}_{r=\mathbf{a}} = 0,$$

is equivalent to

$$B\epsilon^{\iota\lambda\mathbf{a}} + C\epsilon^{-\iota\lambda\mathbf{a}} = 0.$$

Hence we may put, introducing a new constant A,

$$B = A\epsilon^{-\iota\lambda\mathbf{a}},$$

$$C = -A\epsilon^{\iota\lambda\mathbf{a}}.$$

The resultant magnetic force in the dielectric is equal to

$$\{BE_1^+(\lambda r) + CE_1^-(\lambda r)\}\sin\theta\,\epsilon^{\iota pt},$$

or substituting the preceding values of B and C and retaining only the lowest powers of $1/\lambda r$,

$$\frac{A\iota}{\lambda r}\{\epsilon^{\iota\lambda(r-\mathbf{a})} + \epsilon^{-\iota\lambda(r-\mathbf{a})}\}\sin\theta\,\epsilon^{\iota pt},$$

or

$$2\frac{A\iota}{\lambda r}\cos\lambda(r-\mathbf{a})\sin\theta\,\epsilon^{\iota pt}.$$

The tangential electromotive intensity is therefore, by Art. 310,

$$2A\frac{V}{\lambda r}\sin\lambda(r-\mathbf{a})\sin\theta\,\epsilon^{\iota pt},$$

while the normal intensity is

$$4\frac{AV}{\lambda^2 r^2}\cos\lambda(r-\mathbf{a})\cos\theta\,\epsilon^{\iota pt},$$

and is thus, except just at the surface of the spheres, very small compared with the tangential electromotive intensity. The normal intensity changes sign as we go from $r = \mathbf{a}$ to $r = \mathbf{b}$, so that the electrification on the portions of the spheres opposite to each other is of the same sign. In this case the lines of electromotive intensity are approximately tangential; during the vibrations they move backwards and forwards across the short space between the spheres. The case of two parallel planes can be regarded as the limit of that of the two spheres, and the preceding work shows that the wave length of the vibrations will either be a sub-multiple of twice the distance between the planes, or else a length comparable with the dimensions of the plane at right angles to their common normal.

If we arrange two metal surfaces, say two silvered glass plates, so that, as in the experiment for showing Newton's rings, the distance between the plates is comparable with the wave length of the luminous rays, care being taken to insulate one plate from the other, then one of the possible modes of electrical vibration will have a wave length comparable with that of the luminous rays, and so might be expected to affect a photographic plate. These vibrations would doubtless be exceedingly difficult to excite, on account of the difficulty of getting any lines of induction to run down between the plates before discharge took place, but this would to some extent be counterbalanced by the fact that the photographic method would enable us to detect vibrations of exceedingly small intensity.

On the Decay of Electric Currents in Conducting Spheres.

317.] The analysis we have used to determine the electrical oscillations on spheres will also enable us to determine the rate at which a system of currents started in the sphere will decay if left to themselves. Let us first consider the case when, as in the preceding investigation, the lines of magnetic force are circles with a diameter of the sphere for their common axis. Using the same notation as before, when there is only a single sphere of radius $\mathbf{a}$ in the field, we have by equation (107)

$$\frac{\sigma \dfrac{d}{d\mathbf{a}}\{\mathbf{a}S_n(\lambda'\mathbf{a})\}}{S_n(\lambda'\mathbf{a})} = \frac{4\pi}{K\iota p}\,\frac{\dfrac{d}{d\mathbf{a}}\{\mathbf{a}E_n(\lambda\mathbf{a})\}}{E_n(\lambda\mathbf{a})}. \qquad (113)$$

The rate at which the system of currents decay is infinitesimal in comparison with the rate at which a distribution of electricity over the surface changes, so that $\lambda\mathrm{a}$ or $p\mathrm{a}/V$ will in this case be exceedingly small: but when $\lambda\mathrm{a}$ is very small

$$E_n(\lambda\mathrm{a}) = (-1)^n 1.3.5\ldots(2n-1).\frac{\epsilon^{\pm\lambda\mathrm{a}}}{\mathrm{a}^{n+1}},$$

$$\frac{d}{d\mathrm{a}}\{\mathrm{a}E_n(\lambda\mathrm{a})\} = (-1)^{n+1}1.3.5\ldots(2n-1).n\frac{\epsilon^{\pm\lambda\mathrm{a}}}{\mathrm{a}^{n+1}},$$

so that the right-hand side of (113) is equal to

$$-\frac{4\pi n}{K\iota p},$$

or

$$-\frac{4\pi n V^2}{\iota p}.$$

Thus

$$\frac{S_n(\lambda'\mathrm{a})}{\dfrac{d}{d\mathrm{a}}\{\mathrm{a}S_n(\lambda'\mathrm{a})\}} = -\frac{\iota p\sigma}{4\pi n V^2}.$$

Now, since $V^2 = 9\times 10^{20}$ and σ for copper is about 1600, the right-hand side of this equation is excessively small, so that it reduces to

$$S_n(\lambda'\mathrm{a}) = 0.$$

When $n = 1$, since

$$S_1(\lambda'\mathrm{a}) = \frac{\cos\lambda'\mathrm{a}}{\lambda'\mathrm{a}} - \frac{\sin\lambda'\mathrm{a}}{\lambda'^2\mathrm{a}^2},$$

λ' is given by the equation

$$\tan\lambda'\mathrm{a} = \lambda'\mathrm{a};$$

the roots of which are approximately

$$\lambda'\mathrm{a} = 1.4303\pi,\quad 2.4590\pi,\quad 3.4709\pi\ldots.$$

The roots of the equation

$$S_2(\lambda'\mathrm{a}) = 0$$

are approximately

$$\lambda'\mathrm{a} = 1{\cdot}8346\pi,\quad 2{\cdot}8950\pi,\quad 3{\cdot}9225\pi.$$

(See Prof. H. Lamb, 'Electrical Motions on Spherical Conductors,' *Phil. Trans.* Pt. II, p. 530. 1883.)

The value of ιp corresponding to any value of λ' is given by the equation

$$\iota p = -\frac{\sigma\lambda'^2}{4\pi\mu}.$$

The time factors in the expressions for the currents will be of the form $\epsilon^{-\frac{\sigma\lambda'^2}{4\pi\mu}t}$. The most persistent type of current will be that corresponding to the smallest value of λ', i.e.

$$\lambda' = 1{\cdot}4303\,\pi/\mathbf{a}.$$

The time required for a current of this type to sink to $1/\epsilon$ of its original value in a copper sphere when $\sigma = 1600$ is $\cdot 000379\,\mathbf{a}^2$ seconds; for an iron sphere when $\mu = 1000$, $\sigma = 10^4$, it is $\cdot 0622\,\mathbf{a}^2$ seconds, thus the currents will be much more persistent in the iron sphere than in the copper one. The persistence of the vibrations is proportional to the square of the radius of the sphere, thus for very large spheres the rate of decay will be exceedingly slow; for example, it would take nearly 5 million years for currents of this type to sink to $1/\epsilon$ of their original value in a copper sphere as large as the earth.

Since $S_n(\lambda'\mathbf{a}) = 0$, we see from (105) that $B = 0$, and therefore that the magnetic force is zero everywhere outside the sphere. Hence, since these currents produce no magnetic effect outside the sphere they cannot be excited by any external magnetic influence. The current at right angles to the radius inside the sphere is by Art. 310

$$\frac{\sin\theta}{4\pi\mu r}\frac{d}{dr}\{ArS_n(\lambda'r)\}\frac{dQ_n}{d\mu}\epsilon^{\iota pt},$$

or in particular, when $n = 1$

$$\frac{\sin\theta}{4\pi\mu r}\frac{d}{dr}\{rS_1(\lambda'r)\}\epsilon^{\iota pt}.$$

Now $\frac{d}{dr}\{rS_1(\lambda'r)\}$ vanishes when $\lambda'r = 2{\cdot}744$, hence the tangential current will vanish when

$$r = \frac{2{\cdot}744}{1{\cdot}4303\,\pi}\mathbf{a}$$

$$= \cdot 601\,\mathbf{a};$$

thus there is a concentric spherical surface over which the current of this type is entirely radial.

The magnetic force vanishes at the surface and at the centre, and as we travel along a radius attains, when $n = 1$, a maximum when r satisfies the equation

$$\frac{d}{dr}S_1(\lambda'r) = 0.$$

The smallest root of this equation is

$$\lambda' r = \cdot 662\pi,$$

where
$$r = \frac{\cdot 662}{1\cdot 4303}\,\mathbf{a} = \cdot 462\,\mathbf{a}.$$

This is nearer the centre of the sphere than the place where the tangential current vanishes. The lines of flow of the current

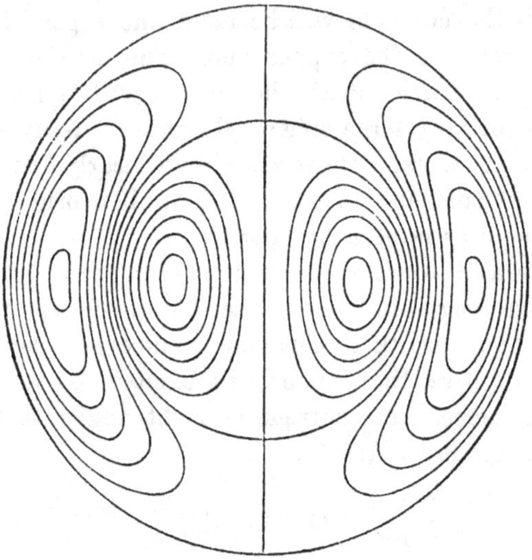

Fig. 112.

in a meridional section of the sphere when $\lambda'\mathbf{a} = 2\cdot 4590\pi$ are given in Fig. 112, which is taken from the paper by Professor Lamb already quoted (p. 378).

Rate of Decay of Currents flowing in Circles which have a Diameter of the Sphere as a Common Axis.

318.] In this case the lines of flow of the current are coincident with the lines of magnetic force of the last example and *vice versa*.

Let P, Q, R denote the components of electromotive intensity, then in the sphere we have

$$\left.\begin{aligned} P &= A\frac{y}{r}S_n(\lambda' r)\frac{dQ_n}{d\mu}\epsilon^{\iota pt}, \\ Q &= -A\frac{x}{r}S_n(\lambda' r)\frac{dQ_n}{d\mu}\epsilon^{\iota pt}, \\ R &= 0; \end{aligned}\right\} \qquad (114)$$

while in the dielectric surrounding the sphere, we have

$$\left.\begin{aligned} P &= B\frac{y}{r}E_n(\lambda r)\frac{dQ_n}{d\mu}\epsilon^{\iota pt}, \\ Q &= -B\frac{x}{r}E_n(\lambda r)\frac{dQ_n}{d\mu}\epsilon^{\iota pt}, \\ R &= 0. \end{aligned}\right\} \qquad (115)$$

Since the electromotive intensity tangential to the sphere is continuous, we have, if $\mathbf{a}$ is the radius of the sphere,

$$AS_n(\lambda'\mathbf{a}) = BE_n(\lambda\mathbf{a}). \qquad (116)$$

If ω is the magnetic induction tangentially to a meridian, then, since the line integral of the electromotive intensity round a circuit is equal to the rate of diminution of the number of lines of magnetic induction passing through it,

$$\frac{d\omega}{dt} = \frac{1}{r}\frac{d}{dr}\{r\{P^2+Q^2\}^{\frac{1}{2}}\}.$$

Since the tangential magnetic force is continuous, we have at the surface

$$\left(\frac{\omega}{\mu}\right) \text{ in the sphere} = \omega \text{ in the dielectric.}$$

Hence

$$\frac{A}{\mu}\frac{d}{d\mathbf{a}}\{\mathbf{a}S_n(\lambda'\mathbf{a})\} = B\frac{d}{d\mathbf{a}}\{\mathbf{a}E_n(\lambda\mathbf{a})\}. \qquad (117)$$

Eliminating A and B from equations (116) and (117), we get

$$\mu\frac{S_n(\lambda'\mathbf{a})}{\dfrac{d}{d\mathbf{a}}\{\mathbf{a}S_n(\lambda'\mathbf{a})\}} = \frac{E_n(\lambda\mathbf{a})}{\dfrac{d}{d\mathbf{a}}\{\mathbf{a}E_n(\lambda\mathbf{a})\}}. \qquad (118)$$

In this case the currents and magnetic forces change so slowly that $\lambda\mathbf{a}$ or $p\mathbf{a}/V$ is an exceedingly small quantity, but when this is the case we have proved Art. 317, that approximately

$$\frac{E_n(\lambda\mathbf{a})}{\dfrac{d}{d\mathbf{a}}\{\mathbf{a}E_n(\lambda\mathbf{a})\}} = -\frac{1}{n},$$

so that equation (118) becomes

$$n\mu S_n(\lambda'\mathbf{a}) + \frac{d}{d\mathbf{a}}\{\mathbf{a}S_n(\lambda'\mathbf{a})\} = 0. \qquad (119)$$

But by equation (100), Art. 309,

$$\mathbf{a}\frac{d}{d\mathbf{a}}S_n(\lambda'\mathbf{a}) + (n+1)S_n(\lambda'\mathbf{a}) = -\lambda'\mathbf{a}S_{n-1}(\lambda'\mathbf{a}),$$

hence (119) may be written

$$n(\mu-1)S_n(\lambda'\mathrm{a})-\lambda'\mathrm{a}\,S_{n-1}(\lambda'\mathrm{a}) = 0. \qquad (120)$$

For non-magnetic metals for which $\mu = 1$ this reduces to

$$S_{n-1}(\lambda'\mathrm{a}) = 0,$$

while for iron, for which μ is very great, the equation approximates very closely to

$$S_n(\lambda'\mathrm{a}) = 0.$$

The smaller roots of the equation

$$S_n(x) = 0,$$

when $n = 0, 1, 2$, are given below;

$$n = 0, \quad x = \pi, \quad 2\pi, \quad 3\pi, \ldots;$$
$$n = 1, \quad x = 1{\cdot}4303\,\pi, \quad 2{\cdot}4590\,\pi, \quad 3{\cdot}4709\,\pi;$$
$$n = 2, \quad x = 1{\cdot}8346\,\pi, \quad 2{\cdot}8950\,\pi, \quad 3{\cdot}9225\,\pi.$$

Thus for a copper sphere for which $\sigma = 1600$, the time the currents of the most permanent type, i.e. those corresponding to the root $\lambda'\mathrm{a} = \pi$, take to fall to $1/\epsilon$ of their original value is $000775\mathrm{a}^2$ seconds, which for a copper sphere as large as the earth is ten million years. These numbers are given by Prof. Horace Lamb in the paper on 'Electrical Motion on a Spherical Conductor,' *Phil. Trans.* 1883, Part II.

319.] As the magnetic force outside the sphere does not vanish in this case, this distribution of currents produces an external magnetic field, and conversely, such a distribution could be induced by changes in such a field. We have supposed that the currents are symmetrical about an axis, but by superposing distributions symmetrical about different axes we could get the most general distribution of this type of current. The most general distribution of this type would however be such that the lines of current flow are on concentric spherical surfaces, it is only distributions of this kind which can be excited in a sphere by variations in the external magnetic field.

We can prove without difficulty that whenever radial currents exist in a sphere the magnetic force outside vanishes, provided displacement currents in the dielectric are neglected.

Let u, v, w be the components of the current inside the sphere, they will, omitting the time factor, be given by equations of the

form
$$u = S_n(\lambda' r) Y',$$
$$v = S_n(\lambda' r) Y'',$$
$$w = S_n(\lambda' r) Y''',$$
where Y', Y'', Y''' are surface harmonics of the nth order.

The radial current is
$$S_n(\lambda' r)\left(\frac{x}{r} Y' + \frac{y}{r} Y'' + \frac{z}{r} Y'''\right),$$
at the surface of the sphere the radial current must vanish, i.e.
$$S_n(\lambda' \mathrm{a})\left\{\frac{x}{\mathrm{a}} Y' + \frac{y}{\mathrm{a}} Y'' + \frac{z}{\mathrm{a}} Y'''\right\}.$$
Now the second factor is a function merely of the angular coordinates, and if it vanished there would not be any radial currents at any point in the sphere, hence, on the hypothesis that there are radial currents in the sphere, we must have
$$S_n(\lambda' \mathrm{a}) = 0,$$
i.e. u, v, w all vanish on the surface of the sphere. But if there are no currents on the surface the electromotive intensity must vanish over the surface, and hence also the radial magnetic induction; for the rate of change of the radial induction through a small area on the surface of the sphere is equal to the electromotive force round that area. But neglecting the displacement current in the dielectric the magnetic force outside the sphere will be derived from a potential; hence, since the radial magnetic force vanishes over the sphere $r = \mathrm{a}$, and over $r = \infty$, and since the space between the two is acyclic, the magnetic force must vanish everywhere in the region between them. Thus the presence of radial currents in the sphere requires the magnetic force due to the currents to be entirely confined to the inside of the sphere.

320.] Returning to the case where the system is symmetrical about an axis, we see from equation (120) that if the sphere is an iron one, λ' is given approximately by the equation
$$S_n(\lambda' \mathrm{a}) = 0.$$
Hence, by equation (114) the electromotive intensity, and therefore the currents, vanish over the surface of the sphere. Since the currents also vanish at the centre, they must attain a maximum at some intermediate position; the distance r of this

position from the centre of the sphere is given by the equation

$$\frac{d}{dr} S_n(\lambda' r) = 0;$$

if $n = 1$, a root of this equation is

$$\lambda' r = \cdot 663\,\pi,$$

and since $$\lambda' \mathrm{a} = 1\cdot 4303\,\pi,$$

we have $$r = \cdot 463\,\mathrm{a}.$$

Currents induced in a Uniform Sphere by the sudden destruction of a Uniform Magnetic Field.

321.] We shall now apply the results we have just obtained to find the currents produced in a sphere placed in a uniform magnetic field which is suddenly destroyed; this problem was solved by Lamb (*Proc. Lond. Math. Soc.* 15, p. 139, 1884). The currents will evidently flow in circles having the diameter of the sphere which is parallel to the magnetic force for axis.

If H is the intensity of the original field at a great distance from the sphere, the lines of force being parallel to z, then inside the sphere the magnetic induction will be parallel to z, and will be equal to $3\mu H/(\mu+2)$. The radial component will thus be proportional to $\cos\theta$. If ρ be the normal component of magnetic induction, a, b, c the components parallel to the axes of x, y, z respectively, then

$$r\rho = xa + yb + zc,$$

$$\nabla^2(r\rho) = x\nabla^2 a + y\nabla^2 b + z\nabla^2 c + 2\left\{\frac{da}{dx} + \frac{db}{dy} + \frac{dc}{dz}\right\}$$

$$= -\lambda'^2 r\rho,$$

since $$\nabla^2 a = -\lambda'^2 a,$$

and $$\frac{da}{dx} + \frac{db}{dy} + \frac{dc}{dz} = 0.$$

Hence, by (97), $$r\rho = C\cos\theta\, S_1(\lambda' r)\,\epsilon^{\iota pt} \qquad (121)$$

where, by (119), λ' is given by the equation

$$(\mu+1)\,S_1(\lambda'\mathrm{a}) + \mathrm{a}\frac{dS_1(\lambda'\mathrm{a})}{d\mathrm{a}} = 0,$$

or by (120) $$(\mu-1)\,S_1(\lambda'\mathrm{a}) - \lambda'\mathrm{a}\,S_0(\lambda'\mathrm{a}) = 0. \qquad (122)$$

When the sphere is non-magnetic $\mu = 1$, and the values of λ' are given by

$$S_0(\lambda'\mathrm{a}) = 0,$$

or

$$\frac{\sin \lambda' \mathbf{a}}{\lambda' \mathbf{a}} = 0,$$

hence $\lambda' = \frac{p\pi}{\mathbf{a}}$, where p is an integer.

When μ is very large, λ' is given approximately by the equation

$$S_1(\lambda' \mathbf{a}) = 0,$$

or

$$\tan \lambda' \mathbf{a} = \lambda' \mathbf{a}.$$

The roots of this equation are given in Art. 317.

We shall for the present not make any assumption as to the magnitude of μ, but suppose that $\lambda_1, \lambda_2 \ldots$ are the values of λ' which satisfy (122). The value of ιp corresponding to λ_s is $-\sigma\lambda_s^2/4\pi\mu$, hence by (121) we have

$$r\rho = \cos\theta \left\{ C_1 S_1(\lambda_1 r)\, \epsilon^{-\frac{\sigma\lambda_1^2}{4\pi\mu}t} + C_2 S_1(\lambda_2 r)\, \epsilon^{-\frac{\sigma\lambda_2^2}{4\pi\mu}t} + \ldots \right\}.$$

To determine $C_1, C_2 \ldots$ we have the condition that when $t = 0$,

$$r\rho = 3r\cos\theta \frac{\mu H}{\mu + 2},$$

hence, for all values of r between 0 and $\mathbf{a}$, we have

$$\frac{3\mu H r}{\mu + 2} = C_1 S_1(\lambda_1 r) + C_2 S_1(\lambda_2 r) + \ldots. \tag{123}$$

Now by Art. 309, if λ_p, λ_q are different roots of (122)

$$\int_0^{\mathbf{a}} r^2 S_1(\lambda_p r)\, S_1(\lambda_q r)\, dr = 0,$$

while

$$\int_0^{\mathbf{a}} r^2 S_1^2(\lambda_p r)\, dr = -\frac{1}{2}\frac{\mathbf{a}^2}{\lambda_p^2}\left\{ S_1(\lambda_p \mathbf{a}) \frac{d}{d\mathbf{a}}\left\{ \mathbf{a}\frac{dS_1(\lambda_p \mathbf{a})}{d\mathbf{a}} \right\} - \mathbf{a}\left(\frac{dS_1(\lambda_p \mathbf{a})}{d\mathbf{a}}\right)^2 \right\}. \tag{124}$$

But $$\frac{d^2}{d\mathbf{a}^2} S_1(\lambda_p \mathbf{a}) + \frac{2}{\mathbf{a}}\frac{d}{d\mathbf{a}} S_1(\lambda_p \mathbf{a}) + \left(\lambda_p^2 - \frac{2}{\mathbf{a}^2}\right) S_1(\lambda_p \mathbf{a}) = 0,$$

and

$$(\mu + 1) S_1(\lambda_p \mathbf{a}) + \mathbf{a}\frac{d}{d\mathbf{a}} S_1(\lambda_p \mathbf{a}) = 0.$$

Substituting in (124) the values of $\frac{d^2}{d\mathbf{a}^2} S_1(\lambda_p \mathbf{a})$ and $\frac{d}{d\mathbf{a}} S_1(\lambda_p \mathbf{a})$ given by these equations, we get

$$\int_0^{\mathbf{a}} r^2 S_1^2(\lambda_p r)\, dr = \frac{\mathbf{a}}{2\lambda_p^2} S_1^2(\lambda_p \mathbf{a}) \{\lambda_p^2 \mathbf{a}^2 + (\mu + 2)(\mu - 1)\}.$$

Hence, multiplying both sides of (123) by $r^2 S_1(\lambda_p r)$ and integrating from 0 to a, we get

$$\frac{3\mu H}{\mu+2}\int_0^{\mathbf{a}} r^3 S_1(\lambda_p r)\,dr$$
$$= \frac{1}{2}\frac{\mathbf{a}C_p}{\lambda_p^2} S_1^2(\lambda_p \mathbf{a})\,\{\lambda_p^2\mathbf{a}^2 + (\mu+2)(\mu-1)\}. \qquad (125)$$

To find the integral on the left-hand side, we notice

$$r^3\frac{d^2 S_1(\lambda_p r)}{dr^2} + 2r^2\frac{d}{dr}S_1(\lambda_p r) - 2rS_1(\lambda_p r) + \lambda_p^2 r^3 S_1(\lambda_p r) = 0,$$

or

$$\frac{d}{dr}\left\{r^3\frac{d}{dr}S_1(\lambda_p r)\right\} - \frac{d}{dr}\{r^2 S_1(\lambda_p r)\} + \lambda_p^2 r^3 S_1(\lambda_p r) = 0;$$

hence, integrating from 0 to a

$$\mathbf{a}^3\frac{d}{d\mathbf{a}}S_1(\lambda_p\mathbf{a}) - \mathbf{a}^2 S_1(\lambda_p\mathbf{a}) + \lambda_p^2\int_0^{\mathbf{a}} r^3 S_1(\lambda_p r)\,dr = 0,$$

which by the use of (122) reduces to

$$\int_0^{\mathbf{a}} r^3 S_1(\lambda_p r)\,dr = \frac{\mathbf{a}^2(\mu+2)}{\lambda_p^2}S_1(\lambda_p\mathbf{a}).$$

Hence, from (125) we get

$$C_p = \frac{6\mu H\mathbf{a}}{S_1(\lambda_p\mathbf{a})} \div \{\lambda_p^2\mathbf{a}^2 + (\mu+2)(\mu-1)\}.$$

322.] When the sphere is non-magnetic $\mu = 1$, and therefore

$$C_p = \frac{6H}{\mathbf{a}S_1(\lambda_p\mathbf{a})\lambda_p^2}.$$

In this case $\lambda_p = \frac{p\pi}{\mathbf{a}}$, and therefore

$$\lambda_p^2 S_1(\lambda_p\mathbf{a}) = \lambda_p\frac{\cos\lambda_p\mathbf{a}}{\mathbf{a}} - \frac{\sin\lambda_p\mathbf{a}}{\mathbf{a}^2}$$
$$= (-1)^p\frac{p\pi}{\mathbf{a}^2}.$$

Thus the normal magnetic induction

$$= \frac{6H\cos\theta.\mathbf{a}^3}{r\pi^3}\sum_{p=1}^{p=\infty}(-1)^p\frac{1}{p^3}\left\{\frac{p\pi}{\mathbf{a}r}\cos p\frac{\pi r}{\mathbf{a}} - \frac{1}{r^2}\sin p\frac{\pi r}{\mathbf{a}}\right\}\epsilon^{-\frac{p^2\pi\sigma}{4\mathbf{a}^2}t}.$$

When $r = \mathbf{a}$, this equals.

$$\frac{6H\cos\theta}{\pi^2}\sum_{p=1}^{p=\infty}\frac{1}{p^2}\epsilon^{-\frac{p^2\pi\sigma}{4\mathbf{a}^2}t}.$$

This summation could be expressed as a theta function, but as the series converges very rapidly it is more convenient to leave it in its present form.

Since we neglect the polarization currents outside the sphere, the magnetic force in that region is derivable from a potential, hence we find that the radial magnetic force is

$$\frac{6H\cos\theta}{\pi^2}\frac{\mathrm{a}^3}{r^3}\sum_{p=1}^{p=\infty}\frac{1}{p^2}\epsilon^{-\frac{p^2\pi\sigma}{4\mathrm{a}^2}t}.$$

The magnetic force at right angles to the radius is

$$\frac{3H\sin\theta\,\mathrm{a}^3}{\pi^2 r^3}\sum_{p=1}^{p=\infty}\frac{1}{p^2}\epsilon^{-\frac{p^2\pi\sigma}{4\mathrm{a}^2}t}.$$

The sphere produces the same effect at an external point as a small magnet whose moment is

$$\frac{3H\mathrm{a}^3}{\pi^2}\sum_{p=1}^{p=\infty}\frac{1}{p^2}\epsilon^{-\frac{p^2\pi\sigma}{4\mathrm{a}^2}t}.$$

323.] When μ is very great

$$C_p = \frac{6H\mathrm{a}}{\mu S_1(\lambda_p\mathrm{a})}.$$

Hence, the normal magnetic force at the surface of the sphere is

$$\frac{6H}{\mu^2}\cos\theta\,\Sigma\epsilon^{-\frac{\lambda p^2\sigma}{4\pi\mu}t}.$$

Outside the sphere the magnetic force is the same as that due to a magnet whose moment is

$$\frac{3H\mathrm{a}^3}{\mu}\Sigma\epsilon^{-\frac{\lambda_p{}^2\sigma}{4\pi\mu}t},$$

placed at its centre. These results are given by Lamb (*l. c.*).

Thus the magnetic effects of the currents induced in a soft iron sphere are less than those which would be produced by a copper sphere of the same size placed in the same field. This is due to the changes of magnetic force proceeding more slowly in the iron sphere on account of its greater self-induction; as the changes in magnetic force are slower, the electromotive forces, and therefore the currents, will be smaller.

Since $S_1(\lambda_p\mathrm{a}) = 0$ when μ is large, the currents on the surface of the sphere vanish, and the currents congregate towards the middle of the sphere.

CHAPTER V.

Experiments on Electromagnetic Waves.

324.] Professor Hertz has recently described a series of experiments which show that waves of electromotive and magnetic force are present in the dielectric medium surrounding an electrical system which is executing very rapid electrical vibrations. A complete account of these will be found in his book *Ausbreitung der elektrischen Kraft*, Leipzig, 1892. The vibrations which Hertz used in his investigations are of the type of those which occur when the inner and outer coatings of a charged Leyden jar are put in electrical connection. The time of vibration of such a system when the resistance of the discharging circuit may be neglected is, as we saw in Art. 296, approximately equal to $2\pi\sqrt{LC}$, where L is the coefficient of self-induction of the discharging circuit for infinitely rapid vibrations and C is the capacity of the jar in electromagnetic measure. If C is the capacity of the jar in electrostatic measure, then, since $C = \mathrm{C}/V^2$, where V is the ratio of the electromagnetic unit of electricity to the electrostatic unit, the time of vibration is equal to $2\pi\sqrt{L\mathrm{C}}/V$. But since V is equal to the velocity of propagation of electrodynamic action through air, the distance the disturbance will travel in the time occupied by a complete oscillation, in other words the wave length in air of these vibrations, will be $2\pi\sqrt{L\mathrm{C}}$. By using electrical systems which had very small capacities and coefficients of self-induction Hertz succeeded in bringing the wave length down to a few metres.

325.] The electrical vibrator which Hertz used in his earlier experiments (*Wied. Ann.* 34, pp. 155, 551, 609, 1888) is represented in Figure 113.

A and B are square zinc plates whose sides are 40 cm. long, copper wires C and D each about 30 cm. long are soldered to the plates, these wires terminate in brass balls E and F. To ensure the success of the experiments it is necessary that these balls should be exceedingly brightly and smoothly polished, and inasmuch as the passage of the sparks from one ball to the other across the air space EF roughens the balls by tearing particles of metal from them, it is necessary to keep repolishing the balls at short intervals during the course of the experiment. It is also advisable to keep the air space EF shaded from the light from any sparks that may be passing in the neighbourhood. In order to excite electrical vibrations in this system the extremities of an induction coil are connected with C and D respectively. When the coil is in action it produces so great a difference of potential between the balls E and F that the electric strength of the air is overcome, sparks pass across the air gap which thus becomes a conductor; the two plates A and B are now connected by a conducting circuit, and the charges on the plates oscillate backwards and forwards from one plate to another just as in the case of the Leyden jar.

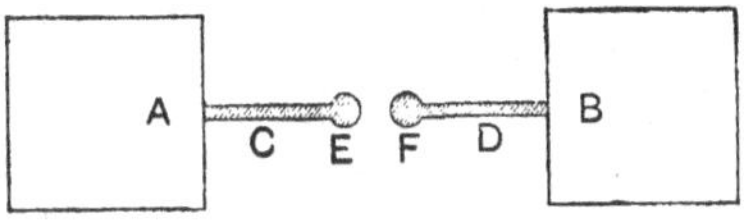

Fig. 113.

326.] As these oscillations are exceedingly rapid they will not be excited unless the electric strength of the air gap breaks down suddenly; if it breaks down so gradually that instead of a spark suddenly rushing across the gap we have an almost continuous glow or brush discharge, hardly any vibrations will be excited. A parallel case to this is that of the vibrations of a simple pendulum, if the bob of such a pendulum is pulled out from the vertical by a string and the string is suddenly cut the pendulum will oscillate; if however the string instead of breaking suddenly gives way gradually, the bob of the pendulum will merely sink to its position of equilibrium and no vibrations will be excited. It is this which makes it necessary to keep the balls E and F well polished, if they are rough there will in all likelihood be sharp points upon them from which the electricity will gradually escape, the constraint of the system will then give way gradually instead of suddenly and no vibrations will be excited.

The necessity of shielding the air gap from light coming from other sparks is due to a similar reason. Ultra-violet light in which these sparks abound possesses, as we saw in Art. 39, the property of producing a gradual discharge of electricity from the negative terminal, so that unless this light is shielded off there will be a tendency to produce a gradual and therefore non-effective discharge instead of an abrupt and therefore effective one.

327.] The presence of the coil does not, as the following calculation of the period of the compound system shows, affect the time of vibration to more than an infinitesimal extent, if, as is practically always the case, the coefficient of self-induction of the secondary of the coil is almost infinite in comparison with that of the vibrator.

Let L be the coefficient of self-induction of the vibrator AB, C its capacity, L' the coefficient of self-induction of the secondary of the coil, M the coefficient of mutual induction between this coil and the vibrator, x the quantity of electricity at any time on either plate of the condenser, $\dot{y}$ the current in the vibrator, $\dot{z}$ that through the secondary of the coil.

Then we have

$$\dot{x} = \dot{y} + \dot{z}$$

or

$$x = y + z.$$

The Kinetic energy of the currents is

$$\tfrac{1}{2}L\dot{y}^2 + \tfrac{1}{2}L'\dot{z}^2 + M\dot{y}\dot{z}.$$

The potential energy is

$$\tfrac{1}{2}\frac{x^2}{C} \quad \text{or} \quad \tfrac{1}{2}\frac{(y+z)^2}{C}.$$

Hence, if we neglect the resistance of the circuit, we have by Lagrange's equations

$$Ly'' + Mz'' + \frac{y+z}{C} = 0,$$

$$L'z'' + My'' + \frac{y+z}{C} = 0.$$

Thus if x and y each vary as $\epsilon^{\iota pt}$, we have

$$y\left(\frac{1}{C} - Lp^2\right) + z\left(\frac{1}{C} - Mp^2\right) = 0,$$

$$z\left(\frac{1}{C} - L'p^2\right) + y\left(\frac{1}{C} - Mp^2\right) = 0.$$

Eliminating y and z we get

$$\left(\frac{1}{C} - Lp^2\right)\left(\frac{1}{C} - L'p^2\right) = \left(\frac{1}{C} - Mp^2\right)^2,$$

or

$$p^2 = \frac{1}{CL}\left\{1 + \frac{L}{L'} - \frac{2M}{L'}\right\} \Big/ \left(1 - \frac{M^2}{LL'}\right).$$

But for a circuit as short as a Hertzian vibrator L/L' and M/L' will be exceedingly small, so that we have as before

$$p^2 = \frac{1}{CL}.$$

The Resonator.

328.] When the electrical oscillations are taking place in the vibrator the space around it will be the seat of electric and magnetic intensities. Hertz found that he could detect these by means of an instrument which is called the Resonator. It consists of a piece of copper wire bent into a circle; the ends of the wire, which are placed very near together, are furnished with two balls or a ball and a point, these are connected by an insulating screw, so that the distance between them admits of very fine adjustment. A resonator without the screw adjustment is shown in Fig. 114. With a vibrator having the dimensions of the one in Art. 325, Hertz used a resonator 35 cm. in radius.

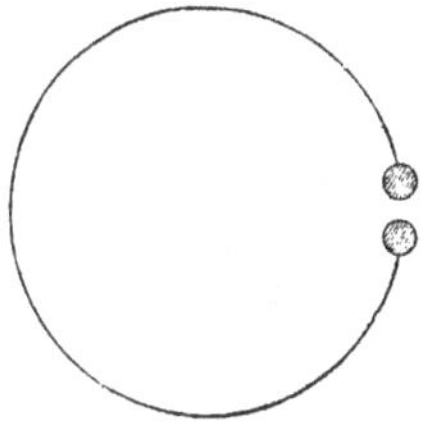

Fig. 114.

329.] When the resonator was held near the vibrator Hertz found that sparks passed across the air space in the resonator and that the length of the air space across which the sparks would pass varied with the position of the resonator. This variation was found by Hertz to be of the following kind:

Let the vibrator be placed so that its axis, the line E F, Fig. 113, is horizontal; let the horizontal line which bisects this axis at right angles, i. e. which passes through the middle point of the air space E F, be called the base line. Then, when the resonator is placed so that its centre is on the base line and its plane at right angles to that line, Hertz found that sparks pass readily in the resonator when its air space is either vertically above or vertically below its centre, but that they cease entirely when the resonator is turned in its own plane round its centre until the air space is in the horizontal plane through that point. Thus the sparks are bright when the line joining the ends of the resonator is parallel to the axis of the vibrator and vanish when it is at right angles to this axis. In intermediate positions of the air gap faint sparks pass between the terminals of the resonator.

When the centre of the resonator is in the base line and its plane at right angles to the axis of the vibrator no sparks pass, whatever may be the position of the air space.

When the centre of the resonator is in the base line and its plane horizontal the sparks are strongest when the air space is nearest to the vibrator, and as the resonator turns about its centre in its own plane the length of the sparks diminishes as the air space recedes from the vibrator and is a minimum when the air gap is at its maximum distance from the axis of the vibrator. They do not however vanish in this case for any position of the air space.

330.] In the preceding experiments the length of the sparks changes as the resonator rotates in its own plane about its centre. Since rotation is not accompanied by any change in the number of lines of magnetic force passing through the resonator circuit, it follows that we cannot estimate the tendency to spark across the air gap by calculating by Faraday's rule the electromotive force round the circuit from the diminution in the number of lines of magnetic force passing through it.

331.] The effects on the spark length are, however, easily explained if we consider the arrangement of the Faraday tubes radiating from the vibrator. The tendency to spark will be proportional to the number of tubes which stretch across the air gap; these tubes may fall directly on the air gap or they may be collected by the wire of the resonator and thrown on the air gap, the resonator acting as a kind of trap for Faraday tubes.

Let us first consider the case when the centre of the resonator is on, and its plane at right angles to, the base line, then in the neighbourhood of the base line the Faraday tubes are approximately parallel to the axis of the vibrator, and their direction of motion is parallel to the base line; thus the Faraday tubes are parallel to the plane of the resonator and are moving at right angles to it. When they strike against the wire of the resonator they will split up into separate pieces as in Fig. 115, which represents a tube moving up to and across the resonator, and after passing the cross-section of the wire of the resonator will join again and go on as if they had not been interrupted. The resonator will thus not catch Faraday tubes and throw them in the air gap, and therefore the tendency to spark across the gap will be due only to those tubes which fall directly upon it. When the air gap is parallel to the tubes, i. e. when it is at the highest or lowest point of the resonator, some of the tubes will be caught and will stretch across the gap and thus tend to produce a spark. When, however, the gap is at right angles to the tubes, i. e. when it is in the horizontal plane through the centre of the resonator, the tubes will pass right through it. None of them will stretch across the gap and there will be consequently no tendency to spark.

When the plane of the resonator is at right angles to the axis of the vibrator, the tubes when they meet the wire of the resonator are, as in the last case, travelling at right angles

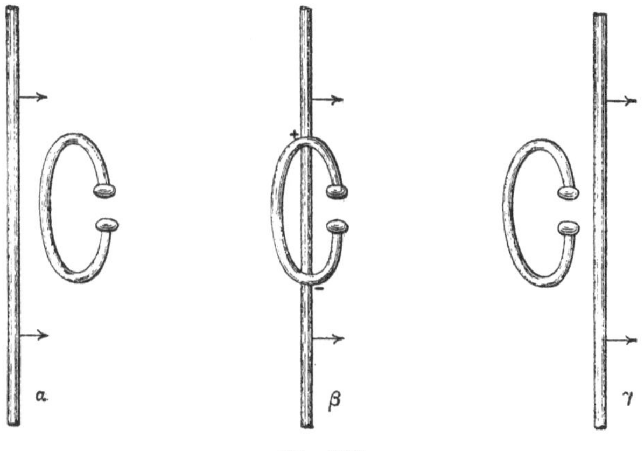

Fig. 115.

to it, so that the wire of the resonator will not collect the tubes and throw them into the air gap. In this case the air gap is always at right angles to the tubes, which will therefore pass right through it, and none of them will stretch across the gap. Thus in this case there is no tendency to spark whatever may be the position of the air space.

Let us now consider the case when the centre of the resonator is on the base line and its plane horizontal. In this case, as we see by the figures Fig. 116, Faraday tubes will be caught by the wire of the resonator and thrown into the air gap wherever that may be; thus, whatever the position of the gap, Faraday tubes will stretch across it, and there will be a tendency to spark. When the gap is as near as possible to the vibrator the Faraday tubes which strike against the resonator will break and a portion of them will stretch right across the gap. When however the gap is a considerable distance from this position the tubes which stretch across it are due to the bending together of two portions of the tubes broken by previously striking against the resonator, the end of one of the portions having travelled along one side of the resonator while the end of the other has travelled along

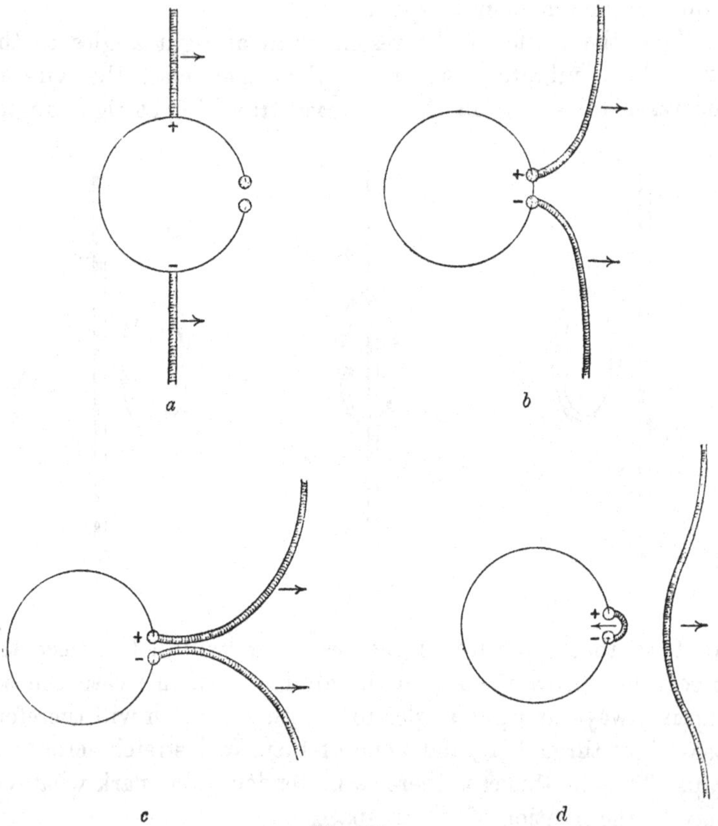

Fig. 116.

the other side, (a); these portions bend together across the gap, (b) and (c); then break up again, one long straight tube travelling outwards, the other shorter one running into the gap, as in (d) Fig. 116. The portion connecting the two sides of the gap diverges more from the shortest distance between the terminals than in the case where the air gap is as near to the vibrator as possible, the field in Fig. 116 will not therefore be so concentrated round the gap, so that there will be less tendency to spark, though this tendency will still remain finite.

Resonance.

332.] Hitherto we have said nothing as to the effect produced by the size of the resonator on the brightness of the sparks, this effect is however often very great, especially when we are using condensers with fairly large capacities which can execute several vibrations before the radiation of their energy reduces the amplitude of the vibration to insignificance.

The cause of this effect is that the resonator is itself an electrical system with a definite period of vibration of its own, hence if we use a resonator the period of whose free vibration is equal to that of the vibrator, the efforts of the vibrator to produce a spark in the resonator will accumulate, and we may be able as the result of this accumulation to get a spark which would not have been produced if the resonator had not been in tune with the vibrator. The case is analogous to the one in which a vibrating tuning fork sets another of the same pitch in vibration, though it does not produce any appreciable effect on another of slightly different pitch.

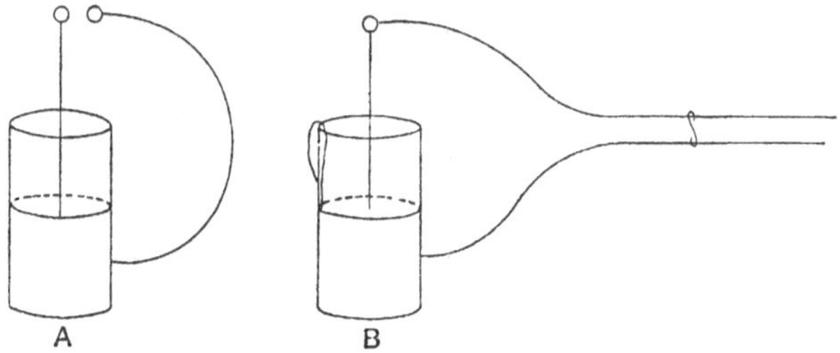

Fig. 117.

333.] Professor Oliver Lodge (*Nature*, Feb. 20, 1890, vol. 41, p. 368) has described an experiment which shows very beautifully the effect of electric resonance. A and B, Fig. 117, represent two Leyden jars whose inner and outer coatings are connected by a wire bent so as to include a considerable area. The circuit connecting the coatings of one of these jars, A, contains an air break. Electrical oscillations are started in this jar by connecting the two coatings with the poles of an electrical machine.

The circuit connecting the coatings of the other jar, B, is provided with a sliding piece by means of which the self-induction of the discharging circuit, and therefore the time of an electrical oscillation of the jar, can be adjusted. The inner and outer coatings of this jar are put almost but not quite into electrical contact by means of a piece of tin-foil bent over the lip of the jar. The jars are placed face to face so that the circuits connecting their coatings are parallel to each other, and approximately at right angles to the line joining the centre of the circuits. When the electrical machine is in action sparks pass across the air break in the circuit in A, and by moving the slider in B about it is possible to find a position for it in which sparks pass by means of the tin-foil from one coating of the jar to the other; as soon however as the slider is moved from this position the sparks cease.

Resonance effects are most clearly marked in cases of this kind, where the system which is vibrating electrically has considerable capacity, since in such cases several complete oscillations have to take place before the radiation of energy from the system has greatly diminished the amplitude of the vibrations. When the capacity is small, the energy radiates so quickly that only a small number of vibrations have any appreciable amplitude; there are thus only a small number of impulses acting on the resonator, and even if the effects of these few conspire, the resonance cannot be expected to be very marked. In the case of the vibrating sphere we saw (Art. 312) that for vibrations about the distribution represented by the first harmonic the amplitude of the second vibration is only about 1/35 of that of the first, in such a case as this the system is practically dead-beat, and there can be no appreciable resonance or interference effects.

The Hertzian vibrator is one in which, as we can see by considering the disposition of the Faraday tubes just before the spark passes across the air, there will be very considerable radiation of energy. Many of the tubes stretch from one plate of the vibrator to the other, and when the insulation of the air space breaks down, closed Faraday tubes will break off from these in the same way as they did from the cylinder; see Fig. 14. These closed tubes will move off from the vibrator with the velocity of light, and will carry the energy of the vibrator away with them. In consequence of this radiation the decay of the oscillations in the vibrator will be very rapid, indeed we should expect the rate of decay to be comparable with its value in the case of the vibrations of electricity over the surfaces of spheres or cylinders, where the Faraday tubes which originally stretched from one part to another of the electrified conductor emit closed tubes which radiate into space in the same way as the similar tubes in the case of the Hertzian vibrator: we have seen, however, that for spheres and cylinders the decay of vibration is so rapid that they may almost be regarded as dead-beat. We should expect a somewhat similar result for the oscillations of the Hertzian vibrator.

334.] On the other hand, the disposition of the Faraday tubes shows us that the electrical vibrations of the resonator will be much more persistent. In this case the Faraday tubes will stretch from side to side across the inside of the resonator as in Fig. 118, and these tubes will oscillate backwards and forwards inside the resonator; they will have no tendency to form closed curves, and consequently there will be little or no radiation of energy. In this case the decay of the vibrations will be chiefly due to the resistance of the resonator, as in the corresponding cases of oscillations in the electrical distribution over spherical or cylindrical cavities in a mass of metal, which are discussed in Arts. 315 and 300.

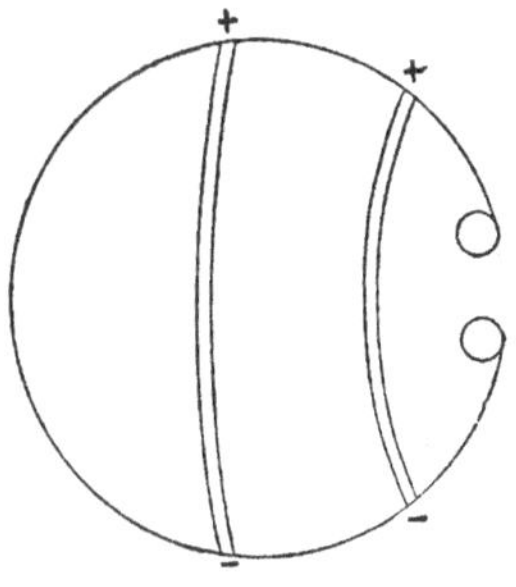

Fig. 118.

335.] The rate at which the vibrations die away for a vibrator and resonator of dimensions not very different from those used by Hertz has been measured by Bjerknes (*Wied. Ann.* 44, p. 74, 1891), who found that in the vibrator the oscillations died away to $1/\epsilon$ of their original value after a time $T/\cdot 26$, where T is the time of oscillation of the vibrator. This rate of decay, though not so rapid as for spheres and cylinders, is still very rapid, as the amplitude of the tenth swing is about 1/14 of that of the first. The amplitudes of the successive vibrations are represented graphically in Fig. 119, which is taken from Bjerknes' paper.

The time taken by the vibrations in the resonator to fade away to $1/\epsilon$ of their original value was found by Bjerknes to be $T'/{\cdot}002$ or $500\,T'$, where T' is the time of the electrical oscillation of the resonator; thus the resonator will make more than 1000 complete oscillations before the amplitude of the vibration falls to 1/10 of its original value. The very slow rate of decay of these oscillations confirms the conclusion we arrived at from

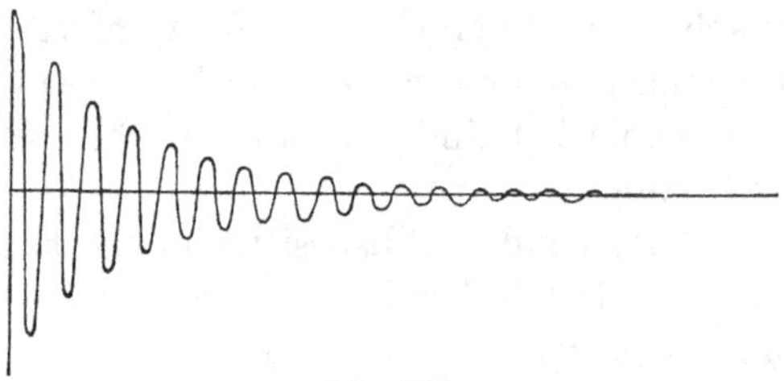

Fig. 119.

the consideration of the Faraday tubes, that there was little or no radiation of energy in this case. The rate of decay of the vibrations in the resonator compares favourably with that of pendulums or tuning-forks, and is in striking contrast to the very rapid fading away of the oscillations of the vibrator. These experiments show that, as the theory led us to expect, we must regard the vibrator as a system having a remarkably large logarithmic decrement, the resonator as one having a remarkably small one.

Reflection of Electromagnetic Waves from a Metal Plate.

336.] We shall now proceed to describe the experiments by which Hertz succeeded in demonstrating, by means of the vibrator and resonator described in Arts. 325 and 328, the existence in the dielectric of waves of electromotive intensity and magnetic force (*Wied. Ann.* 34, p. 610, 1888).

The experiments were made in a large room about 15 metres long, 14 broad, and 6 high. The vibrator was placed 2 m. from one of the main walls, in such a position that its axis was vertical and its base line at right angles to the wall. At all points along the base line the electromotive intensity is vertical, being parallel to the axis of the vibrator. At the further end of the room a piece of sheet zinc 4 metres by 2 was placed vertically against the wall, its plane being thus at right angles to the base line of the vibrator. The zinc plate was connected to earth by means of the gas and water pipes. In one set of experiments the centre of the resonator was on and its plane at right angles to the base line. When it is in this position the Faraday tubes from the vibrator strike the wire of the resonator at right angles; the resonator therefore does not catch the tubes and throw them into the air gap, and the spark will be due to the tubes which fall directly upon the air gap. Thus, as might be expected, the sparks vanish when the gap is at the highest or lowest point of the resonator, when the tubes are at right angles to the direction in which the sparks would pass, and the sparks are brightest when the air gap is in the horizontal plane through the base line, when the incident tubes are parallel to the sparks.

337.] Let the air gap be kept in this plane, and the resonator moved about, its centre remaining on the base line, and its plane at right angles to it. When the resonator is quite close to the zinc plate no sparks pass across the air space; feeble sparks, however, begin to pass as soon as the resonator is moved a short distance away from the plate. They increase rapidly in brightness as the resonator is moved away from the plate until the distance between the two is about 1·8 m., when the brightness of the sparks is a maximum. When the distance is still further increased the brightness of the sparks diminishes, and vanishes again at a distance of about 4 metres from the zinc plate, after which it begins to increase, and attains another maximum, and so on. Thus the sparks exhibit a remarkable periodic character, similar to that which occurs when stationary vibrations are produced by the reflection of wave motion from a surface at right angles to the direction of propagation of the motion.

338.] Let the resonator now be placed so that its plane is the vertical one through the base line, the air gap being at the highest or lowest point; in this position the Faraday tubes which fall directly on the air gap are at right angles to the sparks, so that the latter are due entirely to the Faraday tubes collected by the resonator and thrown into the air gap.

When the resonator is in this position and close to the reflecting plate sparks pass freely. As the resonator recedes from the plate the sparks diminish and vanish when its distance from the plate is about 1·8 metres, the place at which they were a maximum when the resonator was at right angles to the base line; after the resonator passes through this position the sparks increase and attain a maximum 4 metres from the plate, the place where, with the other position of the resonator, they were a minimum; when the resonator is removed still further from the plate the sparks diminish, then vanish, and so on. The sparks in this case show a periodicity of the same wave length as when the resonator was in its former position, the places of minimum intensity for the sparks in one position of the resonator corresponding to those of maximum intensity in the other.

339.] If the zinc reflecting plate is mounted on a movable frame work, so that it can be placed behind the resonator and removed at will, its effect can be very clearly shown by the following experiments:—

Hold the resonator in the position it had in the last experiment at some distance from the vibrator and observe the sparks, the zinc plate being placed on one side out of action: then place the reflector immediately behind the resonator, the sparks will increase in brightness; now push the reflector back, and at about 2 metres from the resonator the sparks will stop. On pushing it still further back the sparks will increase again, and when the reflector is about 4 metres away they will be a little brighter than when it was absent altogether.

340.] Hertz only used one size of resonator, which was selected so as to be in tune with the vibrator. Sarasin and De la Rive (*Comptes Rendus*, March 31, 1891), who repeated this experiment with vibrators and resonators of various sizes, found however that the apparent wave length of the vibrations, that is twice the distance between two adjacent places where the sparks vanish, depended entirely upon the size of the resonator, and not at all upon that of the vibrator. The following table contains the results of their experiments; λ denotes the wave length, a 'loop' means a place where the sparks are at their maximum brightness when the resonator is held in the first position, a 'node' a place where the brightness is a minimum. The line beginning '1/4 λ wire' relates to another series of experiments which we shall consider subsequently. It is included here to avoid the repetition of the table. The distances of the loops and nodes are measured in metres from the reflecting surface.

Diameter of resonator circle (D).	1 metre, stout wire 1 cm. in diameter.	·75 m. stout wire.	·50 m. stout wire.	·35 m. stout wire.	·35 m. fine wire 2 mm. in diameter.
1st Loop . . .	2·11	1·60	1·11	·76	·75
1st Node . . .	4·14	3·01		1·49	1·51
2nd Loop . .				2·30	2·37
2nd Node . .				3·04	3·10
3rd Loop . . .					
3rd Node . .					
$\frac{1}{4}\lambda$ air	2·03	1·41	1·11	·76	·80
$\frac{1}{4}\lambda$ wire . . .	1·92	1·48	·98	·73	
$2D$	2·00	1·50	1·00	·70	·70

Diameter of resonator circle (D).	·25 m. stout wire.	·25 m. fine wire.	·20 m. stout wire.	·20 m. fine wire.	·10 m. stout wire.
1st Loop . . .	·46	·54	.39	·42	·21
1st Node . . .	·94	1·17	·80	·93	·41
2nd Loop . .	1·63	1·89	1·24	1·55	·59
2nd Node . .	2·15	2·40	1·69	2·05	·79
3rd Loop . .	2·71			2·46	·96
3rd Node . .	3·14				
$\frac{1}{4}\lambda$ air		·60	·43	·51	·19
$\frac{1}{4}\lambda$ wire . . .		·56		·45	
$2D$		·50	·40	·40	·20

The most natural interpretation of Hertz's original experiment was to suppose that the vibrator emitted waves of electromotive intensity which, by interference with the waves reflected from the zinc plate, produced standing waves in the region between the vibrator and the reflector, the places in these waves where the electromotive intensity was a maximum being where the sparks were brightest when the resonator was held in the first position.

Sarasin's and De la Rive's discovery of the influence of the size of the resonator on the positions of maximum sparking, and the independence of these positions on the period of the vibrator, compels us, if we retain this explanation, to suppose that any electrical vibrator gives out vibrations of all periods, emitting as it were a continuous electric spectrum.

340*.] This hypothesis appears most improbable, and a more satisfactory explanation seems to be afforded by means of the fact that the oscillations of the vibrator die away with great rapidity, while those of the resonator are extremely persistent. Let us consider what would happen in the extreme case when the oscillations in the vibrator are absolutely dead-beat. Here an electric impulse starts from the vibrator; on its way to the reflector it strikes against the resonator and sets it in electrical vibration; the impulse then travels up to the plate and is reflected, the electromotive intensity in the impulse being reversed by reflection; after reflection the impulse again strikes the resonator, which has maintained the vibrations started by the first impact. If when the reflected impulse reaches the resonator the phase of the vibrations of the latter is opposite to the phase when the impulse passed it on its way to the reflector, the electromotive intensity across the air gap due to the direct and reflected impulses will conspire, so that if the resonator is held in the first position a bright spark will be produced. Now the reflected impulse will strike the resonator the second time when its vibration is in the opposite phase to that which it had just after the first impact if the time which has elapsed between the two impacts is equal to half the time of a complete electrical oscillation of the resonator. The impulse travels at the rate at which electromagnetic action is propagated; hence, if the distance travelled by the impulse between the two impacts is equal to half the wave length of the free electrical vibrations of the resonator, that is, if the distance of the resonator from the reflecting plane is equal to one quarter of the wave length of this vibration, the direct and reflected waves will conspire. If the path travelled by the impulse between the two impacts is equal to a wave length, the electromotive intensity at the air gap due to the incident impulse will be equal and opposite to that due to the reflected one; so that there will in this case, in which the resonator is half a wave length away from the reflector, be no tendency to spark when the resonator is held in this position.

Thus we see that on this view the distances from the reflecting plane of the places where the sparks have their maximum brightness will depend entirely upon the size of the resonator, and not upon that of the vibrator. This, as we have seen, was found by Sarasin and De la Rive to be a very marked feature in their experiments. We have assumed in this explanation that the vibrator does not vibrate. Bjerknes' experiments (l. c.) show that though the vibrations die away very rapidly they are not absolutely dead-beat. The existence of a small number of oscillations in the vibrator will cause the effects to be more vivid with a resonator in tune with it than with any other resonator. Since, however, the rate of decay of the vibrator is infinitely rapid compared with that of the resonator, the positions in which the sparks are brightest will depend much more upon the time of oscillation of the resonator than upon that of the vibrator.

341.] We have still to explain why the places at which the sparks were a maximum when the resonator was in the first position (i.e. with its plane at right angles to the base line) were the places where the sparks vanished when the vibrator was in the second position (i.e. with its plane containing the base line and the axis of the vibrator). When the resonator is in the first position the sparks are wholly due to the Faraday tubes which fall directly upon the air gap, hence the sparks will be a maximum when the state of the resonator corresponds to the incidence upon it of Faraday tubes from the vibrator of the same kind as those which reach it after reflection from the zinc plate. When the resonator is in the second position, having the line joining the terminals of the air gap at right angles to the axis of the vibrator, the sparks are due entirely to the Faraday tubes collected by the resonator and thrown into the air gap, and there would be no tendency to spark in the case just mentioned. For when two Faraday tubes of the same kind moving in opposite directions strike against opposite sides of the resonator, the tubes thrown into the air gap are of opposite signs, and thus do not produce any tendency to spark. When the resonator is in this position the maximum sparks will be produced when the positive tubes strike against one side of the resonator, the negative tubes against the other;

the tubes thrown into the air gap will then be of the same sign and their efforts to produce a spark will conspire: if however the resonator had been held in the first position the positive tubes would have counterbalanced the negative ones, and there would not have been any tendency to spark.

342.] There is one result of Sarasin's and De la Rive's experiments which it is difficult to reconcile with theory. As will be seen from the table they found that the wave length of the vibration was equal to 8 times the diameter of the resonator; theory would lead us to expect that the circumference of the resonator should be half a wave length, since, until the sparks pass, the current in the resonator will vanish at each end of the resonator, as we may neglect the capacity of the knobs. Thus there will be a node at each end of the resonator, and we should expect the wave length to be 2π times the diameter instead of 8 times, as found by Sarasin and De la Rive.

Parabolic Mirrors.

343.] If the vibrator is placed in the focal line of a parabolic cylinder, and if it is of such a kind that the Faraday tubes it emits are parallel to the focal line, then the waves emitted by the vibrator will, if the laws of reflection of these waves are the same as for light, after reflection from the cylinder emerge as a parallel beam and will therefore not diminish in intensity as they recede from the mirror; if such a beam falls on another parabolic mirror whose axis (i.e. the axis of its cross-section) is parallel to the beam, it will be brought to a focus on the focal line of the second mirror. For these reasons the use of parabolic mirrors facilitates very much many experiments on electromagnetic waves.

The parabolic mirrors used by Hertz were made of sheet zinc, and their focal length was about 12·5 cm. The vibrator which was placed in the focal line of one of the mirrors consisted of two equal brass cylinders placed so that their axes were coincident with each other and with the focal line; the length of each of the cylinders was 12 cm. and the diameter 3 cm., their sparking ends being rounded and well polished. The resonator, which was placed in the focal line of an equal parabolic mirror, consisted of two pieces of wire, each had a straight piece 50 cm. long, and was then bent round at right angles so as to pass through the back of the mirror, the length of this bent piece being 15 cm. The ends which came through the mirror were connected with a

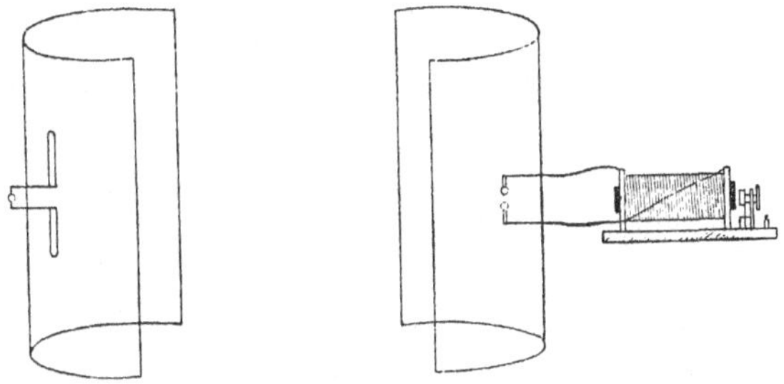

Fig. 120.

spark micrometer and the sparks were observed from behind the mirror. The mirrors are represented in Fig. 120.

Electric Screening.

344.] If the mirrors are placed about 6 or 7 feet apart in such a way that they face each other and have their axes coincident, then when the vibrator is in action vigorous sparks will be observed in the resonator. If a screen of sheet zinc about 2 m. high by 1 broad is placed between the mirrors the sparks in the resonator will immediately cease; they will also cease if a paste-board screen covered with gold-leaf or tin-foil is placed between the mirrors; the interposition of a non-conductor, such as a wooden door, will not however produce any effect. We thus see that a very thin metallic plate acts as a perfect screen and is absolutely opaque to electrical oscillations, while on the other hand a non-conductor allows these radiations to pass through quite freely. The human body is a sufficiently good conductor to produce considerable screening when interposed between the vibrator and resonator.

345.] If wire be wound round a large rectangular framework in such a way that the turns of wire are parallel to one pair of sides of the frame, and if this is interposed between the mirrors, it will stop the sparks when the wires are vertical and thus parallel to the Faraday tubes emitted from the resonator; the sparks however will begin again if the framework is turned through a right angle so that the wires are at right angles to the Faraday tubes.

Reflection of Electric Waves.

346.] To show the reflection of these waves place the mirrors side by side, so that their openings look in the same directions and their axes converge at a point distant about 3 m. from the mirrors. No sparks can be detected at the resonator when the vibrator is in action. If, however, we place at the point of intersection of the axes of the mirrors a metal plate about 2 m. square at right angles to the line which bisects the angle between the axes of the mirrors, sparks will appear at the resonator; they will however disappear if the metal plate is twisted through about 15° on either side. This experiment shows that these waves are reflected and that, approximately at any rate, the angle of incidence is equal to the angle of reflection.

If the framework wound with wire is substituted for the metal plate sparks will appear when the wires are vertical and so parallel to the Faraday tubes, while the sparks will disappear if the framework is turned round until the wires are horizontal. Thus this framework reflects but does not transmit Faraday tubes parallel to the wires, while it transmits but does not reflect Faraday tubes at right angles to them. It behaves in fact towards the electrical waves very much as a plate of tourmaline does to light waves.

Refraction of Electric Waves.

347.] To show the refraction of these waves Hertz used a large prism made of pitch; it was about 1·5 metres in height, had a refracting angle of 30°, and a slant side of 1·2 metres. When the electric waves from the vibrator passed through this prism the sparks in the resonator were not excited when the axes of the two mirrors were parallel, but they were produced when the axis of the mirror of the resonator made a suitable angle with that of the vibrator. When the system was adjusted for minimum deviation the sparks were most vigorous in the resonator when the axis of its mirror made an angle of 22° with that of the vibrator. This shows that the refractive index for pitch is 1·69 for these long electrical waves.

Angle of Polarization.

348.] When light polarized in a plane at right angles to that of incidence falls upon a plate of refracting substance and the normal to the wave front makes with the normal to the surface an angle $\tan^{-1}\mu$, where μ is the refractive index, all the light is refracted and none reflected.

Trouton (*Nature*, February, 21, 1889) has observed a similar effect with these electrical vibrations. From a wall 3 feet thick reflections were obtained when the vibrator, and therefore the Faraday tubes, were perpendicular to the plane of incidence, while there was no reflection when the vibrator was turned through a right angle so that the Faraday tubes were in the plane of incidence. This experiment proves that in the Electromagnetic Theory of Light the Faraday tubes and the electric polarization are at right angles to the plane of polarization.

Before proceeding to describe some other interesting experiments of Mr. Trouton's on the reflection of these waves from slabs of dielectrics, we shall investigate the theory of these phenomena on Maxwell's Theory.

349.] Let us suppose that plane waves are incident on a plate of dielectric bounded by parallel planes, let the plane of the paper be taken as that of incidence and of xy, let the plate be bounded by the parallel planes $x=0$, $x=-h$, the wave being incident on the plane $x=0$. We shall first take the case when the polarization and Faraday tubes are at right angles to the plane of incidence. Let the electromotive intensity in the incident wave be represented by the real part of

$$A\epsilon^{\iota(ax+by+pt)};$$

if i is the angle of incidence, λ the wave length of the vibrations, V their velocity of propagation,

$$a=\frac{2\pi}{\lambda}\cos i, \quad b=\frac{2\pi}{\lambda}\sin i, \quad p=\frac{2\pi}{\lambda}V.$$

Let the intensity in the reflected wave be represented by the real part of

$$A'\epsilon^{\iota(-ax+by+pt)}.$$

The coefficient of y in the exponential in the reflected wave must be the same as that in the incident wave, otherwise the ratio of the reflected to the incident light would depend upon the portion of the plate on which the light fell. The coefficient of x in the expression for the reflected wave can only differ in sign from that in the incident wave: for if E is the electromotive intensity in either the incident or reflected wave, we have

$$\frac{d^2E}{dx^2} + \frac{d^2E}{dy^2} = \frac{1}{V^2}\frac{d^2E}{dt^2},$$

hence the sum of the squares of the coefficients of x and y must be the same for the incident and reflected waves, and since the cofficients of y are the same the coefficients of x can only differ in sign. If E_1, E_2, E_3 are the total electromotive intensities at right angles to the plane of incidence in the air, in the plate, and in the air on the further side of the plate, we may put

$$E_1 = A\epsilon^{\iota(ax+by+pt)} + A'\epsilon^{\iota(-ax+by+pt)},$$
$$E_2 = B\epsilon^{\iota(a'x+by+pt)} + B'\epsilon^{\iota(-a'x+by+pt)},$$
$$E_3 = C\epsilon^{\iota(ax+by+pt)},$$

where

$$a'^2 + b^2 = \frac{p^2}{V'^2},$$

V' being the velocity with which electromagnetic action travels through the plate. The real parts of the preceding expressions only are to be taken.

Since the electromotive intensity is continuous when $x=0$ and when $x=-h$, we have

$$A + A' = B + B', \tag{1}$$
$$C\epsilon^{-\iota ah} = B\epsilon^{-\iota a'h} + B'\epsilon^{\iota a'h}. \tag{2}$$

Since there is no accumulation of Faraday tubes on the surface of the plate the normal flow of these tubes in the air must equal that in the dielectric. Let K be the specific inductive capacity of the plate, that of air being taken as unity, then in the air just above the plate the normal flow of tubes towards the plate is

$$\frac{1}{4\pi}(A-A')V\cos i\,\epsilon^{\iota(by+pt)},$$

the normal flow of tubes in the plate away from the surface $x=0$ is

$$\frac{K}{4\pi}(B-B')V'\cos r\,\epsilon^{\iota(by+pt)},$$

where r is the angle of refraction. Since these must be équal we have

$$(A-A')V\cos i = K(B-B')V'\cos r. \tag{3}$$

The corresponding condition when $x=-h$ gives

$$C\epsilon^{-\iota ah}V\cos i = K(B\epsilon^{-\iota a'h} - B'\epsilon^{\iota a'h})V'\cos r. \tag{4}$$

Equations (3) and (4) are equivalent to the condition that the tangential magnetic force is continuous.

Solving equations (1), (2), (3), (4), we get

$$\left.\begin{aligned} A' &= -A\,(K^2V'^2\cos^2 r - V^2\cos^2 i)\,(\epsilon^{\iota a'h} - \epsilon^{-\iota a'h}) \div \Delta, \\ B &= 2AV\cos i\,(KV'\cos r + V\cos i)\,\epsilon^{\iota a'h} \div \Delta, \\ B' &= 2AV\cos i\,(KV'\cos r - V\cos i)\,\epsilon^{-\iota ah} \div \Delta, \\ C &= 4AKVV'\cos i\cos r\,\epsilon^{\iota ah} \div \Delta, \end{aligned}\right\} \quad (5)$$

where

$$\Delta = (K^2V'^2\cos^2 r + V^2\cos^2 i)\,(\epsilon^{\iota a'h} - \epsilon^{-\iota a'h}) + 2KVV'\cos i\cos r\,(\epsilon^{\iota a'h} + \epsilon^{-\iota a'h}).$$

Thus, corresponding to the incident wave of electromotive intensity

$$\cos\frac{2\pi}{\lambda}(x\cos i + y\sin i + Vt),$$

there will be a reflected wave represented by

$$-(K^2V'^2\cos^2 r - V^2\cos^2 i)\sin\left(\frac{2\pi}{\lambda'}h\cos r\right)\times \cos\left[\frac{2\pi}{\lambda}(-x\cos i + y\sin i + Vt) + \frac{\pi}{2} - \vartheta\right] \div D,$$

where λ' is the wave length in the plate.

$$D^2 = (K^2V'^2\cos^2 r + V^2\cos^2 i)^2\sin^2\left(\frac{2\pi}{\lambda'}h\cos r\right) + 4K^2V'^2V^2\cos^2 i\cos^2 r\cos^2\left(\frac{2\pi}{\lambda'}h\cos r\right),$$

and $$\tan\vartheta = \frac{K^2V'^2\cos^2 r + V^2\cos^2 i}{2KVV'\cos i\cos r}\tan\left(\frac{2\pi}{\lambda'}h\cos r\right).$$

The waves in the plate will be

$$V\cos i\,(KV'\cos r + V\cos i)\times \cos\left[\frac{2\pi}{\lambda'}((x+h)\cos r + y\sin r + V't) - \vartheta\right] \div D,$$

and

$$V\cos i\,(KV'\cos r - V\cos i)\times \cos\left[\frac{2\pi}{\lambda'}(-(x+h)\cos r + y\sin r + V't) - \vartheta\right] \div D;$$

while the wave emerging from the plate will be

$$2KVV'\cos i\cos r\cos\left[\frac{2\pi}{\lambda}((x+h)\cos i + y\sin i + Vt) - \vartheta\right] \div D.$$

Thus we see that when $2\pi h \cos r/\lambda'$ is very small the reflected wave vanishes; this is what we should have expected, as it must require a slab whose thickness is at least comparable with the wave length in the slab to produce any appreciable reflection. When the reflecting surface is too thin we get a result analogous to the blackness of very thin soap films. Trouton has verified that there is no reflection of the electrical waves from window-glass unless this is covered with moisture.

The expression for the amplitude for the reflected wave shows that this will vanish not merely when $2\pi h \cos r/\lambda'$ vanishes but also when this is a multiple of π. Trouton used as the dielectric plate a wall built of paraffin bricks, a method which enabled him to try the effect of altering the thickness of the plate; he found that after reaching the thickness at which the reflected wave became sensible, by making the wall still thicker the reflected wave could be diminished so that its effects were insensible. The case is exactly analogous to that of Newton's rings, where we have darkness whenever $2h \cos r$ is a multiple of a wave length of the light in the plate.

There will be a critical angle in this case if the solution of the equation

$$K^2 V'^2 \cos^2 r - V^2 \cos^2 i = 0 \qquad (6)$$

is real. If the plate is non-magnetic the magnetic permeability is unity, and we have

$$K = \frac{V^2}{V'^2} = \frac{\sin^2 i}{\sin^2 r},$$

so equation (6) becomes

$$\cot^2 r - \cot^2 i = 0,$$

an equation which cannot be satisfied, so that there is no critical angle in this case. This result would not however be true if it were possible to find a magnetic substance which was transparent to electric waves; for if μ' is the magnetic permeability of the substance, we have

$$\mu' K = \frac{V^2}{V'^2},$$

so that equation (6) becomes

$$\frac{\cot^2 r}{\mu'^2} = \cot^2 i,$$

or

$$\frac{\cot r}{\mu'} = \cot i.$$

Since

$$\sqrt{\mu' K} \sin r = \sin i$$

we may transform this equation to

$$\sin^2 i = \frac{\mu'^2 - \mu' K}{\mu'^2 - 1} = \frac{\mu'(\mu' - K)}{\mu'^2 - 1};$$

hence if i is real, μ' must be greater than K. No substance is known which fulfils the conditions of being transparent and having the magnetic permeability greater than the specific inductive capacity, which are the conditions for the existence of a polarizing angle when the Faraday tubes are at right angles to the plane of incidence.

When the plane is infinitely thick, we see that

$$A' = -\frac{KV' \cos r - V \cos i}{KV' \cos r + V \cos i} A,$$

or if the magnetic permeability is unity,

$$A' = -\frac{\sin (i - r)}{\sin (i + r)} A,$$

which is analogous to the expression obtained by Fresnel for the amplitude of the reflected ray when the incident light is polarized in the plane of incidence.

350.] In the preceding investigation the Faraday tubes were at right angles to the plane of incidence, we shall now consider the case when they are in that plane: they are also of course in the planes at right angles to the direction of propagation of the several waves.

Let the electromotive intensity at right angles to the incident ray be

$$A \epsilon^{\iota(ax + by + pt)},$$

that at right angles to the reflected ray

$$A' \epsilon^{\iota(-ax + by + pt)}.$$

Let the electromotive intensity at right angles to the ray which travels in the same sense as the incident one through the plate of dielectric, i.e. in a direction in which x diminishes, be

$$B \epsilon^{\iota(a'x + by + pt)},$$

while that at right angles to the ray travelling in a direction in which x increases is represented by

$$B' \epsilon^{\iota(-a'x + by + pt)}.$$

The electromotive intensity at right angles to the ray emerging from the plate is

$$C \epsilon^{\iota(ax + by + pt)}$$

The conditions at the boundary are (1) that the electromotive intensity parallel to the surface of the plate is continuous; (2) that the electric polarization at right angles to the plate is also continuous.

Hence if i is the angle of incidence, r that of refraction, the boundary conditions at the surface $x=0$ of the plate give

$$\left.\begin{aligned}(A-A')\cos i &= (B-B')\cos r,\\ (A+A')\sin i &= K(B+B')\sin r,\end{aligned}\right\}\qquad(7)$$

where K is the specific inductive capacity of the plate.

The boundary conditions at the lower surface of the plate give

$$\left.\begin{aligned}C\epsilon^{-\iota ah}\cos i &= (B\epsilon^{-\iota a'h}-B'\epsilon^{\iota a'h})\cos r,\\ C\epsilon^{-\iota ah}\sin i &= K(B\epsilon^{-\iota a'h}+B'\epsilon^{\iota a'h})\sin r.\end{aligned}\right\}\qquad(8)$$

Solving equations (7) and (8) we get

$$\begin{aligned}A' &= A\,(K^2\tan^2 r-\tan^2 i)(\epsilon^{\iota a'h}-\epsilon^{-\iota a'h})\div\Delta',\\ B &= 2A\,(\sin i/\cos r)(K\tan r+\tan i)\,\epsilon^{\iota a'h}\div\Delta',\\ B' &= -2A\,(\sin i/\cos r)(K\tan r-\tan i)\,\epsilon^{-\iota a'h}\div\Delta',\\ C &= 4AK\tan i\tan r\,\epsilon^{\iota ah}\div\Delta',\end{aligned}$$

where

$$\Delta'=(K^2\tan^2 r+\tan^2 i)(\epsilon^{\iota a'h}-\epsilon^{-\iota a'h})+2K\tan i\tan r(\epsilon^{\iota a'h}+\epsilon^{-\iota a'h}).$$

From these equations we see that if the incident wave is equal to

$$\cos\frac{2\pi}{\lambda}(x\cos i+y\sin i+Vt)$$

the reflected wave will be

$$(K^2\tan^2 r-\tan^2 i)\sin\left(\frac{2\pi}{\lambda'}h\cos r\right)\times$$
$$\cos\left[\frac{2\pi}{\lambda}(-x\cos i+y\sin i+Vt)+\frac{\pi}{2}-\theta\right]\div D':$$

the waves in the plate will be represented by

$$(\sin i/\cos r)(K\tan r+\tan i)\times$$
$$\cos\left[\frac{2\pi}{\lambda'}((x+h)\cos r+y\sin r+V't)-\theta\right]\div D'$$

and $-(\sin i/\cos r)(K\tan r-\tan i)\times$

$$\cos\left[\frac{2\pi}{\lambda'}(-(x+h)\cos r+y\sin r+V't)-\theta\right]\div D'$$

respectively, while the emergent wave is

$$2K \tan i \tan r \cos\left[\frac{2\pi}{\lambda}\left((x+h)\cos i + y\sin i + Vt\right) - \theta\right] \div D',$$

where

$$D'^2 = (K^2\tan^2 r + \tan^2 i)^2 \sin^2\left(\frac{2\pi}{\lambda'} h\cos r\right)$$
$$+ 4K^2 \tan^2 r \tan^2 i \cos^2\left(\frac{2\pi}{\lambda'} h\cos r\right),$$

and
$$\tan\theta = \frac{K^2\tan^2 r + \tan^2 i}{2K\tan r\tan i}\tan\left(\frac{2\pi}{\lambda'} h\cos r\right).$$

From these expressions we see that, as before, there is no reflected wave when h is very small compared with λ' and when $h\cos r$ is a multiple of $\lambda'/2$; these results are the same whether the Faraday tubes are in or at right angles to the plane of incidence. We see now, however, that in addition to this the reflected wave vanishes, whatever the thickness of the plate, when $K\tan r = \tan i$, or since $\sqrt{\mu' K}\sin r = \sin i$ where μ' is the magnetic permeability, the reflected wave vanishes when

$$\tan^2 i = \frac{K(K-\mu')}{\mu' K - 1};$$

if the plate is non-magnetic $\mu' = 1$, and we have

$$\tan i = \sqrt{K}.$$

When $K\tan r = \tan i$ the reflected wave and one of the waves in the plate vanish; the electromotive intensity in the other wave in the plate is equal to

$$\sqrt{\frac{\mu'}{K}}\cos\frac{2\pi}{\lambda'}(x\cos r + y\sin r + V't),$$

and the emergent wave is

$$\cos\frac{2\pi}{\lambda}\left((x+h)\cos i + y\sin i + Vt - \frac{h\lambda}{\lambda'}\cos r\right).$$

The intensity of all these waves are independent of the thickness of the plate.

If the plate is infinitely thick we must put $B' = 0$ in equations (7); doing this we find from these equations that

$$A' = A\,\frac{(K\tan r - \tan i)}{K\tan r + \tan i},$$

$$B = A\,\frac{\sin 2i}{\sin i\cos r + K\cos i\sin r}.$$

If the plate is made of a non-magnetic material $K = \sin^2 i / \sin^2 r$, and in this case we have

$$A' = A \frac{\tan(i-r)}{\tan(i+r)},$$

$$B = 4A \frac{\sin r \cos i}{\sin 2i + \sin 2r}.$$

Reflection from a Metal Plate.

351.] The very important case when the plate is made of a metal instead of an insulator can be solved in a similar way. The expressions for the electromotive intensities in the various media will be of the same type as before; in the case of metallic reflection however the quantity a', which occurs in the expression for the electromotive intensity in the plate, will no longer be real. In a conductor whose specific resistance is σ the electromotive intensity will satisfy a differential equation of the form

$$\frac{d^2E}{dx^2} + \frac{d^2E}{dy^2} = \frac{4\pi\mu}{\sigma}\frac{dE}{dt},$$

or, since E varies as $\epsilon^{\iota pt}$,

$$\frac{d^2E}{dx^2} + \frac{d^2E}{dy^2} = \frac{4\pi\mu\iota p}{\sigma} E.$$

Hence, since in the metal plate E varies as $\epsilon^{\iota(\pm a'x + by + pt)}$, we see that

$$a'^2 + b^2 = -4\pi\mu\iota p/\sigma. \qquad (9)$$

To compare the magnitude of the terms in this equation, let us suppose that we are dealing with a wave whose wave length is 10^q centimetres. Then since $2\pi/p$ is the time of a vibration, if V is the velocity of propagation of electromagnetic action in air,

$$V 2\pi/p = \lambda,$$

but V is equal to 3×10^{10}, hence

$$p = 6\pi\, 10^{10-q}.$$

If the plate is made of zinc σ is about 10^4, so that the modulus of $4\pi\mu\iota p/\sigma$ is about $24\pi^2 10^{6-q}$. Now b^2 is less than $4\pi^2/\lambda^2$, i.e. $4\pi^2 \times 10^{-2q}$, hence the ratio of the modulus of $4\pi\mu\iota p/\sigma$ to b^2 is of the order $6 \times 10^{6+q}$, and is therefore exceedingly large unless q is less than -6, that is, unless the wave length of the electrical oscillation is much less than that of green light. Thus for waves appreciably longer than this we may for a zinc plate neglect b^2 in equation (9), which then becomes

$$a'^2 = -4\pi\mu\iota p/\sigma,$$

or

$$a' = \pm\sqrt{2\pi\mu p/\sigma}\,(1-\iota),$$

thus a' is exceedingly large compared with a.

We shall first consider the case when the Faraday tubes are at right angles to the plane of incidence as in Art. 349. The condition that the electromotive intensity parallel to the surface of the plate is continuous will still be true, but since there is no real angle of refraction in metals it is convenient to recognize the second condition of that article as expressing the condition that the tangential magnetic force is continuous. The tangential magnetic force is parallel to y and is equal to

$$\frac{1}{\mu\iota p}\frac{dE}{dx},$$

where μ is the magnetic permeability. By means of this and the previous condition we find, using the notation of Art. (349),

$$\left.\begin{aligned}
A' &= -A\,(a'^2/\mu^2 - a^2)\,(\epsilon^{\iota ha'} - \epsilon^{-\iota ha'}) \div D,\\
B &= 2\,A\,a\,(a'/\mu + a)\,\epsilon^{\iota ha'} \div D,\\
B' &= 2\,A\,a\,(a'/\mu - a)\,\epsilon^{-\iota ha'} \div D,\\
C &= A\,4\,a\,(a'/\mu)\,\epsilon^{\iota ha} \div D,\\
&\text{where}\\
D &= (a'^2/\mu^2 + a^2)\,(\epsilon^{\iota ha'} - \epsilon^{-\iota ha'}) + 2\,a\,(a'/\mu)\,(\epsilon^{\iota ha'} + \epsilon^{-\iota ha'}).
\end{aligned}\right\}\quad(10)$$

Since $\epsilon^{\iota(a'x+by+pt)}$ represents a wave travelling in the plate in the direction of the incident wave, i. e. so that x is increasingly negative; the real part of $\iota a'$ must be positive, otherwise the amplitude of the wave would continually increase as the wave travelled onwards; hence if ha' is very large, equations (10) become approximately, remembering that a'/a is also very large,

$$A' = -A,$$

$$B = \frac{2\,a\mu}{a'}A,$$

$$C = B' = 0.$$

Hence in this case there is complete reflection from the metal plate, and since $A' + A = 0$ we see that the electromotive intensity vanishes at the surface of the plate, and since $C = 0$ there is no electromotive intensity on the far side of the plate.

The condition that the plate should act as a perfect reflector or, which is the same thing, as a perfect screen, is that $\{4\pi\mu p h^2/\sigma\}^{\frac{1}{2}}$ should be large. In the case of zinc plates the value of this quantity for vibrations whose wave length is 10^q centimetres is equal to $1\cdot5\times10^{4-q/2}h$, so that for waves 1 metre long it is equal to $1500h$; thus, if h were as great as $\frac{1}{15}$ of a millimetre, $a'h$ would be equal to 10, and since ϵ^{10} is very large the reflection in this case would be practically perfect. We see from this result the reason why gold-leaf and tin-foil are able to reflect these very rapid oscillations almost completely. If however the conductor is an electrolyte σ may be of the order 10^{10}, so that $a'h$ will now be only $1\cdot5h$ for waves 1 metre in length, in this case it will require a slab of electrolyte several millimetres in thickness to produce complete reflection. We shall consider a little more fully the wave emergent from the metallic plate. We have by equations (10)

$$C = \frac{4Aaa'\epsilon^{\iota ha}}{\mu\{(a'^2/\mu^2+a^2)(\epsilon^{\iota ha'}-\epsilon^{-\iota ha'})+(2aa'/\mu)(\epsilon^{\iota ha'}+\epsilon^{-\iota ha'})\}}. \quad (11)$$

If ha' is very small this may be written

$$C = \frac{2Aaa'\epsilon^{\iota ha}}{\mu\{(a'^2/\mu^2+a^2)ha'\iota+(2aa'/\mu)\}},$$

or, since a'^2/μ^2 is very large compared with a^2,

$$C = \frac{A\epsilon^{\iota ha}}{1+\dfrac{\iota ha'^2}{2\mu a}+\frac{1}{2}\iota\mu ha}$$

$$= \frac{A\epsilon^{\iota ha}}{1+(2\pi Vh/\sigma)+\frac{1}{2}\iota\mu ha}.$$

Thus, corresponding to the incident wave

$$\cos\frac{2\pi}{\lambda}(x+Vt),$$

we have, since ha is very small, an emergent wave

$$\frac{1}{1+(2\pi hV/\sigma)}\cos\frac{2\pi}{\lambda}(x+h'+Vt),$$

where
$$h' = h\left\{1-\frac{1}{2}\frac{\mu}{1+2\pi hV/\sigma}\right\}.$$

Since V is equal to 3×10^{10} and σ for electrolytes is rarely greater than 10^9, we see that for very moderate thicknesses $(2\pi hV/\sigma)$ will be large compared with unity, so that the expression for the emergent wave becomes

$$\frac{1}{(2\pi h V/\sigma)} \cos \frac{2\pi}{\lambda}(x+h+Vt).$$

The thickness of the conducting material which, when interposed in the path of the wave, produces a given diminution in the electric intensity is thus proportional to the specific resistance of the material; this result has been applied to measure the specific resistance of electrolytes under very rapidly alternating currents (see J. J. Thomson, *Proc. Roy. Soc.* 45, p. 269, 1889).

The preceding investigation applies to the case when the Faraday tubes are at right angles to the plane of incidence, the same results will apply when the Faraday tubes are in the plane of incidence: the proof of these results for this case we shall however leave as an exercise for the student.

Reflection of Light from Metals.

352.] The assumption that a'/a is very large is legitimate when we are dealing with waves as long as those produced by Hertz's apparatus, it ceases however to be so when the length of the wave is as small as it is in the electrical vibrations we call light. We shall therefore consider separately the theory of the reflection of such waves from metallic surfaces. With the view of making our equations more general we shall not in this case neglect the effects of the polarization currents in the metal; when we include these, the components of the magnetic force and electromotive intensity in the metal satisfy differential equations of the form

$$\mu K' \frac{d^2 f}{dt^2} + \frac{4\pi\mu}{\sigma}\frac{df}{dt} = \frac{d^2 f}{dx^2} + \frac{d^2 f}{dy^2} + \frac{d^2 f}{dz^2}. \qquad (1)$$

See Maxwell's *Electricity and Magnetism*, Art. 783; here K' is the specific inductive capacity of the metal.

353.] Let us first consider the case when the incident wave is polarized in the plane of incidence, which we take as the plane of xy, the reflecting surface being given by the equation $x = 0$. In this case the electromotive intensity Z is parallel to the axis of z; let the incident wave be

$$Z = \epsilon^{\iota(ax+by+pt)},$$

the reflected wave

$$Z = A\epsilon^{\iota(-ax+by+pt)},$$

where $$a^2+b^2=Kp^2, \qquad (2)$$

K being the specific inductive capacity of the dielectric, and the magnetic permeability of this dielectric being assumed to be unity.

Let the wave in the metal be given by the equation

$$Z=B\epsilon^{\iota(a'x+by+pt)},$$

where $$a'^2+b^2=p^2\mu K'-\frac{4\pi\mu\iota p}{\sigma}. \qquad (3)$$

Thus in the dielectric we have

$$Z=\epsilon^{\iota(ax+by+pt)}+A\epsilon^{\iota(-ax+by+pt)},$$

and in the metal $$Z=B\epsilon^{\iota(a'x+by+pt)}.$$

Since Z, the electromotive intensity, is continuous when $x=0$, we have $$1+A=B.$$

By equation (2) of Art. 256 the magnetic induction parallel to y is equal to

$$\frac{1}{\iota p}\frac{dZ}{dx},$$

and since the magnetic force parallel to y is continuous when $x=0$, we have

$$a(1-A)=\frac{a'}{\mu}B.$$

From these equations we find

$$A=\frac{1-\dfrac{a'}{\mu a}}{1+\dfrac{a'}{\mu a}}. \qquad (3^*)$$

Let us for the present confine our attention to the non-magnetic metals for which $\mu=1$, in this case the preceding equation becomes

$$A=\frac{1-\dfrac{a'}{a}}{1+\dfrac{a'}{a}}.$$

The expression given by Fresnel for the amplitude of the wave reflected from a transparent substance is of exactly the same form as this result, the only difference being that for a transparent substance a' is real, while in the case of metals it is complex.

Now for transparent substances the relation between a' and a is

$$\frac{a'^2+b^2}{a^2+b^2} = \mu'^2,$$

where μ' is the refractive index of the substance.

In the case of metals however the relation between a' and a is

$$\frac{a'^2+b^2}{a^2+b^2} = \mu\frac{K'}{K} - \frac{4\pi\mu\iota}{Kp\sigma} = R^2\epsilon^{2\iota\alpha}, \text{ say,} \tag{4}$$

which is of exactly the same form as the preceding, with $R\epsilon^{\iota\alpha}$ written instead of μ', the refractive index of the transparent substance.

Thus, if in Fresnel's formula for the reflected light we suppose that the refractive index is complex and equal to $R\epsilon^{\iota\alpha}$, where R and α are defined by equation (4), we shall arrive at the results given by the preceding theory of the reflection of light by metals.

354.] Let us now consider the case when the plane of polarization is perpendicular to the plane of incidence; in this case the electromotive intensity is in the plane of incidence and the magnetic force γ at right angles to it. If the incident wave is expressed by the equation

$$\gamma = \epsilon^{\iota(ax+by+pt)},$$

then in the dielectric we may put

$$\gamma = \epsilon^{\iota(ax+by+pt)} + A'\epsilon^{\iota(-ax+by+pt)},$$

while in the metal we have

$$\gamma = B'\epsilon^{\iota(a'x+by+pt)}.$$

Since the magnetic force parallel to the surface is continuous, we have

$$1 + A' = B'. \tag{5}$$

The other boundary condition we shall employ is that Q, the tangential electromotive intensity parallel to the axis of y, is continuous. Now if g is the electric polarization parallel to y, and v the conduction current in the same direction, then in the dielectric above the metal

$$4\pi\frac{dg}{dt} = -\frac{d\gamma}{dx},$$

or since

$$g = \frac{K}{4\pi}Q = \frac{a^2+b^2}{4\pi p^2}Q$$

by equation (2) we have

$$\frac{\iota(a^2+b^2)}{p}Q = -\frac{d\gamma}{dx}.$$

In the metal

$$4\pi\frac{dg}{dt} + 4\pi v = -\frac{d\gamma}{dx},$$

or

$$\left(K'\iota p + \frac{4\pi}{\sigma}\right)Q = -\frac{d\gamma}{dx},$$

this by equation (3) becomes

$$\frac{\iota}{p\mu}(a'^2+b^2)\,Q = -\frac{d\gamma}{dx};$$

hence, since Q is continuous when $x = 0$, we have

$$\frac{a}{a^2+b^2}(1-A') = \frac{\mu a'}{(a'^2+b^2)}B'. \qquad (6)$$

Equations (5) and (6) give

$$A' = \frac{1-\mu\dfrac{a'}{a}\dfrac{a^2+b^2}{a'^2+b^2}}{1+\mu\dfrac{a'}{a}\dfrac{a^2+b^2}{a'^2+b^2}},$$

which is again, for non-magnetic metals for which $\mu = 1$, of the same form as Fresnel's expression for the amplitude of the reflected wave from a transparent substance. So that in this case, as in the previous one, we see that we can get the results of this theory of metallic reflection by substituting in Fresnel's expression a complex quantity for the refractive index.

355.] This result leads to a difficulty similar to the one which was pointed out by Lord Rayleigh (*Phil. Mag.* [4], 43, p. 321, 1872) in the theory of metallic reflection on the elastic solid theory of light. The result of substituting in Fresnel's expressions a complex quantity for the refractive index has been compared with the result of experiments on metallic reflection by Eisenlohr (*Pogg. Ann.* 104, p. 368, 1858) and Drude (*Wied. Ann.* 39, p. 481, 1890). The latter writer finds that if the real part of $R^2\epsilon^{2\iota\alpha}$, the quantity which for metals replaces the square of the refractive index for transparent substances, is written as $n^2(1-k^2)$, the imaginary part as $-2\iota n^2 k$; then n and k have the following values, where the accented letters refer to the values for red light, the unaccented to sodium light.

	n	n'	k	k'
Bismuth	1·90	2·07	1·93	1·90
Lead, pure	2·01	1·97	1·73	1·74
Lead, impure	1·97		1·74	
Mercury, pure	1·73	1·87	2·87	2·78
Mercury, impure	1.55		3·14	
Platinum, pure	2·06	2·16	2·06	2·06
Platinum, impure	2·15		1·92	
Gold, pure	·366	·306	7·71	10·2
Gold, impure	·570		5·31	
Antimony	3·04	3·17	1·63	1·56
Tin, solid	1·48	1·66	3·55	3·30
Tin, liquid	2·10		2·15	
Cadmium	1·13	1·31	4·43	4·05
Silver	·181	·203	20·3	19·5
Zinc	2·12	2·36	2·60	2·34
Copper, pure	·641	·580	4·09	5·24
Copper, impure	·686		3·85	
Copper—Nickel alloy	1·55		2·14	
Nickel	1·79	1·89	1·86	1·88
Iron	2·36		1·36	
Steel	2·41	2·62	1·38	1·32
Aluminium	1·44	1·62	3·63	3·36
Magnesium	·37	·40	11·8	11·5

It will be seen that for all these metals without exception the value of k is greater than unity, so that the real part of $R^2 \epsilon^{2\iota\alpha}$ or $n^2(1-k^2)$ is negative. Equation (4), Art. 353, shows, however, that the real part of $R^2 \epsilon^{2\iota\alpha}$ is equal to $\mu K'/K$, an essentially positive quantity. This shows that the electromagnetic theory of metallic reflection is not general enough to cover the facts. In this respect, however, it is in no worse position than any other existing theory of light, while it possesses the advantage over other theories of explaining why metals are opaque.

356.] The direction in which to look for an improvement of the theory seems pretty obvious. The preceding table shows how rapidly the effects vary with the frequency of the light vibrations; they are in this respect analogous to the effects of 'anomalous dispersion' (see Glazebrook, *Report on Optical Theories*, *B. A. Report*, 1885), which have been accounted for by assuming that the molecules of the substance through which the light passes have free periods of vibration comparable with the frequency of the light vibrations. The energy absorbed by such molecules is then a function of the frequency of the light vibrations, and the optical character of the medium cannot be fixed by one or two constants, such as the specific inductive capacity or the specific resistance; we require to know in addition the free periods of the molecules.

357.] We now return to the case of the magnetic metals; the question arises whether or not these substances retain their magnetic properties under magnetic forces which oscillate as rapidly as those in a wave of light. We have seen (Art. 286) that iron retains its magnetic properties when the magnetic forces make about one million vibrations per second; in the light waves, however, the magnetic forces are vibrating more than five hundred million times faster than this, and the only means we have of testing whether magnetic substances retain their properties under such circumstances is to examine the light reflected from or transmitted through such bodies. When we do this, however, we labour under the disadvantage that, as the preceding investigation shows, the theory of metallic reflection is incomplete, so that the conclusions we may come to as the results of this theory are not conclusive. Such evidence as we have, however, tends to show that iron does not retain its magnetic properties under such rapidly alternating magnetic forces. An example of such evidence is furnished by equation (3*), Art. 353. We see from that equation that if μ for light waves in iron were very large, the intensity of the light reflected from iron would be very nearly the same as that of the incident light, in other words iron would have a very high reflecting power. The reverse, however, seems to be true; thus Drude (*Wied. Ann.* 39, p. 549, 1890) gives the following numbers as representing the reflective powers of some metals for yellow light:—

Silver.	Gold.	Copper.	Iron.	Steel.	Nickel.
95·3	85·1	73·2	56·1	58·5	62·0

Rubens (*Wied. Ann.* 37, p. 265, 1889) gives for the same metals the following numbers:—

Silver.	Gold.	Copper.	Iron.	Nickel.
90·3	71·1	70·0	56·1	62·1

The near agreement of the numbers found by these two experimenters seems to show that the smallness of the reflection observed from iron could not be due to any accidental cause such as want of polish. Another reason for believing that iron does not manifest magnetic properties under the action of light waves, is that there is nothing exceptional in the position of iron with respect to the optical constants of metals in the table given in Art. 353. The theory of metallic reflection is however so far from accounting for the facts that we cannot attach much weight to considerations based on it. The only conclusion we can come to is the negative one, that there is no evidence to show that iron does retain its magnetic properties for the light vibrations.

The change in Phase produced by the Transmission of Light through thin Films of Metal.

358.] Quincke (*Pogg., Ann.* 120, p. 599, 1863) investigated the change in phase produced when light passed through thin silver plates, and found that in many cases the phase was accelerated, the effect being the same as if the velocity of light through silver was greater than that through air. Kundt (*Phil. Mag.* [5], 26, p. 1, 1888), in a most beautiful series of experiments, measured the deviation of a ray passing through a small metal *prism*, and found that when the prism was made of silver, gold, or copper, the deviation was towards the thin end. With platinum, nickel, bismuth, and iron prisms the deviation was, on the other hand, towards the thick end. We can readily find on the electromagnetic theory of light the change in phase produced when the light passes through a thin film of metal. The equation (11) of Art. 351 shows, that if the incident wave (supposed for simplicity to be travelling at right angles to the film) is represented by

$$\epsilon^{\iota(ax+pt)},$$

the emergent wave will be

$$\frac{4a(a'/\mu)\,\epsilon^{\iota ha}\,\epsilon^{\iota(ax+pt)}}{(a'^2/\mu^2+a^2)(\epsilon^{\iota ha'}-\epsilon^{-\iota ha'})+2a(a'/\mu)(\epsilon^{\iota ha'}+\epsilon^{-\iota ha'})},$$

or if the film is so thin that ha' is a small quantity, the emergent wave is equal to

$$\frac{\epsilon^{\iota ha}\,\epsilon^{\iota(ax+pt)}}{1+\frac{1}{2}\frac{\iota h\mu}{a}\left(\frac{a'^2}{\mu^2}+a^2\right)}.$$

Now, since in this case $b=0$, we have by equation (4) of Art. 353 $\frac{a'^2}{a^2}=R^2\epsilon^{2\iota\alpha}$, hence the emergent wave is equal to

$$\frac{\epsilon^{\iota ha}\,\epsilon^{\iota(ax+pt)}}{1+\frac{1}{2}\iota h\mu a\left\{\frac{R^2\epsilon^{2\iota\alpha}}{\mu^2}+1\right\}},$$

or, neglecting squares and higher powers of h, this is equal to

$$\epsilon^{\frac{1}{2}haR^2 \sin 2a\mu^{-1}} \epsilon^{\iota ha} \epsilon^{-\frac{1}{2}\iota h\mu a(1+R^2\cos 2a/\mu^2)} \epsilon^{\iota(ax+pt)}$$

$$= \epsilon^{\frac{1}{2}haR^2 \sin 2a\mu^{-1}} \epsilon^{\iota ha\left(1-\frac{\mu}{2}\{1+R^2\cos 2a/\mu^2\}\right)} \epsilon^{\iota(ax+pt)},$$

hence the acceleration of phase expressed as a length is equal to

$$h\left(1-\frac{\mu}{2}\{1+R^2\cos 2a/\mu^2\}\right),$$

or for non-magnetic substances to

$$\tfrac{1}{2}h(1-R^2\cos 2a).$$

In the interpretation of this result we are beset with difficulties, whether we take $R^2\cos 2a$ as determined by the electromagnetic theory, or whether we take it as given by Drude's experiments. In the former case $R^2\cos 2a$ is positive, so that the acceleration cannot be greater than $h/2$, or the apparent speed of light through the metal cannot be greater than twice that through air; this is not in accordance with Kundt's experiments on silver and gold. If, on the other hand, we take Drude's values for $R^2\cos 2a$, since these are negative for all metals, the apparent velocity of light through a film of any metal ought to be more than double that through air; this again is not in accordance with Kundt's observations, according to which the apparent velocity of light through films of metals other than gold, silver, or copper is less than that through air. We might have anticipated that such a discrepancy would arise, for we have assumed in deducing the expression for the transmitted ray that the electromotive intensity parallel to the surface of the metal is continuous. Now if we suppose that the light vibrations have periods comparable with periods of the molecules of the metal, the electromotive intensity in the metal will arise from two causes. The first is due to magnetic induction, this will be continuous with that due to the same cause in the air; the second is due to the reaction of the molecules of the metal on the medium conveying the light. Now there does not seem to be any reason to assume that this part of the electromotive intensity should be continuous as we pass from the air which does not exhibit anomalous dispersion to the metal which does. The electromotive intensity parallel to the boundary is thus probably discontinuous, and we could not therefore expect a formula obtained by the condition that this intensity was continuous to be in accordance with experiment.

Reflection of Electromagnetic Waves from Wires.

Reflection from a Grating.

359.] We shall now consider the reflection of electromagnetic waves from a grating consisting of similar and parallel metallic wires, whose cross-sections we leave for the present indeterminate, arranged at equal intervals, the axes of all the wires being in one plane, which we shall take as the plane of yz, the axis of z being parallel to the wires: the distance between the axes of two adjacent wires is a. We shall suppose that a wave in which the electromotive intensity is parallel to the wires, and whose front is parallel to the plane of the grating, falls upon the wires. The electromotive intensity in the incident wave may be represented by the real part of $A\epsilon^{\frac{\iota 2\pi}{\lambda}(Vt+x)}$, x being measured from the plane of the grating towards the advancing wave. The incidence of this wave will induce currents in the wires, and these currents will themselves produce electromotive intensities parallel to z in the region surrounding them; these intensities will evidently be expressed by a periodic function of y of such a character that when y is increased by a the value of the function remains unchanged. If we make the axis of z coincide with the axis of one of the wires, the electromotive intensity will evidently be an even function of y. Thus E_2, the electromotive intensity due to the currents in the wire, will be given by an equation of the form

$$E_2 = \Sigma A_m \cos\frac{2m\pi y}{a}\,\epsilon^{\iota nx}\,\epsilon^{\frac{\iota 2\pi}{\lambda}Vt},$$

where m is an integer.

Since the electromotive intensity satisfies the equation

$$\frac{d^2E}{dx^2}+\frac{d^2E}{dy^2}=\frac{1}{V^2}\frac{d^2E}{dt^2},$$

we have

$$n^2 = -\frac{4\pi^2 m^2}{a^2}+\frac{4\pi^2}{\lambda^2}.$$

We shall assume that the distance between the wires of the grating is very small compared with the length of the wave; thus, unless m is zero, the first term on the right-hand side of the above equation will be very large compared with the second, so that when m is not zero we may put

$$n = \pm \frac{\iota 2\pi m}{a},$$

while when m is zero

$$n = -\frac{2\pi}{\lambda},$$

the minus sign being taken so as to represent a wave diverging from the wires. Substituting these values we find that when x is positive,

$$E_2 = A_0 \epsilon^{\frac{\iota 2\pi}{\lambda}(Vt-(x+a))} + \sum_{m=1}^{m=\infty} A_m \epsilon^{-\frac{2\pi m}{a}x} \cos\frac{2\pi m y}{a} \epsilon^{\frac{\iota 2\pi}{\lambda}Vt},$$

where a is a constant.

When the rate of alternation is so rapid that the waves are only a few metres in length the electromotive intensity at the surface of the metal wire must vanish, see Arts. 300 and 301; hence if E_1 is the electromotive intensity in the incident wave, $E_1 + E_2$ must vanish at the surface of the wire. Near the grating however x/λ will be small; hence we may put, writing

$$A_m' \cos\frac{2\pi}{\lambda}Vt + B_m' \sin\frac{2\pi}{\lambda}Vt \quad \text{for} \quad A_m \epsilon^{\iota 2\pi Vt/\lambda},$$

$$E_1 + E_2 = (A + A_0)\cos\frac{2\pi}{\lambda}Vt + (A_0(x+a) - Ax)\frac{2\pi}{\lambda}\sin\frac{2\pi}{\lambda}Vt$$

$$+ \Sigma \epsilon^{-\frac{2\pi m x}{a}} \cos\frac{2\pi m y}{a}\left(A_m' \cos\frac{2\pi}{\lambda}Vt + B_m' \sin\frac{2\pi}{\lambda}Vt\right).$$

Now in Maxwell's *Electricity and Magnetism*, Vol. i. Art. 203, it is shown that the expression

$$C \log\left\{1 - 2\epsilon^{-\frac{2\pi x}{a}} \cos\frac{2\pi y}{a} + \epsilon^{-\frac{4\pi x}{a}}\right\} + Dx,$$

where C and D are constants, is constant over a series of equidistant parallel wires, whose axes are at a distance a apart and whose cross-section is approximately circular. The logarithm can be expanded in the form

$$-2C\Sigma \frac{1}{m} \epsilon^{-\frac{2m\pi x}{a}} \cos\frac{2m\pi y}{a}.$$

Now in the expression for $E_1 + E_2$ put

$$A + A_0 = 0, \quad A_m = 0, \quad B_m = -\frac{2C}{m},$$

then

$$E_1+E_2 = A\cos\frac{2\pi}{\lambda}(Vt+x)-A\cos\frac{2\pi}{\lambda}(Vt-(x+a))$$
$$+C\log\left(1-2\epsilon^{-\frac{2\pi x}{a}}\cos\frac{2\pi y}{a}+\epsilon^{-\frac{4\pi x}{a}}\right)\sin\frac{2\pi}{\lambda}Vt,$$

hence near the grating where x/λ is small

$$E_1+E_2 = \sin\frac{2\pi}{\lambda}Vt\left\{-A\frac{2\pi x}{\lambda}-A\frac{2\pi}{\lambda}(x+a)\right.$$
$$\left.+C\log\left(1-2\epsilon^{-\frac{2\pi x}{a}}\cos\frac{2\pi y}{a}+\epsilon^{-\frac{4\pi x}{a}}\right)\right\},$$

and we see by Maxwell's result that the quantity inside the bracket has a constant value over the surface of the wires; hence, if we make this value zero, we shall have satisfied the conditions of the problem. Let $2c$ be the diameter of any one of the wires in the plane of the grating, then when $x=0$ and $y=c$ the expression inside the bracket must vanish, hence

$$-A\frac{2\pi}{\lambda}a+C\log 4\sin^2\frac{\pi c}{a}=0.$$

To find another relation between A, C, and a we must consider the equation to the cross-section of the wire at the origin, viz.,

$$-A\frac{2\pi}{\lambda}(2x+a)+C\log\left(1-2\epsilon^{-\frac{2\pi x}{a}}\cos\frac{2\pi y}{a}+\epsilon^{-\frac{4\pi x}{a}}\right)=0,$$

or substituting for C its value in terms of A,

$$\left(\frac{2x}{a}+1\right)\log\left\{4\sin^2\frac{\pi c}{a}\right\}=\log\left(1-2\epsilon^{-\frac{2\pi x}{a}}\cos\frac{2\pi y}{a}+\epsilon^{-\frac{4\pi x}{a}}\right). \quad (1)$$

If d is the value of x when $y=0$,

$$a=2d\frac{\log 2\sin\frac{\pi c}{a}}{\log\left\{\dfrac{1-\epsilon^{-\frac{2\pi d}{a}}}{2\sin\frac{\pi c}{a}}\right\}}. \quad (2)$$

When $c=d$, this equation becomes, since c/a is small,

$$a=-\frac{2a}{\pi}\log 2\sin\frac{\pi c}{a}.$$

The expression for E_2 consists of two parts, one of which is

$$-A\epsilon^{\frac{\iota 2\pi}{\lambda}(Vt-(x+a))},$$

which represents a reflected wave equal in intensity to the incident one, but whose phase is changed by reflection by $(\frac{1}{2}\lambda - a)$, where a is given by (2) and depends upon the size of the wires and their distance apart. The other part of the expression for E_2 is

$$C \log\left(1 - 2\epsilon^{-\frac{2\pi x}{a}} \cos\frac{2\pi y}{a} + \epsilon^{-\frac{4\pi x}{a}}\right).$$

This is inappreciable at a distance from the grating 4 or 5 times the distance between the wires, hence the reflection, at some distance from the grating, is the same, except for the alteration in phase as from a continuous metallic surface.

360.] If the electromotive intensity had been at right angles to the wires the reflection would have been very small; thus a grating of this kind will act like a polariscope, changing either by reflection or transmission an unpolarised set of electrical vibrations into a polarised one. When used to produce polarisation by transmission we may regard it as the electrical analogue of a plate of tourmaline crystal.

Scattering of Electromagnetic Waves by a Metallic Wire.

361.] The scattering produced when a train of plane electromagnetic waves impinges on an infinitely long metal cylinder, whose axis is at right angles to the direction of propagation of the waves and whose diameter is small compared with the wave length, can easily be found as follows:—

We shall begin with the case where the electromotive intensity in the incident wave is parallel to the axis of the cylinder, which we take as the axis of z; the axis of x being at right angles to the fronts of the incident waves.

Let λ be the wave length, then E_1, the electromotive intensity in the incident waves, may be represented by the equation

$$E_1 = \epsilon^{\frac{\iota 2\pi}{\lambda}(Vt + x)},$$

where the real part of the right-hand side is to be taken. The positive direction of x is opposite to that in which the waves are travelling. In the neighbourhood of the cylinder x/λ is small, so that we may put

$$E_1 = \epsilon^{\frac{\iota 2\pi}{\lambda}Vt}\left(1 + \iota x\frac{2\pi}{\lambda}\right)$$

approximately, or if r and θ are the polar coordinates of the point where the intensity is E_1,

$$E_1 = \epsilon^{\frac{\iota 2\pi}{\lambda}Vt}\left(1+\iota\frac{2\pi}{\lambda}r\cos\theta\right).$$

Let E_2 be the electromotive intensity due to the currents induced in the cylinder, then E_2 satisfies the differential equation

$$\frac{d^2E_2}{dr^2}+\frac{1}{r}\frac{dE_2}{dr}+\frac{1}{r^2}\frac{d^2E_2}{d\theta^2}=\frac{1}{V^2}\frac{d^2E_2}{dt^2}$$
$$=-\frac{4\pi^2}{\lambda^2}E_2,$$

or if E_2 varies as $\cos n\theta$,

$$\frac{d^2E_2}{dr^2}+\frac{1}{r}\frac{dE_2}{dr}+\left(\frac{4\pi^2}{\lambda^2}-\frac{n^2}{r^2}\right)E_2=0.$$

The solution of which outside the cylinder is

$$E_2 = A_n\cos n\theta K_n\left(\frac{2\pi}{\lambda}r\right)\epsilon^{\frac{\iota 2\pi}{\lambda}Vt},$$

where K_n represents the 'external' Bessel's function of the nth order.

Thus

$$E_2=\left\{A_0K_0\left(\frac{2\pi}{\lambda}r\right)+A_1\cos\theta K_1\left(\frac{2\pi}{\lambda}r\right)\right.$$
$$\left.+A_2\cos 2\theta K_2\left(\frac{2\pi}{\lambda}r\right)+\ldots\right\}\epsilon^{\frac{\iota 2\pi}{\lambda}Vt}.$$

Now since the cylinder is a good conductor, the total tangential electromotive intensity must vanish over its surface, see Arts. 300 and 301. Hence if c is the radius of the cylinder, $E_1+E_2=0$ when $r=c$; from this condition we get

$$A_0=-\frac{1}{K_0\left(\frac{2\pi}{\lambda}c\right)},\quad A_1=-\frac{\iota 2\pi c}{\lambda K_1\left(\frac{2\pi}{\lambda}c\right)},\quad A_2=A_3=\ldots=0.$$

Thus

$$E_2=\left\{-\frac{K_0\left(\frac{2\pi}{\lambda}r\right)}{K_0\left(\frac{2\pi}{\lambda}c\right)}-\frac{\iota 2\pi c}{\lambda}\cos\theta\frac{K_1\left(\frac{2\pi}{\lambda}r\right)}{K_1\left(\frac{2\pi}{\lambda}c\right)}\right\}\epsilon^{\frac{\iota 2\pi}{\lambda}Vt}.$$

362.] Let us first consider the effect of the cylinder on the lines of magnetic force in its neighbourhood. If α, β are the components of the magnetic force parallel to the axes of x and y respectively, E the total electromotive intensity, then

$$\frac{dE}{dx} = \frac{d\beta}{dt} = \iota \frac{2\pi}{\lambda} V\beta,$$

$$\frac{dE}{dy} = -\frac{d\alpha}{dt} = -\iota \frac{2\pi}{\lambda} V\alpha.$$

Thus the direction of the magnetic force will be tangential to the curves over which E is constant, the equations to the lines of magnetic force in the neighbourhood of the cylinder are therefore

$$\left\{\left(1 - \frac{K_0\left(\frac{2\pi}{\lambda}r\right)}{K_0\left(\frac{2\pi}{\lambda}c\right)}\right) + \iota\frac{2\pi}{\lambda}\cos\theta\left(r - c\frac{K_1\left(\frac{2\pi}{\lambda}r\right)}{K_1\left(\frac{2\pi}{\lambda}c\right)}\right)\right\}\epsilon^{\iota\frac{2\pi}{\lambda}Vt} = C,$$

where C is independent of r and θ.

Now $2\pi c/\lambda$ is by hypothesis very small, and when x is small then, by Art. 261, the values of K_0 and K_1 are given approximately by the equations

$$K_0(x) = \log(2\gamma/x),$$

$$K_1(x) = -K_0'(x) = \frac{1}{x},$$

where γ is Euler's constant and $\log\gamma$ is equal to ·5772157.

In the neighbourhood of the cylinder r/λ is small as well as c/λ, so that in this region the equations to the lines of magnetic force are, approximately,

$$\frac{\log(r/c)}{\log(\gamma\lambda/\pi c)}\cos\frac{2\pi}{\lambda}Vt + \frac{2\pi}{\lambda}\cos\theta\frac{(c^2-r^2)}{r}\sin\frac{2\pi}{\lambda}Vt = C.$$

In this expression the coefficient of $\cos(2\pi Vt/\lambda)$ is very large compared with that of $\sin(2\pi Vt/\lambda)$, so that unless $2\pi Vt/\lambda$ is an odd multiple of $\pi/2$, that is, unless the intensity in the incident wave at the axis of the cylinder vanishes, the equations to the lines of magnetic force are

$$\log(c/r) = \text{a constant},$$

so that these lines are circles concentric with the cylinders.

When $2\pi Vt/\lambda$ is an odd multiple of $\pi/2$, the lines of magnetic force are given by the equation

$$\cos\theta\frac{(c^2-r^2)}{r} = C,$$

or in Cartesian coordinates

$$x\{c^2-(x^2+y^2)\} = C(x^2+y^2);$$

these curves are shown in Fig. 121.

363.] Since the direction of motion of the Faraday tubes is at right angles to themselves and to the magnetic force, when the lines of magnetic force near the cylinder are circles, these tubes will, in the neighbourhood of the cylinder, move radially, the positive tubes (i.e. those parallel to the tubes in the incident wave) moving inwards, the negative ones outwards. In the special case where the electromotive intensity vanishes at the

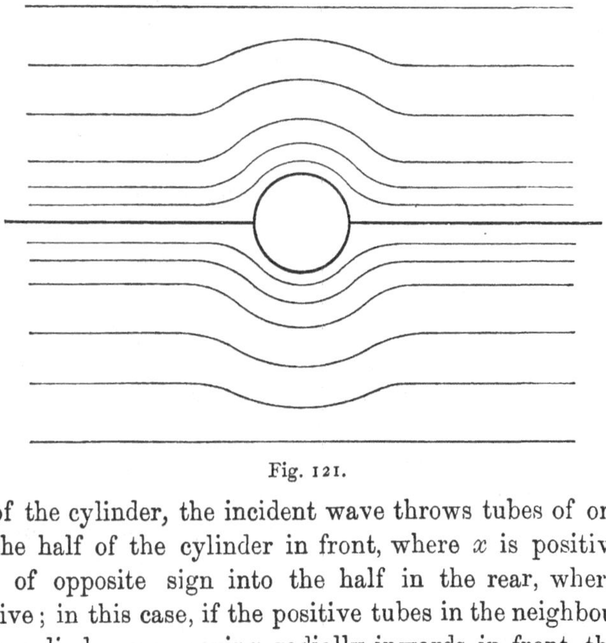

Fig. 121.

axis of the cylinder, the incident wave throws tubes of one sign into the half of the cylinder in front, where x is positive, and tubes of opposite sign into the half in the rear, where x is negative; in this case, if the positive tubes in the neighbourhood of the cylinder are moving radially inwards in front, they are moving radially outwards in the rear and *vice versâ*; there are in this case but few tubes near the equatorial plane, and the motion of these is no longer radial.

364.] When the distance from the cylinder is large compared with the wave length, we have

$$K_0\left(\frac{2\pi}{\lambda}r\right) = \tfrac{1}{2}\iota^{\frac{1}{2}}\frac{\epsilon^{-\iota 2\pi r/\lambda}}{(r/\lambda)^{\frac{1}{2}}},$$

$$K_1\left(\frac{2\pi}{\lambda}r\right) = -\tfrac{1}{2}\iota^{\frac{3}{2}}\frac{\epsilon^{-\iota 2\pi r/\lambda}}{(r/\lambda)^{\frac{1}{2}}}.$$

Thus in the wave 'scattered' by the cylinder

$$E_2 = -\frac{\epsilon^{-\iota\frac{2\pi}{\lambda}\left(r - Vt - \frac{\lambda}{8}\right)}}{2\,(r/\lambda)^{\frac{1}{2}}}\left\{\frac{1}{\log(\gamma\pi\lambda/c)} + \frac{4\pi^2c^2}{\lambda^2}\cos\theta\right\}.$$

Thus in this case, as we should expect, the part of the scattered wave which is independent of the azimuth is very much larger than the part which varies with θ, so that there is no direction in which the intensity of the scattered light vanishes. In this respect the metal cylinder resembles one made of a non-conductor, the effect of which on a train of waves has been investigated by Lord Rayleigh (*Phil. Mag.* [5], 12, p. 98, 1881): there are however some important differences between the two cases; in the first place we see that since c occurs in the leading term only as a logarithm, the amount of light scattered by the cylinder changes very slowly with the dimensions of the cylinder, while in the light scattered from a dielectric cylinder the electromotive intensity in the scattered wave is proportional to the area of the cross-section of the cylinder. Again, when the cylinder is a good conductor the electromotive intensity in the scattered wave, if we regard the logarithmic term as approximately constant, varies as $\lambda^{\frac{1}{2}}$ and so increases with the wave length, while when the cylinder is an insulator the electromotive intensity varies as $\lambda^{-\frac{3}{2}}$, so that the scattering *decreases* rapidly as the length of the wave increases. The most interesting case of this kind is when the wave incident on the cylinder is a wave of light; in this case the theory indicates that the light scattered by the metallic cylinder would be slightly reddish, while that from the insulating cylinder would be distinctly blue; the blue in the latter case would be much more decided than the red of the previous one, since the variation of the intensity of the scattered light with the wave length is much more rapid when the cylinder is an insulator than when it is a good conductor.

365.] We shall now proceed to consider the case when the electromotive intensity in the incident wave is at right angles to the axis of the cylinder. This case is of more interest than the preceding because the general features of the results obtained will apply to the scattering of light by particles limited in every direction; it is thus representative of the scattering by small particles in general, while the peculiarities of the case discussed in the preceding article were due to the cylindrical shape of the obstacle. The only case to which the results of this article would not be applicable without further investigation is that in which the particles are highly magnetic, and we shall find that even this case constitutes no exception since our results do not involve the magnetic permeability of the cylinder.

As the electromotive intensity is at right angles to the axis of the cylinder, the magnetic force will be parallel to the axis.

Let the magnetic force H_1 in the incident wave be expressed by the equation

$$H_1 = \epsilon^{\frac{\iota 2\pi}{\lambda}(Vt+x)}.$$

When x which is equal to $r\cos\theta$ is small compared with λ, this is approximately

$$H_1 = \epsilon^{\frac{\iota 2\pi}{\lambda}Vt}\left\{1 - \frac{\pi^2}{\lambda^2}r^2 + \frac{\iota 2\pi}{\lambda}r\cos\theta - \frac{\pi^2}{\lambda^2}r^2\cos 2\theta\right\}.$$

Since H, the magnetic force, satisfies the differential equation

$$\frac{d^2H}{dx^2} + \frac{d^2H}{dy^2} = \frac{1}{V^2}\frac{d^2H}{dt^2},$$

the magnetic force H_2 due to the currents induced in the cylinder may be expressed by the equation

$$H_2 = \epsilon^{\frac{\iota 2\pi}{\lambda}Vt}\left\{A_0K_0\left(\frac{2\pi}{\lambda}r\right) + A_1\cos\theta\, K_1\left(\frac{2\pi}{\lambda}r\right) + A_2\cos 2\theta\, K_2\left(\frac{2\pi}{\lambda}r\right)\right\},$$

where A_0, A_1 and A_2 are arbitrary constants.

The condition to be satisfied at the boundary of the cylinder is that the tangential electromotive intensity at its surface should vanish. In this case we have, however,

$$\frac{d}{dr}(H_1 + H_2) = 4\pi \text{ (intensity of current at right angles to } r).$$

The current in the dielectric is a polarization current, and if E is the tangential electromotive intensity, the intensity of this current at right angles to r is

$$\frac{K}{4\pi}\frac{dE}{dt},$$

which is equal to

$$\frac{K}{4\pi}\frac{\iota 2\pi}{\lambda}VE.$$

Thus the condition that E should vanish at the surface is

equivalent to the condition that

$$\frac{d}{dr}(H_1+H_2)=0$$

when $r=c$, c being the radius of the cylinder.

From this condition we get

$$-2c\frac{\pi^2}{\lambda^2}+A_0\frac{d}{dc}K_0\left(\frac{2\pi}{\lambda}c\right)=0,$$

$$\frac{\iota 2\pi}{\lambda}+A_1\frac{d}{dc}K_1\left(\frac{2\pi}{\lambda}c\right)=0,$$

$$-2c\frac{\pi^2}{\lambda^2}+A_2\frac{d}{dc}K_2\left(\frac{2\pi}{\lambda}c\right)=0.$$

Since $2\pi c/\lambda$ is very small and therefore approximately

$$K_0\left(\frac{2\pi}{\lambda}c\right)=\log\left(2\gamma/\frac{2\pi c}{\lambda}\right),$$

$$K_1\left(\frac{2\pi}{\lambda}c\right)=\frac{\lambda}{2\pi c},$$

$$K_2\left(\frac{2\pi}{\lambda}c\right)=\frac{\lambda^2}{2\pi^2c^2},$$

we get

$$A_0=-2\pi^2\frac{c^2}{\lambda^2},$$

$$A_1=\iota 4\pi^2\frac{c^2}{\lambda^2},$$

$$A_2=-2\pi^4\frac{c^4}{\lambda^4}.$$

Thus the magnetic force due to the currents induced in the cylinder is given by the equation

$$H_2=2\pi^2\frac{c^2}{\lambda^2}\epsilon^{\frac{\iota 2\pi}{\lambda}Vt}\left\{-K_0\left(\frac{2\pi}{\lambda}r\right)+2\iota\cos\theta K_1\left(\frac{2\pi}{\lambda}r\right)-\frac{\pi^2c^2}{\lambda^2}\cos 2\theta K_2\left(\frac{2\pi}{\lambda}r\right)\right\}.$$

366.] To draw the lines of electromotive intensity, we notice that if ds is an element of a curve in the dielectric, $d(H_1+H_2)/ds$ is proportional to the electromotive intensity at right angles to ds, so that the lines of electromotive intensity will be the lines

$$H_1+H_2=\text{a constant}.$$

When r/λ is small, this condition leads to the equation

$$\epsilon^{\frac{\iota 2\pi}{\lambda}Vt}\Big[1-\frac{\pi^2}{\lambda^2}r^2-\frac{2\pi^2c^2}{\lambda^2}K_0\Big(\frac{2\pi}{\lambda}r\Big)+\frac{2\iota\pi}{\lambda}\cos\theta\Big\{r+\frac{2\pi}{\lambda}c^2K_1\Big(\frac{2\pi}{\lambda}r\Big)\Big\}$$
$$-\frac{\pi^2}{\lambda^2}\cos 2\theta\Big\{r^2+\frac{2\pi^2c^4}{\lambda^2}K_2\Big(\frac{2\pi}{\lambda}r\Big)\Big\}\Big]=C,$$

where C is a constant.

Substituting the approximate values of K_0, K_1 and K_2 this becomes

$$\epsilon^{\frac{\iota 2\pi}{\lambda}Vt}\Big[1-\frac{\pi^2}{\lambda^2}r^2+\frac{2\pi^2c^2}{\lambda^2}\log(\pi r/\gamma\lambda)+\frac{2\iota\pi}{\lambda}\cos\theta\frac{(r^2+c^2)}{r}$$
$$-\frac{\pi^2}{\lambda^2}\cos 2\theta\Big(r^2+\frac{c^4}{r^2}\Big)\Big]=C.$$

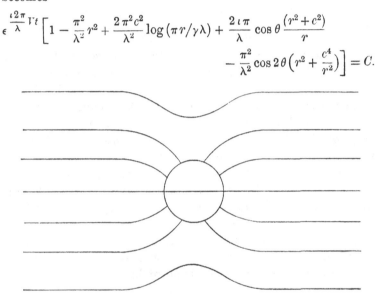

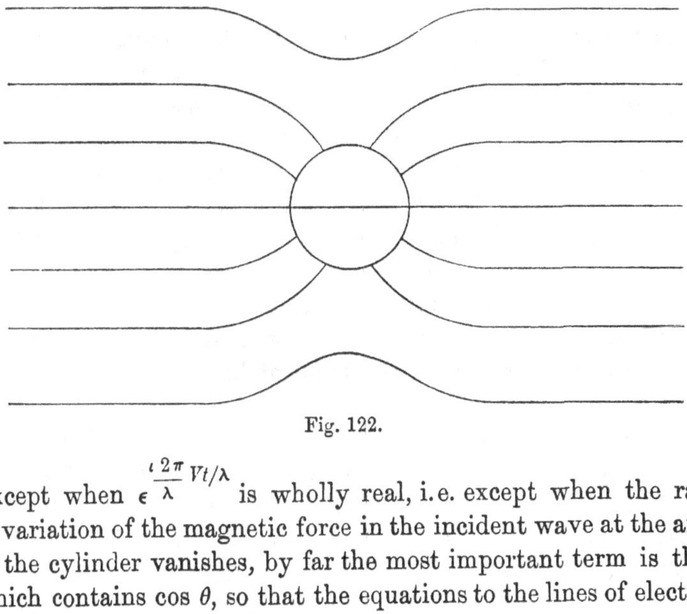

Fig. 122.

Except when $\epsilon^{\frac{\iota 2\pi}{\lambda}Vt/\lambda}$ is wholly real, i.e. except when the rate of variation of the magnetic force in the incident wave at the axis of the cylinder vanishes, by far the most important term is that which contains $\cos\theta$, so that the equations to the lines of electromotive intensity are

$$\frac{c^2+r^2}{r}\cos\theta = \text{a constant} = C', \text{ say.}$$

The lines of electromotive intensity are represented in Fig. 122.

At the times when $\epsilon^{\iota 2\pi Vt/\lambda}$ is wholly real, the lines are approximately circles concentric with the cross-section of the cylinder, since in this case the term involving the logarithm is the most important of the variable terms.

367.] When r is large compared with λ, we find by introducing the values of the K functions when the argument is very large, viz.

$$K_0(x) = \iota^{\frac{1}{2}} \left(\frac{\pi}{2x}\right)^{\frac{1}{2}} \epsilon^{-\iota x},$$

$$K_1(x) = -\iota^{\frac{3}{2}} \left(\frac{\pi}{2x}\right)^{\frac{1}{2}} \epsilon^{-\iota x},$$

$$H_2 = -\frac{\pi^2 c^2}{r^{\frac{1}{2}} \lambda^{\frac{3}{2}}} \epsilon^{\frac{\iota 2\pi}{\lambda}\left(Vt - r + \frac{\lambda}{8}\right)} (1 + 2\cos\theta),$$

retaining only the lowest powers of c/λ.

Thus the magnetic force in the scattered wave vanishes when $2\cos\theta = -1$, or in a direction making an angle of 120° with the incident ray. When the wave is scattered by an insulating cylinder Lord Rayleigh (l. c.) found that the magnetic intensity in the scattered ray was expressed by a similar formula *with the exception that the factor* $(1 + 2\cos\theta)$ *was replaced by* $\cos\theta$. Thus, if we take the case where the incident wave is a luminous one, the scattered light will vanish in the direction of the electric displacement when the particles are insulators, while it will vanish in a direction making an angle of 30° with this direction if the particles are metallic. If the incident light is not polarized, then with metallic particles the scattered light will be completely polarized in a direction making 120° with the direction of propagation of the incident light, while if the particles are insulators the direction in which the polarization is complete is at right angles to the direction of the incident light. The observations of Tyndall, Brücke, Stokes, and Lord Rayleigh afford abundant proof of the truth of the last statement: but no experiments seem to have been published on the results of the reflection of light from small metallic particles.

368.] The preceding results have also an important application to the consideration of the influence of the size of the reflector on the intensity of reflected electromagnetic waves. When the electromotive intensity is parallel to the axis of the cylinder, the most important term in the expression for the reflected wave only involves the radius of the cylinder as a logarithm, it will thus only vary slowly with the radius, so that in this case the size of the cylinder is of comparatively little importance: hence we may conclude that we shall get good reflection if the length of the reflector measured in the direction of the electromotive intensity is considerable, whatever may be the breadth of the reflector at right angles to the electromotive intensity. On the other hand, when the electromotive intensity is at right angles to the axis of the cylinder, the electromotive intensity in the scattered wave increases as the square of the radius of the cylinder, so that in this case the size of the reflector is all important. These results are confirmed by Trouton's experiments on 'The Influence the Size of the Reflector exerts in Hertz's Experiment,' *Phil. Mag.* [5], 32, p. 80, 1891.

On the Scattering of Electric Waves by Metallic Spheres.

369.] We shall proceed to discuss in some detail the problem of the incidence of a plane electric wave upon a metal sphere *.

If α, β, γ; f, g, h are respectively the components of the magnetic force and of the polarization in the dielectric which are radiated from the sphere, then if ψ stands for any one of these quantities it satisfies a differential equation of the form

$$\frac{d^2\psi}{dx^2}+\frac{d^2\psi}{dy^2}+\frac{d^2\psi}{dz^2}=\frac{1}{V^2}\frac{d^2\psi}{dt^2}, \qquad (1)$$

where V is the velocity with which electric action is propagated through the dielectric surrounding the sphere. If λ is the wave length of the disturbance incident upon the sphere, then the components of magnetic induction and of electric polarization will all vary as $\epsilon^{\iota\frac{2\pi}{\lambda}Vt}$; thus $V^{-2}d^2\psi/dt^2$ may be replaced by $-4\pi^2\psi/\lambda^2$, so that writing k for $2\pi/\lambda$, equation (1) may be written

$$\frac{d^2\psi}{dx^2}+\frac{d^2\psi}{dy^2}+\frac{d^2\psi}{dz^2}+k^2\psi=0,$$

a solution of which is by Art. 308,

$$\psi=\epsilon^{\iota kVt}\Sigma f_n(kr)S_n,$$

where r is the distance from the centre of the sphere. Since the waves of magnetic force and dielectric polarization are radiating outwards from the sphere

$$f_n(kr)=\left(\frac{1}{kr}\frac{d}{d(kr)}\right)^n\frac{\epsilon^{-\iota kr}}{kr},$$

* The scattering by an insulating sphere is discussed by Lord Rayleigh (*Phil. Mag.* 12, p. 98, 1881). The incidence of a plane wave on a sphere was the subject of a dissertation sent in to Trinity College, Cambridge, by Professor Michell in 1890. I do not know of any papers which discuss the special problem of the scattering by *metal* spheres.

S_n is a solid spherical harmonic of degree n. It should be noted that $f_n(kr)$ of this article is $(kr)^{-n}f(kr)$ of article 308.

370.] We shall now prove a theorem due to Professor Lamb (*Proc. Lond. Math. Soc.* 13, p. 189, 1881), that if α, β, γ satisfy equations of the form (1), and if

$$\frac{d\alpha}{dx} + \frac{d\beta}{dy} + \frac{d\gamma}{dz} = 0;$$

then the most general solution of these equations is given by

$$\left.\begin{aligned}
\alpha &= \Sigma\left\{(n+1)f_{n-1}(kr)\frac{d\omega_n}{dx} - nk^2r^{2n+3}f_{n+1}(kr)\frac{d}{dx}\frac{\omega_n}{r^{2n+1}}\right\} \\
&\qquad + \Sigma f_n(kr)\left(y\frac{d}{dz} - z\frac{d}{dy}\right)\omega'_n, \\
\beta &= \Sigma\left\{(n+1)f_{n-1}(kr)\frac{d\omega_n}{dy} - nk^2r^{2n+3}f_{n+1}(kr)\frac{d}{dy}\frac{\omega_n}{r^{2n+1}}\right\} \\
&\qquad + \Sigma f_n(kr)\left(z\frac{d}{dx} - x\frac{d}{dz}\right)\omega'_n, \\
\gamma &= \Sigma\left\{(n+1)f_{n-1}(kr)\frac{d\omega_n}{dz} - nk^2r^{2n+3}f_{n+1}(kr)\frac{d}{dz}\frac{\omega_n}{r^{2n+1}}\right\} \\
&\qquad + \Sigma f_n(kr)\left\{x\frac{d}{dy} - y\frac{d}{dx}\right\}\omega'_n.
\end{aligned}\right\} \quad (2)$$

where ω_n, ω'_n represent arbitrary solid spherical harmonics of degree n.

Since $$\frac{d\omega_n}{dx}, \qquad \frac{d}{dx}\frac{\omega_n}{r^{2n+1}}, \qquad \left(y\frac{d}{dz} - z\frac{d}{dy}\right)\omega'_n$$

are solid spherical harmonics of degrees $(n-1)$, $-(n+1)$, n respectively, we see that the expression given for α satisfies the differential equation (1); similarly this equation is satisfied by the values of β and γ.

Let us now find the value of $d\alpha/dx + d\beta/dy + d\gamma/dz$; we notice that the terms involving ω'_n vanish identically, and since

$$\nabla^2(\omega_n) = 0, \qquad \nabla^2\frac{\omega_n}{r^{2n+1}} = 0,$$

we have

$$\begin{aligned}
\frac{d\alpha}{dx} + \frac{d\beta}{dy} + \frac{d\gamma}{dz} &= \Sigma(n+1)\frac{k}{r}f'_{n-1}(kr)\left\{x\frac{d}{dx} + y\frac{d}{dy} + z\frac{d}{dz}\right\}\omega_n \\
&\quad - \Sigma\left[nk^3r^{2n+2}f'_{n+1}(kr) + n(2n+3)k^2r^{2n+1}f_{n+1}(kr)\right] \times \\
&\qquad\qquad \left\{x\frac{d}{dx} + y\frac{d}{dy} + z\frac{d}{dz}\right\}\frac{\omega_n}{r^{2n+1}} \\
&= \Sigma n.\overline{n+1}.\frac{k}{r}\left\{f'_{n-1}(kr) + k^2r^2f'_{n+1}(kr) + (2n+3)krf_{n+1}(kr)\right\}\omega_n.
\end{aligned}$$

Now

$$f_n(kr) = \left(\frac{1}{kr}\frac{d}{d(kr)}\right)^n \frac{\epsilon^{-\iota kr}}{kr},$$

hence

$$f'_{n-1}(kr) = krf_n(kr). \tag{3}$$

We have also

$$f''_n(kr) + \frac{2(n+1)}{kr} f_n'(kr) + f_n(kr) = 0,$$

which may be written as

$$\frac{d}{d(kr)}\{krf'_n(kr) + (2n+1)f_n(kr)\} = -krf_n(kr)$$
$$= -f'_{n-1}(kr) \text{ by (3)};$$

hence, since the constant of integration must vanish since all the f's involve $\epsilon^{-\iota kr}$,

$$krf'_n(kr) + (2n+1)f_n(kr) = -f_{n-1}(kr), \tag{4}$$

and by (101), Art. 308,

$$(2n+1)f_n(kr) = -\{f_{n-1}(kr) + k^2r^2f_{n+1}(kr)\}. \tag{5}$$

Writing $(n+1)$ for n in (4), we have

$$krf'_{n+1}(kr) + (2n+3)f_{n+1}(kr) = -f_n(kr)$$
$$= -\frac{f'_{n-1}(kr)}{kr}. \tag{6}$$

From this equation we see that

$$\frac{d\alpha}{dx} + \frac{d\beta}{dy} + \frac{d\gamma}{dz} = 0.$$

To prove that equation (2) gives the most general expressions for α, β, γ, we notice that the values of α, β may be written

$$\left.\begin{aligned} \alpha &= \Sigma f_n(kr)\left[\left\{(n+2)\frac{d\omega_{n+1}}{dx} - (n-1)k^2r^{2n+1}\frac{d}{dx}\frac{\omega_{n-1}}{r^{2n-1}}\right\}\right. \\ &\qquad\qquad\left. + \left(y\frac{d}{dz} - z\frac{d}{dy}\right)\omega'_n\right], \\ \beta &= \Sigma f_n(kr)\left[\left\{(n+2)\frac{d\omega_{n+1}}{dy} - (n-1)k^2r^{2n+1}\frac{d}{dy}\frac{\omega_{n-1}}{r^{2n-1}}\right\}\right. \\ &\qquad\qquad\left. + \left(z\frac{d}{dx} - x\frac{d}{dz}\right)\omega'_n\right]. \end{aligned}\right\} \tag{7}$$

The most general expressions for α, β, when they represent radiation outwards from the sphere, may however, Art. 308, be expressed in the form

$$\left.\begin{aligned} \alpha &= \Sigma f_n(kr)\,U_n, \\ \beta &= \Sigma f_n(kr)\,V_n, \end{aligned}\right\} \tag{8}$$

where U_n, V_n are solid spherical harmonics of degree n. Since

ω_n and ω'_n are arbitrary, we may determine them so as to make the values of α and β given by (7) agree with those given by (8). Thus (7) are sufficiently general expressions for α, β, and when α and β are given γ follows from the equation

$$\frac{d\alpha}{dx}+\frac{d\beta}{dy}+\frac{d\gamma}{dz}=0.$$

371.] If α, β, γ represent the components of the magnetic force, f, g, h the components of the electric polarization are, in a dielectric, given by the equations

$$4\pi\frac{df}{dt}=\frac{d\gamma}{dy}-\frac{d\beta}{dz},$$

$$4\pi\frac{dg}{dt}=\frac{d\alpha}{dz}-\frac{d\gamma}{dx},$$

$$4\pi\frac{dh}{dt}=\frac{d\beta}{dx}-\frac{d\alpha}{dy}.$$

Taking the values of β and γ given in (2), we see that the term in $4\pi df/dt$ involving ω_n is equal to

$$\left\{(n+1)\frac{k}{r}f'_{n-1}(kr)-nk^3rf'_{n+1}(kr)-n(2n+3)k^2f_{n+1}(kr)\right\}\times\left\{y\frac{d\omega_n}{dz}-z\frac{d\omega_n}{dy}\right\},$$

and this by equations (4) and (6) is equal to

$$(2n+1)k^2f_n(kr)\left\{y\frac{d\omega_n}{dz}-z\frac{d\omega_n}{dy}\right\}.$$

Let us now consider the term in $4\pi df/dt$ involving ω_n'; this equals

$$f_n(kr)\left\{-2\frac{d}{dx}-\left(x\frac{d}{dx}+y\frac{d}{dy}+z\frac{d}{dz}\right)\frac{d}{dx}\right\}\omega'_n$$
$$+f'_n(kr)k\left(x\frac{d}{dr}-r\frac{d}{dx}\right)\omega'_n$$

$$=-(n+1)f_n(kr)\frac{d\omega'_n}{dx}+rkf_n'(kr)\frac{nx}{r^2}\omega_n'-krf_n'(kr)\frac{d\omega_n'}{dx},$$

this by equations (4) and (6) equals

$$-\frac{n+1}{2n+1}\left\{(k^2r^2f_{n+1}(kr)+f_{n-1}(kr))-k^2r^2f_{n+1}(kr)\right\}\frac{d\omega_n'}{dx}$$
$$+rkf_n'(kr)\frac{nx}{r^2}\omega_n'$$

$$=\frac{1}{(2n+1)}\left\{(n+1)f_{n-1}(kr)\frac{d\omega_n'}{dx}-nk^2r^{2n+3}f_{n+1}(kr)\frac{d}{dx}\frac{\omega_n'}{r^{2n+1}}\right\}.$$

Thus if α, β, γ are given by (2), then we have

$$4\pi\frac{df}{dt}=\Sigma\frac{1}{2n+1}\left\{(n+1)f_{n-1}(kr)\frac{d\omega_n'}{dx}-nk^2r^{2n+3}f_{n+1}(kr)\frac{d}{dx}\frac{\omega_n'}{r^{2n+1}}\right\}$$
$$+\Sigma(2n+1)k^2f_n(kr)\left(y\frac{d\omega_n}{dz}-z\frac{d\omega_n}{dy}\right),$$

$$4\pi\frac{dg}{dt}=\Sigma\frac{1}{2n+1}\left\{(n+1)f_{n-1}(kr)\frac{d\omega_n'}{dy}-nk^2r^{2n+3}f_{n+1}(kr)\frac{d}{dy}\frac{\omega_n'}{r^{2n+1}}\right\}$$
$$+\Sigma(2n+1)k^2f_n(kr)\left(z\frac{d\omega_n}{dx}-x\frac{d\omega_n}{dz}\right),$$

$$4\pi\frac{dh}{dt}=\Sigma\frac{1}{2n+1}\left\{(n+1)f_{n-1}(kr)\frac{d\omega_n'}{dz}-nk^2r^{2n+3}f_{n+1}(kr)\frac{d}{dz}\frac{\omega_n'}{r^{2n+1}}\right\}$$
$$+\Sigma(2n+1)k^2f_n(kr)\left(x\frac{d\omega_n}{dy}-y\frac{d\omega_n}{dx}\right).$$

372.] In the plane electrical wave incident on the sphere, let us suppose that the electric polarization h_0 in the wave front is parallel to z and expressed by the equation

$$h_0=\epsilon^{\frac{\iota 2\pi}{\lambda}(Vt+x)}=\epsilon^{\iota k(Vt+x)},$$

where the axis of x is at right angles to the wave front.

We have to expand h_0 in the form

$$\epsilon^{\iota kVt}\,\Sigma A_nQ_n,$$

where Q_n is a zonal harmonic of degree n whose axis is the axis of x and A_n is a function of r which we have to determine.

Since
$$\epsilon^{\iota kx}=\Sigma A_nQ_n,$$

and since it satisfies the equation

$$\frac{d^2\psi}{dx^2}+\frac{d^2\psi}{dy^2}+\frac{d^2\psi}{dz^2}+k^2\psi=0,$$

and is finite when $r=0$, we see by Art. 308 that

$$A_n=A_n'S_n(kr)=A_n'(kr)^n\left\{\frac{1}{kr}\frac{d}{d(kr)}\right\}^n\frac{\sin kr}{kr},$$

where A_n' is independent of r.

Since
$$\frac{\sin kr}{kr}=1-\frac{k^2r^2}{3!}+\frac{k^4r^4}{5!}-\ldots,$$

we see that when kr is very small

$$A_n=(-1)^nA_n'\frac{(kr)^n}{(2n+1)(2n-1)\ldots 1}. \tag{9}$$

But if $x/r = \mu$, we have

$$\epsilon^{\iota kr\mu} = \Sigma A_n Q_n,$$

$$\therefore \quad \int_{-1}^{+1} \epsilon^{\iota kr\mu} Q_n \, d\mu = A_n \int_{-1}^{+1} Q_n^2 \, d\mu = \frac{2 A_n}{2n+1}.$$

The lowest power of kr on the left-hand side of this equation is the nth, the coefficient of this is equal to

$$\frac{\iota^n}{\underline{|n}} \int_{-1}^{+1} \mu^n Q_n \, d\mu = \frac{2\iota^n}{(2n+1)(2n-1)(2n-3)\ldots 1};$$

hence when kr is small we have

$$\frac{2\iota^n (kr)^n}{(2n+1)(2n-1)\ldots 1} = \frac{2 A_n}{2n+1}.$$

Comparing this equation with (9) we see that

$$A'_n = \frac{(2n+1)}{\iota^n},$$

$$A_n = \frac{2n+1}{\iota^n} S_n(kr),$$

so that

$$\epsilon^{\iota kr\mu} = \Sigma \frac{2n+1}{\iota^n} S_n(kr) Q_n.$$

This expression is given by Lord Rayleigh (*Theory of Sound*, ii. p. 239).

By equation (101) of Art. 308 we have

$$\frac{A_{n-1}}{2n-1} - \frac{A_{n+1}}{2n+3} = \frac{1}{\iota kr} A_n. \tag{10}$$

This can also be proved directly thus,

$$\frac{A_{n-1}}{2n-1} = \tfrac{1}{2} \int_{-1}^{+1} \epsilon^{\iota kr\mu} Q_{n-1} \, d\mu,$$

$$\frac{A_{n+1}}{2n+3} = \tfrac{1}{2} \int_{-1}^{+1} \epsilon^{\iota kr\mu} Q_{n+1} \, d\mu,$$

$$\frac{A_{n-1}}{2n-1} - \frac{A_{n+1}}{2n+3} = \left[\tfrac{1}{2} \frac{\epsilon^{\iota kr\mu}}{\iota kr} (Q_{n-1} - Q_{n+1})\right]_{-1}^{+1}$$
$$- \tfrac{1}{2} \frac{1}{\iota kr} \int_{-1}^{+1} \epsilon^{\iota kr\mu} \left(\frac{dQ_{n-1}}{d\mu} - \frac{dQ_{n+1}}{d\mu}\right) d\mu.$$

The terms within square brackets vanish, and since

$$\frac{dQ_{n-1}}{d\mu} - \frac{dQ_{n+1}}{d\mu} = -(2n+1) Q_n,$$

we have

$$\frac{A_{n-1}}{2n-1} - \frac{A_{n+1}}{2n+3} = \frac{1}{2\iota kr}\int_{-1}^{+1}(2n+1)\,\epsilon^{\iota kr\mu}\,Q_n\,d\mu$$

$$= \frac{A_n}{\iota kr}.$$

373.] It will be convenient to collect together the results we have obtained.

In the incident wave,

$$f_0 = 0, \qquad g_0 = 0, \qquad h_0 = \epsilon^{\iota kVt}\,\Sigma\,\frac{2n+1}{\iota^n}\,Q_n S_n(kr),$$

and therefore by Art. 9,

$$a_0 = 0, \qquad \gamma_0 = 0, \qquad \beta_0 = 4\pi h_0 V = 4\pi V\epsilon^{\iota kVt}\,\Sigma\,\frac{2n+1}{\iota^n}\,Q_n S_n(kr).$$

For the wave scattered by the sphere, omitting the time factor, we have since $d/dt = \iota kV$

$$4\pi\iota kVf = \Sigma\frac{1}{2n+1}\left\{(n+1)f_{n-1}(kr)\frac{d\omega'_n}{dx} - nk^2r^{2n+3}f_{n+1}(kr)\frac{d}{dx}\frac{\omega'_n}{r^{2n+1}}\right\}$$
$$+\Sigma(2n+1)k^2f_n(kr)\left(y\frac{d\omega_n}{dz} - z\frac{d\omega_n}{dy}\right),$$

$$4\pi\iota kVg = \Sigma\frac{1}{2n+1}\left\{(n+1)f_{n-1}(kr)\frac{d\omega'_n}{dy} - nk^2r^{2n+3}f_{n+1}(kr)\frac{d}{dy}\frac{\omega'_n}{r^{2n+1}}\right\}$$
$$+\Sigma(2n+1)k^2f_n(kr)\left(z\frac{d\omega_n}{dx} - x\frac{d\omega_n}{dz}\right),$$

$$4\pi\iota kVh = \Sigma\frac{1}{2n+1}\left\{(n+1)f_{n-1}(kr)\frac{d\omega'_n}{dz} - nk^2r^{2n+3}f_{n+1}(kr)\frac{d}{dz}\frac{\omega'_n}{r^{2n+1}}\right\}$$
$$+\Sigma(2n+1)k^2f_n(kr)\left(x\frac{d\omega_n}{dy} - y\frac{d\omega_n}{dx}\right).$$

$$a = \Sigma\left\{(n+1)f_{n-1}(kr)\frac{d\omega_n}{dx} - nk^2r^{2n+3}f_{n+1}(kr)\frac{d}{dx}\frac{\omega_n}{r^{2n+1}}\right\}$$
$$+\Sigma f_n(kr)\left(y\frac{d\omega'_n}{dz} - z\frac{d\omega'_n}{dy}\right),$$

$$\beta = \Sigma\left\{(n+1)f_{n-1}(kr)\frac{d\omega_n}{dy} - nk^2r^{2n+3}f_{n+1}(kr)\frac{d}{dy}\frac{\omega_n}{r^{2n+1}}\right\}$$
$$+\Sigma f_n(kr)\left(z\frac{d\omega'_n}{dx} - x\frac{d\omega'_n}{dz}\right),$$

$$\gamma = \Sigma\left\{(n+1)f_{n-1}(kr)\frac{d\omega_n}{dz} - nk^2r^{2n+3}f_{n+1}(kr)\frac{d}{dz}\frac{\omega_n}{r^{2n+1}}\right\}$$
$$+\Sigma f_n(kr)\left(x\frac{d\omega_n'}{dy} - y\frac{d\omega_n'}{dx}\right).$$

374.] To determine ω_n, ω_n', we shall assume that the sphere is a perfect conductor and therefore that the electromotive intensity, and therefore the electric polarization, is at right angles to the sphere. This condition is satisfied whatever the resistance of the sphere if the frequency is so great that $kV\mu a^2/\sigma$ is large; a being the radius of the sphere, σ its specific resistance, and μ its magnetic permeability. If R is the normal electromotive polarization, Θ that along a tangent to a meridian, Φ that along a parallel of latitude, then the condition

$$\frac{df}{dx} + \frac{dg}{dy} + \frac{dh}{dz} = 0$$

is equivalent to

$$\frac{d}{dr}(r^2 R) + \frac{1}{\sin\theta}\frac{d}{d\theta}(r\sin\theta\,\Theta) + \frac{1}{\sin\theta}\frac{d}{d\phi}(r\Phi) = 0;$$

but since Θ and Φ vanish all over the sphere, this, if a is the radius of the sphere, gives the condition

$$\frac{d}{dr}(r^2 R) = 0 \text{ when } r = a.$$

Now $$rR = x(f+f_0) + y(g+g_0) + z(h+h_0);$$

but

$$4\pi\iota kV(xf+yg+zh) = \Sigma\frac{n.(n+1)}{2n+1}\left(f_{n-1}(kr) + k^2r^2 f_{n+1}(kr)\right)\omega_n$$

$$= -\Sigma n.\overline{n+1}.f_n(kr)\,\omega_n', \text{ by equation (6).}$$

$$xf_0 + yg_0 + zh_0 = z\Sigma A_n Q_n, \text{ omitting the time factor.}$$

But if r, θ, ϕ are the polar coordinates of the point whose Cartesian coordinates are x, y, z,

$$z = r\sin\theta\sin\phi,$$

and $$Q_n = \frac{1}{2n+1}\left\{\frac{dQ_{n+1}}{d\mu} - \frac{dQ_{n-1}}{d\mu}\right\};$$

hence, if $\omega_n' = r^n Y_n'$ where Y_n' is a surface harmonic of degree n, the condition

$$\frac{d}{dr}(r^2 R) = 0 \text{ when } r = a$$

becomes

$$\frac{1}{4\pi\iota KV}\Sigma n.(n+1)Y_n'\frac{d}{da}(a^{n+1}f_n(ka)) = \sin\theta\sin\phi\,\Sigma Q_n\frac{d}{da}(a^2A_n)$$

$$= \sin\theta\sin\phi\,\Sigma\frac{dQ_n}{d\mu}\frac{d}{da}\left\{\frac{a^2A_{n-1}}{2n-1} - \frac{a^2A_{n+1}}{2n+3}\right\};$$

but $\sin\theta\sin\phi\dfrac{dQ_n}{d\mu}$ is a surface harmonic of degrees n, hence

$$Y_n' = \frac{\sin\theta\sin\phi\dfrac{dQ_n}{d\mu}\dfrac{d}{da}\left\{\dfrac{a^2A_{n-1}}{2n-1}-\dfrac{a^2A_{n+1}}{2n+3}\right\}}{\dfrac{n.n+1}{4\pi\iota kV}\cdot\dfrac{d}{da}(a^{n+1}f_n(ka))},$$

or by (10)

$$Y_n' = 4\pi V\sin\theta\sin\phi\frac{dQ_n}{d\mu}\frac{\dfrac{d(aA_n)}{da}}{n.n+1.\dfrac{d}{da}(a^{n+1}f_n(ka))},$$

and $\omega_n' = r^n Y_n'$.

375.] We now proceed to find ω_n. The line integral of the electromotive intensity taken round any closed curve is equal to the rate of diminution of the number of lines of magnetic induction passing through it: if we take as our closed curve one drawn on the surface of the sphere, we see, since the tangential electromotive intensity over the surface of the sphere vanishes, that the rate of diminution of the normal magnetic induction also vanishes; this condition, since the induction varies harmonically, is equivalent to the condition that the normal magnetic induction vanishes over the surface of the surface; hence when $r=a$, we have

$$x(a+a_0)+y(\beta+\beta_0)+z(\gamma+\gamma_0) = 0. \tag{1}$$

But when $r=a$,

$$xa+y\beta+z\gamma = \Sigma n.(n+1)(2n+1)f_n(ka)\,\omega_n,$$

$$\begin{aligned} xa_0+y\beta_0+z\gamma_0 &= 4\pi Vy\Sigma A_nQ_n \\ &= 4\pi Va\sin\theta\cos\phi\,\Sigma\left(\frac{A_{n-1}}{2n-1}-\frac{A_{n+1}}{2n+3}\right)\frac{dQ_n}{d\mu} \\ &= \frac{4\pi V}{\iota k}\sin\theta\cos\phi\,\Sigma A_n\frac{dQ_n}{d\mu}. \end{aligned}$$

Let $\omega_n = r^n Y_n$, where Y_n is a surface harmonic of degree n. Then we have

$$Y_n = \frac{4\pi Va^{-n}}{n.n+1.2n+1.\iota k}\sin\theta\cos\phi\frac{dQ_n}{d\mu}\frac{A_n}{f_n(ka)}.$$

376.] Substituting the values just found for ω_n, ω_n', we find that the values of $f, g, h, a, \beta, \gamma$ in the wave scattered by the sphere are, omitting the time factor, given by the equations

$$f = \Sigma \frac{1}{n \,.\, n+1 \,.} \frac{1}{\iota^n} \frac{1}{\iota k} \Bigg\{ (n+1) f_{n-1}(kr) \frac{d}{dx} \left(r^n \sin\theta \sin\phi \frac{dQ_n}{d\mu} \right)$$

$$- n k^2 r^{2n+3} f_{n+1}(kr) \frac{d}{dx} \left(\frac{\sin\theta \sin\phi \dfrac{dQ_n}{d\mu}}{r^{n+1}} \right) \Bigg\} \frac{\dfrac{d}{da}(aS_n(ka))}{\dfrac{d}{da}(a^{n+1} f_n(ka))}$$

$$- \Sigma \frac{2n+1}{\iota^n a^n . n . n+1 .} \cdot \frac{S_n(ka)}{f_n(ka)} f_n(kr) \left(y \frac{d}{dz} - z \frac{d}{dy} \right) \left(r^n \sin\theta \cos\phi \frac{dQ_n}{d\mu} \right),$$

$$g = \Sigma \frac{1}{n \,.\, n+1} \cdot \frac{1}{\iota^n} \frac{1}{\iota k} \Bigg\{ (n+1) f_{n-1}(kr) \frac{d}{dy} \left(r^n \sin\theta \sin\phi \frac{dQ_n}{d\mu} \right)$$

$$- n k^2 r^{2n+3} f_{n+1}(kr) \frac{d}{dy} \left(\frac{\sin\theta \sin\phi \dfrac{dQ_n}{d\mu}}{r^{n+1}} \right) \Bigg\} \frac{\dfrac{d}{da}(aS_n(ka))}{\dfrac{d}{da}(a^{n+1} f_n(ka))}$$

$$- \Sigma \frac{2n+1}{\iota^n a^n . n . n+1 .} \cdot \frac{S_n(ka)}{f_n(ka)} f_n(kr) \left(z \frac{d}{dx} - x \frac{d}{dz} \right) \left(r^n \sin\theta \cos\phi \frac{dQ_n}{d\mu} \right),$$

$$h = \Sigma \frac{1}{n \,.\, n+1 \,.} \frac{1}{\iota^n} \frac{1}{\iota k} \Bigg\{ (n+1) f_{n-1}(kr) \frac{d}{dz} \left(r^n \sin\theta \sin\phi \frac{dQ_n}{d\mu} \right)$$

$$- n k^2 r^{2n+3} f_{n+1}(kr) \frac{d}{dz} \left(\frac{\sin\theta \sin\phi \dfrac{dQ_n}{d\mu}}{r^{n+1}} \right) \Bigg\} \frac{\dfrac{d}{da}(aS_n(ka))}{\dfrac{d}{da}(a^{n+1} f_n(ka))}$$

$$- \Sigma \frac{2n+1}{\iota^n a^n . n . n+1 .} \cdot \frac{S_n(ka)}{f_n(ka)} f_n(kr) \left(x \frac{d}{dy} - y \frac{d}{dx} \right) \left(r^n \sin\theta \cos\phi \frac{dQ_n}{d\mu} \right);$$

$$a = 4\pi V \Sigma \frac{a^{-n}}{n \,.\, n+1 \,.} \frac{1}{\iota^n \iota k} \frac{S_n(ka)}{f_n(ka)} \Bigg\{ (n+1) f_{n-1}(kr) \frac{d}{dx} \left(r^n \sin\theta \cos\phi \frac{dQ_n}{d\mu} \right)$$

$$- n k^2 f_{n+1}(kr) r^{2n+3} \frac{d}{dx} \left(\frac{\sin\theta \cos\phi \dfrac{dQ_n}{d\mu}}{r^{n+1}} \right) \Bigg\}$$

$$+ 4\pi V \Sigma \cdot \frac{2n+1}{n \,.\, n+1 .} \cdot \frac{1}{\iota^n} \frac{\dfrac{d}{da}(aS_n(ka))}{\dfrac{d}{da}(a^{n+1} f_n(ka))} f_n(kr) \left(y \frac{d}{dz} - z \frac{d}{dy} \right) \left(r^n \sin\theta \sin\phi \frac{dQ_n}{d\mu} \right)$$

$$\beta = 4\pi V\Sigma \frac{a^{-n}}{n.n+1.\iota^n.\iota k} \frac{1}{f_n(ka)} \frac{S_n(ka)}{f_n(ka)} \Big\{(n+1)f_{n-1}(kr)\frac{d}{dy}\Big(r^n \sin\theta\cos\phi \frac{dQ_n}{d\mu}\Big)$$

$$-nk^2 f_{n+1}(kr)\, r^{2n+3} \frac{d}{dy}\left(\frac{\sin\theta\cos\phi\dfrac{dQ_n}{d\mu}}{r^{n+1}}\right)\Big\}$$

$$\pi V\Sigma \cdot \frac{2n+1}{n.n+1.} \cdot \frac{1}{\iota^n} \frac{\dfrac{d}{da}(aS_n(ka))}{\dfrac{d}{da}(a^{n+1}f_n(ka))} f_n(kr)\Big(z\frac{d}{dx} - x\frac{d}{dz}\Big)\Big(r^n \sin\theta\sin\phi\frac{dQ_n}{d\mu}\Big),$$

$$\gamma = 4\pi V\Sigma \frac{a^{-n}}{n.n+1} \frac{1}{\iota^n \iota k} \frac{S_n(ka)}{f_n(ka)} \Big\{(n+1)f_{n-1}(kr)\frac{d}{dz}\Big(r^n\sin\theta\cos\phi\frac{dQ_n}{d\mu}\Big)$$

$$-nk^2 f_{n+1}(kr)\, r^{2n+3}\frac{d}{dz}\left(\frac{\sin\theta\cos\phi\dfrac{dQ_n}{d\mu}}{r^{n+1}}\right)\Big\}$$

$$+4\pi V\Sigma \frac{2n+1}{n.n+1} \cdot \frac{1}{\iota^n} \frac{\dfrac{d}{da}(aS_n(ka))}{\dfrac{d}{da}(a^{n+1}f_n(ka)} \Big(x\frac{d}{dy} - y\frac{d}{dx}\Big)\Big(r^n\sin\theta\sin\phi\frac{dQ_n}{d\mu}\Big).$$

377.] These expressions give the solution of the problem of the scattering of a plane wave by a sphere of any size. The particular case when the radius of the sphere is very small compared with the wave length of the incident wave is of great importance. In this case ka is very small, and the approximate values of $S_n(ka)$, $f_n(ka)$ are, Art. 308, expressed by the equations

$$S_n(ka) = \frac{(-1)^n (ka)^n}{2n+1.2n-1\ldots 1},$$

$$f_n(ka) = (-1)^n\, 2n-1.2n-3\ldots 1\, \frac{\epsilon^{-\iota ka}}{(ka)^{2n+1}}.$$

Substituting these values in the preceding equations and retaining only the lowest powers of ka, we find, omitting the time factor,

$$f = k^5 a^3 \epsilon^{\iota ka} f_2(kr)\, xz + \frac{1}{2\iota} k^4 a^3 \epsilon^{\iota ka} f_1(kr)\, z,$$

$$g = k^5 a^3 \epsilon^{\iota ka} f_2(kr)\, yz,$$

$$h = \tfrac{1}{3} k^3 a^3 \epsilon^{\iota ka}\{2f_0(kr) + k^2(3z^2 - r^2) f_2(kr)\} - \frac{1}{2\iota} k^4 a^3 \epsilon^{\iota ka} f_1(kr)\, x;$$

$$\left.\begin{aligned}\alpha &= -4\pi V \tfrac{1}{2} k^5 a^3 \epsilon^{\iota k a} f_2(kr)\, xy + 4\pi V \iota k^4 a^3 \epsilon^{\iota k a} f_1(kr)\, y,\\ \beta &= -4\pi V \tfrac{1}{6} k^3 a^3 \epsilon^{\iota k a}\, \{2 f_0(kr) + k^2(3y^2 - r^2) f_2(kr)\}\\ &\qquad -4\pi V \iota k^4 a^3 \epsilon^{\iota k a} f_1(kr)\, x,\\ \gamma &= -4\pi V \tfrac{1}{2} k^5 a^3 \epsilon^{\iota k a} f_2(kr)\, yz.\end{aligned}\right\}$$

At a distance from the sphere, which is large compared with the wave length, kr is very large; we then have approximately

$$f_2(kr) = -\frac{\epsilon^{-\iota kr}}{k^3 r^3}, \qquad f_1(kr) = -\frac{\iota\epsilon^{-\iota kr}}{k^2 r^2}, \qquad f_0(kr) = \frac{\epsilon^{-\iota kr}}{kr}.$$

Substituting their value and introducing the time factor, we get

$$\left.\begin{aligned} f &= -\epsilon^{\iota k(Vt-(r-a))} \frac{k^2 a^3}{r} \left(\frac{xz}{r^2} + \tfrac{1}{2}\frac{z}{r}\right),\\ g &= -\epsilon^{\iota k(Vt-(r-a))} \frac{k^2 a^3}{r} \frac{yz}{r^2},\\ h &= \epsilon^{\iota k(Vt-(r-a))} \frac{k^2 a^3}{r} \left(1 - \frac{z^2}{r^2} + \tfrac{1}{2}\frac{x}{r}\right);\end{aligned}\right\}$$

$$\left.\begin{aligned} \alpha &= 4\pi V \epsilon^{\iota k(Vt-(r-a))} \frac{k^2 a^3}{r} \left\{\tfrac{1}{2}\frac{xy}{r^2} + \frac{y}{r}\right\},\\ \beta &= 4\pi V \epsilon^{\iota k(Vt-(r-a))} \frac{k^2 a^3}{r} \left\{\frac{y^2 - r^2}{2r^2} - \frac{x}{r}\right\},\\ \gamma &= 4\pi V \epsilon^{\iota k(Vt-(r-a))} \frac{k^2 a^3}{r} \tfrac{1}{2} yz.\end{aligned}\right\}$$

From these expressions we see that

$$xf + yg + zh = 0, \qquad x\alpha + y\beta + z\gamma = 0;$$

so that both the electric polarization and the magnetic induction are at right angles to the radius. We have also

$$f\alpha + g\beta + h\gamma = 0,$$

so that the electric polarization is at right angles to the magnetic induction. Taking the real part of the preceding expressions, we find

$$f^2 + g^2 + h^2 = \cos^2 k(Vt - (r-a)) \frac{k^4 a^6}{r^2} \left\{\left(\frac{x}{r} + \tfrac{1}{2}\right)^2 + \tfrac{3}{4}\frac{y^2}{r^2}\right\},$$

$$\alpha^2 + \beta^2 + \gamma^2 = (4\pi V)^2 \cos^2 k(Vt - (r-a)) \frac{k^4 a^6}{r^2} \left\{\left(\frac{x}{r} + \tfrac{1}{2}\right)^2 + \tfrac{3}{4}\frac{y^2}{r^2}\right\}.$$

Thus we see that the resultant magnetic induction is equal to $4\pi V$ times the resultant electric displacement. We could have deduced this result directly from Art. 9, since the Faraday tubes are moving outwards at right angles to themselves with the velocity V.

378.] We see from the expressions for the resultant electric polarization and the magnetic force that at the places where the scattered wave vanishes

$$x/r = -\tfrac{1}{2}, \qquad y = 0.$$

Thus the scattered light produced by the incidence of a plane polarized wave vanishes in the plane through the centre at right angles to the magnetic induction in the incident wave along a line, making an angle of 120° with the radius to the point at which the wave first strikes the sphere, and it does not vanish in any direction other than this. Thus if non-polarized waves of light or of electric displacement are incident upon a sphere, whose radius is small compared with the wave length of the incident vibration, the direction in which the scattered light is plane polarized will be inclined at an angle of 120° to the direction of the incident light. The scattering of light by small metallic spheres thus follows laws which are quite different from those which hold when the scattering is produced by non-conducting particles. In the latter case (see Lord Rayleigh, *Phil. Mag.* [5], 12, p. 81, 1881), when a ray of plane polarized light falls upon a small sphere, the scattered light vanishes at all points in the plane normal to the magnetic induction, where the radius vector makes an angle of 90°, and not 120°, with the direction of the incident light. Thus, when non-polarized light falls upon a small non-conducting sphere, the scattered light will be completely polarized at any point in a plane through the centre of the sphere at right angles to the direction of the incident light. When the light is scattered by a conducting sphere, the points at which the light is completely polarized are on the surface of a cone whose axis is the direction of propagation of the incident light and whose semi-vertical angle is 120°. The Faraday tubes given off by the conducting sphere form two sets of closed curves, which are separated by the surface of this cone. The momentum of these tubes being at right angles both to the magnetic induction and the electric polarization is radial, so that the energy emitted by the conducting sphere is, when we are considering a point whose distance from the centre is a large number of wave lengths, travelling radially outwards from the sphere.

At a point close to the sphere kr is very small, so that we have approximately

$$f_0(kr) = \frac{\epsilon^{-\iota kr}}{kr}, \qquad f_1(kr) = -\frac{\epsilon^{-\iota kr}}{k^3 r^3}, \qquad f_2(kr) = \frac{3\epsilon^{-\iota kr}}{k^5 r^5}.$$

Substituting these values in the expressions in Art. 377, we find that the components of the total electric polarization and magnetic induction, i.e. the polarization and induction scattered from the sphere plus that due to the incident wave, are given approximately by the equations

$$f = \frac{3a^3}{r^5} xz \cos kVt,$$

$$g = \frac{3a^3}{r^5} yz \cos kVt,$$

$$h = \left\{\frac{a^3}{r^5}(3z^2 - r^2) + 1\right\} \cos kVt;$$

$$a = -6\pi V \frac{a^3}{r^5} xy \cos kVt,$$

$$\beta = -2\pi V \left\{\frac{a^3}{r^5}(3y^2 - r^2) - 2\right\} \cos kVt,$$

$$\gamma = -6\pi V \frac{a^3}{r^5} yz \cos kVt.$$

Thus when $r = a$,

$$f = \frac{3xz}{a^2} \cos kVt, \qquad g = \frac{3yz}{a^2} \cos kVt, \qquad h = \frac{3z^2}{a^2} \cos kVt;$$

$$a = -6\pi V \frac{xy}{a^2} \cos kVt, \qquad \beta = 6\pi V \frac{(x^2 + z^2)}{a^2} \cos kVt,$$

$$\gamma = -6\pi V \frac{yz}{a^2} \cos kVt.$$

Thus at the surface of the sphere the resultant electric polarization is radial and proportional to z; there is thus a distribution of electricity over the sphere whose surface density varies as the distance of the point on the sphere from a plane through its centre parallel to the plane of polarization of the incident wave, —the plane of polarization being the plane at right angles to the electric polarization.

The magnetic induction at the surface of the sphere is tangential to the sphere and equal to

$$6\pi V \frac{1}{a} \{x^2+z^2\}^{\frac{1}{2}} \cos kVt;$$

it is thus proportional to the distance of a point on the surface of the sphere from the diameter of the sphere parallel to the magnetic force in the incident wave. The lines of magnetic force on the sphere are great circles all passing through this diameter.

Since the electric polarization is radial and the magnetic induction is tangential, the momentum due to the Faraday tubes which is at right angles to each of these quantities is tangential. The direction of the momentum is tangential to a series of small circles on the sphere whose planes are at right angles to the diameter of the sphere parallel to the magnetic induction in the incident wave.

Waves along Wires.

379.] If the electric potential at one end of a wire be made to vary harmonically so as at any time to be represented by $\cos pt$, the electromotive intensity, as we proceed along the wire, will be a harmonic function of the distance from the end of the wire; if the wave length of this harmonic distribution is λ, the velocity of propagation of the disturbance along the wire is defined to be $\lambda p/2\pi$. This velocity ought, if Maxwell's theory is true, to be equal to V, the velocity with which electrodynamic disturbances are propagated through air (see Art. 267). Indeed on this theory the effects observed do in reality travel through the air even though the wire is present, so that the introduction of the wire does not materially alter the physical conditions. The electrical vibrations considered in this chapter are all of very high frequency, being produced by the discharge of condensers through short discharging circuits. In this case (see Art. 269) the electromotive intensity in the region around the wire is at right angles to it, and we may suppose that the phenomena near the wire are due to radial Faraday tubes, with their ends on the wire travelling along it with the velocity of light.

380.] Considerable interest attaches to some experiments made by Hertz, which seemed to indicate that the velocity along the wire was considerably less than that through the air; and though later experiments have shown that this conclusion is erroneous, and that, as Maxwell's theory indicates, the two velocities are identical, Hertz's experiments are of great interest both from the methods used and the points they illustrate.

In these experiments Hertz (*Wied. Ann.* 34, p. 551, 1888) used the vibrator described in Art. 325. This was placed in a vertical plane; behind and parallel to one of the metal plates A, and insulated from it, was a metal plate B of equal area (see Fig. 123). A long wire was soldered to B and bent round so

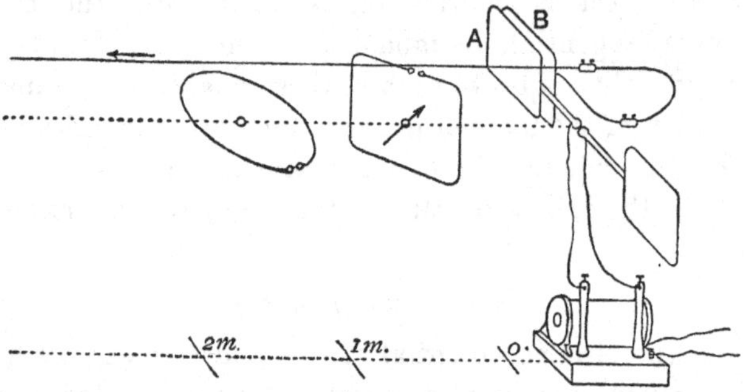

Fig. 123.

as to come in front of the vibrator and lie in the vertical plane of symmetry of the vibrator about a foot above the base line. The wire, which was above sixty metres long, was taken through a window, and was kept as far as possible from walls, &c., so as to avoid disturbances arising from reflected waves. In the first set of experiments the free end of the wire was insulated. The resonator used was the circular coil of wire 35 cm. in radius previously described. When the plane of the resonator was at right angles to the axis of the vibrator, the electromotive intensity due to the vibrator (apart from the action of the wire) did not (Art. 331) produce any tendency to spark in the resonator, so that the sparks in this position of the resonator must have been entirely due to the disturbance produced by the wire. To observe the effects due to the wire, the resonator was turned round in its own plane until the air gap was at the highest point, and therefore parallel to the wire. When the resonator was moved along the wire the following effects were observed. At the free end of the wire (which was insulated) the sparks in the resonator were extremely small, as the resonator was moved towards the vibrator the sparks increased and attained a maximum; they then decreased again until they almost vanished. If we call such a place a node, then, as the resonator moved along the wire, such nodes were found to occur at approximately equal intervals.

381.] Similar periodic effects were observed when the plane of the resonator was at right angles to the wire, the air gap being vertical; in such a position there would have been no sparks unless the wire had been present. On moving the resonator along the wire the brightness of the sparks changed in a periodic way: the positions however in which the sparks were brightest with the resonator in this position were those in which they had been dullest when the resonator was in its previous position.

This result is what we should expect from theoretical considerations. For when the resonator is in the first position, with its plane passing through the wire, the air gap is placed parallel to the wire. Now the Faraday tubes travelling along the wire are, as we saw Art. 269, at right angles to it and therefore to the air gap: thus the tubes which fall directly on the air gap do not tend to produce a spark; the sparks must be due to the tubes collected by the resonator and thrown by it into the air gap. The tubes which travel with their ends on the wire will be reflected from the insulated extremity of it, so that there will be tubes travelling in opposite directions along the wire; incident tubes travelling from the vibrator to the free end of the wire, and reflected tubes travelling back from the free end to the vibrator.

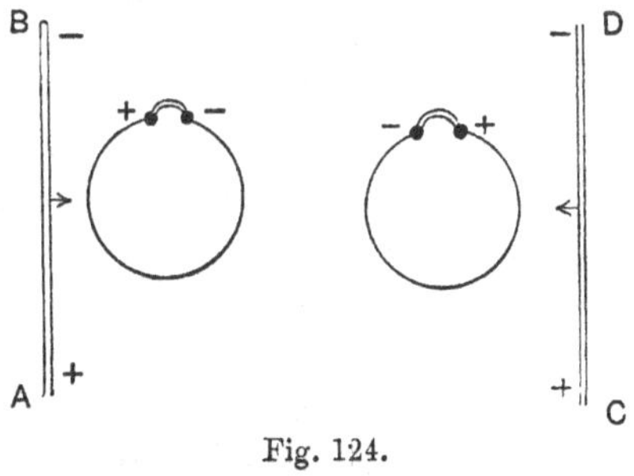

Fig. 124.

Let us now consider what will happen when the vibrator is in such a position as that represented in Fig. 124. The tube thrown into the air gap by a positive tube, such as *CD* proceeding from the vibrator, will be of opposite sign to that thrown by a positive tube, such as *AB* proceeding from the free end: thus in this position of the vibrator the positive tubes moving in opposite directions will neutralize each other's effects in producing sparks, though they increase the resultant electromotive intensity: thus, in this case, at the places where the electromotive intensity is greatest there will be no sparks in the resonator, for this maximum intensity will be due to two sets of tubes of the same sign, one set moving in one direction, the other in the opposite.

Since the free end of the wire has little or no capacity, no electricity can accumulate there, so that when one set of positive tubes arrives at the free end from the vibrator an equal number of positive tubes must start from the free end and move towards the vibrator; thus at the free end we have equal numbers of positive (or negative) tubes travelling in opposite directions. We should expect therefore that no sparks would be produced when the resonator was placed close to the free end; this, as we have seen, was found by Hertz to be the case.

When however the resonator is placed in the second position, with its plane at right angles to the wire, the conditions are very different; for the tubes which though they strike the resonator yet miss the air gap, are not hampered by the resonator in their passage through it; thus the resonator does not in this case collect tubes and throw them into the air gap. The sparks are now entirely due to the tubes which strike the air gap itself, and thus will be brightest at those points on the wire where the electromotive intensity is a maximum, while at such places, as we have seen, the sparks vanish when the resonator is in the former position.

382.] Hertz found that when the wire was cut at a node the nodes in the portion of the wire which remained were not altered in position, but that they were displaced when the wire was cut at any place other than a node.

Hertz also found that the distance between the nodes was independent of the diameter of the wire and of the material of which it was made, and that in particular the positions of the nodes were not affected by substituting an iron wire for a copper one.

The distance between the nodes is half the wave length along the wire; thus, if we know the period of the electrical vibrations of the system we can determine the velocity of propagation along the wire. Hertz, by using the formula $2\pi\sqrt{LC}$ for the wave length of the vibrations emitted by a condenser of capacity C, whose plates are connected by a discharging circuit whose coefficient of self-induction is L, came to the conclusion that the velocity of propagation along the wire was only about 2/3 of that through the dielectric; there are however many difficulties and doubtful points in the theoretical calculation of the period of vibration of such a system as Hertz's.

383.] Before discussing these we shall consider another method which Hertz used to compare directly the velocity of propagation along a wire with that through the air.

In this method interference was produced in the following way between the waves travelling out from the vibrator through the air and those travelling along the wire. The free end of the wire was put to earth so as to get rid of reflected waves along the wire, and as there were no metallic reflectors in the way of the waves proceeding directly through the air from the vibrator, the only reflected waves of this kind must have come from the floors or walls of the room; we shall assume for the present that there were no reflected air waves. The resonator was placed so that the air gap was at the highest point and vertically under the wire, and the plane of the resonator could rotate about a vertical axis passing through the middle of the air gap. When the plane of the resonator was at right angles to the wire, the waves proceeding along the latter had no tendency to produce a spark; any sparks that passed across the resonator must have been entirely due to the waves travelling from the vibrator through the air independently of the wire. In Hertz's experiments when the resonator was in this position the sparks were about 2 mm. long. On the other hand, when the resonator was twisted about the axis so that its plane passed through the wire and was at right angles to the axis of the vibrator, the direct waves through the air from the vibrator would have no tendency to produce sparks; which in this case must have been entirely due to the waves travelling along the wire. In Hertz's experiments when the resonator was in this position the sparks were again about 2 mm. long. When the resonator was in a position intermediate between these two, the sparks were due to the combined action of the waves travelling along the wire and those coming directly through the air. In such a case the brightness of the sparks would, in general, change when the plane of the vibrator was twisted through a considerable angle. If now the fronts of the two sets of waves were parallel and moving forward with the same velocity, then the effect of turning the plane of the vibrator through a definite angle in a definite direction would be the same at all points on the wire: if however the two waves were travelling at different rates the effect of turning the resonator would vary as it is moved from place to place along the wire.

384.] To prove this, let the electromotive intensity in the air gap due to the wave travelling along the wire be

$$A \cos \frac{2\pi}{\lambda}(Vt-z),$$

when the plane of the resonator passes through the wire; here the wire is taken as the axis of z, and λ is the wave length of the waves travelling along it. Then, when the plane of the resonator is twisted through an angle ϕ from this position, the electromotive intensity in the air gap due to the wire waves will be

$$A \cos \phi \cos \frac{2\pi}{\lambda}(Vt-z),$$

since the electromotive intensity is approximately proportional to the projection of the resonator on the plane through the wire and the base line of the vibrator.

Let the electromotive intensity in the air gap due to the waves coming from the vibrator independently of the wire be, when the plane of the resonator is at right angles to the wire,

$$B \cos \frac{2\pi}{\lambda'}(V't-(z-a)),$$

where λ' is the wave length and V' the velocity of the air waves; then, if the plane of the resonator is turned until it makes an angle ϕ with the plane through the wire and the base line, the electromotive intensity resolved parallel to the air gap is equal to

$$B \sin \phi \cos \frac{2\pi}{\lambda'}(V't-(z-a)).$$

Thus, considering both the air waves and those along the wire, the electromotive intensity when the resonator is in this position is equal to

$$A \cos \phi \cos \frac{2\pi}{\lambda}(Vt-z) + B \sin \phi \cos \frac{2\pi}{\lambda'}(V't-(z-a)),$$

which may, since V/λ is equal to V'/λ', be written as

$$R \cos \left\{\frac{2\pi}{\lambda} V(t+\epsilon)\right\},$$

where

$$R^2 = A^2 \cos^2\phi + B^2 \sin^2\phi$$
$$+ 2AB\cos\phi\sin\phi\cos\left\{\left(\frac{2\pi}{\lambda} - \frac{2\pi}{\lambda'}\right)z + \frac{2\pi}{\lambda'}a\right\}.$$

Now R is the maximum electromotive intensity acting on the air gap, and will be measured by the brightness of the spark. We see from the preceding expression that if $\lambda = \lambda'$, that is, if the velocity of the waves along the wire is the same as that of the air waves which are not affected by the wire, the last term in the expression for R^2 will cease to be a periodic function of z, so that in this case there will be no periodic change in the effect produced by a given rotation as we move the resonator along the wire. When however λ is not equal to λ', the effect on the spark length of a given rotation of the resonator will vary harmonically along the wire. Since in Hertz's experiments the sparks were about equally long in the two extreme positions, $\phi = 0$ and $\phi = \pi/2$, we may in discussing these experiments put $A = B$, and therefore

$$R^2 = A^2\left(1 + 2\cos\phi\sin\phi\cos\left\{\left(\frac{2\pi}{\lambda} - \frac{2\pi}{\lambda'}\right)z + \frac{2\pi}{\lambda'}a\right\}\right);$$

thus, if the resonator is rotated so that ϕ changes from $+\beta$ to $-\beta$, R^2 is diminished by

$$2A^2\sin 2\beta\cos\left\{\left(\frac{2\pi}{\lambda} - \frac{2\pi}{\lambda'}\right)z + \frac{2\pi}{\lambda'}a\right\}.$$

Thus when

$$\left(\frac{2\pi}{\lambda} - \frac{2\pi}{\lambda'}\right)z + \frac{2\pi}{\lambda'}a = (2n+1)\frac{\pi}{2},$$

that is, at places separated by the intervals

$$\tfrac{1}{2}\Big/\left\{\frac{1}{\lambda} - \frac{1}{\lambda'}\right\}$$

along the wire the rotation of the resonator will produce no effect upon the sparks, while on one side of one of these positions it will increase, on the other side diminish the brightness of the sparks. If λ' were very large compared with λ, that is, if the velocity of the waves travelling freely through the air were very much greater than that of those travelling along the wire, the distance between the places where rotation produces no effect would be $\frac{1}{2}\lambda$, which is the distance between the nodes observed in the experiments described in Art. 380. Hertz, however, came to the conclusion that the places where rotation produced no effect were separated by a much greater interval than the nodes. These he had determined to be about 2·8 metres apart, whereas the places where rotation produced no effect seemed to be separated by about 7·5 metres. Assuming these numbers we have

$$\lambda = 5{\cdot}6,$$

$$\tfrac{1}{2}\Big/\left(\frac{1}{\lambda} - \frac{1}{\lambda'}\right) = 7{\cdot}5\,;$$

hence $\lambda' = 8{\cdot}94$. Thus from these experiments the velocity of the free air waves would appear to be greater than those along the wire in the proportion of 8·94 to 5·6 or 1·6 to 1; or the velocity of the air waves is about half as large again as that of the wire waves.

We have, however, in the preceding investigations made several assumptions which it would be difficult to realise in practice; we have assumed, for example, that in the neighbourhood of the resonator the front of the air waves was at right angles to the wire. Since the resonator was close to the axis of the vibrator this assumption would be justifiable if there had been no reflection of the air waves from the walls or floors of the room. Since the thickness of the walls was small compared with the wave length it is not likely, unless they were very damp, that there would be much reflection from them; the case of the floor is however very different, and it is difficult to see how reflection from it could have been entirely avoided. Reflection from the floor would however introduce waves, the normals to whose fronts would make a finite angle with the wire. The electromotive intensity in the spark gap due to such waves would no longer be represented by a term of the form

$$\cos\left(2\pi(V't - z)/\lambda'\right),$$

but by one of the form

$$\cos\left(2\pi(V't - z\cos\theta)/\lambda'\right),$$

where θ is the angle between the normal to the wave front and the wire. Thus in the preceding investigation we must, for such waves, replace λ' by $\lambda'\sec\theta$, and their apparent wave length along the wire would be $\lambda'\sec\theta$ and not λ', so that the reflection would have the effect of increasing the apparent wave length of the air waves. The result then of Hertz's experiments that the wave length of the air waves, measured parallel to the wire, was greater than that of the wire waves, may perhaps be explained by the reflection of the waves from the floor of the room, without supposing that the velocity of the free air waves is different from that of those guided by the wire.

385.] The experiments of Sarasin and De la Rive (*Archives des Sciences Physiques et Naturelles Genève*, 1890, t. xxiii, p. 113) on the distance between the nodes (1) along a wire, (2) when produced by interference between direct air waves and waves reflected from a large metallic plate, seem to prove conclusively that the velocity of the waves guided by a wire is the same as that of free air waves. The experiments on the air waves have already been described in Art. 339; those on the wire waves were made in a slightly different way from Hertz's experiments.

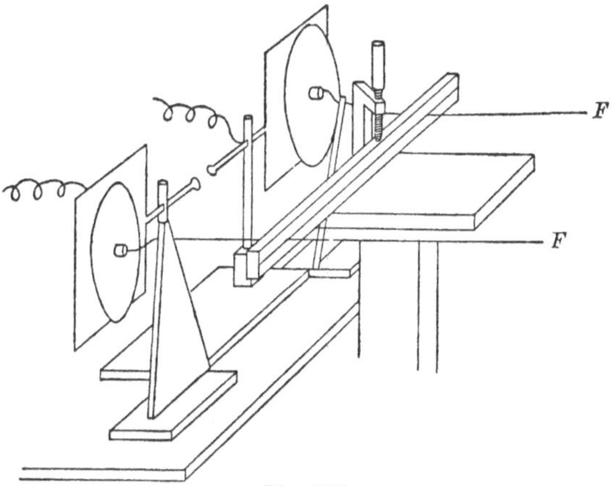

Fig. 125.

The method used by Sarasin and De la Rive is indicated in Fig. 125. Two metallic plates placed in front of the plates of the vibrator have parallel wires *F*, *F* soldered to them, the wires being of equal length and insulated. The plane of the resonator is at right angles to the wires, and the air gap is at the highest point, so that the air gap is parallel to the shortest distance between the wires. The resonator is mounted on a wagon by means of which it can be moved to and fro along the wires, while a scale on the bench along which the wagon slides enables the position of the latter to be determined. The resonator with its mounting is shown in Fig. 126. Sarasin and De la Rive found that as long as the same resonator was used the distance between the nodes as determined by this apparatus was the same as when the nodes were produced by the interference of direct air waves and those reflected from a metallic plate. The relative distances are given in the table in Art. 340, where 'λ for wire' indicates twice the distance between the nodes measured along the wire. They found with the wires, as later on they found for the air waves, that the distance between the nodes depended entirely upon the size of the resonator and not upon that of the vibrator; in fact the distance between the nodes was directly proportional to the diameter of the resonator; while it did not seem to depend to any appreciable extent upon the size of the vibrator. These peculiarities can be explained in the same way as the corresponding ones for the air waves, see Art. 341.

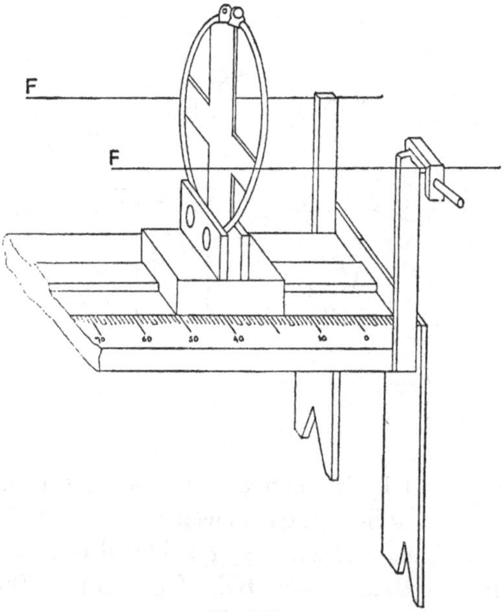

Fig. 126.

When the extremities of the wires remote from the vibrator are attached to large metallic plates, instead of being free, the electromotive intensity parallel to the plates at the ends must vanish; hence, whenever a bundle of positive Faraday tubes from the vibrator arrives at a plate an equal number of negative tubes must start from the plate and travel towards the vibrator, while, when the end of the wire is free, the tubes starting from the end of the wire in response to those coming from the vibrator are of the same sign as those arriving. Thus, when the end is free, the current vanishes and the electromotive intensity is a maximum, while when the end is attached to a large plate the electromotive intensity vanishes and the current is a maximum. Since the sparks in the resonator, when used as in Sarasin and De la Rive's experiments, are due to the tubes falling directly on the air gap, the sparks will be brightest when the electromotive intensity is a maximum, and will vanish when it vanishes; thus the loops when the ends are free will coincide with the nodes when the wires are attached to large plates. This was found by Sarasin and De la Rive to be the case.

A similar point arises in connection with the experiments with wires to that which was mentioned in Art. 342 in connection with the experiments on the air waves. The distance between the nodes, which is half the wave length of the vibration of the resonator, is, as is seen from the table in Art. 340, very approximately four times the diameter; if the resonator were a straight wire the half wave length would be equal to the length of the wire, and we should expect that bending the wire into a circle would tend to shorten the period, we should therefore have expected the distance between the nodes to have been a little less than the circumference of the resonator. Sarasin and De la Rive's experiments show however that it was 80 per cent. greater than this: it is remarkable however that the distance of the first node from the end of the wire, which is a loop, was always equal to half the circumference of the resonator, which is the value it would have had if the wave length of the vibration emitted by the resonator had been equal to twice its circumference.

386.] The experiments of Sarasin and De la Rive show that when vibrators of the kind shown in Fig. 113 are used, the oscillations which are detected by a circular resonator are those in the resonator rather than the vibrator.

Rubens, Paalzow, Ritter, and Arons (*Wied. Ann.* 37, p. 529, 1889; 40, p. 55, 1890; 42, pp. 154, 581, 1891) have used another method of measuring wave lengths, which though it certainly requires great care and labour, yet when used in a particular way would seem to give very accurate results. The method depends upon the change which takes place in the resistance of a wire when it is heated by the passage of a current through it. Rubens finds that the rapidly alternating currents induced by the vibrator can produce heat sufficient to increase the resistance of a fine wire by an amount which can be made to cause a considerable deflection in a delicate galvanometer.

387.] Rubens' apparatus, which is really a bolometer, is arranged as follows. Rapidly alternating currents pass through a very fine iron wire *L*. This wire forms one of the arms of a Wheatstone's Bridge provided with a battery and a galvanometer. When the rapidly alternating currents do not pass through *L* this bridge is balanced, and there is no deflection of the galvanometer. When however a rapidly alternating discharge passes

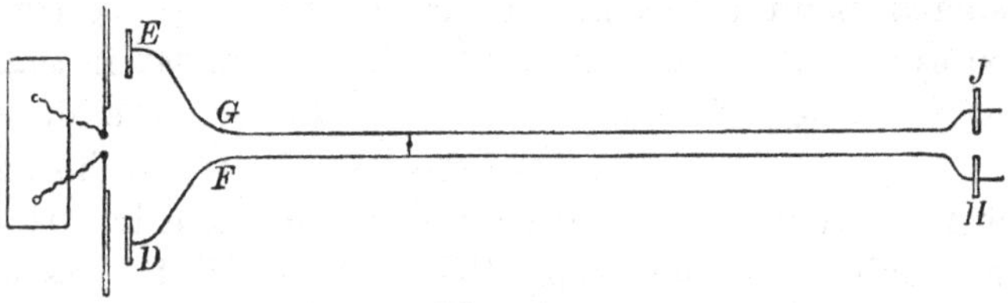

Fig. 127.

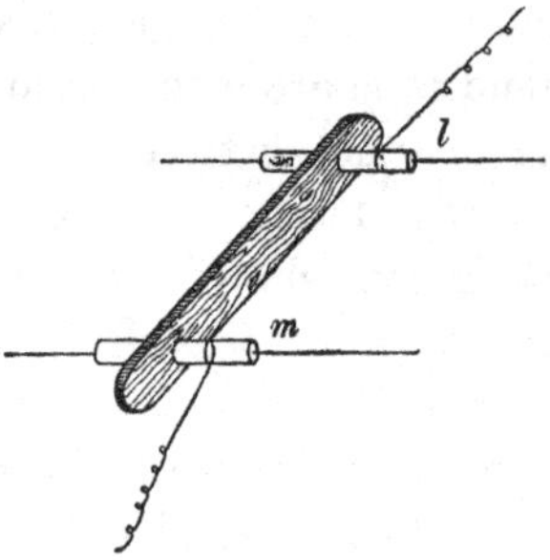

Fig. 128.

through the fine wire it heats it and so alters its resistance, and as the Bridge is no longer balanced the galvanometer is deflected. This arrangement is so sensitive that it is not necessary to place *L* in series with the wires connected with the plates of the vibrating system. Rubens found if a wire in series with *L* encircled, without touching, one of the wires *EJ*, *DH* in the experiment figured in Fig. 127 (Rubens, *Wied. Ann.* 42, p. 154, 1871), the deflection of the galvanometer was large enough to be easily measured. The apparatus was so delicate that a rise in temperature of 1/10,000 of a degree in the wire produced a deflection of a millimetre on the galvanometer scale. In one of his experiments the wire joined in series with L was bent round two pieces of glass tubing through which the wires EJ, DH passed, the plane of the turns round the glass tube being at right angles to the wires. In this case each turn of the wire and the wire it surrounds acted like a little Leyden jar, and the electricity which flowed through the wire L and disturbed the balance in the Bridge was due to the charging and discharging of these jars.

The pieces of glass tube were attached to a frame work, see Fig. 128, which was moved along the wire, and the deflection of the galvanometer observed as it moved along the wire. The

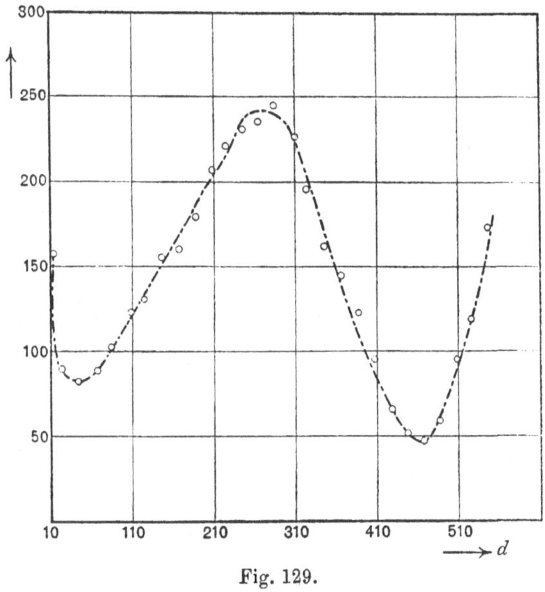

Fig. 129.

relation between the galvanometer deflection and the position of the tubes is shown in Fig. 129, where the ordinates represent the deflection of the galvanometer and the abscissae, the distance of the turns in the bolometer circuit from the point F in the wire. The curve shows very clearly the harmonic character of the disturbance along the wire.

388.] The results however of experiments of this kind were not very accordant, and in the majority of his experiments Rubens used another method which had previously been used by Lecher, who instead of a bolometer employed the brightness of the discharge through an exhausted tube as a measure of the intensity of the waves.

In these experiments the turns l, m (Fig. 128) in the bolometer circuit were kept at the ends J and H of the main wire (Fig. 127), while a metallic wire forming a bridge between the two parallel wires was moved along from one end of the wires to the other. The deflection of the bolometer depended on the position of the bridge, in the manner represented in Fig. 130, where the ordinates represent the deflection of the galvanometer, the abscissae the position of the bridge.

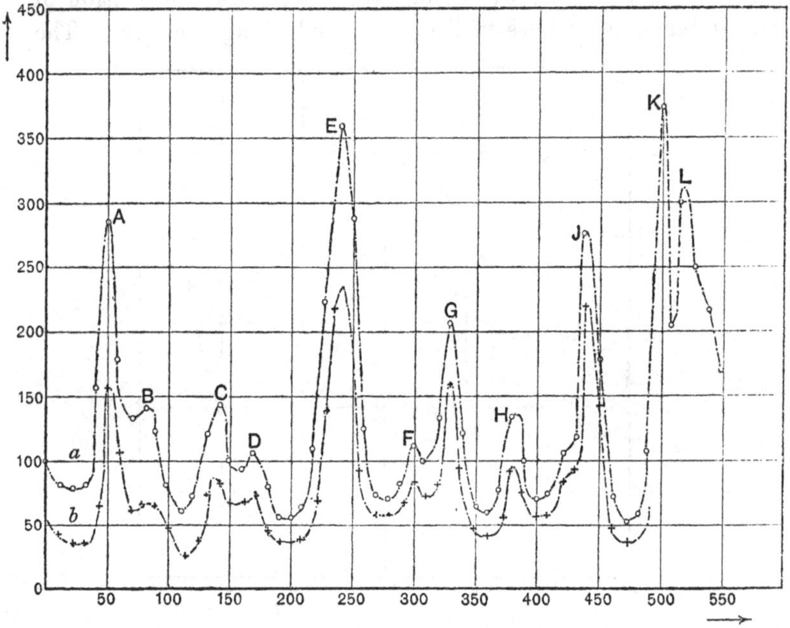

Fig. 130.

Rubens found that the positions of the bridge, in which the deflection of the galvanometer was a maximum, were independent of the length of the wire connecting the plates of the vibrator to the balls between which the sparks passed, and therefore of the period of vibration of the vibrator. This result shows that the vibrations in the wires which are detected by the bolometer cannot be 'forced' by the vibrator; for though, if this were the case, the deflection of the bolometer would vary with the position of the bridge, the places where the bridge produced a maximum deflection would depend upon the period of the vibrator. We can see this in the following way, if the bridge was at a place where the electromotive intensity at right angles to the wire vanished—which, if there were no capacity at the ends J, H, would be an odd number of quarter wave lengths from these ends—the introduction of the bridge would, since no current would flow through it, produce no diminution in the electromotive intensity at the ends J, H; in other positions of the bridge some of the current, which in its absence would go to the ends, would be diverted by the bridge, so that the electromotive intensity at the ends would be weakened. Thus, when the deflection of the bolometer was a maximum, the distances of the bridge from the ends J, H would be an odd multiple of a quarter of the wave length of the vibration travelling along the wire; thus, if these vibrations were 'forced' by the vibrator, the positions of the bridge which give a maximum deflection in the bolometer would depend upon the period of the vibrator. Rubens' experiments show that this was not the case.

We may therefore, as the result of these experiments, assume that the effect of the sparks in the vibrator is to give an electrical impulse to the wires and start the 'free' vibrations proper to them. The capacity of the plates at the ends of the wire makes the investigation of the free periods troublesome; we may however avail ourselves of the results of some experiments of Lecher's (*Wied. Ann.* 41, p. 850, 1890), who found that the addition of capacity to the ends might be represented by supposing the wires prolonged to an extent depending upon this additional capacity.

389.] Let AB, CD, Fig. 131, be the original wires, $A\alpha$, $B\beta$, $C\gamma$, $D\delta$ the amount by which they have to be prolonged to represent the capacity at the ends, we shall call the wires $\alpha\beta$, $\gamma\delta$ the 'equivalent' wires. Let PQ represent the position of the bridge.

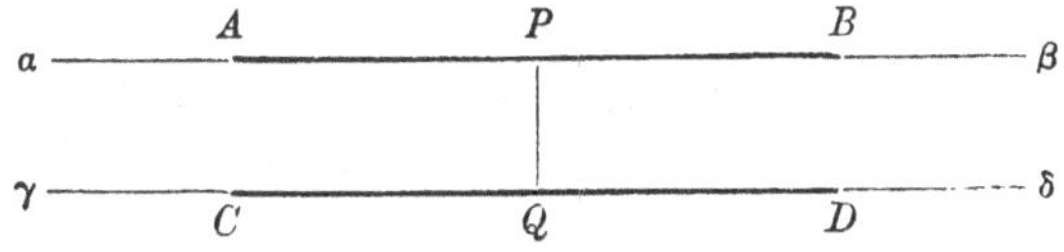

Fig. 131.

The electrical disturbance produced by the coil may start several systems of currents in the wires $\alpha\beta$, $\gamma\delta$. Then there may be a system of longitudinal currents along $\alpha\beta$, $\gamma\delta$ determined by the condition that the currents must vanish at α, β, and at γ, δ. Another system might flow round $\alpha PQ\gamma$, their wave length being determined by the condition that the currents along the wire must vanish at α and γ, and that by symmetry the electrification at these points must be equal and opposite. A third system of currents might flow round $\beta PQ\delta$, the flow vanishing at β and δ. If the bridge PQ were near the ends α, γ, we might expect, *a priori*, that the current in the circuit $\alpha PQ\gamma$ would be the most intense. Since the currents induced in the wires by the coil would tend to distribute themselves so that their self-induction should be as small as possible they would therefore tend to take the shortest course, i.e. that round the circuit $\alpha PQ\gamma$: these currents would induce currents round the circuit $\beta PQ\delta$. Lecher's experiments (*Wied. Ann.* 41, p. 850, 1890) show that the currents circulating round $\alpha PQ\gamma$, $\beta PQ\delta$ are much more efficacious in producing the electrical disturbance at the ends than the longitudinal ones along $\alpha\beta$, $\gamma\delta$. As a test of the magnitude of the disturbance at the ends, Lecher used an exhausted tube containing nitrogen and a little turpentine vapour; this was placed across the wires at the ends, and the brilliancy of the luminosity in the tube served as an indication of the magnitude of the electromotive intensity across $\beta\delta$. In one of his experiments Lecher used a bridge formed of two wires, PQ, $P'Q'$ in parallel, and moved this about until the luminosity in the tube was a maximum; he then cut the wires $\alpha\beta$, $\gamma\delta$ between PQ and $P'Q'$, so that the two circuits $\alpha PQ\gamma$, $\beta P'Q'\delta$ were no longer in metallic connection. Lecher found that this division of the circuit produced very little diminution in the brilliancy of the luminosity in the tube, though the longitudinal flow of the currents from α to β and from γ to δ must have been almost entirely destroyed by it. Lecher also found that the position of the bridge in which the luminosity of the tube was a maximum depended upon the length of the bridge; if the bridge were lengthened it had to be pushed towards, and if shortened away from the coil, to maintain the luminosity of the tube at its maximum value. He also found that, as might be expected, if the bridge were very short the tube at the end remained dark wherever the bridge was placed, while if the bridge were very long the tube was always bright whatever the position of the bridge. These experiments show that it is the currents round the circuits $\alpha PQ\gamma$, $\beta PQ\delta$ which chiefly cause the luminosity in the tube. Since the currents in the circuit $\beta PQ\delta$ are induced by those in the circuit $\alpha PQ\gamma$, they will be greatest when the time of the electrical vibration of the system $\alpha PQ\gamma$ is the same as that of $\beta PQ\delta$. The periods of vibration of these circuits are determined by the conditions that the current must vanish at their extremities and that these must be in opposite electrical conditions; these conditions entail that the wave lengths must be odd submultiples of the lengths of the circuit. If the two circuits are in unison the wave lengths must be the same, hence the ratio of the lengths of the two circuits must be of the form $(2n-1)/(2m-1)$, where n and m are integers.

This conclusion is verified in a remarkable way by Rubens' experiments with the bolometer. The relation between the deflections of the bolometer (the ordinates) and the distances of the bridge from G in Fig. 127 (the abscissae) is represented in Fig. 130. The length of the bridge in these experiments was 14 cm., that of the curved piece of the wire EG was 83 cm., and that of the straight portion GJ was 570 cm. The lengths Aa, $B\beta$ which had to be added to the wires to represent the effects of the capacity at the ends were assumed to be 55 cm. for the end of the wire next the coil, and 60 cm. for the end next the bolometer. These two lengths were chosen so as best to fit in with the observations, and were thus really determined by the measurements given in the following table; in spite of this, so many maxima were observed that the observations furnish satisfactory evidence of the truth of the theory just described.

m.	*n.*	$2m-1$.	$2n-1$.	*Distance of point of maximum deflection from G.*		*Corresponding point in Fig.* 130.
				Calculated.	*Observed.*	
2	1	3	1	51	50	*A*
4	2	7	3	89	86	*B*
3	2	5	3	148	143	*C*
4	3	7	5	181	182	*D*
2	2	3	3	246	245	*E*
3	4	5	7	311	305	*F*
2	3	3	5	343	334	*G*
2	4	3	7	402	386	*H*
1	2	1	3	441	443	*J*
1	3	1	5	506	503	*K*
1	4	1	7	529	523	*L*

Specific Inductive Capacity of Dielectrics in rapidly alternating Electric Fields.

390.] Methods analogous to those we have just described have been applied to determine the specific inductive capacities of dielectrics when transmitting electrical waves a few metres long.

One of the most striking results of Maxwell's *Electromagnetic Theory of Light* is the connection which it entails between the specific inductive capacity and the refractive index of a transparent body. On this theory the refractive index for infinitely long waves is (Maxwell's *Electricity and Magnetism*, vol. ii, Art. 786) equal to the square root of the specific inductive capacity of the dielectric under a steady electric field.

391.] Some determinations of K, the specific inductive capacity of various dielectrics in slowly varying fields, are given in the following table, which also contains the value of μ^2, the square of the refractive index for such dielectrics as are transparent. The letter following the value of μ^2 denotes the Frauenhofer line for which the refractive index is measured; when ∞ is affixed to the value of μ^2 the number denotes the square of the refractive index for infinitely long waves deduced from Cauchy's formula.

When μ is given by the observer of the specific inductive capacity this value has been used, in other cases μ has been taken from Landolt's and Börnstein's 'Physicalisch-Chemische Tabellen.'

Substance.	Observer.	K.	Temperature.	μ^2.
Glass, very light flint. . . .	Hopkinson [1]	6·57	...	2·375 D
„ light flint.	„	6·85	...	2·478 D
„ dense flint	„	7·4	...	2·631 D
„ extra dense flint . . .	„	10·1	...	2·924 D
„ hard crown	„	6·96		
„ plate	„	8·45		
Paraffin	„	2·29	...	2·022 ∞
Sulphur, along greatest axis	Boltzmann [2]	4·73	...	4·89 B
„ „ mean axis .	„	3·970	...	4·154 B
„ „ least axis .	„	3·811	...	3·748 B
„ non-crystalline . .	„	3·84		
Calcite, perpendicular to axis	Romich & Nowak [3]	7·7	...	2·734 A
„ along axis	„ „	7·5	...	2·197 A

[1] Hopkinson, *Phil. Trans.* 1878, Part I, p. 17, and *Phil. Trans.* 1881, Part II, p. 355.

[2] Boltzmann, *Wien. Berichte* 70, 2nd abth. p. 342, 1874.

[3] Romich and Nowak, *Wien. Berichte* 70, 2nd abth. p. 380, 1874.

Substance.	Observer.	*K*.	Temperature.	μ^2.
Fluor Spar	Romich & Nowak	6·7	...	2·050 *B*
Mica	Klemenčič[1]	6·64	...	2·526 *D*
Ebonite	Boltzmann[2]	3·15		
Resin	,,	2·55		
Quartz along optic axis	Curie[3]	4·55	...	2·41 *D*
,, perpendicular to axis	,,	4·49	...	2·38 *D*
Tourmaline along axis	,,	6·05	...	2·63 *D*
,, perpendicular to axis	,,	7·10	...	2·70 *D*
Beryl along axis	,,	6·24	...	2·48 *D*
,, perpendicular to axis	,,	7·58	...	2·50 *D*
Topaz	,,	6·56	...	2·61 *D*
Gypsum	,,	6·33	...	2·32 *D*
Alum	,,	6·4	...	2·2 *D*
Rock Salt	,,	5·85	...	2·36 *D*
Petroleum Spirit	Hopkinson[4]	1·92	...	1·922 ∞
Petroleum Oil, Field's	,,	2·07	...	2·075 ∞
,, ,, Common	,,	2·10	...	2·078 ∞
Ozokerite	,,	2·13	...	2·086 ∞
Turpentine, commercial	,,	2·23	...	2·128 ∞
Castor Oil	,,	4·78	...	2·153 ∞
Sperm Oil	,,	3·02	...	2·135 ∞
Olive Oil	,,	3·16	...	2·131 ∞
Neat's-foot Oil	,,	3·07	...	2·125 ∞
Benzene C_6H_6	Hopkinson[5]	2·38	...	2·2614 *D*
,, ,,	Negreano[6]	2·2988	25	2·2434 *D*
,, ,,	,,	2·2921	14	2·2686 *D*
Toluene C_7H_8	,,	2·242	27	2·224 *D*
,, ,,	,,	2·3013	14	2·245 *D*
Toluene	Hopkinson[5]	2·42	...	2·2470 *D*
Xylene C_8H_{10}	,,	2·39	...	2·2238 *D*
,, ,,	Negreano[6]	2·2679	27	2·219 *D*
Metaxylene C_8H_{10}	,,	2·3781	12	2·243 *D*
Pseudocumene C_9H_{12}	,,	2·4310	14	2·201 *D*
Cymene $C_{10}H_{14}$	,,	2·4706	19	2·201 *D*
,, ,,	Hopkinson[5]	2·25	...	2·2254 *D*
Terebenthine $C_{10}H_{16}$	Negreano[6]	2·2618	20	2·168 *D*
Carbon bisulphide	Hopkinson[5]	2·67	...	2·673 *D* (at 10°)
Ether	,,	4·75	...	1·8055 ∞
Amylene	,,	2·05	...	1·9044 *D*
Distilled Water	Cohn and Arons[7]	76·	15°?	1·779 *D*
,, ,,	Rosa[8]	75·7	25°	

[1] Klemenčič, *Wien. Berichte* 96, 2nd abth. p. 807, 1887.

[2] Boltzmann, *Wien. Berichte* 70, 2nd abth. p. 342, 1874.

[3] Curie, *Annales de Chimie et de Physique*, 6, 17, p. 385, 1889.

[4] Hopkinson, *Phil. Trans.* 1878, Part I, p. 17, and *Phil. Trans.* 1881, Part II, p. 355.

[5] Hopkinson, *Proc. Roy. Soc.* 43, p. 161, 1887.

[6] Negreano, *Compt. rend.* 104, p. 425, 1887.

[7] Cohn and Arons, *Wied. Ann.* 33, p. 13, 1888.

[8] Rosa, *Phil. Mag.* [5], 31, p. 188, 1891.

Substance.	Observer.	K.	Temperature.	μ^2.
Ethyl alcohol (98%)	Cohn and Arons [1]	26·5	...	1·831 ∞
Amyl alcohol	,, ,,	15·	...	1·951 ∞
Mixture of Xylene and Ethyl alcohol containing x parts of alcohol in unit volume				
x = ·00	,, ,,	2·36		
= ·09	,, ,,	3·08		
= ·17	,, ,,	3·98		
= ·30	,, ,,	7·08		
= ·40	,, ,,	9·53		
= ·50	,, ,,	13·0		
= 1·	,, ,,	26·5		

The values of K for the following gases at the pressure of 760 mm. of mercury are expressed in terms of that for a vacuum. In deducing them it has been assumed that for air at different pressures the changes in K are proportional to the changes in the pressure.

Gas.	Observer.	K.	Temperature.	μ^2.
Air	Boltzmann [2]	1·000590	0°	1·000588 D
,,	Klemenčič [3]	1·000586	0°	
Hydrogen	Boltzmann [2]	1·000264	0°	1·000278 D
,,	Klemenčič [3]	1·000264	0°	
Carbonic acid . . .	Boltzmann [2]	1·000946	0°	1·000908 D
,, ,,	Klemenčič [3]	1·000984	0°	
Carbonic oxide . .	Boltzmann [2]	1·00069	0°	1·00067 D
,, ,,	Klemenčič [3]	1·000694	0°	
Nitrous oxide . .	Boltzmann [2]	1·000994	0°	1·001032 D
,, ,,	Klemenčič [3]	1·001158	0°	
Olefiant gas	Boltzmann [2]	1 001312	0°	1·001356 D
Marsh gas	Boltzmann [2]	1·000944	0°	1·000886
Methyl alcohol . .	Lebedew [4]	1·0057	100°	
Ethyl alcohol . . .	,,	1·0065	100°	1·001745 D (at 0°)
Methyl formate . .	,,	1·0069	100°	
Ethyl formate . .	,,	1·0083	100°	
Methyl acetate . .	,,	1·0073	100°	
Ethyl ether	,,	1·0045	100°	
,, ,,	Klemenčič [3]	1·0074	0°	1·003048 D
Carbon bisulphide	,,	1·0029	0°	1·00296 D
Toluene	Lebedew [4]	1·0043	126°	
Benzene	,,	1·0027	100°	

[1] Cohn and Arons, *Wied. Ann.* 33, p. 13, 1888.
[2] Boltzmann, *Pogg. Ann.* 155, p. 403, 1875.
[3] Klemenčič, *Wien. Berichte* 91, 2nd abth. p. 712, 1885.
[4] Lebedew, *Wied. Ann.* 44, p. 288, 1891.

Ayrton and Perry (*Practical Electricity*, p. 310) found that the specific inductive capacity of a vacuum in which they estimated the pressure to be ·001 mm. was about ·994. This would make K for air referred to this vacuum as the unit about 1·006, while μ^2 from a vacuum to air is about 1·000588, there is thus a serious discrepancy between these values.

392.] We see from the above table that for some substances, such as sulphur, paraffin, liquid hydrocarbons, and the permanent gases, the relation $K = \mu^2$ is very approximately fulfilled; while for most other substances the divergence between K and μ^2 is considerable. When, however, we remember (1) that even when μ is estimated for infinitely long waves this is done by Cauchy's formula, and that the values so deduced would be completely invalidated if there were any anomalous dispersion below the visible rays, (2) that Maxwell's equations do not profess to contain any terms which would account for dispersion, the marvel is not that there should be substances for which the relation $K = \mu^2$ does not hold, but that there should be any for which it does. To give the theory a fair trial we ought to measure the specific inductive capacity for electrical waves whose wave length is the same as the luminous waves we use to determine the refractive index.

393.] Though we are as yet unable to construct an electrical system which emits electrical waves whose lengths approach those of the luminous rays, it is still interesting to measure the values of the specific inductive capacity for the shortest electrical waves we can produce.

We can do this by a method used by von Bezold (*Pogg. Ann.* 140, p. 541, 1870) twenty years ago to prove that the velocity with which an electric pulse travels along a wire is independent of the material of wire, it was also used by Hertz in his experiments on electric waves.

This method is as follows. Let $ABCD$ be a rectangle of wires with an air space at EF in the middle of CD; this rectangle is connected to one of the poles of an induction coil by a wire attached to a point K in AB, then if K is at the middle of AB the pulse coming along the wire from the induction coil will divide at K and will travel to E and F, reaching these points simultaneously; thus E and F will be in similar electric states and there will be no tendency to spark across the air gap EF.

If now we move K to a position which is not symmetrical with respect to E and F, then, when a pulse travels along the rectangle, it will reach one of these points before the other; their electric states will therefore be different and there will be a tendency to spark.

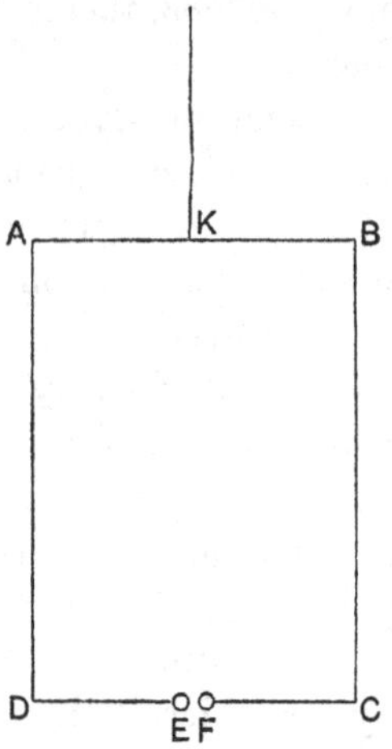

Fig. 132.

Suppose that with K at the middle point of AB, we insert BC into a dielectric through which electromagnetic disturbances travel more slowly than they do through air, then the pulse which goes round AD will arrive at E before the pulse which goes round BC arrives at F; thus E and F will not be in the same electrical state and sparks will therefore pass across the air space. To get rid of the sparks we must either move K towards B or else keep K fixed and, as the waves travel more slowly through the dielectric than through air, lengthen the side AD of the figure. If we do this until the sparks disappear we may conclude that E and F are in similar electric states, and therefore that the time taken by the pulse to travel round one arm of the circuit is the same as that round the other. By seeing how much the length of the one arm exceeds that of the other we can compare the velocity of electromagnetic action through the dielectric in which BC is immersed with that through air.

394.] I have used (*Phil. Mag.* [5], 30, p. 129, 1890) this method to determine the velocity of propagation of electromagnetic action through paraffin and sulphur. This was done by leading one of the wires, say BC, through a long metal tube filled with either paraffin or sulphur, the wire being insulated from the tube which was connected to earth. By measuring the length of wire it was necessary to insert in AD to stop the sparks, I found that the velocities with which electromagnetic action travels through sulphur and paraffin are respectively 1/1·7 and 1/1·35 of the velocity through air. The corresponding values of the specific inductive capacities would be about 2·9 and 1·8.

395.] Rubens and Arons (*Wied. Ann.* 42, p. 581; 44, p. 206), while employing a method based on the same principles, have made it very much more sensitive by using a bolometer instead of observing the sparks and by using two quadrilaterals instead of one. The arrangement they used is represented in Fig. 133 (*Wied. Ann.* 42, p. 584).

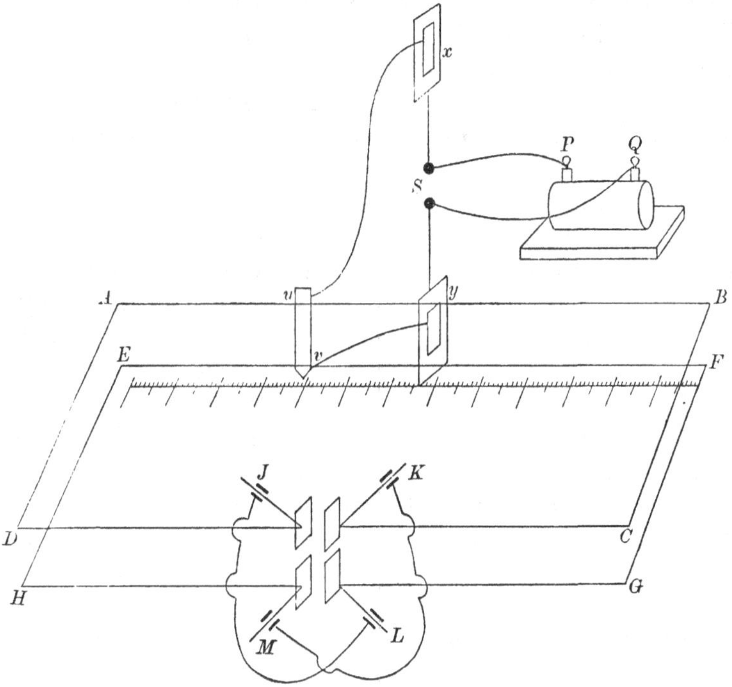

Fig. 133.

The poles P and Q of an induction coil are connected to the balls of a spark gap S, to each of these balls a metal plate, 40 cm. square, was attached by vertical brass rods 15 cm. long.

Two small tin plates x, y, 8 cm. square, were placed at a distance of between 3 and 4 cm. from the large plates. Then wires connected to these plates made sliding contacts at u and v with the wire rectangles $ABCD$, $EFGH$ ·230 cm. by 35 cm. One of these rectangles was placed vertically over the other, the distance between them being 8 cm. The points u, v were connected with each other by a vertical wooden rod, ending in a pointer which moved over a millimetre scale. The direct action of the coil on the rectangles was screened off by interposing a wire grating through which the wires $u\,x$, $v\,y$ were led. The wires CD, GH were cut in the middle and the free ends were attached to small metal plates 5.5 cm. square; metal pieces attached to these plates went between the plates of the little condensers J, K, L, M, the plates of these condensers were attached cross-wise to each other as in the figure. The two wires connecting the plates were attached to a bolometer circuit similar to that described in Art. 387. By means of a sliding coil attached to the bolometer circuit, Arons and Rubens investigated the electrical condition of the circuits $uADJ$, $uBCK$, &c., and found that approximately there was a node in the middle and a loop at each end; these circuits then may be regarded as executing electrical vibrations whose wave lengths are twice the lengths of the circuits. If the times of vibrations of the circuits on the left of u, v are the same as those on the right, the plates J and K will be in similar electrical states, as will also L and M, and there will be no deflection of the galvanometer in the bolometer circuit. When the wires are surrounded by air this will be when u, v are at the middle points of AB, EF. In practice Arons and Rubens found that the deflection of the galvanometer never actually vanished, but attained a very decided minimum when u, v were in the middle, and that the effect produced by sliding u, v through 1 cm. could easily be detected.

To determine the velocity of propagation of electromagnetic action through different dielectrics, one of the short sides of the rectangles was made so that the wires passed through a zinc box, 18 cm. long, 13 cm. broad, and 14 cm. high; the wires were carefully insulated from the box; the wires outside the box were straight, but the part inside was sometimes straight and sometimes zigzag. This box could be filled with the dielectric under observation, and the velocity of propagation of the electromagnetic action through the dielectric was deduced from the alteration made in the null position (i. e. the position in which the deflection of the galvanometer in the bolometer circuit was a minimum) of uv by filling the box with the dielectric.

Let p_1 and p_2 be the readings of the pointer attached to uv when a straight wire of length D_g and a zigzag of length D_k are respectively inserted in the box, the box in this case being empty. Then since in each case the lengths of the circuits on the right and left of uv must be the same, the difference in the lengths of the circuits on the left, when the straight wire and the zigzag respectively are inserted, must be equal to the difference in the lengths of the circuits on the right. The length of the circuit on the left when the zigzag is in exceeds that when the straight wire is in by

$$(D_k - p_2) - (D_g - p_1),$$

while the difference in the length of the circuits on the right is

$$p_2 - p_1;$$

hence $$D_k - D_g - (p_2 - p_1) = p_2 - p_1,$$

or $$D_k - D_g = 2(p_2 - p_1).$$

When the wires are surrounded by the dielectric, Arons and Rubens regard them as equivalent to wires in air, whose lengths are nD_g and nD_k, where n is the ratio of the velocity of transmission of electromagnetic action through air to that through the dielectric; for the time taken by a pulse to travel over a wire of length nD_g in air, is the same as that required for the pulse to travel over the length D_g in the dielectric. We shall return to this point after describing the results of these experiments. If p_3 and p_4 are the readings for the null positions of uv when the box is filled with the dielectric, then we have, on Arons and Rubens' hypothesis,

$$n(D_k - D_g) = 2(p_4 - p_3);$$

or, eliminating $D_k - D_g$, $$n = \frac{p_4 - p_3}{p_2 - p_1};$$

hence, if p_1, p_2, p_3, p_4 are determined, the value of n follows immediately.

In this way Arons and Rubens found as the values of n for the following substances:—

	n.	$\sqrt{K}$.
Castor Oil . . .	2·05	2·16
Olive Oil . . .	1·71	1·75
Xylol	1·50	1·53
Petroleum . . .	1·40	1·44

The values of K, the specific inductive capacity in a slowly varying field, were determined by Arons and Rubens for the same samples as they used in their bolometer experiments.

396.] The method used by Arons and Rubens to reduce their observations leads to values of the specific inductive capacity which are in accordance with those found by other methods. It is however very difficult to see, using any theory of the action of the divided rectangle that has been suggested, why the values of the specific inductive capacity should be accurately deduced from the observations by this method, except in the particular case when the wires outside the box are very short compared with the wave length of the electrical vibrations.

Considering the case of the single divided rectangle, there seem to be three ways in which it might be supposed to act. We may suppose that a single electrical impulse comes to K (Fig. 132), and there splits up into two equal parts, one travelling round AD to E, the other round BC to F. If these impulses arrived at E and F simultaneously they would, if they were of equal intensity, cause the electric states of E and F to be similar, so that there would be no tendency to spark across the gap EF. Thus, if the pulses arrived at E and F undiminished in intensity, the condition for there to be no spark would be that the time taken by a pulse to travel from K to E should be equal to that from K to F. This reasoning is not applicable however when the pulse in its way round one side of the circuit passes through regions in which its velocity is not the same as when passing through air, because in this case the pulse will be partly reflected as it passes from one medium to another, and will therefore proceed with diminished intensity. Thus, though this pulse may arrive at the air gap at the same time as the pulse which has travelled round the other side of the rectangle, it will not have the same intensity as that pulse; the electrical conditions of the knobs will therefore be different, and there will therefore be a tendency to spark. When the pulse has to travel through media of high specific inductive capacity the reflection must be very considerable, and the inequality in the pulses on the two sides of the air gap so great that we should not expect to get under any circumstances such a diminution in the intensity of the sparks as we know from experience actually takes place. We conclude therefore that this method of regarding the action of the rectangle is not tenable.

397.] Another method of regarding the action is to look on the rectangle as the seat of vibrations, whose period is determined by the electrical system with which it is connected. Thus we may regard the potential at K as expressed by $\phi_0 \cos pt$;

then the condition that there should be no sparks is that the potentials at E and F should be the same. We can deduce the expressions for the potentials at E and F from that at K when E and F are nodes or loops. Let us consider the case when the capacity of the knobs E and F is so small that the current at E and F vanishes. Then we can easily show by the method of Art. 298 that if there is no discontinuity in the current along the wire, and if the self-induction per unit length of the wire is the same at all points in $KADE$, and if the portions AK, DF are in air while AD is immersed in a dielectric in which the velocity of propagation of electromagnetic action is V', that through air being V, then if the potential at K is $\phi_0 \cos pt$, that at F is equal to

$$\frac{\phi_0 \cos pt}{\Delta},$$

where $\Delta = \cos\left(\frac{p}{V'}AD\right)\cos\frac{p}{V}(KA+DF) - \sin\left(\frac{p}{V'}AD\right)\times$

$$\left\{\mu \sin\left(\frac{p}{V}DF\right)\cos\left(\frac{p}{V}KA\right) + \frac{1}{\mu}\sin\left(\frac{p}{V}KA\right)\cos\left(\frac{p}{V}DF\right)\right\},$$

and $\mu = V/V'$.

The potential at E is

$$\frac{\phi_0 \cos pt}{\cos\frac{p}{V}KE}$$

if KE represents the total length $KB+BC+CE$, the whole of which is supposed to be surrounded by air. Hence, if the potentials at E and F are the same, we have

$$\cos\left(\frac{p}{V'}AD\right)\cos\frac{p}{V}(KA+DF) - \sin\left(\frac{p}{V'}AD\right)\times$$
$$\left\{\mu \sin\left(\frac{p}{V}DF\right)\cos\left(\frac{p}{V}KA\right) + \frac{1}{\mu}\sin\left(\frac{p}{V}KA\right)\cos\left(\frac{p}{V}DF\right)\right\}$$
$$= \cos\frac{p}{V}KE. \qquad (1)$$

To make the interpretation of this equation as simple as possible, suppose $KA = DF$, equation (1) then becomes

$$\cos\left(\frac{p}{V'}AD\right)\cos\left(\frac{2p}{V}KA\right) - \left(\mu+\frac{1}{\mu}\right)\tfrac{1}{2}\sin\left(\frac{p}{V'}AD\right)\sin\left(\frac{2p}{V}KA\right)$$
$$= \cos\left(\frac{p}{V}KE\right). \qquad (2)$$

Let us now consider one or two special cases of this equation. Let us suppose that AD is so small that $\left(\mu+\frac{1}{\mu}\right)\sin\left(\frac{p}{V'}AD\right)$ is a small quantity, then equation (2) may be written approximately

$$\cos\left\{\frac{2p}{V}KA+\tfrac{1}{2}\left(\mu+\frac{1}{\mu}\right)\frac{p}{V'}AD\right\}=\cos\left(\frac{p}{V}KE\right);$$

hence
$$2KA+\tfrac{1}{2}(\mu^2+1)AD=KE,$$

therefore
$$\frac{\delta KE}{\delta AD}=\tfrac{1}{2}(\mu^2+1),$$

so that in this case the process which Arons and Rubens applied to their measurements would give $(\mu^2+1)/2$ and not μ.

If, on the other hand, KA is so small that $\left(\mu+\frac{1}{\mu}\right)\sin\frac{2p}{V}KA$ is small, equation (2) may be written approximately

$$\cos\left\{\frac{p}{V'}AD+\left(\mu+\frac{1}{\mu}\right)\frac{p}{V}KA\right\}=\cos\left(\frac{p}{V}KE\right),$$

or
$$\mu AD+\left(\mu+\frac{1}{\mu}\right)KA=KE,$$

so that
$$\frac{\delta KE}{\delta AD}=\mu,$$

and in this case Arons and Rubens' process gives the correct result.

398.] A third view of the action of the rectangle, which seems to be that taken by Arons and Rubens, is that the vibrations are not forced, but that each side of the rectangle executes its natural vibrations independently of the other. If the extremities are to keep in the same electrical states, then the times of vibration of the two sides must be equal.

Arons and Rubens' measurements with the bolometer show that there is a loop at K and nodes at E and F.

Now if $2\pi/p$ is the time of vibration of a wire such as $KADF$ with a node at F and a loop at K, surrounded by air along KA, DF, and along AD by a medium through which electromagnetic action travels with the velocity V', then we can show by a process similar to that in Art. 298 that p is given by the equation

$$\frac{1}{\mu}\cot\left(\frac{p}{V'}AD\right)-\frac{1}{\mu}\cot\left(\frac{p}{V'}AD\right)\cot\left(\frac{p}{V}KA\right)\cot\left(\frac{p}{V}DF\right)$$
$$+\frac{1}{\mu^2}\cot\left(\frac{p}{V}KA\right)+\cot\left(\frac{p}{V}DF\right)=0. \qquad (3)$$

Let us take the case when $KA = DF$, then this equation becomes

$$\cot\left(\frac{p}{V'}AD\right) = \left(\mu + \frac{1}{\mu}\right)\frac{\cot\left(\frac{p}{V}KA\right)}{\cot^2\left(\frac{p}{V}KA\right)-1}$$

$$= \tfrac{1}{2}\left(\mu + \frac{1}{\mu}\right)\tan\left(\frac{2p}{V}KA\right),$$

or

$$\cot\left(\frac{2p}{V}KA\right) = \tfrac{1}{2}\left(\mu + \frac{1}{\mu}\right)\tan\frac{p}{V'}AD. \quad (4)$$

Let us consider the special case when $p.AD/V'$ is small, the solution of (4) is then

$$\frac{2p}{V}KA = \frac{\pi}{2} - \tfrac{1}{2}\left(\mu + \frac{1}{\mu}\right)\frac{p}{V'}AD,$$

or

$$\frac{p}{V}\left\{2KA + \frac{\mu^2+1}{2}AD\right\} = \frac{\pi}{2}.$$

If p' is the time of vibration of $KBCE$ with a loop at K and a node at E, this wire being entirely surrounded by air, then

$$\frac{p'}{V}(KE) = \frac{\pi}{2};$$

hence if $p' = p$,

$$2KA + \frac{\mu^2+1}{2}AD = KE,$$

so that

$$\frac{\delta KE}{\delta AD} = \frac{\mu^2+1}{2}.$$

Arons and Rubens when reducing their observations took the ratio $\delta KE/\delta AD$ to be always equal to μ. The above investigation shows that this is not the case when pAD/V' is small. We might show that $\delta KE/\delta AD$ is equal to μ when KA/AD is small.

The results given on the third view of the electrical vibrations of the compound wire seem parallel to those which hold for vibrating strings and bars. Thus if we have three strings of different materials stretched in series between two points, the time of longitudinal vibration of this system is not proportional to the sum of the times a pulse would take to travel over the strings separately (see Routh's *Advanced Rigid Dynamics*, p. 397), but is given by an equation somewhat resembling (3).

399.] The discrepancy between the results of the preceding theory of the action of the divided rectangle and the method employed by Arons and Rubens to reduce their observations, may perhaps explain to some extent the difference between the values of the specific inductive capacity of glass in rapidly alternating electric fields obtained by these observers and those obtained by M. Blondlot and myself for the same quantity.

Arons and Rubens (*Wied. Ann.* 44, p. 206, 1891) determined the ratio of the velocity of electromagnetic action through air to that through glass by filling with glass blocks a box through which the wires on one side of their rectangle passed. Employing the same method of reduction as for liquid dielectrics, they found μ (the ratio of the velocities) to be 2·33, whence $K = \mu^2$ is 5·43; while the value of K for the same glass, in slowly varying fields, was 5·37, which is practically identical with the preceding value. If, however, we adopted the method of reduction indicated by the preceding theory we should get a considerably smaller value of K. In order to see what kind of diminution we might expect, let us suppose that the circuit through the glass is so short that the relation expressed by (4) holds. This gives the same value for $(K+1)/2$ as Arons and Rubens get for μ; hence we find $K = 3{\cdot}66$, a value considerably less than under steady fields.

400.] Arons and Rubens checked their method by finding by means of it the specific inductive capacity of paraffin. This substance happens to be one for which either method of reduction leads to very much the same result. For example, for fluid paraffin their method of reduction gave $\mu = \sqrt{K} = 1{\cdot}47$, $K = 2{\cdot}16$; if we suppose that we ought to have $(K+1)/2$ instead of μ we get $K = 1{\cdot}94$, while the value in slowly varying fields is 1·98; so that the result for this substance is not decisive between the methods of reduction.

Both M. Blondlot and myself found that the specific inductive capacity of glass was smaller under rapidly changing fields than in steady ones. The following is the method used by M. Blondlot (*Comptes Rendus*, May 11, 1891, p. 1058; *Phil. Mag.* [5], 32, p. 230, 1891). A large rectangular plate of copper AA', Fig. 134, is fixed vertically, and a second parallel and smaller plate BB' forms a condenser with the first. This condenser discharges itself by means of the knobs a, b; a is connected with the gas pipes, b with one pole of an induction-coil, the other pole of which is connected to the gas pipes. When the coil is working, electrical oscillations take place in the condenser, the period of which is of the order 1/25,000,000 of a second. There is thus on the side of AA' a periodic electric field which has xx as the plane of symmetry. Two square plates, CD, $C'D'$, are placed in this field parallel to AA' and symmetrical with respect to xx; two wires terminating in EE' are soldered at DD' to the middle points of the sides of these plates. The wires are connected at EE' to two carbon points kept facing each other at a very small distance apart.

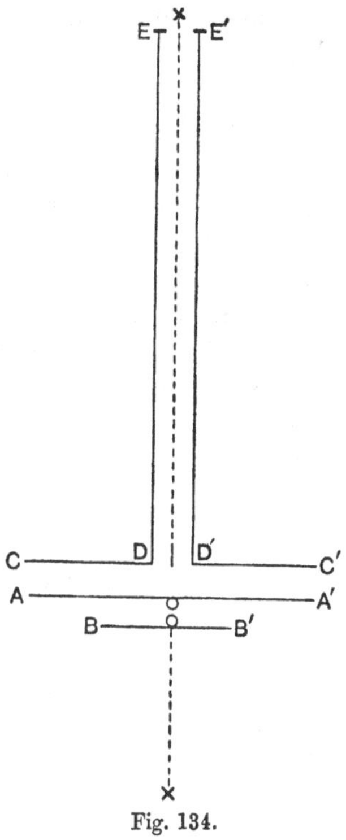

Fig. 134.

When the coil is working no sparks are observed between E and E', this is due to the symmetry of the apparatus. When, however, a glass plate is placed between AA' and CD sparks immediately pass between E and E'; these are caused by the induction received by CD differing from that received by $C'D'$. By interposing between AA' and C' $'$ a sheet of sulphur of suitable thickness the sparks can be made to disappear again. We can thus find the relative thicknesses of plates of glass and sulphur which produce the same effect on the electromagnetic waves passing through them, and we can therefore compare the specific inductive capacity of glass and sulphur under similar electrical conditions. M. Blondlot found the specific inductive capacity of the sulphur he employed by Curie's method (*Annales de Chimie et de Physique*, [6], 17, p. 385, 1889), and assuming that its inductive capacity was the same in rapidly alternating fields as in steady ones, he found the specific inductive capacity of the glass to be 2·84, which is considerably less than its value in steady fields.

401.] I had previously (*Proc. Roy. Soc.* 46, p. 292) arrived at the same conclusion by measuring the lengths of the electrical

I i

waves emitted by a parallel plate condenser, (1) when the plates were separated by air, (2) when they were separated by glass. The period of vibration of the condenser depends upon its capacity, and this again upon the dielectric between the plates, so that the determination of the periods gives us the means of determining the specific inductive capacity of the glass. The parallel plate condenser loses its energy by radiation slowly, and will thus force the vibration of its own period upon any electrical system under its influence. It differs in this respect from the condenser in Fig. 113, which radiates its energy away so rapidly that its action on neighbouring electrical conductors approximates to an impulse which starts the free vibrations of such systems.

The wave lengths in those observations were determined by observations on sparks. This is not comparable in delicacy with the bolometric method of Arons and Rubens; the method was however sufficiently sensitive to show a considerable falling off in the specific inductive capacity of the glass, for which I obtained the value 2·7, almost coincident with that obtained by M. Blondlot. Sulphur and ebonite on the other hand, when tested in the same way, showed no appreciable change in their specific inductive capacity.

The Effects produced by a Magnetic Field on Light.

402.] The connection between optical and electromagnetic phenomena is illustrated by the effects produced by a magnetic field on light passing through it. Faraday was the first to discover the action of magnetism on light; he found (*Experimental Researches*, vol. 3, p. 1) that when plane polarized light passes through certain substances, such as bisulphide of carbon or heavy glass, placed in a magnetic field where the lines of force are parallel to the direction of propagation of the light, the plane of polarization is twisted round the direction of the magnetic force. The laws of this phenomenon are described in Maxwell's *Electricity and Magnetism*, Chapter XXI.

403.] Subsequent investigations have shown that a magnetic field produces other effects upon light, which, though they probably have their origin in the same cause as that which produces the rotation of the plane of polarization in the magnetic field, manifest themselves in a different way.

Thus Kerr (*Phil. Mag.* [5], 3, p. 321, 1877), whose experiments have been verified and extended by Righi (*Annales de Chimie et de Physique*, [6], 4, p. 433, 1885; 9, p. 65, 1886; 10, p. 200, 1887), Kundt (*Wied. Ann.* 23, p. 228, 1884), Du Bois (*Wied. Ann.* 39, p. 25, 1890), and Sissingh (*Wied. Ann.* 42, p. 115, 1891) found that when plane polarized light is incident on the pole of an electromagnet, polished so as to act like a mirror, the plane of polarization of the reflected light is not the same when the magnet is 'on' as when it is 'off.'

The simplest case is when the incident plane polarized light falls normally on the pole of an electromagnet. In this case, when the magnet is not excited, the reflected ray is plane polarized, and can be completely stopped by an analyser placed in a suitable position. If the analyser is kept in this position and the electromagnet excited, the field, as seen through the analyser, is no longer quite dark, but becomes so, or very nearly so, when the analyser is turned through a small angle, showing that the plane of polarization has been twisted through a small angle by reflection from the magnetized iron. Righi (l. c.) has shown that the reflected light is not quite plane polarized, but that it is elliptically polarized, the axes of the ellipse being of very unequal magnitude. These axes are not respectively in and at right angles to the plane of incidence. If we regard for a moment the reflected elliptically polarised light as approximately plane polarized, the plane of polarization being that through the major axis of the ellipse, the direction of rotation of the plane of polarization depends upon whether the pole from which the light is reflected is a north or south pole. Kerr found that the direction of rotation was opposite to that of the currents exciting the pole from which the light was reflected.

The rotation produced is small. Kerr, who used a small electromagnet, had to concentrate the lines of magnetic force in the neighbourhood of the mirror by placing near to this a large mass of soft iron, before he could get any appreciable effects. By the use of more powerful magnets Gordon and Righi have succeeded in getting a difference of about half a degree between the positions of the analyser for maximum darkness with the magnetizing current flowing first in one direction and then in the opposite.

A piece of gold-leaf placed over the pole entirely stops the magnetic rotation, thus proving that the rotation of the plane of polarization is not produced in the air.

Hall (*Phil. Mag.* [5], 12, p. 157, 1881) found that the rotation takes place when the light is reflected from nickel or cobalt, instead of from iron, and is in the same direction as for iron.

Righi (l. c.) showed that the amount of rotation depends on the nature of the light; the longer the wave length the greater (at least within the limits of the luminous spectrum) the rotation.

Oblique Incidence on a Magnetic Pole.

404.] When the light is incident obliquely and not normally on the polished pole of an electromagnet it is necessary, in order to be able to measure the rotation, that the incident light should be polarized either in or at right angles to the plane of incidence, since it is only in these two cases that plane polarized light remains plane polarized after reflection from a metallic surface, even though this is not in a magnetic field. When light polarized in either of these planes is incident on the polished pole of an electromagnet, the light, when the magnet is on, is elliptically polarized after reflection, and the major and minor axes of the ellipse are not respectively in and at right angles to the plane of incidence. The ellipticity of the reflected light is very small. If we regard the light as consisting of two plane polarized waves of unequal amplitudes and complementary phases, then the rotation from the plane of polarization of the incident wave to that of the plane in which the amplitude of the reflected wave is greatest is in the direction opposite to that of the currents which circulate round the poles of the electromagnet.

According to Righi the amount of this rotation when the incident light is polarized in a plane perpendicular to that of incidence reaches a maximum when the angle of incidence is between 44° and 68°; while when the light is polarized in the plane of incidence the rotation steadily decreases as the angle of incidence is increased. The rotation when the light is polarized in the plane of incidence is always less than when it is polarized at right angles to that plane, except when the incidence is normal, when of course the two rotations are equal.

These results of Righi's differ in some respects from those of some preceding investigations by Kundt, who, when the light was polarized at right angles to the plane of incidence, obtained a reversal of the sign of the rotation of the plane of polarization near grazing incidence.

Reflection from Tangentially Magnetized Iron.

405.] In the preceding experiments the lines of magnetic force were at right angles to the reflecting surface; somewhat similar effects are however produced when the mirror is magnetized tangentially. In this case Kerr (*Phil. Mag.* [5], 5, p. 161, 1878) found :—

1. That when the plane of incidence is perpendicular to the lines of magnetic force no change is produced by the magnetization on the reflected light.

2. No change is produced at normal incidence.

3. When the incidence is oblique, the lines of magnetic force being in the plane of incidence, the reflected light is elliptically polarized after reflection from the magnetized surface, and the axes of the ellipse are not in and at right angles to the plane of incidence. When the light is polarized in the plane of incidence. the rotation of the plane of polarization (that is the rotation from the original plane to the plane through the major axis of the ellipse) is for all angles of incidence in the opposite direction to that of currents which would produce a magnetic field of the same sign as the magnet. When the light is polarized at right angles to the plane of incidence, the rotation is in the same direction as these currents when the angle of incidence is between $0°$ and $75°$ according to Kerr, between $0°$ and $80°$ according to Kundt, and between $0°$ and $78° 54'$ according to Righi. When the incidence is more oblique than this, the rotation of the plane of polarization is in the opposite direction to the electric currents which would produce a magnetic field of the same sign.

406.] Kerr's experiments were confined to the case of light reflected from metallic surfaces. Kundt (*Phil. Mag.* [5], 18, p. 308, 1884) has made a most interesting series of observations of the effect of thin plates of the magnetic metals iron, nickel and cobalt, on the plane of polarization of light passing through these plates in a strong magnetic field where the lines of force are at right angles to the surface of the plates.

Kundt found that in these circumstances the magnetic metals possess to an extraordinary degree the power of rotating the plane of polarization of the light. The rotation due to an iron plate is for the mean rays of the spectrum more than 30,000 times that of a glass plate of the same thickness in the same magnetic field, and nearly 1,500 times the natural rotation (i. e. the rotation independent of magnetic force) due to a plate of quartz of the same thickness. The rotation of the plane of polarization is with all three substances in the direction of the currents which would produce a magnetic field of the same sign as the one producing the rotation. The rotation under similar circumstances is nearly the same for iron and cobalt, while for nickel it is decidedly weaker. The rotation is greater for the red rays than for the blue.

407.] The phenomena discovered by Kerr show that when the rapidly alternating currents which accompany light waves are flowing through iron, nickel, or cobalt in a magnetic field, electromotive intensities are produced which are at right angles both to the current and the magnetic force. Let us take, for example, the simple case when light is incident normally on the pole of an electromagnet. Let us suppose that the incident light is polarized in the plane of zx, where $z=0$ is the equation to the reflecting surface, so that in the incident wave the electromotive intensities and the currents are at right angles to this plane; Kerr found, however, that the reflected wave had a component polarized in the plane of yz; thus after reflection there are electromotive intensities and currents parallel to x, that is at right angles to both the direction of the external magnetic field which is parallel to z and to the intensities in the incident wave which are parallel to y.

The Hall Effect.

408.] In the *Philosophical Magazine* for November, 1880, Hall published an account of some experiments, which show that when a steady current is flowing in a steady magnetic field electromotive intensities are developed which are at right angles both to the magnetic force and to the current, and are proportional to the product of the intensity of the current, the magnetic force and the sine of the angle between the directions of these quantities. The nature of the experiments by which this effect was demonstrated was as follows: A thin film of metal was deposited on a glass plate; this plate was placed over the pole of an electromagnet and a steady current sent through the film from two electrodes. The distribution of the current was indicated by finding two places in the film which were at the same potential; this was done by finding two points such that if they were placed in electrical connection with the terminals of a delicate galvanometer (G) they produced no current through it when the electromagnet was 'off.' If now the current was sent through an electromagnet a deflection of the galvanometer (G) was produced, and this continued as long as the electromagnet was 'on,' showing that the distribution of current in the film was altered by the magnetic field. The method used by Hall to measure this effect is described in the following extract taken from one of his papers on this subject (*Phil. Mag.* [5], 19, p. 419, 1885). 'In most cases, when possible, the metal was used in the form of a thin strip about 1·1 centim. wide and about 3 centim. long between the two pieces of brass B, B (Fig. 135), which, soldered to the ends of the strip, served as electrodes for the entrance and escape of the main current. To the arms a, a, about 2 millim. wide and perhaps 7 millim. long, were soldered the wires w, w, which led to a Thomson galvanometer. The notches c, c show how adjustment was secured. The strip thus prepared was fastened to a plate of glass by means of a cement of beeswax and rosin, all the parts shown in the figure being imbedded in and covered by this cement, which was so hard and stiff as to be quite brittle at the ordinary temperature of the air.

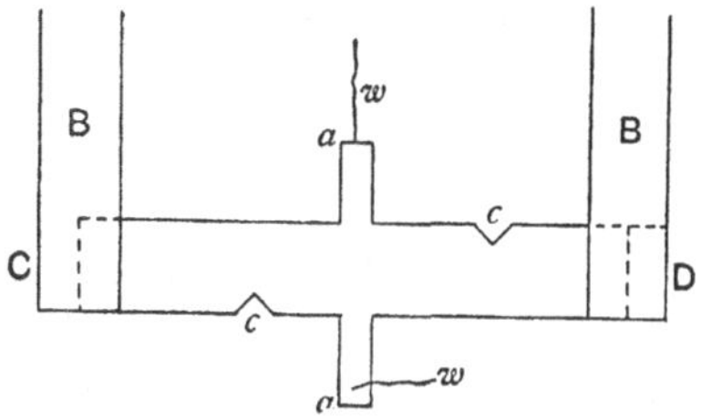

Fig. 135.

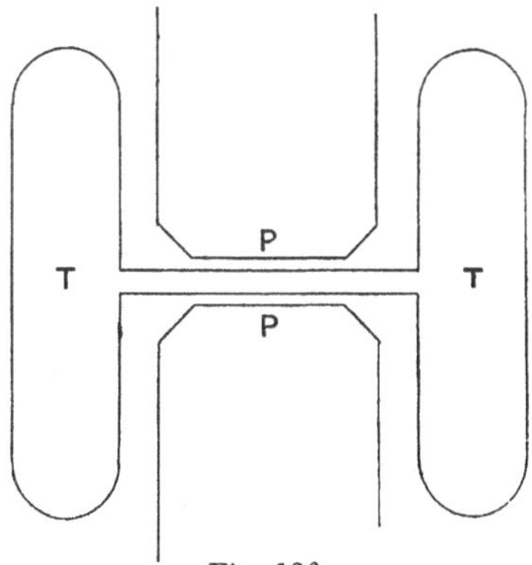

Fig. 136.

'The plate of glass bearing the strip of metal so embedded was, when about to be tested, placed with B, B vertical in the narrow part of a tank whose horizontal section is shown in Fig. 136. This tank, *TT*, containing the plate of glass with the metal strip was placed between the poles *PP* of the electromagnet. The tank was filled with water which was sometimes at rest and sometimes flowing. By this means the temperature of the strip of metal was under tolerable control, and the inconvenience from thermoelectric effects at *a* and *a* considerably lessened. The diameter of the plane circular ends of the pole pieces *PP* were about 3·7 centim.'

By means of experiments of this kind Hall arrived at the conclusion that if α, β, γ; u, v, w denote respectively the components of the magnetic force and the intensity of the current, electromotive intensities are set up whose components parallel to the axes of x, y, z are respectively

$$C(\beta w-\gamma v),\quad C(\gamma u-\alpha w),\quad C(\alpha v-\beta u).$$

The values of C in electromagnetic units for some metals at 20° C, as determined by Hall (*Phil. Mag.* [5], 19, p. 419, 1885), are given in the following table (l. c. p. 436):—

Metal.	$C\times 10^{15}$.
Copper	-520
Zinc	$+820$
Iron	$+7850$
Steel, soft	$+12060$
„ tempered	$+33000$
Cobalt	$+2460$
Nickel	-14740
Bismuth	-8580000
Antimony	$+114000$
Gold	-660

With regard to the magnetic metals, it is not certain that the quantity primarily involved in the Hall effect is the magnetic force rather than the magnetic induction, or the intensity of magnetization. Hall's experiments with nickel seem to point to its being the last of these three, as he found, using strong magnetic fields, that the effect ceased to be proportional to the external magnetic field, and fell off in a way similar to that in which the magnetization falls off when the field is increased. We must remember, if we use Hall's value of C for iron and the other magnetic metals, to use in the expression for the electromotive intensities the magnetic induction instead of the magnetic force. For in Hall's experiments the magnetic force measured was the normal magnetic force outside the iron. Since the plate was very thin the normal magnetic force outside the iron would be large compared with that inside; the normal magnetic induction inside would however be equal to the normal magnetic force outside, so that Hall in this case measured the relation between the electromotive intensity produced and the magnetic induction producing it.

Hall has thus established for steady currents the existence of an effect of the same nature as that which Kerr's experiments proved (assuming the electromagnetic theory of light) to exist for the rapidly alternating currents which constitute light. Here however the resemblance ends; the values of the coefficient C deduced by Hall from his experiments on steady currents do not apply to rapidly alternating light currents. Thus Hall found that for steady currents the sign of C was positive for iron, negative for nickel; the magneto-optical properties of these bodies are however quite similar. Again, both Hall and Righi found that the C for bismuth was enormously larger than that for iron or nickel. Righi, however, was unable to find any traces of magneto-optical effects in bismuth.

The optical experiments previously described show that there is an electromotive intensity at right angles both to the magnetic force and to the electromotive intensity; they do not however show without further investigation on what function of the electromotive intensity the magnitude of the transverse intensity depends. Thus, for example, the complete current in the metal is the sum of the polarization and conduction currents. Thus, if the electromotive intensity is X, the total current u is given by the equation

$$u = \left(\frac{K'}{4\pi}\frac{d}{dt} + \frac{1}{\sigma}\right)X,$$

or if the effects are periodic and proportional to $\epsilon^{\iota pt}$,

$$u = \left(\frac{K'}{4\pi}\iota p + \frac{1}{\sigma}\right)X,$$

where K' is the specific inductive capacity of the metal and σ its specific resistance.

We do not know from the experiments, without further discussion, whether the transverse electromotive intensity is proportional to u, the total current, or only to $K'\iota pX/4\pi$, the polarization part of it, or to X/σ, the conduction current.

We shall assume that the components of the transverse electromotive intensity are given by the expressions

$$k(bw-cv),$$
$$k(cu-aw),$$
$$k(av-bu);$$

where a, b, c are the components of the magnetic induction, u, v, w those of the total current.

This form, if k is a real constant, makes the transverse intensity proportional to the total current; the form is however sufficiently general analytically to cover the cases where the transverse intensity is proportional to the polarization current alone or to the conduction one. Thus, if we put

$$k = \left(\frac{K'\iota p/4\pi}{K'\iota p/4\pi + 1/\sigma}\right)k',$$

where k' is a real constant, the transverse intensity will be proportional to the displacement current; while if we put

$$k = \frac{k''}{K'\iota p/4\pi + 1/\sigma},$$

where k'' is a real constant, the transverse intensity will be proportional to the conduction current. We shall now proceed to investigate which, if any, of these hypotheses will explain the results observed by Kerr.

409.] Let P, Q, R be the components of the electromotive intensity in a conductor, P', Q', R' the parts of these which arise from electromagnetic induction, a, b, c the components of the magnetic induction, α, β, γ those of the magnetic force, u, v, w the components of the current. K', μ', σ are respectively the specific inductive capacity, the magnetic permeability, and the specific resistance of the metal.

Then we have in the metal

$$P = P' + k(bw-cv),$$
$$Q = Q' + k(cu-aw),$$
$$R = R' + k(av-bu),$$

where k is a coefficient which bears the same relation to rapidly alternating currents as C (Art. 408) does to steady currents. If the external field is very strong, we may without appreciable error substitute for a, b, c, in the terms multiplied by k, a_0, b_0, c_0, the components of the external field. We shall suppose that this field is uniform, so that a_0, b_0, c_0 are independent of x, y, z.

By equation (2) of Art. 256

$$\frac{da}{dt} = \frac{dQ'}{dz} - \frac{dR'}{dy}$$

$$= \frac{dQ}{dz} - \frac{dR}{dy} - k\left(a_0 \frac{d}{dx} + b_0 \frac{d}{dy} + c_0 \frac{d}{dz}\right) u, \qquad (1)$$

since $\frac{du}{dx} + \frac{dv}{dy} + \frac{dw}{dz} = 0$ on Maxwell's hypothesis that all the currents are closed. Now since u is the component of the total current parallel to x, it is equal to the sum of the components of the polarization and conduction currents in that direction. The polarization current is equal to

$$\frac{K'}{4\pi} \frac{dP}{dt},$$

the conduction current to P/σ, hence

$$4\pi u = K' \frac{dP}{dt} + \frac{4\pi}{\sigma} P.$$

We shall confine our attention to periodic currents and suppose that the variables are proportional to $\epsilon^{\iota pt}$; in this case the preceding equation becomes

$$4\pi u = (K'\iota p + 4\pi/\sigma) P,$$

but

$$4\pi u = \frac{d\gamma}{dy} - \frac{d\beta}{dz};$$

hence we have

$$(K'\iota p + 4\pi/\sigma) P = \frac{d\gamma}{dy} - \frac{d\beta}{dz};$$

similarly

$$(K'\iota p + 4\pi/\sigma) Q = \frac{da}{dz} - \frac{d\gamma}{dx},$$

$$(K'\iota p + 4\pi/\sigma) R = \frac{d\beta}{dx} - \frac{da}{dy};$$

and therefore since

$$\frac{da}{dx} + \frac{d\beta}{dy} + \frac{d\gamma}{dz} = 0,$$

$$(K'\iota p + 4\pi/\sigma)\left(\frac{dQ}{dz} - \frac{dR}{dy}\right) = \frac{d^2 a}{dx^2} + \frac{d^2 a}{dy^2} + \frac{d^2 a}{dz^2};$$

and hence equation (1) becomes

$$(K'\iota p+4\pi/\sigma)\mu'\frac{d\alpha}{dt}=\frac{d^2\alpha}{dx^2}+\frac{d^2\alpha}{dy^2}+\frac{d^2\alpha}{dz^2}$$
$$-\frac{k}{4\pi}(K'\iota p+4\pi/\sigma)\left(a_0\frac{d}{dx}+b_0\frac{d}{dy}+c_0\frac{d}{dz}\right)\left(\frac{d\gamma}{dy}-\frac{d\beta}{dz}\right).$$

Similarly we have

$$\left.\begin{aligned}(K'\iota p+4\pi/\sigma)\mu'\frac{d\beta}{dt}&=\frac{d^2\beta}{dx^2}+\frac{d^2\beta}{dy^2}+\frac{d^2\beta}{dz^2}\\&\quad-\frac{k}{4\pi}(K'\iota p+4\pi/\sigma)\left(a_0\frac{d}{dx}+b_0\frac{d}{dy}+c_0\frac{d}{dz}\right)\left(\frac{d\alpha}{dz}-\frac{d\gamma}{dx}\right),\\(K'\iota p+4\pi/\sigma)\mu'\frac{d\gamma}{dt}&=\frac{d^2\gamma}{dx^2}+\frac{d^2\gamma}{dy^2}+\frac{d^2\gamma}{dz^2}\\&\quad-\frac{k}{4\pi}(K'\iota p+4\pi/\sigma)\left(a_0\frac{d}{dx}+b_0\frac{d}{dy}+c_0\frac{d}{dz}\right)\left(\frac{d\beta}{dx}-\frac{d\alpha}{dy}\right).\end{aligned}\right\}\quad(2)$$

We are now in a position to discuss the reflection of waves of light from a plane metallic surface. Let us take the plane separating the metal from the air as the plane of xy, the plane of incidence as the plane of xz; the positive direction along z is from the metal to the air.

Let us suppose that waves of magnetic force are incidence on the metal, these incident waves may be expressed by equations of the form

$$\alpha=A_0\epsilon^{\iota(lx+mz+pt)},$$
$$\beta=B_0\epsilon^{\iota(lx+mz+pt)};$$

and since
$$\frac{d\alpha}{dx}+\frac{d\beta}{dy}+\frac{d\gamma}{dz}=0,$$

$$\gamma=-\frac{l}{m}A_0\epsilon^{\iota(lx+mz+pt)},$$

where
$$l^2+m^2=\frac{p^2}{V^2},$$

and A_0 and B_0 are constants.

V is the velocity of propagation of electromagnetic action through the air and so is equal to $1/K$, where K is the electromagnetic measure of the specific inductive capacity of the air, whose magnetic permeability is taken as unity. Waves will be reflected from the surface of the metal, and the amplitudes of these waves will be proportional to $\epsilon^{\iota(lx-mz+pt)}$, so that α, β, γ, the components of the total magnetic force in the air, will, since it is due to both the incident and reflected waves, be represented by the equations

$$\alpha = A_0 \epsilon^{\iota(lx+mz+pt)} + A\epsilon^{\iota(lx-mz+pt)},$$
$$\beta = B_0 \epsilon^{\iota(lx+mz+pt)} + B\epsilon^{\iota(lx-mz+pt)},$$
$$\gamma = -\frac{l}{m} A_0 \epsilon^{\iota(lx+mz+pt)} + \frac{l}{m} A\epsilon^{\iota(lx-mz+pt)},$$

where A and B are constants.

We shall suppose that the metal is so thick that there is no reflection except from the face $z = 0$; in this case the waves in the metal will travel in the negative direction of z.

Thus in the metal we may put

$$\alpha = A'\epsilon^{\iota(lx+m'z+pt)},$$
$$\beta = B'\epsilon^{\iota(lx+m'z+pt)},$$
$$\gamma = -\frac{l}{m'} A'\epsilon^{\iota(lx+m'z+pt)},$$

where if m' is complex the real part must be positive in order that the equations should represent a wave travelling in the negative direction of z; the imaginary part of m' must be negative, otherwise the amplitude of the wave of magnetic force would increase indefinitely as the wave travelled along.

Substituting these values of α, β, γ in equations (2), we get

$$A'(-p^2\mu' K' + 4\pi\mu'\iota p/\sigma + l^2 + m'^2)$$
$$= -\frac{k}{4\pi}(K'\iota p + 4\pi/\sigma)\, m'(la_0 + m'c_0) B', \qquad (3)$$

$$B'(-p^2\mu' K' + 4\pi\mu'\iota p/\sigma + l^2 + m'^2)$$
$$= \frac{k}{4\pi}(K'\iota p + 4\pi/\sigma)\frac{l^2+m'^2}{m'}(la_0 + m'c_0) A'. \qquad (4)$$

Eliminating A' and B' from these equations, we get

$$-p^2\mu' K' + 4\pi\mu'\iota p/\sigma + l^2 + m'^2$$
$$= \pm\frac{\iota k}{4\pi}(K'\iota p + 4\pi/\sigma)(l^2+m'^2)^{\frac{1}{2}}(la_0 + m'c_0). \qquad (5)$$

There are only two values of m' which satisfy this equation and which have their real parts positive and their imaginary parts negative. We shall denote these two roots by m_1, m_2; m_1 being the root when the plus sign is taken in the ambiguity in sign in equation (4), m_2 the root when the minus sign is taken.

We have from equation (3), if A_1 and B_1 are the values of A' and B' corresponding to the root m_1,

$$A_1 \iota (l^2 + m_1^2)^{\frac{1}{2}} = -B_1 m_1;$$

or if

$$l^2 + m_1^2 = \omega_1^2,$$

$$A_1 \iota \omega_1 = -B_1 m_1.$$

If A_2 and B_2 are the values of A' and B' corresponding to the root m_2, we have

$$A_2 \iota \omega_2 = B_2 m_2,$$

where

$$l^2 + m_2^2 = \omega_2^2.$$

Thus in the metal we have

$$a = A_1 \epsilon^{\iota(lx + m_1 z + pt)} + A_2 \epsilon^{\iota(lx + m_2 z + pt)},$$

$$\beta = -\frac{\iota \omega_1}{m_1} A_1 \epsilon^{\iota(lx + m_1 z + pt)} + \frac{\iota \omega_2}{m_2} A_2 \epsilon^{\iota(lx + m_2 z + pt)},$$

$$\gamma = -\frac{l}{m_1} A_1 \epsilon^{\iota(lx + m_1 z + pt)} - \frac{l}{m_2} A_2 \epsilon^{\iota(lx + m_2 z + pt)}.$$

We thus see that the original plane wave is in the metal split up into two plane waves travelling with the velocities p/ω_1, p/ω_2 respectively. We also see from the equations for a, β, γ that the waves are two circularly polarized ones travelling with different velocities. Starting from this result Prof. G. F. Fitzgerald (*Phil. Trans.* 1880, p. 691) has calculated the rotation of the plane of polarization produced by reflection from the surface of a *transparent* medium which under the action of magnetic force splits up a plane wave into two circularly polarized ones; some of the results which he has arrived at are not in accordance with the results of Kerr's and Righi's experiments on the reflection from metallic surfaces placed in a magnetic field, proving that in them we must take into account the opacity of the medium if we wish to completely explain the results of these experiments.

410.] In order to determine the reflected and transmitted waves we must introduce the boundary conditions. We assume (1) that a and β, the tangential components of the magnetic force, are continuous; (2) that the normal magnetic induction is continuous; and (3) that the part of the tangential electromotive intensity which is due to magnetic induction is continuous. It should be noticed that condition (3) makes the total tangential electromotive intensity discontinuous, for the total electromotive intensity is made up of two parts, one due to electromagnetic induction, the other due to the causes which produce the Hall effect; it is only the first of these parts which we assume to be continuous.

If P is the component parallel to x of the total electromotive intensity, P' the part of it due to electromagnetic induction, then

$$P = P' + k(b_0 w - c_0 v);$$

but

$$P = \frac{1}{K'\iota p + 4\pi/\sigma}\left(\frac{d\gamma}{dy} - \frac{d\beta}{dz}\right)$$

$$= -\frac{1}{K'\iota p + 4\pi/\sigma}\frac{d\beta}{dz},$$

since in the present case γ does not depend upon y.

Hence, substituting the values of w and v in terms of the magnetic force, the condition that P' is continuous is equivalent to that of

$$-\frac{1}{K'\iota p + 4\pi/\sigma}\frac{d\beta}{dz} - \frac{k}{4\pi}\left(b_0\frac{d\beta}{dx} \quad c_0\left(\frac{d\alpha}{dz} - \frac{d\gamma}{dx}\right)\right)$$

being continuous.

We shall suppose that in the air $k = 0$.

The condition that α is continuous gives

$$A_0 + A = A_1 + A_2; \tag{6}$$

the condition that β is continuous gives

$$B_0 + B = -\frac{\iota\omega_1}{m_1}A_1 + \frac{\iota\omega_2}{m_2}A_2; \tag{7}$$

the condition that the normal magnetic induction is continuous gives

$$-\frac{l}{m}(A_0 - A) = -\mu'\left(\frac{l}{m_1}A_1 + \frac{l}{m_2}A_2\right),$$

or dividing by l,

$$-\frac{1}{m}(A_0 - A) = -\mu'\left(\frac{1}{m_1}A_1 + \frac{1}{m_2}A_2\right). \tag{8}$$

We can easily prove independently that this equation is true when $l = 0$, though in that case it cannot be legitimately deduced from the preceding equation.

The condition that P' is continuous gives, since $k=0$ and $\sigma=\infty$ for air,

$$-\frac{m}{K\iota p}(B_0-B)=\frac{\iota(\omega_1 A_1-\omega_2 A_2)}{K'\iota p+4\pi/\sigma}$$

$$-\frac{k}{4\pi}\left\{b_0 l\left(-\frac{\iota\omega_1}{m_1}A_1+\frac{\iota\omega_2}{m_2}A_2\right)-c_0\left(\frac{\omega_1^2}{m_1}A_1+\frac{\omega_2^2}{m_2}A_2\right)\right\}. \quad (9)$$

The equations (6), (7), (8), and (9) are sufficient to determine the four quantities A, A_1, A_2, B, and thus to determine the amplitudes and phases of the reflected and transmitted waves.

411.] We shall now proceed to apply these equations to the case of reflection from a tangentially magnetized reflecting surface, as the peculiar reversal of the direction of rotation of the plane of polarization which (Art. 405) Kerr found to take place when the angle of incidence passes through 75° seems to indicate that this case is the one which is best fitted to distinguish between rival hypotheses.

Since in this case the magnetic force is tangential $c_0=0$; hence, referring to equation (5), we see that there will only be one value of m' if la_0 vanishes, i. e. if $l=0$, in which case the incidence is normal, or if $a_0=0$, in which case the magnetic force is at right angles to the plane of incidence; hence, since there is only one value of m', there will not be any rotation of the plane of polarization in either of these cases; this agrees with Kerr's experiments (see Art. 405).

Let us suppose that the light is polarized perpendicularly to the plane of incidence and that the mirror is magnetized in that plane. In the incident wave the magnetic force is at right angles to the plane of incidence, so that the A_0 of equations (6), (7), (8), and (9) vanishes. Putting

$$A_0=0,\quad b_0=0,\quad c_0=0,$$

we get from these equations

$$A=A_1+A_2,$$

$$B_0+B=-\iota\left(\frac{\omega_1}{m_1}A_1-\frac{\omega_2}{m_2}A_2\right),$$

$$A=-\mu' m\left(\frac{1}{m_1}A_1+\frac{1}{m_2}A_2\right)$$

$$-\frac{m}{K\iota p}(B_0-B)=\frac{\iota(\omega_1 A_1-\omega_2 A_2)}{K'\iota p+4\pi/\sigma}.$$

Since $(K'\iota p+4\pi/\sigma)/K\iota p = \frac{1}{\mu'}R^2\epsilon^{2\iota\alpha}$, see Art. 353, the last equation may be written

$$-(B_0-B) = \frac{\iota\mu'}{R^2\epsilon^{2\iota\alpha}m}(\omega_1 A_1-\omega_2 A_2).$$

The rotation observed is small, we shall therefore neglect the squares and higher powers of (m_1-m_2); doing this we find from the preceding equations that

$$\frac{A}{B} = \frac{\iota\mu' m\left(\frac{1}{m_1}-\frac{1}{m_2}\right)}{\omega\left(\frac{\mu'}{R^2\epsilon^{2\iota\alpha}m}-\frac{1}{M}\right)\left(1+\mu'\frac{m}{M}\right)}, \tag{10}$$

where M is the value of m_1 or m_2, when $k=0$, and $\omega^2 = l^2+M^2$.

From equation (5) we have, when $c_0 = 0$,

$$-p^2\mu' K'+4\pi\mu'\iota p/\sigma+l^2+m_1^2 = \frac{\iota k}{4\pi}(K'\iota p+4\pi/\sigma)(l^2+m_1^2)^{\frac{1}{2}}la_0,$$

$$-p^2\mu' K'+4\pi\mu'\iota p/\sigma+l^2+m_2^2 = -\frac{\iota k}{4\pi}(K'\iota p+4\pi/\sigma)(l^2+m_2^2)^{\frac{1}{2}}la_0.$$

Hence, when m_1-m_2 is small, we have approximately

$$(m_1-m_2)M = \frac{\iota k}{4\pi}(K'\iota p+4\pi/\sigma)\,\omega l a_0$$
$$= \frac{\iota k}{4\pi\mu'}R^2\epsilon^{2\iota\alpha}\,\iota p V_0^{-2}\,\omega l a_0,$$

where V_0 denotes the velocity of propagation of electromagnetic action through air. Substituting this value of m_1-m_2 in equation (10) we get

$$\frac{A}{B} = \frac{\iota kp}{4\pi}\,\frac{R^2\epsilon^{2\iota\alpha}lmV_0^{-2}a_0\omega}{M\left(\frac{\mu' M}{R^2\epsilon^{2\iota\alpha}m}-1\right)(M+\mu' m)}. \tag{11}$$

If
$$\frac{A}{B} = \theta+\iota\phi,$$

where θ and ϕ are real quantities, then if the reflected light polarized perpendicularly to the plane of incidence is represented by

$$\beta = \cos(pt+lx-mz),$$

the reflected light polarized in the plane of incidence will be represented by

$$a = \theta\cos(pt+lx-mz)-\phi\sin(pt+lx-mz);$$

thus, unless ϕ vanishes, the reflected light will be elliptically polarized. If however θ and ϕ are small, then the angle between the major axis of the ellipse for the reflected light and that of the incident light (regarding this which is plane polarized as the limit of elliptically polarized light when the minor axis of the ellipse vanishes) will be approximately θ. Hence if the analysing prism is set so as to extinguish the light reflected from the mirror when it is not magnetized, the field after magnetization will be darkest when the analyser is turned through an angle θ, though even in this case it will not be absolutely dark. We proceed now to find θ from equation (11).

We have by Art. (353)

$$l^2 + M^2 = R^2 \epsilon^{2\iota a} (l^2 + m^2),$$

or

$$M^2 = (R^2 \epsilon^{2\iota a} - 1) l^2 + R^2 \epsilon^{2\iota a} m^2.$$

Now for metals the modulus of $R^2 \epsilon^{2\iota a}$ is large, the table in Art. 355 showing that for steel it is about 17; hence we have approximately

$$M^2 = R^2 \epsilon^{2\iota a} (l^2 + m^2)$$

$$= R^2 \epsilon^{2\iota a} \frac{p^2}{V_0^2}.$$

We shall put $\mu' = 1$ in the denominator of the right-hand side of equation (11), since there is no evidence that iron and steel retain their magnetic properties for magnetic forces alternating as rapidly as those in the light waves. Making this substitution and putting $m = (p/V_0) \cos i$, where i is the angle of incidence, we find

$$\left(\frac{M}{R^2 \epsilon^{2\iota a} m} - 1\right)(M + m) = \frac{p}{V_0}\left(\frac{1}{\cos i} - R\epsilon^{\iota a} + \frac{1}{R\epsilon^{\iota a}} - \cos i\right)$$

$$= \frac{p}{V_0 \cos i}(1 - \cos^2 i - R\epsilon^{\iota a} \cos i)$$

approximately, since the modulus of $R\epsilon^{\iota a}$ is large.

Hence we see

$$\frac{A}{B} = \frac{\iota k}{4\pi} \frac{p a_0 V_0^{-2} R\epsilon^{\iota a} \sin i \cos^2 i}{\sin^2 i - R\epsilon^{\iota a} \cos i},$$

so that if k is real,

$$\theta = -\frac{k}{4\pi} \frac{p a_0 V_0^{-2} \sin^3 i \cos^2 i \, R \sin a}{\sin^4 i - 2 \sin^2 i \cos i \, R \cos a + R^2 \cos^2 i}.$$

This does not change sign for any value of i between 0 and $\pi/2$; this result is therefore inconsistent with Kerr's and Kundt's experiments, and we may conclude that the hypothesis on which it is founded—that the transverse intensity is proportional to the *total* current—is erroneous.

As Kerr's and Kundt's experiments were made with magnetic metals it seems desirable to consider the results of supposing these metals to retain their magnetic properties. When μ' is not put equal to unity, θ is proportional to

$$\cos^2 i \sin i \sin a \left(\mu' \sin^2 i + \frac{2\mu'^2}{R} \cos a \cos i\right);$$

this does not change sign for any value of i between 0 and $\pi/2$, so that the preceding hypothesis cannot be made to agree with the facts by supposing the metals to retain their magnetic properties.

412.] Let us now consider the consequence of supposing that the transverse electromotive intensity is proportional not to the total current but to the polarization current; we can do this by putting

$$k = \frac{K'\iota p/4\pi}{K'\iota p/4\pi + 1/\sigma} k',$$

where k' is a real quantity.

This equation may be written

$$k = \frac{K'V_0^2}{R^2 \epsilon^{2\iota a}} k'.$$

Substituting this value of k in equation (11) we find

$$\frac{A}{B} = \frac{\iota k' K' p a_0}{4\pi R \epsilon^{\iota a}} \frac{\sin i \cos^2 i}{\sin^2 i - R\epsilon^{\iota a} \cos i}.$$

If we write this in the form

$$\frac{A}{B} = \theta' + \iota\phi',$$

where θ' and ϕ' are real, we find

$$\theta' = \frac{k'K'pa_0}{4\pi R} \frac{\sin i \cos^2 i (\sin a \sin^2 i - R \sin 2a \cos i)}{\sin^4 i - 2R \sin^2 i \cos i \cos a + R^2 \cos^2 i}. \qquad (12)$$

The angle through which the analyser has to be twisted in order to produce the greatest darkness is, as we have seen, equal to θ' the real part of A/B. Equation (12) shows that this changes sign when i passes through the value given by the equation

$$\sin a \sin^2 i - R \sin 2a \cos i = 0,$$

or

$$\sin^2 i = 2R \cos a \cos i;$$

with the notation of the table in Art. 355 this is

$$\sin^2 i = 2n \cos i.$$

If μ' is not equal to unity the corresponding equation may easily be shown to be

$$\mu' \sin^2 i = 2n \cos i.$$

From the table in Art. 355 we see that for steel $n = 2{\cdot}41$, the corresponding value of i when $\mu' = 1$ is about 78°, which agrees well with the results of Kerr's experiments. Hence we see that the consequences of the hypotheses, that the transverse electromotive intensity is proportional to the polarization current, and that $\mu' = 1$, agree with the results of experiments.

We shall now consider the consequences of supposing that the transverse electromotive intensity is proportional to the conduction current. We can do this by putting

$$k = \frac{k''}{K' \iota p / 4\pi + 1/\sigma},$$

where k'' is a constant real quantity.

This equation may be written

$$k = \frac{4\pi k'' V_0^2}{\iota p R^2 \epsilon^{2\iota a}}.$$

Substituting this value of k in equation (11) we find

$$\frac{A}{B} = \frac{k'' \sin i \cos^2 i}{R \epsilon^{\iota a} (\sin^2 i - R \epsilon^{\iota a} \cos i)},$$

the real part of which is

$$\frac{k'' \sin i \cos^2 i}{R} \frac{(\cos a \sin^2 i - R \cos 2a \cos i)}{\sin^4 i - 2 \sin^2 i \cos i R \cos a + R^2 \cos^2 i}.$$

This is the angle through which the analyser must be twisted in order to quench the reflected light as much as possible. The rotation of the analyser will change sign when i passes through the value given by the equation

$$\cos a \sin^2 i = R \cos 2a \cos i,$$

or

$$R \cos a \sin^2 i = R^2 \cos 2a \cos i.$$

With the notation of Art. 355 this may be written

$$n \sin^2 \iota = n^2 (1-k^2) \cos \iota.$$

From the table in Art. 355 we see that $1-k^2$ is negative, hence, since n is positive there is no real value of i less than $\pi/2$ which satisfies this equation, so that if this hypothesis were correct there would be no reversal of the direction of rotation of the analyser.

Hence of the three hypotheses, (1) that the transverse electromotive intensity concerned in these magnetic optical effects is proportional to the total current, (2) that it is proportional to the polarization current, (3) that it is proportional to the conduction current, we see that (1) and (3) are inconsistent with Kerr's experiments on the reflection from tangentially magnetized mirrors, while (2) is completely in accordance with them.

413.] The transverse electromotive intensity indicated by hypothesis (2) is of a totally different character from that discovered by Hall. In Hall's experiments the electromotive intensities, and therefore the currents through the metallic plates, were constant; when however this is the case the 'polarization' current vanishes. Thus in Hall's experiments there could have been no electromotive intensity of the kind assumed in hypothesis (2); there is therefore no reason to expect that the order of the metals with respect to Kerr's effect should be the same as that with respect to Hall's.

It is worth noting that reflection from a transparent body placed in a magnetic field can be deduced from the preceding equations by putting $\alpha = 0$, since this makes the refractive index real. In this case we see, by equation (12), that the real part of A/B vanishes, so that the reflected light is elliptically polarized, with the major axis of the ellipse in the plane of incidence; any small rotation of the analyser would therefore in this case increase the brightness of the field.

414.] We now proceed to consider the case of reflection from a normally magnetized mirror. We shall confine ourselves to the case of normal incidence.

If the incident light is plane polarized we may (using the notation of Art. 409) put $B_0 = 0$; we have also $l = 0$, $\omega_1 = m_1$, $\omega_2 = m_2$, and since the mirror is magnetized normally, $a_0 = 0$,

$b_0 = 0$. Making these substitutions, equations (6), (7), (8), and (9) of Art. 410 become, putting $\mu' = 1$,

$$A_0 + A = A_1 + A_2, \tag{13}$$

$$B = -\iota(A_1 - A_2), \tag{14}$$

$$A_0 - A = m\left(\frac{A_1}{m_1} + \frac{A_2}{m_2}\right), \tag{15}$$

$$\frac{m}{K\iota p}B = \frac{\iota(m_1 A_1 - m_2 A_2)}{K'\iota p + 4\pi/\sigma} + \frac{k}{4\pi}c_0(m_1 A_1 + m_2 A_2), \tag{16}$$

where K is the specific inductive capacity of air. The last equation by means of (5) reduces to

$$\frac{m}{K\iota p}B = p\left(\frac{A_1}{m_1} - \frac{A_2}{m_2}\right),$$

or since

$$Kp^2 = m^2,$$

$$B = \iota m\left(\frac{A_1}{m_1} - \frac{A_2}{m_2}\right). \tag{17}$$

Solving these equations we find

$$\frac{B}{A} = -\frac{\iota m(m_1 - m_2)}{m_1 m_2 - m^2}.$$

Now $m_1 - m_2$ is small, and we may therefore, if we neglect the squares of small quantities, in the denominator of the expression for B/A, put M for either m_1 or m_2, where M is the value of these quantities when the magnetic field vanishes.

We have by equation (5)

$$-p^2\mu'K' + 4\pi\mu'\iota p/\sigma + m_1^2 = \frac{\iota k}{4\pi}(K'\iota p + 4\pi/\sigma)m_1^2 c_0,$$

$$-p^2\mu'K' + 4\pi\mu'\iota p/\sigma + m_2^2 = -\frac{\iota k}{4\pi}(K'\iota p + 4\pi/\sigma)m_2^2 c_0;$$

hence

$$m_1 - m_2 = \frac{\iota k}{4\pi}(K'\iota p + 4\pi/\sigma)Mc_0$$

approximately.

Since the transverse electromotive intensity is proportional to the polarization current we have

$$k = \frac{K'\iota p/4\pi}{K'\iota p/4\pi + 1/\sigma}k',$$

where k' is a real quantity. Substituting this value of k in the expression for $m_1 - m_2$, we get

$$m_1 - m_2 = -\frac{pK'k'Mc_0}{4\pi};$$

but $M = R\epsilon^{\iota a}m$, so that

$$\frac{B}{A} = \frac{\iota p K' k' R \epsilon^{\iota a} c_0}{4\pi (R^2 \epsilon^{2\iota a} - 1)},$$

or, since the modulus of $R^2\epsilon^{2\iota a}$ is large compared with unity,

$$\frac{B}{A} = \frac{\iota p K' k' \epsilon^{-\iota a}}{4\pi R} c_0$$

$$= \frac{pK'k'}{4\pi R} c_0 (\sin a + \iota \cos a)$$

approximately.

Hence, if the magnetic force in the reflected wave, which is polarized in the same plane as the incident wave, is represented by

$$\cos (pt + mz),$$

the magnetic force in the reflected wave polarized in the plane at right angles to this will be represented by

$$\frac{pK'k'}{4\pi R} c_0 \sin a \cos (pt + mz) - \frac{pK'k'}{4\pi R} c_0 \cos a \sin (pt + mz).$$

Thus in the expression for the light polarized in this plane one term represents a component in the same phase as the constituent in the original plane, while the phase of the component represented by the other term differs from this by quarter of a wave length. The resultant reflected light will thus be slightly elliptically polarized. As in Art. (411) however, we may show that the field can be darkened by twisting the analyser through a small angle from the position in which it completely quenched the light when the mirror was not magnetized. The angle for which the darkening is as great as possible is equal to the real term in the expression for B/A, i. e. to

$$\frac{pK'k'}{4\pi R} c_0 \sin a.$$

Thus though the reflected light cannot be completely quenched by rotating the analyser, its intensity can be very considerably reduced; this agrees with the results of Righi's experiments, see Art. 403.

We can deduce from this case that of reflection from a transparent substance by putting $a = 0$, as this assumption makes the refractive index wholly real; in this case the reflected light is elliptically polarized, but as the axes of the ellipse are respectively in and at right angles to the plane of the original polarization any small rotation of the analyser will increase the brightness of the field.

We can solve by similar means the case of oblique reflection from a normally magnetized mirror; the results agree with Kerr's experiments; want of space compels us however to pass on to apply the same principles to the case where light, as in Kundt's experiments Art. 406, passes through thin metallic films placed in a magnetic field.

On the Effect produced by a thin Magnetized Plate on Light passing through it.

415.] We shall assume that the plate is bounded by the planes $z=0$, $z=-h$, the incident light falling normally on the plane $z=0$. The external magnetic field is supposed to be parallel to the axis of z.

Let the incident light be plane polarized, the magnetic force in it being parallel to the axis of x. The reflected light will consist of two portions, one polarized in the same plane as the incident light, the other polarized in the plane at right angles to this: the magnetic force in the latter part of the light will therefore be parallel to the axis of y.

If α, β are the components of the magnetic force parallel to the axes of x and y respectively, then in the region for which z is positive we have

$$\alpha = A_0\epsilon^{\iota(mz+pt)} + A\epsilon^{\iota(-mz+pt)},$$
$$\beta = B\epsilon^{\iota(-mz+pt)},$$

where $A_0\epsilon^{\iota(mz+pt)}$ represents the magnetic force in the incident wave and A and B are constants.

In the plate we have

$$\alpha = A_1\epsilon^{\iota(m_1z+pt)} + A_1'\epsilon^{\iota(-m_1z+pt)} + A_2\epsilon^{\iota(m_2z+pt)} + A_2'\epsilon^{\iota(-m_2z+pt)},$$

and therefore, as in Art. 409, as $l=0$,

$$\beta = -\iota A_1\epsilon^{\iota(m_1z+pt)} - \iota A_1'\epsilon^{\iota(-m_1z+pt)} + \iota A_2\epsilon^{\iota(m_2z+pt)} + \iota A_2'\epsilon^{\iota(-m_2z+pt)},$$

where m_1, m_2 are the roots of equation (5) and A_1, A_1', A_2, A_2' are constants.

After the light has passed through the plate, the components of the magnetic force will be given by equations of the form

$$\alpha = C\epsilon^{\iota(mz+pt)},$$
$$\beta = D\epsilon^{\iota(mz+pt)}.$$

The four boundary conditions at the surface $z = 0$ give, if $\mu' = 1$,

$$\left.\begin{aligned} A_0 + A &= A_1 + A_1' + A_2 + A_2', \\ A_0 - A &= m\left(\frac{A_1}{m_1} - \frac{A_1'}{m_1} + \frac{A_2}{m_2} - \frac{A_2'}{m_2}\right); \end{aligned}\right\} \quad (18)$$

$$\left.\begin{aligned} B &= -\iota\,(A_1 + A_1' - A_2 - A_2'), \\ B &= \iota m\left(\frac{A_1}{m_1} - \frac{A_1'}{m_1} - \frac{A_2}{m_2} + \frac{A_2'}{m_2}\right). \end{aligned}\right\} \quad (19)$$

The boundary conditions when $z = -h$ give, writing θ and ϕ for $-\iota m_1 h$, $-\iota m_2 h$ respectively,

$$\left.\begin{aligned} C\epsilon^{-\iota mh} &= A_1\epsilon^{\theta} + A_1'\epsilon^{-\theta} + A_2\epsilon^{\phi} + A_2'\epsilon^{-\phi}, \\ C\epsilon^{-\iota mh} &= m\left(\frac{A_1\epsilon^{\theta}}{m_1} - \frac{A_1'\epsilon^{-\theta}}{m_1} + \frac{A_2\epsilon^{\phi}}{m_2} - \frac{A_2'\epsilon^{-\phi}}{m_2}\right); \end{aligned}\right\} \quad (20)$$

$$\left.\begin{aligned} D\epsilon^{-\iota mh} &= -\iota\,(A_1\epsilon^{\theta} + A_1'\epsilon^{-\theta} - A_2\epsilon^{\phi} - A_2'\epsilon^{-\phi}), \\ D\epsilon^{-\iota mh} &= \iota m\left(\frac{A_1\epsilon^{\theta}}{m_1} - \frac{A_1'\epsilon^{-\theta}}{m_1} - \frac{A_2\epsilon^{\phi}}{m_2} + \frac{A_2'\epsilon^{-\phi}}{m_2}\right). \end{aligned}\right\} \quad (21)$$

From equations (19), (20), and (21) we get

$$A_1\epsilon^{\theta}\left(1 - \frac{m}{m_1}\right) + A_1'\epsilon^{-\theta}\left(1 + \frac{m}{m_1}\right) + A_2\epsilon^{\phi}\left(1 - \frac{m}{m_2}\right) + A_2'\epsilon^{-\phi}\left(1 + \frac{m}{m_2}\right) = 0.$$

$$A_1\epsilon^{\theta}\left(1 + \frac{m}{m_1}\right) + A_1'\epsilon^{-\theta}\left(1 - \frac{m}{m_1}\right) - A_2\epsilon^{\phi}\left(1 + \frac{m}{m_2}\right) - A_2'\epsilon^{-\phi}\left(1 - \frac{m}{m_2}\right) = 0,$$

$$A_1\left(1 + \frac{m}{m_1}\right) + A_1'\left(1 - \frac{m}{m_1}\right) - A_2\left(1 + \frac{m}{m_2}\right) - A_2'\left(1 - \frac{m}{m_2}\right) = 0.$$

The solution of these equations may be expressed in the form

$$= \frac{A_1\epsilon^{\theta}}{\epsilon^{\phi}\left(1 - \frac{m}{m_2}\right)\left(1 + \frac{m^2}{m_1 m_2}\right) - \epsilon^{-\phi}\left(1 + \frac{m}{m_2}\right)\left(1 - \frac{m^2}{m_1 m_2}\right) + 2\epsilon^{\theta}\left(1 - \frac{m}{m_1}\right)\frac{m}{m_2}},$$

$$= \frac{-A_1'\epsilon^{-\theta}}{\epsilon^{\phi}\left(1 - \frac{m}{m_2}\right)\left(1 - \frac{m^2}{m_1 m_2}\right) - \epsilon^{-\phi}\left(1 + \frac{m}{m_2}\right)\left(1 + \frac{m^2}{m_1 m_2}\right) + 2\epsilon^{-\theta}\left(1 + \frac{m}{m_1}\right)\frac{m}{m_2}},$$

$$= \frac{A_2\epsilon^{\phi}}{\epsilon^{\theta}\left(1 - \frac{m}{m_1}\right)\left(1 + \frac{m^2}{m_1 m_2}\right) - \epsilon^{-\theta}\left(1 + \frac{m}{m_1}\right)\left(1 - \frac{m^2}{m_1 m_2}\right) + 2\epsilon^{\phi}\left(1 - \frac{m}{m_2}\right)\frac{m}{m_1}},$$

$$= \frac{-A_2'\epsilon^{-\phi}}{\epsilon^{\theta}\left(1 - \frac{m}{m_1}\right)\left(1 - \frac{m^2}{m_1 m_2}\right) - \epsilon^{-\theta}\left(1 + \frac{m}{m_1}\right)\left(1 + \frac{m^2}{m_1 m_2}\right) + 2\epsilon^{-\phi}\left(1 + \frac{m}{m_2}\right)\frac{m}{m_1}}.$$

Now by equations (20) and (21) we have

$$\frac{D}{C} = -\frac{\iota(A_1\epsilon^{\theta}+A_1'\epsilon^{-\theta}-A_2\epsilon^{\phi}-A_2'\epsilon^{-\phi})}{A_1\epsilon^{\theta}+A_1'\epsilon^{-\theta}+A_2\epsilon^{\phi}+A_2'\epsilon^{-\phi}}.$$

Substituting the ratios of A_1, A_1', A_2, A_2' just found, we get

$$-\frac{D}{C} =$$

$$\frac{\iota\left\{\frac{1}{m_2}\left(1+\frac{m^2}{m_1^2}\right)(\epsilon^{\theta}-\epsilon^{-\theta})-\frac{1}{m_1}\left(1+\frac{m^2}{m_2^2}\right)(\epsilon^{\phi}-\epsilon^{-\phi})+\frac{2m}{m_1m_2}(\epsilon^{\phi}+\epsilon^{-\phi}-\epsilon^{\theta}-\epsilon^{-\theta})\right\}}{\frac{1}{m_2}\left(1-\frac{m^2}{m_1^2}\right)(\epsilon^{\theta}-\epsilon^{-\theta})+\frac{1}{m_1}\left(1-\frac{m^2}{m_2^2}\right)(\epsilon^{\phi}-\epsilon^{-\phi})}$$

We notice that the numerator vanishes when $m_1 = m_2$, in which case $\theta = \phi$: it therefore contains the factor $m_1 - m_2$; hence, if we neglect the squares and higher powers of $(m_1 - m_2)$, we may in the denominator put $m_1 = m_2 = M$ and $\phi = \theta$.

If the thickness of the film is so small that θ and ϕ are small quantities, then neglecting powers of h higher than the second, we find

$$\frac{D}{C} = \frac{1}{2}\frac{m_2^2 - m_1^2}{M^2 - m^2}(\iota - mh).$$

Substituting the value of $m_2^2 - m_1^2$ from equation (5), and putting $M = R\epsilon^{\iota a}m$, we see that

$$\frac{D}{C} = \frac{pK'k'c_0}{4\pi}\frac{(\iota - mh)}{1-\frac{\epsilon^{-2\iota a}}{R^2}}.$$

Since R^2 is large for metals we may, as a first approximation, put

$$\frac{D}{C} = \frac{pK'k'c_0}{4\pi}(\iota - mh).$$

The angle through which the plane of polarization is twisted is equal to the real part of D/C, and is therefore equal to

$$-pK'k'c_0 mh/4\pi;$$

it is thus to our order of approximation independent of the opacity of the plate. We see from Art. 414 that when light is incident normally on a magnetized mirror the rotation of the plane of polarization of the reflected light is proportional to sin a, and thus depends primarily on the opacity of the mirror, vanishing when the mirror is transparent.

The imaginary part of D/C remains finite though h is made indefinitely small, we therefore infer that the transmitted light is elliptically polarized, and that the ratio of the axes of the ellipse is approximately independent of the thickness of the plate.

Let us now consider the light reflected from the plate. We have by equations (18) and (19)

$$\frac{B}{2A} = -\frac{\iota(A_1 + A_1' - A_2 - A_2')}{A_1\left(1 - \frac{m}{m_1}\right) + A_1'\left(1 + \frac{m}{m_1}\right) + A_2\left(1 - \frac{m}{m_2}\right) + A_2'\left(1 + \frac{m}{m_2}\right)}.$$

Substituting the values of A_1, A_1', A_2, A_2' previously given, we find, neglecting squares and higher powers of $m_1 - m_2$,

$$-\frac{B}{2A} =$$

$$\frac{\left\{\left(\epsilon^{\theta-\phi} - \epsilon^{-(\theta-\phi)}\right)\frac{2m}{M}\left(1 + \frac{m^2}{M^2}\right) + \left(\epsilon^{-(\theta+\phi)} - \epsilon^{\theta+\phi}\right)m\left(\frac{1}{m_1} - \frac{1}{m_2}\right)\left(1 - \frac{m^2}{M^2}\right)\right\}}{2\left(1 - \frac{m^2}{M^2}\right)^2\left\{2 - \left(\epsilon^{2\theta} + \epsilon^{-2\theta}\right)\right\}}.$$

If the plate is so thin that θ and ϕ are small, we have approximately

$$\frac{B}{A} = \frac{(m_1 - m_2)\frac{m}{M}\left\{1 + \frac{m^2}{M^2} + \left(1 - \frac{m^2}{M^2}\right)\right\}}{\left(1 - \frac{m^2}{M^2}\right)^2 M^2 h}$$

$$= (m_1 - m_2)\frac{2m}{M^3 h\left(1 - \frac{m^2}{M^2}\right)^2}$$

$$= \frac{2(m_1 - m_2)m}{M^3 h},$$

since m/M is small for metals.

Substituting the value of $m_1 - m_2$ from equation (5) we get, putting $M = R\epsilon^{\iota a} m$,

$$\frac{B}{A} = -\frac{pK'k'c_0}{2\pi}\frac{1}{mR^2\epsilon^{2\iota a}h};$$

the rotation of the plane of polarization is equal to the real part of B/A, and hence to

$$-\frac{K'k'c_0}{2\pi}\frac{p}{mh}\frac{\cos 2a}{R^2}.$$

Since this is proportional to $1/h$ we see that the rotation increases as the thickness of the plate diminishes. The explanation of this is that while the intensities of the two components reflected light, viz. the component polarized in the same plane as the incident wave and the component polarized in the plane at right angles to this, both diminish as the thickness of the plate diminishes; the first component diminishes much more rapidly than the second; thus the ratio of the second component to the first and therefore the angle of rotation of the plane of polarization increases as the thickness of the plate diminishes.

416.] The effect of a magnetic field in producing rotation of the plane of polarization thus seems to afford strong evidence of the existence of a transverse electromotive intensity in a conductor placed in a magnetic field, this intensity being quite distinct from that discovered by Hall, inasmuch as the former is proportional to the rate of variation of the electromotive intensity, whereas the Hall effect is proportional to the electromotive intensity itself. We shall now endeavour to form some estimate of the magnitude of this transverse intensity revealed to us by optical phenomena.

Kundt (*Wied. Ann.* 23, p. 238, 1884) found from his experiments that if ϕ, the rotation of the plane of polarization produced by the passage of light of wave length λ through a magnetized plate of thickness h, is given by an equation of the form

$$\phi = \frac{\pi h}{\lambda}(n - n'),$$

then $\phi = 1^{\circ}.48'$ when

$$\lambda = 5.8 \times 10^{-5}, \text{ and } h = 5.5 \times 10^{-6},$$

thus

$$n - n' = \cdot 1.$$

But we have seen that the rotation in this case is equal to

$$\frac{pK'k'c_0}{2\pi}\frac{\pi h}{\lambda};$$

hence, comparing this with Kundt's result, we find

$$\frac{pK'k'c_0}{2\pi} = \cdot 1,$$

but if $\lambda = 5.8 \times 10^{-5}$, $p = 2\pi \times 3 \times 10^{10} \times 10^{5}/5.8 = 3.2 \times 10^{15}$.

Substituting these values, we find

$$\frac{K'k'}{2\pi} c_0 = 3.1 \times 10^{-17}.$$

Now if f is the electric polarization parallel to x, the transverse electromotive intensity is equal to

$$k'c_0 \frac{df}{dt} = k'\iota p c_0 f$$
$$= k' \frac{K'}{4\pi} \iota p c_0 X,$$

where X is the electromotive intensity parallel to x. Hence $k'K'pc_0/4\pi$ is the ratio of the magnitude of the transverse intensity to that producing the current; this ratio is for iron therefore equal to

$$1.6 \times 10^{-17} p$$

for magnetic fields of the strength used by Kundt. The factor multiplying p is so small as to make it probable that the effects of this transverse force are insensible except when the electromotive intensity is changing with a rapidity comparable with the rate of change in light waves, in other words, that it is only in optical phenomena that this transverse electromotive intensity produces any measurable effect.

CHAPTER VI.

THE DISTRIBUTION OF RAPIDLY ALTERNATING CURRENTS.

417.] PROBLEMS concerning alternating currents have become in recent years of much greater importance than they were at the time when Maxwell's Treatise was published; this is due to the extensive use of such currents for electric lighting, and to the important part which the much more rapidly oscillating currents produced by the discharge of Leyden jars now play in electrical researches. It is therefore desirable to consider more fully than is done in the *Electricity and Magnetism* the application of Maxwell's principles to such currents. In doing this we shall follow the methods used by Lord Rayleigh in his papers on 'The Reaction upon the Driving-Point of a System executing Forced Harmonic Oscillations of various Periods, with Applications to Electricity,' *Phil. Mag.* [5], 21, p. 369, 1886, and on 'The Sensitiveness of the Bridge Method in its Application to Periodic Electric Currents,' *Proc. Roy. Soc.* 49, p. 203, 1891.

418.] When the currents are steady their distribution among a net-work of conductors is determined by the condition that the rate of heat production must be a minimum, see Maxwell's *Electricity and Magnetism*, vol. i. p. 408. Thus, if F is the Dissipation Function (*Electricity and Magnetism*, vol. i. p. 408), $\dot{x}_1$, $\dot{x}_2$, $\dot{x}_3$ ... the currents flowing through the circuits, these variables being chosen so that they are sufficient but not more than sufficient to determine the currents flowing through each branch of the net-work, then $\dot{x}_1$, $\dot{x}_2$, &c. are determined by the equations

$$\frac{dF}{d\dot{x}_1} = \frac{dF}{d\dot{x}_2} = \frac{dF}{d\dot{x}_3} = \dots = 0.$$

When, however, the currents are variable these equations are no longer true; we have instead of them the equations

$$\frac{d}{dt}\frac{dT}{d\dot{x}_1} + \frac{dF}{d\dot{x}_1} - \frac{dV}{dx_1} = 0,$$

. . . .

where T is the Kinetic Energy due to the Self and Mutual induction of the circuits, F as before is the Dissipation Function, and V is the Potential Energy arising from the charges that may be in any condensers in the system.

If the currents are periodic and proportional to $\epsilon^{\iota pt}$, the preceding equation may be written as

$$\iota p\frac{dT}{d\dot{x}_1} + \frac{dF}{d\dot{x}_1} - \frac{dV}{dx_1} = 0;$$

and thus when p increases indefinitely the preceding equation approximates to

$$\frac{dT}{d\dot{x}_1} = 0;$$

we have similarly

$$\frac{dT}{d\dot{x}_2} = \frac{dT}{d\dot{x}_3} = \ldots = 0.$$

Thus in this case the distribution of currents is independent of the resistances, and is determined by the condition that the Kinetic Energy and not the Dissipation Function is a minimum.

419.] We have already considered several instances of this effect. Thus, when a rapidly alternating current travels along a wire, the currents fly to the outside of the wire, since by doing this the mean distance between the parts of the current is a maximum and the Kinetic Energy therefore a minimum. Again, when two currents in opposite directions flow through two parallel plates the currents congregate on the adjacent surfaces of the plates, since by so doing the average distance between the opposite currents, and therefore the Kinetic Energy, is a minimum.

Mr. G. F. C. Searle has devised an experiment which shows this tendency of the currents in a very striking way. AB, Fig. 123. is an exhausted tube through which the periodic currents produced by the discharge of a Leyden jar are sent. When none of the wires leading from the jar to the tube passes parallel to it in its neighbourhood, the glow produced by the currents fills the tube uniformly. When however one of the leads is bent, as in Fig. 137, so as to pass near the tube in such a way that the current through the lead is in the opposite direction to that through the

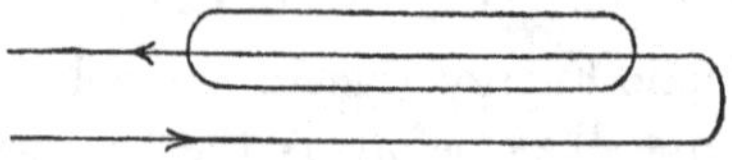

Fig. 137.

tube, the glow no longer fills the tube but concentrates itself on the side of the tube next the wire, thus getting as near as possible to the current in the opposite direction through the wire. When however the wire is bent, as in Fig. 138, so that the current through the lead is in the same direction as that through the

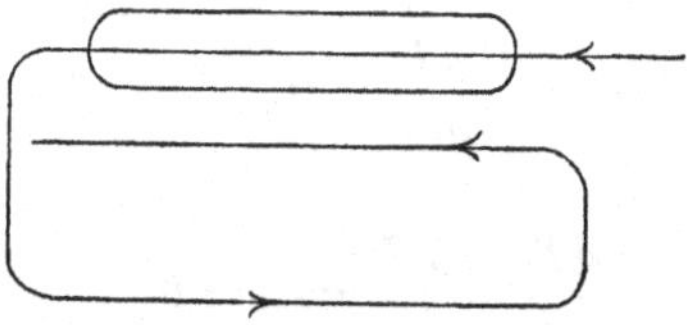

Fig. 138.

tube, the glow flies to the part of the tube most remote from the wire.

420.] We shall now proceed to consider the distribution of alternating currents among various systems of conductors. The first case we shall consider is the distribution of an alternating current between two conductors *ACB*, *ADB* in parallel. Let the resistance and self-induction in the arm *ACB* be respectively

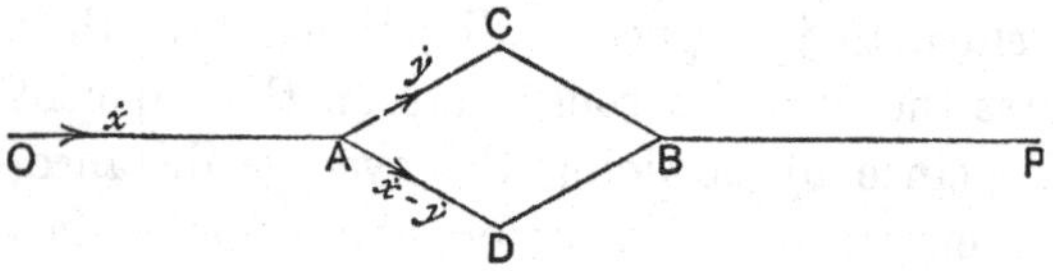

Fig. 139.

R, *L*, the corresponding quantities in the arm *ADB* being denoted by *S*, *N*, and let *M* be the coefficient of mutual induction between the circuits *ACB*, *ADB*. We shall suppose that the rate of alternation of the current is not so rapid as to produce any appreciable variation in the intensity of the current from one end of ACB or ADB to the other, in other words, that the wave length corresponding to the rate of alternation of the current is large compared with the length ACB or ADB; the case when this wave length is comparable with the length of the circuit is considered separately in Art. 298. Let the current flowing in along OA and out along BP be denoted by $\dot{x}$; we shall assume that $\dot{x}$ varies as $\epsilon^{\iota pt}$. Let the current in ACB be $\dot{y}$, that in ADB will be $\dot{x}-\dot{y}$. Then T, the Kinetic Energy in the branch $ACDB$ of the circuit, is expressed by the equation

$$T = \tfrac{1}{2}\{L\dot{y}^2 + 2M(\dot{x}-\dot{y})\dot{y} + N(\dot{x}-\dot{y})^2\}.$$

The dissipation function F is given by

$$F = \tfrac{1}{2}\{R\dot{y}^2 + S(\dot{x}-\dot{y})^2\},$$

and we have

$$\frac{d}{dt}\frac{dT}{d\dot{y}} + \frac{dF}{d\dot{y}} = 0,$$

or
$$(L+N-2M)\frac{d\dot{y}}{dt} + (R+S)\dot{y} - (N-M)\frac{d\dot{x}}{dt} - S\dot{x} = 0.$$

Let $\dot{x} = \epsilon^{\iota pt}$, then from this equation we have

$$\dot{y} = \frac{(N-M)\,\iota p + S}{(L+N-2M)\,\iota p + (R+S)}\,\epsilon^{\iota pt},$$

or, taking the real part of this, corresponding to the current $\cos pt$ along OA, we find

$$\dot{y} = \frac{\{S(R+S)+(L+N-2M)(N-M)p^2\}\cos pt - p\{R(N-M)-S(L-M)\}\sin pt}{(L+N-2M)^2p^2+(R+S)^2}, \quad (1)$$

$$\dot{x}-\dot{y} = \frac{\{R(S+R)+(L+N-2M)(L-M)p^2\}\cos pt + p\{R(N-M)-S(L-M)\}\sin pt}{(L+N-2M)^2p^2+(R+S)^2}. \quad (2)$$

These expressions may be written in the forms

$$\dot{y} = \left\{\frac{S^2+(N-M)^2p^2}{(L+N-2M)^2p^2+(R+S)^2}\right\}^{\frac{1}{2}}\cos(pt+\epsilon) = A\cos(pt+\epsilon), \text{ say,}$$

$$\dot{x}-\dot{y} = \left\{\frac{R^2+(L-M)^2p^2}{(L+N-2M)^2p^2+(R+S)^2}\right\}^{\frac{1}{2}}\cos(pt+\epsilon') = B\cos(pt+\epsilon'),$$

where
$$\tan\epsilon = \frac{p\{R(N-M)-S(L-M)\}}{S(R+S)+(L+N-2M)(N-M)p^2},$$

and
$$\tan\epsilon' = -\frac{p\{R(N-M)-S(L-M)\}}{R(R+S)+(L+N-2M)(L-M)p^2}.$$

The maximum currents through ACB, ADB are proportional to A and B, and we see from the preceding equations that

$$\frac{A}{\{S^2+(N-M)^2p^2\}^{\frac{1}{2}}}=\frac{B}{\{R^2+(L-M)^2p^2\}^{\frac{1}{2}}}.$$

When p is very large, this equation becomes

$$\frac{A}{N-M}=\frac{B}{L-M},$$

so that in this case the distribution of the currents is governed entirely by the induction in the circuits, and not at all by their resistances. Referring to equations (1) and (2) we see that when p is infinite

$$\dot{y}=\frac{N-M}{L+N-2M}\cos pt, \tag{3}$$

$$\dot{x}-\dot{y}=\frac{L-M}{L+N-2M}\cos pt. \tag{4}$$

An inspection of these equations leads to the interesting result that when the alternations are very rapid the maximum current in one or both of the branches may be greater than that in the leads. Consider the case when the two circuits ACB, ADB are wound close together. Suppose, for example, that they are parts of a circular coil, and that there are m turns in the circuit ACB, and n turns in ADB, then if the coils are close together we may put

$$L=Km^2,\quad M=Knm,\quad N=Kn^2,$$

where K is a constant.

Substituting these values for L, M, N in equations (3) and (4) we find

$$\dot{y}=\frac{n^2-nm}{(n-m)^2}\cos pt=\frac{n}{n-m}\cos pt, \tag{5}$$

$$\dot{x}-\dot{y}=\frac{m^2-nm}{(n-m)^2}=-\frac{m}{n-m}\cos pt. \tag{6}$$

Thus the currents are of opposite signs in the two coils, the current in the coil with the smallest number of turns flows in the same direction as the current in the leads. When $n-m$ is very small both currents become large, being now much greater than the current in the leads whose maximum value was taken as unity; thus by introducing an alternating current of small intensity into a divided circuit, we can produce in the arms of this circuit currents of very much greater intensity. The reason of this becomes clear when we consider the energy in the loop, when the rate of alternation is exceedingly rapid. The effects of the inertia of the system become all important, and the distribution of currents is that which would result if we considered merely the Kinetic Energy of the system. In this case, in accordance with dynamical principles, the actual solution is that which makes the Kinetic Energy as small as possible consistent with the condition that the algebraical sum of the currents in ACB, ADB shall be equal to $\dot{x}$.

Thus, as the Kinetic Energy is to be as small as possible, and this energy is in the field around the loop and proportional at each place to the square of the magnetic force, the currents will distribute themselves in the wires so as to neutralize as much as possible each other's magnetic effect. Thus if the wires are wound close together the currents will flow in opposite directions, the branch having the smallest number of turns having the largest current, so as to be on equal terms as far as magnetic force is concerned with the branch with the larger number of turns. In fact we see from equations (5) and (6) that the current in each branch is inversely proportional to the number of turns. If the two branches are exactly equal in all respects the current in each will be in the same direction, but this distribution will be unstable, the slightest difference of the coefficients of induction in the two branches being sufficient to make the current in the branch of least inductance flow in the direction of that in the leads, and the current in the other branch in the opposite direction, the intensity in either branch at the same time increasing largely.

When the currents are distributed in accordance with equations (3) and (4), the Kinetic Energy in the loop is

$$\frac{1}{2}\frac{LN-M^2}{L+N-2M}p^2\cos^2 pt.$$

We notice that $(LN-M^2)/(L+N-2M)$ is always less than L or N. $L+N-2M$ is always positive, since it is proportional to the Kinetic Energy in the loop when the currents are equal and opposite.

We see from equations (1) and (2) that when

$$R(N-M) = S(L-M),$$

$$\dot{y} = \frac{S}{R+S}\cos pt,$$

$$\dot{x}-\dot{y} = \frac{R}{R+S}\cos pt.$$

So that in this case the distribution of alternating currents of any frequency is the same as when the currents are steady.

421.] We shall now consider the self-induction and resistance of the two wires in parallel. Let L_0 and r be respectively the self-induction and resistance of the leads, and suppose that there is no mutual induction between the leads and the branches *ACB*, *ADB*.

Then we have

$$(L_0+N)\frac{d\dot{x}}{dt} - (N-M)\frac{d\dot{y}}{dt} + (r+S)\dot{x} - S\dot{y}$$

$$= \text{external electromotive force tending to increase } x.$$

Substituting in this expression the value of $\dot{y}$ in terms of $\dot{x}$ previously obtained in Art. 420, we find

$$(L_0+N)\frac{d\dot{x}}{dt} - \frac{\{(N-M)\iota p+S\}^2}{(L+N-2M)\iota p+R+S}\dot{x} + (r+S)\dot{x}$$

$$= \text{external electromotive force tending to increase } x.$$

Remembering that $\iota p\dot{x} = d\dot{x}/dt$, we see that the left-hand side of this equation may be written

$$\left\{L_0 + \frac{NR^2+LS^2+2MRS+p^2(LN-M^2)(L+N-2M)}{(R+S)^2+p^2(L+N-2M)^2}\right\}\frac{d\dot{x}}{dt}$$

$$+\left\{r + \frac{RS(R+S)+p^2\{R(N-M)^2+S(L-M)^2\}}{(R+S)^2+p^2(L+N-2M)^2}\right\}\dot{x}.$$

From the form of this equation we see that the self-induction of the two wires in parallel is

$$\frac{NR^2+LS^2+2MRS+p^2(LN-M^2)(L+N-2M)}{(R+S)^2+p^2(L+N-2M)^2},$$

which may be written as

$$\frac{NR^2+LS^2+2MRS}{(R+S)^2}$$

$$-\frac{p^2(L+N-2M)}{(R+S)^2+p^2(L+N-2M)^2}\{R(N-M)-S(L-M)\}^2.$$

The impedance of the loop is

$$\frac{RS(R+S)+p^2\{R(N-M)^2+S(L-M)^2\}}{(R+S)^2+p^2(L+N-2M)^2},$$

which is equal to

$$\frac{RS}{R+S}+\frac{p^2\{R(N-M)-S(L-M)\}^2}{(R+S)\{(R+S)^2+p^2(L+N-2M)^2\}}.$$

We see from the expression for the self-induction of the loop that it is greatest when $p = 0$, when its value is

$$\frac{NR^2+2MRS+LS^2}{(R+S)^2},$$

and least when p is infinite when it is equal to

$$\frac{LN-M^2}{L+N-2M}.$$

If $$R(N-M) = S(L-M),$$

the self-induction of the loop is independent of the period.

From the expression for the impedance of the loop we see that it is least when $p = 0$ when its value is

$$\frac{RS}{R+S},$$

and greatest when p is infinite when it is equal to

$$\frac{R(N-M)^2+S(L-M)^2}{(L+N-2M)^2};$$

and if $$R(N-M) = S(L-M),$$

the impedance is independent of the period. Thus in this case the self-induction and the impedance are unaltered, whatever the frequency of the currents. In all other cases the self-induction diminishes and the impedance increases as the frequency of the currents increases.

422.] We shall now proceed to investigate the general case when there are any number of wires in parallel. Let $\dot{x}_0$ be the current in the leads, $\dot{x}_1, \dot{x}_2, \ldots \dot{x}_n$ the currents in the n wires in parallel; we shall assume, as before, that there is no induction between these wires and the leads. Let a_{rr} be the self-induction and r_r the resistance of the wire through which the current is $\dot{x}_r$, a_{rs} the coefficient of mutual induction between this wire and the wire through which the current is $\dot{x}_s$. Let a_0 be the self-induction, r_0 the resistance of the leads, E_0 the electromotive force in the external circuit; we shall suppose that this varies as

$\epsilon^{\iota pt}$. The current through the leads and those through the wires in parallel are connected by the relation

$$\dot{x}_0 - (\dot{x}_1 + \dot{x}_2 + \dots \dot{x}_n) = 0\,;$$

we shall denote this by $\phi = 0$.

Then T being the Kinetic Energy, F the Dissipation function, and λ an arbitrary multiplier, the equations determining the currents are of the form

$$\frac{d}{dt}\frac{dT}{d\dot{x}_s} + \frac{dF}{d\dot{x}_s} + \lambda\frac{d\phi}{d\dot{x}_s}$$

$$= \text{external electromotive force tending to increase } \dot{x}_s.$$

From these equations we get

$$(a_0\,\iota p + r_0)\,\dot{x}_0 + \lambda = E_0, \tag{7}$$

$$\left.\begin{aligned} (a_{11}\,\iota p + r_1)\,\dot{x}_1 + a_{12}\,\iota p\dot{x}_2 + \dots - \lambda &= 0,\\ a_{12}\,\iota p\dot{x}_1 + (a_{22}\,\iota p + r_2)\,\dot{x}_2 + \dots - \lambda &= 0,\\ \cdots\cdots\cdots\cdots\cdots\cdots&\\ a_{1n}\,\iota p\dot{x}_1 + a_{2n}\,\iota p\dot{x}_2 + \dots \qquad - \lambda &= 0. \end{aligned}\right\} \tag{8}$$

Solving equations (8) we find

$$\frac{\dot{x}_1}{A_{11} + A_{12} + \dots A_{1n}} = \frac{\dot{x}_2}{A_{12} + A_{22} + \dots A_{2n}}$$

$$= \frac{\dot{x}_n}{A_{1n} + A_{2n} + \dots A_{nn}} = \frac{\lambda}{\Delta}, \tag{9}$$

where

$$\Delta = \begin{vmatrix} a_{11}\,\iota p + r_1, & a_{12}\,\iota p, & \dots & a_{1n}\,\iota p\\ a_{12}\,\iota p, & a_{22}\,\iota p + r_2, & \dots & a_{2n}\,\iota p\\ \cdot & \cdot & \cdot & \cdot\\ a_{1n}\,\iota p, & a_{2n}\,\iota p, & \dots & a_{nn}\,\iota p + r_n \end{vmatrix},$$

and A_{pq} denotes the minor of Δ corresponding to the constituent $a_{pq}\,\iota p$.

Since $\dot{x}_0 = \dot{x}_1 + \dot{x}_2 + \dots,$

we have from the above equations

$$\frac{\dot{x}_0}{A_{11} + A_{22} + \dots A_{nn} + 2A_{12} + 2A_{13} + 2A_{23} + \dots} = \frac{\lambda}{\Delta}.$$

Substituting this value of λ in equation (7) we find

$$\left(a_0\,\iota p + r_0 + \frac{\Delta}{S}\right)\dot{x}_0 = E_0, \tag{10}$$

where S is written for

$$A_{11} + A_{22} + \dots A_{nn} + 2A_{12} + 2A_{13} + 2A_{23} + \dots.$$

The self-induction and impedance of the leads can be deduced from (10); the expressions for them are however in general very complicated, but they take comparatively simple forms when ιp is either very large or very small.

When ιp is very large,

$$\frac{\Delta}{S} = \iota p \frac{D}{S'} + \frac{r_1 (A'_{11} + A'_{12} + \ldots A'_{1n})^2 + r_2 (A'_{12} + A'_{22} + \ldots A'_{2n})^2 + \ldots}{S'^2},$$

where

$$D = \begin{vmatrix} a_{11}, & a_{12}, & a_{1n} \\ a_{12}, & a_{22}, & a_{2n} \\ \cdot & \cdot & \cdot \\ a_{1n}, & a_{2n}, & a_{nn} \end{vmatrix},$$

and A'_{pq} is the minor of D corresponding to the constituent a_{pq}, while

$$S' = A'_{11} + A'_{22} + \ldots A'_{nn} + 2A'_{12} + 2A'_{13} + 2A'_{23} + \ldots.$$

Thus the self-induction of the wires in parallel is in this case

$$\frac{D}{S'},$$

while the impedance is

$$\{r_1 (A'_{11} + A'_{12} + \ldots A'_{1n})^2 + r_2 (A'_{12} + A'_{22} + \ldots A'_{2n})^2 + \ldots\}/S'^2.$$

When ιp is very small,

$$\frac{\Delta}{S} = \iota p \frac{\left(\frac{a_{11}}{r_1^2} + \frac{a_{22}}{r_2^2} + \ldots + \frac{2a_{12}}{r_1 r_2} + \frac{2a_{13}}{r_1 r_3} + \ldots\right)}{\left(\frac{1}{r_1} + \frac{1}{r_2} + \ldots \frac{1}{r_n}\right)^2} + \frac{1}{\frac{1}{r_1} + \frac{1}{r_2} + \ldots \frac{1}{r_n}}.$$

So that in this case the self-induction of the wires in parallel is

$$\frac{\frac{a_{11}}{r_1^2} + \frac{a_{22}}{r_2^2} + \ldots + \frac{2a_{12}}{r_1 r_2} + \frac{2a_{13}}{r_1 r_3} + \ldots}{\left(\frac{1}{r_1} + \frac{1}{r_2} + \ldots \frac{1}{r_n}\right)^2},$$

and the resistance is

$$\frac{1}{\frac{1}{r_1} + \frac{1}{r_2} + \ldots \frac{1}{r_n}}.$$

When there is no induction between the wires in parallel,

a_{12}, a_{13}, a_{23}, &c. all vanish; hence, when ιp is infinite, the self-induction is

$$\frac{1}{\frac{1}{a_{11}}+\frac{1}{a_{22}}+\dots\frac{1}{a_{nn}}},$$

and the impedance

$$\frac{\frac{r_1}{a^2_{11}}+\frac{r_2}{a^2_{22}}+\dots}{\left(\frac{1}{a_{11}}+\frac{1}{a_{22}}+\dots\right)^2}.$$

423.] We shall now consider the case of any number of circuits; the investigation will apply whether the circuits are arranged so as to form separate circuits or whether some or all of them are metallically connected so as to form a net-work of conductors.

Let $\dot{x}_1, \dot{x}_2, \dots \dot{x}_n$ be the variables required to fix the distribution of currents through the circuits; let T, the Kinetic Energy due to these currents, be expressed by the equation

$$T = \tfrac{1}{2}\{a_{11}\dot{x}_1{}^2 + a_{22}\dot{x}_2{}^2 + \dots + 2a_{12}\dot{x}_1\dot{x}_2 + \dots\},$$

while the Dissipation Function F is given by

$$F = \tfrac{1}{2}\{r_{11}\dot{x}_1{}^2 + r_{22}\dot{x}_2{}^2 + \dots + 2r_{12}\dot{x}_1\dot{x}_2 + \dots\}.$$

Let us suppose that there are no external forces of types $\dot{x}_2$, $\dot{x}_3$, &c., and that X_1, the external force of type x_1, is proportional to $\epsilon^{\iota pt}$.

The equations giving the currents are

$$\begin{aligned}(a_{11}\,\iota p + r_{11})\,\dot{x}_1 + (a_{12}\,\iota p + r_{12})\,\dot{x}_2 + \dots &= X,\\ (a_{12}\,\iota p + r_{12})\,\dot{x}_1 + (a_{22}\,\iota p + r_{22})\,\dot{x}_2 + \dots &= 0,\\ (a_{13}\,\iota p + r_{13})\,\dot{x}_1 + (a_{23}\,\iota p + r_{23})\,\dot{x}_2 + \dots &= 0,\\ \cdots\cdots\cdots\cdots\cdots\cdots\end{aligned}$$

From the last $(n-1)$ of these equations we have

$$\frac{\dot{x}_1}{B_{11}} = \frac{\dot{x}_2}{B_{12}} = \frac{\dot{x}_3}{B_{13}} = \dots, \qquad (11)$$

where B_{pq} denotes the minor of the determinant

$$\begin{vmatrix} a_{11}\,\iota p + r_{11}, & a_{12}\,\iota p + r_{12}, & \dots \\ a_{12}\,\iota p + r_{12}, & a_{22}\,\iota p + r_{22}, & \dots \\ \cdots & \cdots & \cdots \end{vmatrix}$$

corresponding to the constituent $a_{pq}\,\iota p + r_{pq}$; we shall denote the determinant by Δ.

Substituting the values of $\dot{x}_2, \dot{x}_3, \ldots$ in the first equation, we have

$$(a_{11}\iota p + r_{11})\dot{x}_1 + \frac{1}{B_{11}}\{(a_{12}\iota p + r_{12})B_{12} + (a_{13}\iota p + r_{13})B_{13} + \ldots\}\dot{x}_1 = X_1,$$

which may be written

$$\frac{\Delta}{B_{11}}\dot{x}_1 = X_1. \qquad (12)$$

If Δ/B_{11} be written in the form $L\iota p + R$, where L and R are real quantities, then L is the effective self-induction of the circuit and R the impedance.

By equation (11) we have

$$\frac{\Delta}{B_{12}}\dot{x}_2 = X_1.$$

If an electromotive force X_2 of the same period as X_1 acted on the second circuit, then the current $\dot{x}_1$ induced in the first circuit would be given by

$$\frac{\Delta}{B_{12}}\dot{x}_1 = X_2.$$

Comparing these results we get Lord Rayleigh's theorem, that when a periodic electromotive force F acts on a circuit A the current induced in another circuit B is the same in amplitude and phase as the current induced in A when an electromotive force equal in amplitude and phase to F acts on the circuit B.

When there are only two circuits in the field,

$$\frac{\Delta}{B_{11}} = a_{11}\iota p + r_{11} - \frac{(a_{12}\iota p + r_{12})^2}{a_{22}\iota p + r_{22}};$$

if the circuits are not in metallic connection $r_{12} = 0$, and we have

$$\frac{\Delta}{B_{11}} = \left(a_{11} - \frac{p^2 a_{22} a^2_{12}}{a^2_{22}p^2 + r^2_{22}}\right)\iota p + r_{11} + \frac{p^2 r_{22} a^2_{12}}{a^2_{22}p^2 + r^2_{22}}.$$

Thus the presence of the second circuit diminishes the self-induction of the first by

$$\frac{p^2 a_{22} a^2_{12}}{a^2_{22}p^2 + r^2_{22}},$$

while it increases the impedance by

$$\frac{p^2 r_{22} a^2_{12}}{a^2_{22}p^2 + r^2_{22}}.$$

These results were given by Maxwell in his paper 'A Dynamical Theory of the Electromagnetic Field' (*Phil. Trans.* 155, p. 459, 1865). We see from these expressions that the diminution in the self-induction and the increase in the impedance increase continuously as the frequency of the electromotive force increases.

424.] Lord Rayleigh has shown that this result is true whatever may be the number of circuits. We have by (12)

$$\frac{\Delta}{B_{11}}\dot{x}_1 = X_1.$$

Now while keeping $\dot{x}_1$ the same we can choose $\dot{x}_2$, $\dot{x}_3$, &c., so that the two quadratic expressions

$$a_{22}\dot{x}^2_2 + a_{33}\dot{x}_3^2 + \dots 2a_{23}\dot{x}_2\dot{x}_3 + \dots,$$
$$r_{22}\dot{x}^2_2 + r_{33}\dot{x}^2_3 + \dots 2r_{23}\dot{x}_2\dot{x}_3 + \dots,$$

i.e. the expressions got by putting $\dot{x}_1 = 0$ in $2T$ and $2F$ respectively, reduce to the sums of squares of $\dot{x}_2$, $\dot{x}_3$, &c.; when $\dot{x}_2$, $\dot{x}_3$, &c. are chosen in this way,

$$a_{23} = a_{24} = a_{pq} = 0,$$

when p is not equal to q and both are greater than unity.

In this case

$$\Delta = \begin{vmatrix} a_{11}\iota p + r_{11}, & a_{12}\iota p + r_{12}, & a_{13}\iota p + r_{13}, & \dots & a_{1n}\iota p + r_{1n} \\ a_{12}\iota p + r_{12}, & a_{22}\iota p + r_{22}, & 0, & \dots & 0 \\ a_{13}\iota p + r_{13}, & 0, & a_{33}\iota p + r_{33}, & \dots & 0 \\ \cdot & \cdot & \cdot & \cdot & \cdot \\ a_{1n}\iota p + r_{1n}, & 0, & 0, & \dots & a_{nn}\iota p + r_{nn} \end{vmatrix}$$

$$= (a_{11}\iota p + r_{11})(a_{22}\iota p + r_{22})\dots(a_{nn}\iota p + r_{nn}) \times$$
$$\left\{1 - \frac{(a_{12}\iota p + r_{12})^2}{(a_{11}\iota p + r_{11})(a_{22}\iota p + r_{22})} - \frac{(a_{13}\iota p + r_{13})^2}{(a_{11}\iota p + r_{11})(a_{33}\iota p + r_{33})} - \dots - \frac{(a_{1n}\iota p + r_{1n})^2}{(a_{11}\iota p + r_{11})(a_{nn}\iota p + r_{nn})}\right\},$$

$$B_{11} = (a_{22}\iota p + r_{22})\dots(a_{nn}\iota p + r_{nn}).$$

Hence

$$\frac{\Delta}{B_{11}} = a_{11}\iota p + r_{11} - \frac{(a_{12}\iota p + r_{12})^2}{a_{22}\iota p + r_{22}} - \frac{(a_{13}\iota p + r_{13})^2}{a_{33}\iota p + r_{33}} - \dots - \frac{(a_{1n}\iota p + r_{1n})^2}{a_{nn}\iota p + r_{nn}}$$

$$= \left\{a_{11} + \Sigma\left(\frac{a_{nn}r_{1n}{}^2 - 2a_{1n}r_{1n}r_{nn}}{r_{nn}{}^2}\right) - \Sigma\left(\frac{a_{nn}p^2(a_{1n}r_{nn} - a_{nn}r_{1n})^2}{r^2_{nn}(a^2_{nn}p^2 + r^2_{nn})}\right)\right\}\iota p$$
$$+ r_{11} - \Sigma\frac{r^2_{1n}}{r_{nn}} + \Sigma\left(\frac{p^2(a_{1n}r_{nn} - a_{nn}r_{1n})^2}{r_{nn}(a^2_{nn}p^2 + r^2_{nn})}\right).$$

The coefficient of ιp in the first line is the coefficient of self-induction of the first circuit,—we see that it is diminished by any increase in p; the second line is the impedance, and we see that this is increased by any increase in p.

425.] We shall now return to the general case. The reduction of Δ/B_{11} to the form $L\iota p+R$ without any limitation as to the value of p would usually lead to very complicated expressions; we can, however, obtain without difficulty the values of L and R, (1) when p is very large, (2) when it is very small.

When ιp is very large we see that

$$L=\frac{D}{A_{11}},$$

where

$$D=\begin{vmatrix} a_{11}, & a_{12}\dots a_{1n} \\ a_{12}, & a_{22}\dots a_{2n} \\ \cdot & \cdot \quad \cdot \quad \cdot \\ a_{1n}, & a_{2n}\dots a_{nn} \end{vmatrix},$$

and A_{11} is the minor of D corresponding to the constituent a_{11}. If A_{pq} denotes the minor of D corresponding to the constituent a_{pq}, then we have by (11)

$$\frac{\dot{x}_1}{A_{11}}=\frac{\dot{x}_2}{A_{12}}=\dots=\frac{\dot{x}_n}{A_{1n}}. \tag{13}$$

Substituting these values of $\dot{x}_2$, $\dot{x}_3$, &c. in terms of $\dot{x}_1$, in the Dissipation Function, we find that

$$R=\frac{1}{A_{11}^{\,2}}\{r_{11}A_{11}^{\,2}+r_{22}A_{12}^{\,2}+\dots r_{nn}A_{1n}^{\,2}+2r_{12}A_{11}A_{12}+2r_{pq}A_{1p}A_{1q}+\dots\};$$

we might of course have deduced this value directly from that of Δ/B_{11}.

When ιp is very small, we see by putting $\iota p=0$ in Δ/B_{11} that

$$R=\frac{C}{R_{11}},$$

where

$$C=\begin{vmatrix} r_{11}, & r_{12}\dots r_{1n} \\ r_{12}, & r_{22}\dots r_{2n} \\ r_{1n}, & r_{2n}\dots r_{nn} \end{vmatrix},$$

and R_{11} is the minor of C corresponding to the constituent r_{11}; if R_{pq} denotes the minor of C corresponding to the constituent r_{pq}, then we have by (11)

$$\frac{\dot{x}_1}{R_{11}}=\frac{\dot{x}_2}{R_{12}}=\dots=\frac{\dot{x}_n}{R_{1n}}.$$

Substituting these values of $\dot{x}_1$, $\dot{x}_2$, $\dot{x}_3$, ... in the expression for the Kinetic Energy, we see that

$$L = \frac{1}{R_{11}^2}\{a_{11}R_{11}^2 + a_{22}R_{12}^2 + \dots 2a_{pq}R_{1p}R_{1q} + \dots\}.$$

426.] Suppose we have a series of circuits arranged so that each circuit acts by induction only on the two adjacent ones; this is expressed by the condition that a_{12} is finite but that a_{1p} vanishes when $p > 2$; again, a_{12}, a_{23} are finite, but a_{2p} vanishes if p differs from 2 by more than unity. Substituting these values of a_{1p}, a_{2p}, a_{3p}..., we easily find

$$A_{12} = -a_{12}\frac{dA_{11}}{da_{22}},$$

$$A_{13} = a_{12}a_{23}\frac{d^2A_{11}}{da_{22}da_{33}},$$

$$A_{14} = -a_{12}a_{23}a_{34}\frac{d^3A_{11}}{da_{22}da_{33}da_{44}},$$

. . . .

$$A_{1n} = (-1)^{n-1}a_{12}a_{23}a_{34}\dots a_{n-1\,n}.$$

Now T, the Kinetic Energy, is always positive, but the condition for this is (Maxwell's *Electricity and Magnetism*, vol. i. p. 111) that

$$D,\ A_{11},\ \frac{dA_{11}}{da_{22}},\ \frac{d^2A_{11}}{da_{22}da_{33}}\dots$$

should all be positive; hence we see if we take a_{12}, a_{23}..., &c. all positive, A_{11}, A_{12}, A_{13} will be alternately plus and minus, but when the frequency of the electromotive force is very great, $\dot{x}_1$, $\dot{x}_2$, ... are by (13) respectively proportional to A_{11}, A_{12}...; hence we see that in this case the adjacent currents are flowing in opposite directions: a result given by Lord Rayleigh. Another way of stating this result is to say that the direction of the currents is such that all the terms involving the product of two currents in the expression for the Kinetic Energy of the system of currents are negative, and in this form we recognise it as a consequence of the principle that the distribution of the currents must be such as to make the Kinetic Energy a minimum.

427.] We shall now apply these results to the case when the circuits are a series of m co-axial right circular solenoids of equal length, which act inductively on each other but which are not in metallic connection. We shall suppose that a is the radius of the first solenoid, b that of the second, c that of the third, and so on, a, b, c being in ascending order of magnitude; and that $n_1, n_2, n_3 \ldots$ are the numbers of turns of wire per unit length of the first, second, and third circuits. Then if l is the length of the solenoids, we have

$$\begin{array}{lll} a_{11} = 4\pi^2 n_1^2 l a^2, & a_{22} = 4\pi^2 n_2^2 l b^2, & a_{33} = 4\pi^2 n_3^2 l c^2, \\ a_{12} = 4\pi^2 n_1 n_2 l a^2, & a_{23} = 4\pi^2 n_2 n_3 l b^2, & a_{34} = 4\pi^2 n_3 n_4 l c^2, \\ a_{13} = 4\pi^2 n_1 n_3 l a^2, & a_{24} = 4\pi^2 n_2 n_4 l b^2, & \ldots\ldots \\ \ldots\ldots & \ldots\ldots & \end{array}$$

Hence

$$D = \begin{vmatrix} a_{11}, & a_{12}, & a_{13} \ldots \\ a_{12}, & a_{22}, & a_{23} \ldots \\ a_{13}, & a_{23}, & a_{33} \ldots \\ \cdot & \cdot & \cdot \end{vmatrix}$$

$$= (4\pi^2 l)^m n_1^2 n_2^2 n_3^2 \ldots a^2 (b^2 - a^2)(c^2 - b^2)(d^2 - c^2) \ldots,$$

and $$A_{11} = \frac{dD}{da_{11}}$$

$$= (4\pi^2 l)^{m-1} n_2^2 n_3^2 \ldots b^2 (c^2 - b^2)(d^2 - c^2) \ldots.$$

Now the coefficient of self-induction of the first circuit for very rapidly alternating current is

$$\frac{D}{A_{11}}.$$

Substituting the preceding expressions for D and A_{11}, we find that the self-induction equals

$$4\pi^2 l n_1^2 a^2 \left(1 - \frac{a^2}{b^2}\right).$$

Thus the only one of the circuits which affects the self-induction of the first is the one immediately adjacent to it. We can at once see the reason for this if we notice that

$$\frac{a_{12}}{a_{22}} = \frac{a_{13}}{a_{23}} = \frac{a_{14}}{a_{24}} = \ldots,$$

and therefore $$A_{13} = A_{14} = A_{15} = \ldots = 0.$$

Now when the rate of alternation is very rapid, $\dot{x}_3, \dot{x}_4, \dot{x}_5 \ldots$, the currents in the third, fourth, and fifth circuits, &c. are by equation (13) Art. (425) proportional to $A_{13}, A_{14}, A_{15} \ldots$; hence we see that in this case these currents all vanish, in other words the second solenoid forms a perfect electric screen, and screens off all induction from the solenoids outside it.

428.] Let us consider the case of three solenoids each of length l when the frequency is not infinitely rapid; we shall suppose that the primary coil is inside and has a radius a, number of turns per unit length n_1, resistance r; next to this is the secondary, radius b, turns per unit length n_2, resistance s; and outside this is the tertiary, radius c, turns per unit length n_3, resistance t. Since the circuits are not in metallic connection $r_{12} = r_{13} = r_{23} = 0$. If X_1, the electromotive force acting on the primary, is proportional to $\epsilon^{\iota pt}$, then we have by equations (11) and (12)

$$\dot{x}_3 = -\frac{n_1 n_3 a^2 s \iota p}{n_1^2 n_2^2 n_3^2 (4\pi^2 l)^2} \frac{X_1}{\begin{vmatrix} a^2 \iota p + \dfrac{r}{4\pi^2 l n_1^2}, & a^2 \iota p, & a^2 \iota p \\ a^2 \iota p, & b^2 \iota p + \dfrac{s}{4\pi^2 l n_2^2}, & b^2 \iota p \\ a^2 \iota p, & b^2 \iota p, & c^2 \iota p + \dfrac{t}{4\pi^2 l n_3^2} \end{vmatrix}}$$

We see from this expression that as long as the radius and length of the secondary remain the same, the effect produced by it on the current in the tertiary circuit depends on the ratio s/n_2^2, since s and n_2 only enter into the expression for $\dot{x}_3$ as constituents of the factor s/n_2^2. Thus all secondaries of radius b and length l will produce the same effect if s/n_2^2 remains constant.

We can apply this result to compare resistances in the following way: take two similar systems A and B each consisting of three co-axial solenoids, the primaries of A and B being exactly equal, as are also the two tertiaries, while the two secondaries are of the same size but differ as to the materials of which they are made. Let us use A and B as a Hughes' Induction Balance, putting the two primaries in series and connecting the tertiaries so that the currents generated in them by their respective primaries tend to circulate in opposite directions; then if, by altering if necessary the resistance in one of the secondaries, we make the resultant current in the combined tertiaries vanish, we know that s/n_2^2 is the same for A and B. Suppose that the secondary in B is a thin tube of thickness τ

and specific resistance σ, then considering the tube as a solenoid wound with wire of square section a packed close together, we see that for the tube

$$s = 2\pi b l n_2 \frac{\sigma}{a} = 2\pi b l n_2{}^2 \frac{\sigma}{\tau}.$$

Now $s/n_2{}^2$ for the tube is equal to $s/n_2{}^2$ for the secondary of A, which may be an ordinary solenoid. We thus have

$$\frac{s}{n_2{}^2} = 2\pi b l \sigma/\tau,$$

a relation by which we can deduce σ.

In order that this method should be sensitive the interposition of the secondary ought to produce a considerable effect on the currents induced in the tertiary. If the resistance of the secondary is large this will not happen unless the frequency of the electromotive force is very great; for ordinary metals a frequency of about a thousand is sufficient, but this would be useless if the specific resistance of the tube were comparable with that of electrolytes.

On the other hand, if the frequency is infinite, there will not be any current in the tertiaries whatever the resistance of the secondaries may be.

Wheatstone's Bridge with Self-Induction in the Arms.

429.] The preceding investigation can be applied to find the effect of self-induction in the arms of a Wheatstone's Bridge. Let $ABCO$ represent the bridge, let an electromotive force X proportional to $\epsilon^{\iota pt}$ act in the arm CB. Let x be the current in CB, y that in BA, z that in AO, then the currents along BO, AC, OC are respectively $x-y$, $y-z$, and $x-y+z$.

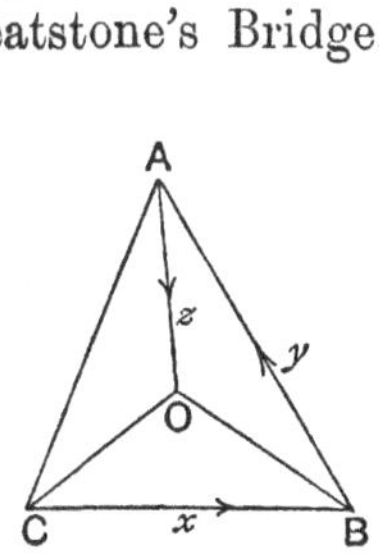

Fig. 140.

Let the self-induction in CB, BA, AC, AO, BO, CO be respectively A, C, B, L, M, N, while the resistance in these arms are respectively a, c, b, a, β, γ. We suppose, moreover, that there is no mutual induction between the various arms of the Bridge. Then the Kinetic Energy T of the system of currents is expressed by the equation

$$2T = Ax^2 + Cy^2 + B(y-z)^2 + Lz^2 + M(x-y)^2 + N(x-y+z)^2.$$

The Dissipation Function F is given by the expression

$$2F = ax^2 + cy^2 + b(y-z)^2 + az^2 + \beta(x-y)^2 + \gamma(x-y+z)^2.$$

Comparing this with our previous notation, we must put

$$a_{11} = A+M+N, \quad a_{12} = -(M+N),$$
$$a_{22} = B+C+M+N, \quad a_{13} = N,$$
$$a_{33} = B+L+N, \quad a_{23} = -(N+B);$$

$$r_{11} = a+\beta+\gamma, \quad r_{12} = -(\beta+\gamma),$$
$$r_{22} = b+c+\beta+\gamma, \quad r_{13} = \gamma,$$
$$r_{33} = b+a+\gamma, \quad r_{23} = -(\gamma+b).$$

Now by equations (11) and (12)

$$z = \frac{B_{13}}{\Delta} X,$$

where B_{13} is the minor of Δ corresponding to the constituent $a_{13}\iota p + r_{13}$, i.e.

$$B_{13} = (a_{12}\iota p + r_{12})(a_{23}\iota p + r_{23}) - (a_{22}\iota p + r_{22})(a_{13}\iota p + r_{13}).$$

Substituting the preceding values for the a's and the r's, we find

$$B_{13} = -p^2(MB-NC) + \iota p(Mb + B\beta - Nc - C\gamma) + b\beta - c\gamma.$$

Now if z vanishes B_{13} must vanish; hence if the Bridge is balanced for all values of p we must have

$$MB - NC = 0,$$
$$Mb + B\beta - Nc - C\gamma = 0,$$
$$b\beta - c\gamma = 0;$$

while if the Bridge is only balanced for a particular value of p, we have

$$b\beta - c\gamma = p^2(MB-NC),$$
$$p(Mb + B\beta - Nc - C\gamma) = 0.$$

When the frequency is very great the most important term in the expression for B_{13} is $-p^2(MB-NC)$, so that the most important condition to be fulfilled when the Bridge is balanced is

$$MB - NC = 0;$$

thus for high frequencies the Bridge tests the self-induction rather than the resistances of its arms.

Combination of Self-Induction and Capacity.

430.] We have supposed in the preceding investigations that the circuits were closed and devoid of capacity; very interesting results, however, occur when some or all of the circuits are cut and their free ends connected to condensers of suitable capacity. We can by properly adjusting the capacity inserted in a circuit in relation to the frequency of the electromotive force and the self-induction of the circuit, make the circuit behave under the action of an electromotive force of given frequency as if it possessed no apparent self-induction.

The explanation of this will, perhaps, be clear if we consider the behaviour of a simple mechanical system under the action of a periodic force. The system we shall take is that of the rectilinear motion of a mass attached to a spring and resisted by a frictional force proportional to its velocity.

Suppose that an external periodic force X acts on the system, then at any instant X must be in equilibrium with the resultant of (1) minus the rate of change of momentum of the system, (2) the force due to the compression or extension of the spring, (3) the resistance. If the frequency of X is very great, then for a given momentum (1) will be very large, so that unless it is counterbalanced by (2) a finite force of infinite frequency would produce an infinitely small momentum. Let us, however, suppose that the frequency of the force is the same as that of the free vibrations of the system when the friction is zero. When the mass vibrates with this frequency (1) and (2) will balance each other, thus all the external force has to do is to balance the resistance. The system will thus behave like one without either mass or stiffness resisted by a frictional force.

In the corresponding electrical system, self-induction corresponds to mass, the reciprocal of the capacity to the stiffness of the spring, and the electric resistance to the frictional resistance. If now we choose the capacity so that the period of the electrical vibrations, calculated on the supposition that the resistance of the circuit vanishes, is the same as that of the external electromotive force, the system will behave as if it had neither self-induction nor capacity but only resistance. Hence, if L is the self-induction of a circuit whose ends are connected to the plates of a condenser whose capacity in electromagnetic measure is C, the system will behave as if it had no self-induction under an electromotive force whose frequency is $p/2\pi$ if $LCp^2=1$.

431.] We shall now consider the case represented in the figure, where we have two circuits in parallel, one of the circuits being cut and its ends connected to the plates of a condenser. Let Λ be the self-induction of the leads, r their resistance; L, N the

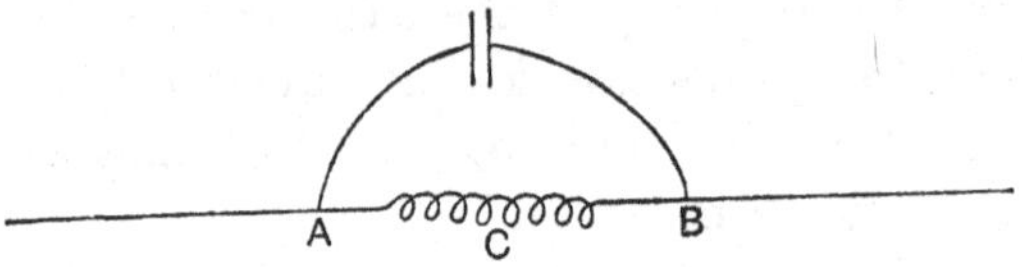

Fig. 141.

coefficients of self-induction of ACB and the condenser circuit respectively, M the coefficient of mutual induction between these circuits. Let R, S be the resistances respectively of ACB and the condenser circuit, C the capacity of the condenser. Let $\dot{x}$ be the current in the leads, $\dot{y}$ that in the condenser circuit, then that in the circuit ACB will be $\dot{x}-\dot{y}$. Let X, the electromotive force in the leads, be proportional to $\epsilon^{\iota pt}$. If there is no mutual induction between the leads and the wires in parallel, the equations giving $\dot{x}$, $\dot{y}$ are

$$(\Lambda+L)\frac{d\dot{x}}{dt}-(L-M)\frac{d\dot{y}}{dt}+(r+R)\dot{x}-R\dot{y}=X,$$

$$(L+N-2M)\frac{d\dot{y}}{dt}-(L-M)\frac{d\dot{x}}{dt}+(S+R)\dot{y}-R\dot{x}+\frac{y}{C}=0.$$

Substituting the value of $\dot{y}$ in terms of $\dot{x}$ and remembering that $d/dt=\iota p$, we get

$$\left\{\Lambda+L+\frac{\xi\{R^2-(L-M)^2p^2\}-2R(R+S)(L-M)}{p^2\xi^2+(R+S)^2}\right\}\iota p\dot{x}$$
$$+\left\{r+R-\frac{(R+S)\{R^2-(L-M)^2p^2\}+2p^2\xi R(L-M)}{p^2\xi^2+(R+S)^2}\right\}\dot{x}=X. \quad (14$$

where $$\xi=(L+N-2M)-\frac{1}{Cp^2}.$$

From the form of this equation we see that the self-induction of the two circuits in parallel is

$$L+\frac{\xi\{R^2-(L-M)^2p^2\}-2R(R+S)(L-M)}{p^2\xi^2+(R+S)^2},$$

this will vanish if

$$Lp^2\xi^2+\xi\{R^2-(L-M)^2p^2\}$$
$$+(R+S)\{L(R+S)-2R(L-M)\}=0. \qquad (15)$$

If the roots of this quadratic are real, then it is possible to choose C so that the self-induction of the loop vanishes. An important special case is when $S=0$, $M=0$, when the quadratic reduces to

$$Lp^2\xi^2+\xi(R^2-L^2p^2)-LR^2=0;$$

thus $$\xi=-\frac{R^2}{Lp^2} \text{ or } L;$$

the first root gives

$$\frac{1}{C}=(L+N)p^2+\frac{R^2}{L},$$

the second

$$\frac{1}{C}=Np^2;$$

this last value of $1/C$ makes $\dot{x}=\dot{y}$, so that none of the current goes through ACB.

When ξ satisfies (15) the self-induction of the loop vanishes. If in that equation we substitute $L+\Lambda$ for L and $M+\Lambda$ for M, the values of ξ which satisfy the new equation will make the self-induction of the whole circuit vanish.

432.] We shall next consider the case of an induction coil or transformer, the primary of which is cut and its free ends connected to the plates of a condenser whose capacity is C. Let L, N be the self-induction of the primary and secondary respectively, M the coefficient of mutual induction between the two, R the resistance of the primary, S that of the secondary, $\dot{x}$, $\dot{y}$ the currents in the primary and secondary respectively; then if X is the electromotive force acting on the primary, we have

$$L\frac{d\dot{x}}{dt}+M\frac{d\dot{y}}{dt}+R\dot{x}+\frac{x}{C}=X,$$

$$M\frac{d\dot{x}}{dt}+N\frac{d\dot{y}}{dt}+S\dot{y}=0.$$

Hence if X varies as $\epsilon^{\iota pt}$, we find

$$\dot{y}=\frac{-M\iota pX}{-p^2(\xi N-M^2)+RS+\iota p(RN+S\xi)},$$

where $$\xi=L-\frac{1}{Cp^2}.$$

The amplitude of $\dot{y}$ for a given amplitude of X is proportional to

$$\frac{XMp}{\{(RS-p^2(\xi N-M^2))^2+(RN+S\xi)^2p^2\}^{\frac{1}{2}}}.$$

This vanishes when $p=0$, because in this case the current in the primary is steady; it also vanishes in general when p is infinite, because in consequence of the self-induction of the primary only an indefinitely small current passes through it in this case. If however

$$\xi N = M^2,$$

or

$$\frac{1}{Cp^2} = L - \frac{M^2}{N},$$

then the amplitude of the current in the secondary is finite when p is infinite, and is equal to

$$\frac{MNX}{RN^2+SM^2};$$

thus when the frequency of the electromotive force is very high the amplitude of the current in the secondary may be increased enormously by cutting the primary circuit and connecting its ends to a condenser of suitable capacity.

432*.] We can apply a method similar to that of Art. 424 to determine the effect of placing a vibrating electrical system near a number of other such systems.

We shall suppose that the systems are not in electrical connection, and neglect the resistances of the circuits. Let T be the Kinetic, V the Potential Energy of the system of currents; let $\dot{x}_1$ denote the current in the first circuit, and let $\dot{x}_2, \dot{x}_3, \ldots$, the currents in the other circuits, be so chosen that when x_1 is put equal to zero the expressions for T and V reduce to the sums of squares of $\dot{x}_2, \dot{x}_3, \ldots;\ x_2, x_3, \ldots$ respectively.

Let T be given by the same expression as in Article 424, while

$$V = \tfrac{1}{2}\left\{\frac{x_1^2}{c_1} + \frac{x_2^2}{c_2} + \frac{x_3^2}{c_3} + \ldots\right\}.$$

Then the equations of the type

$$\frac{d}{dt}\frac{dT}{d\dot{x}} + \frac{dV}{dx} = 0$$

give, if all the variables are proportional to $\epsilon^{\iota pt}$,

$$\left(-a_{11}p^2+\frac{1}{c_1}\right)x_1-a_{12}p^2x_2-a_{13}p^2x_3-\ldots=0$$

$$-a_{12}p^2x_1+\left(-a_{22}p^2+\frac{1}{c_2}\right)x_2 \qquad =0$$

$$-a_{13}p^2x_1+\left(-a_{33}p^2+\frac{1}{c_3}\right)x_3 \qquad =0.$$

$$\cdots\cdots\cdots\cdots\cdots$$

Hence substituting for x_2, x_3 in terms of x_1, we get

$$-a_{11}p^2+\frac{1}{c_1}=\frac{a_{12}^2p^4}{\dfrac{1}{c_2}-a_{22}p^2}+\frac{a_{13}^2p^4}{\dfrac{1}{c_3}-a_{33}p^2}+\ldots.$$

Let us suppose that the period of the first system is only slightly changed, so that we may in the right-hand side of this equation write p_1 for p, where p_1 is the value of p when the first vibrator is alone in the field.

Let $p_2, p_3, \ldots$ be the values of p for the other vibrators when the first one is absent, then

$$\frac{1}{c_2}=p_2^2a_{22}$$

$$\frac{1}{c_3}=p_3^2a_{33}.$$

Thus if δp_1^2 denotes the increase in p_1^2 due to the presence of the other vibrators, we have

$$-a_{11}\delta p_1^2=p_1^4\left\{\frac{a_{12}^2}{a_{22}(p_2^2-p_1^2)}+\frac{a_{13}^2}{a_{33}(p_3^2-p_1^2)}+\ldots\right\}.$$

Thus we see that if p_2 is greater than p_1 the effect of the proximity of the circuit whose period is p_2 is to diminish p_1, while if p_2 is less than p_1 the proximity of this circuit increases p_1. Similar remarks apply to the other circuits. Thus the first system, if its free period is slower than that of the second, is made to vibrate still more slowly by the presence of the latter; while if its free period is faster than that of the second the presence of the latter makes it vibrate still more quickly. In other words, the effect of putting two vibrators near together is to make the difference between their periods greater than it is when the vibrators are free from each other's influence; the quicker period is accelerated, the slower one retarded.

CHAPTER VII.

ELECTROMOTIVE INTENSITY IN MOVING BODIES.

433.] THE equations (B) given in Art. 598 of Maxwell's *Electricity and Magnetism* for the components of the electromotive intensity in a moving body involve a quantity Ψ, whose physical meaning it is desirable to consider more fully. The investigation by which the equations themselves are deduced tells us nothing about Ψ; it is introduced after the investigation is finished, so as to make the expressions for the electromotive intensity as general as it is possible for them to be and yet be consistent with Faraday's Law of the induction of currents in a variable magnetic field.

Let u, v, w denote the components of the velocity of the medium; a, b, c the components of the magnetic induction; F, G, H those of the vector potential; X, Y, Z those of the electromotive intensity.

In the course of Maxwell's investigation of the values of X, Y, Z due to induction, the terms

$$-\frac{d}{dx}(Fu+Gv+Hw), \quad -\frac{d}{dy}(Fu+Gv+Hw),$$
$$-\frac{d}{dz}(Fu+Gv+Hw)$$

respectively in the final expressions for X, Y, Z are included under the Ψ terms. We shall find it clearer to keep these terms separate and write the expressions for X, Y, Z as

$$\left.\begin{aligned} X &= cv-bw-\frac{dF}{dt}-\frac{d}{dx}(Fu+Gv+Hw)-\frac{d\phi}{dx}, \\ Y &= aw-cu-\frac{dG}{dt}-\frac{d}{dy}(Fu+Gv+Hw)-\frac{d\phi}{dy}, \\ Z &= bu-av-\frac{dH}{dt}-\frac{d}{dz}(Fu+Gv+Hw)-\frac{d\phi}{dz}. \end{aligned}\right\} \quad (1)$$

For Faraday's law to hold, the line integral of the electromotive intensity taken round any closed curve must be independent of ϕ, hence ϕ must be a continuous function.

When there is no free electricity

$$\frac{dX}{dx}+\frac{dY}{dy}+\frac{dZ}{dz}=0.$$

Substituting the values of X, Y, Z just given, we find, using

$$\frac{dF}{dx}+\frac{dG}{dy}+\frac{dH}{dz}=0,$$

$$F\nabla^2 u+G\nabla^2 v+H\nabla^2 w+2\left(\frac{dF}{dx}\frac{du}{dx}+\frac{dG}{dy}\frac{dv}{dy}+\frac{dH}{dz}\frac{dw}{dz}\right)$$
$$+\left(\frac{dH}{dy}+\frac{dG}{dz}\right)\left(\frac{dw}{dy}+\frac{dv}{dz}\right)+\left(\frac{dF}{dz}+\frac{dH}{dx}\right)\left(\frac{du}{dz}+\frac{dw}{dx}\right)$$
$$+\left(\frac{dG}{dx}+\frac{dF}{dy}\right)\left(\frac{dv}{dx}+\frac{du}{dy}\right)=-\nabla^2\phi.$$

If the medium is moving like a rigid body, then

$$u=p+\omega_2 z-\omega_3 y,$$
$$v=q+\omega_3 x-\omega_1 z,$$
$$w=r+\omega_1 y-\omega_2 x;$$

where p, q, r are the components of the velocity of the origin and ω_1, ω_2, ω_3 the rotations about the axes of x, y, z respectively. Substituting these values we see that whenever the system moves as a rigid body

$$\nabla^2\phi=0.$$

434.] In order to see the meaning of ϕ we shall take the case of a solid sphere rotating with uniform angular velocity ω about the axis of z in a uniform magnetic field where the magnetic induction is parallel to the axis z and is equal to c. We may suppose that the magnetic induction is produced by a large cylindrical solenoid with the axis of z for its axis; in this case

$$F=-\tfrac{1}{2}cy,\quad G=\tfrac{1}{2}cx,\quad H=0.$$

In the rotating sphere

$$u=-\omega y,\quad v=\omega x,\quad w=0.$$

If the system is in a steady state, dF/dt, dG/dt, dH/dt all vanish.

Thus in the sphere

$$X = c\omega x - \tfrac{1}{2}\frac{d}{dx}\{c\omega(x^2+y^2)\} - \frac{d\phi}{dx},$$

$$Y = c\omega y - \tfrac{1}{2}\frac{d}{dy}\{c\omega(x^2+y^2)\} - \frac{d\phi}{dy},$$

$$Z = \qquad\qquad\qquad\qquad - \frac{d\phi}{dz};$$

these equations reduce to

$$X = -\frac{d\phi}{dx},$$

$$Y = -\frac{d\phi}{dy},$$

$$Z = -\frac{d\phi}{dz},$$

and we have also $\nabla^2\phi = 0$.

In the space outside the sphere the medium does not move as a rigid body. The process by which the equations (1) were obtained could not without further investigation be held to justify us in applying them to cases where the velocity is discontinuous, for in the investigation, see Maxwell, Art. 598, it is assumed that the variations δx, δy, δz are continuous, and that these are proportional to the components of the velocity. To avoid any discontinuity in the velocity at the surface of the sphere we shall suppose that the medium in contact with the sphere moves at the same rate as the sphere, but that as we recede from the surface of the sphere the velocity diminishes in the same way as it does in a viscous fluid surrounding a rotating sphere. Thus we shall suppose that the rotating sphere whose radius is $\mathbf{a}$ is surrounded by a fixed sphere whose radius is $\mathbf{b}$, and that between the spheres the components of the velocity are given by the expressions

$$u = -\left(A\frac{d}{dy}\frac{1}{r} + By\right), \quad v = \left(A\frac{d}{dx}\frac{1}{r} + Bx\right), \quad w = 0,$$

where r is the distance from the centre of the rotating sphere.

When $r = \mathbf{b}$, $u = 0$, $v = 0$, hence

$$-\frac{A}{\mathbf{b}^3} + B = 0;$$

when $r = \mathbf{a}$, $u = -\omega y$, $v = \omega x$, hence

$$-\frac{A}{\mathbf{a}^3} + B = \omega,$$

hence

$$A = -\frac{\omega\,\mathbf{a}^3\mathbf{b}^3}{\mathbf{b}^3 - \mathbf{a}^3}.$$

Substituting these values of u, v in equation (1), we find that when $\mathbf{a} < r < \mathbf{b}$,

$$X = \tfrac{1}{2} c A\,(x^2 + y^2)\frac{d}{dx}\frac{1}{r^3} - \frac{d\phi}{dx},$$

$$Y = \tfrac{1}{2} c A\,(x^2 + y^2)\frac{d}{dy}\frac{1}{r^3} - \frac{d\phi}{dy},$$

$$Z = \tfrac{1}{2} c A\,(x^2 + y^2)\frac{d}{dz}\frac{1}{r^3} - \frac{d\phi}{dz};$$

hence, since

$$\frac{dX}{dx} + \frac{dY}{dy} + \frac{dZ}{dz} = 0,$$

we have

$$\nabla^2\phi = 0.$$

Again, when $r > \mathbf{b}$ the medium is at rest, here we have

$$X = -\frac{d\phi}{dx},$$

$$Y = -\frac{d\phi}{dy},$$

$$Z = -\frac{d\phi}{dz},$$

and $\nabla^2\phi = 0$.

The boundary conditions satisfied by ϕ and its differential coefficients will depend upon whether the sphere is a conductor or an insulator. We shall first consider the case when it is an insulated conductor. In this case, when the system is in a steady state, the radial currents in the sphere must vanish, otherwise the electrical condition of the surface of the sphere could not be constant.

Thus at any point on the surface of the sphere

$$xX + yY + zZ = 0,$$

this is equivalent to

$$\frac{d\phi_1}{dr} = 0,$$

where ϕ_1 is the value of ϕ inside the rotating sphere; hence we have

$$\phi_1 = K,$$

where K is a constant.

If ϕ_2, ϕ_3 are the values of ϕ in the region between the fixed and moving spheres, and in the fixed sphere respectively, then we may put

$$\phi_2 = L + \frac{M}{r} + NQ_2\left(\frac{r^2}{\mathrm{a}^2} - \frac{\mathrm{a}^3}{r^3}\right),$$

$$\phi_3 = \frac{PQ_2}{r^3},$$

where L, M, N, P are constants, and Q_2 is the second zonal harmonic with z for its axis.

The continuity of ϕ gives

$$K = L + \frac{M}{\mathrm{a}}, \quad 0 = L + \frac{M}{\mathrm{b}},$$

$$P = \frac{N(\mathrm{b}^5 - \mathrm{a}^5)}{\mathrm{a}^2}.$$

If K_1 is the specific inductive capacity of the medium between the two spheres, K_2 that of the medium beyond the outer sphere; then, since the normal electric polarization must be continuous when $r = \mathrm{b}$, we have

$$3K_2P\frac{Q_2}{\mathrm{b}^4} = K_1\left\{cA\frac{(Q_2 - 1)}{\mathrm{b}^2} + \frac{M}{\mathrm{b}^2} - NQ_2\left(\frac{2\mathrm{b}}{\mathrm{a}^2} + \frac{3\mathrm{a}^3}{\mathrm{b}^4}\right)\right\}.$$

Solving these equations we find

$$\left.\begin{aligned} P &= \frac{cAK_1\mathrm{b}^2(\mathrm{b}^5 - \mathrm{a}^5)}{3K_2(\mathrm{b}^5 - \mathrm{a}^5) + K_1(2\mathrm{b}^5 + 3\mathrm{a}^5)}, \\ N &= \frac{cAK_1\mathrm{a}^2\mathrm{b}^2}{3K_2(\mathrm{b}^5 - \mathrm{a}^5) + K_1(2\mathrm{b}^5 + 3\mathrm{a}^5)}, \\ M &= cA, \quad L = -cA/\mathrm{b}, \quad K = cA(\mathrm{b} - \mathrm{a})/\mathrm{ab}, \end{aligned}\right\} \quad (2)$$

where $$A = -\omega\mathrm{a}^3\mathrm{b}^3/(\mathrm{b}^3 - \mathrm{a}^3).$$

The surface density of the electricity on the moving sphere is

$$\frac{K_1cA}{4\pi\mathrm{a}^2}\left\{\frac{K_1(2\mathrm{b}^5 + 3\mathrm{a}^5 - 5\mathrm{a}^3\mathrm{b}^2) + 3K_2(\mathrm{b}^5 - \mathrm{a}^5)}{3K_2(\mathrm{b}^5 - \mathrm{a}^5) + K_1(2\mathrm{b}^5 + 3\mathrm{a}^5)}\right\}Q_2.$$

The preceding formulæ are general; we shall now consider some particular cases.

435.] The first we shall consider is when $\mathrm{b} - \mathrm{a} = \delta$ is small compared with either b or a. In this case we have approximately, when K_2 is not infinite,

$$P = -\tfrac{1}{3}c\omega\mathrm{a}^5, \quad N = -\tfrac{1}{15}\frac{c\omega\mathrm{a}^3}{\delta}, \quad M = -\tfrac{1}{3}\frac{c\omega\mathrm{a}^4}{\delta},$$

$$L = \tfrac{1}{3}\frac{c\omega\mathrm{a}^3}{\delta}, \quad K = -\tfrac{1}{3}c\omega\mathrm{a}^2.$$

Thus in the outer fixed sphere the components of the electromotive intensity are equal to the differential coefficients with respect to x, y, z of the function

$$\tfrac{1}{3} c\, \omega\, \mathbf{a}^5 \frac{Q_2}{r^3}.$$

Thus the radial electromotive intensity close to the surface of the rotating sphere is

$$-c\omega\mathbf{a}Q_2,$$

while the tangential intensity is

$$-c\omega\mathbf{a}\sin\theta\cos\theta.$$

These results show that the effects produced by rotating uncharged spheres in a strong magnetic field ought to be quite large enough to be measurable. Thus if the sphere is rotating so fast that a point on its equator moves with the velocity 3×10^3, which is about 100 feet per second, and if $c = 10^3$, then the maximum radial intensity is about 1/33 of a volt per centimetre, and the maximum tangential intensity about 1/2 of this: these are quite measurable quantities, and if it were necessary to increase the effect both c and ω might be made considerably greater than the values we have assumed.

The surface density of the electricity on the rotating sphere when $(\mathbf{b}-\mathbf{a})/\mathbf{a}$ is small is

$$-\frac{1}{4\pi} K_2 c\,\omega\,\mathbf{a}\,Q_2.$$

436.] If the outer fixed sphere is a conductor, the electromotive intensity must vanish when $r > \mathbf{b}$, hence $P=0$, so that $N=0$, while M, L, K have the same values as before. In this case the surface density of the electricity on the surface of the rotating sphere is

$$\frac{K_1}{4\pi\mathbf{a}^2} c A Q_2,$$

and when $\mathbf{b}-\mathbf{a}$ is small, this is equal to

$$-\frac{K_1}{12\pi\delta} c\,\omega\,\mathbf{a}^2 Q_2.$$

Since this expression is proportional to $1/\delta$, the surface density can be increased to any extent by diminishing the distance between the rotating and fixed surfaces.

In the general case, when $\mathbf{b}-\mathbf{a}$ is not necessarily small, the surface density of the electricity on the rotating sphere is

$$\frac{K_1}{4\pi \mathbf{a}^2} c A Q_2,$$

the surface density on the fixed sphere is

$$-\frac{K_1}{4\pi \mathbf{b}^2} c A Q_2.$$

The electrostatic potential due to this distribution of electricity at a distance r from the centre of the rotating sphere is, when $r > \mathbf{b}$,

$$-\frac{cA}{5}(\mathbf{b}^2-\mathbf{a}^2)\frac{Q_2}{r^3},$$

while when $r < \mathbf{a}$ it is

$$-\frac{cA}{5}\left(\frac{1}{\mathbf{b}^3}-\frac{1}{\mathbf{a}^3}\right) r^2 Q_2.$$

The values of ϕ in these regions are respectively zero and a constant. Hence this example is sufficient to show us that ϕ is not equal to the electrostatic potential due to the free electricity on the surface of the conductors.

437.] We may (though there does not seem to be any advantage gained by so doing) regard ϕ as the sum of two parts, one of which, ϕ_e, is the electrostatic potential due to the distribution of free electricity over the surfaces separating the different media; the other, ϕ_m, being regarded as peculiarly due to electromagnetic induction.

Let us consider the case of a body moving in any manner, then we must have, since there is no volume distribution of electricity,

$$\nabla^2 \phi_e = 0.$$

If σ is the surface density of the electricity over any surface of separation at a point where the direction cosines of the outward drawn normal are l, m, n, then if K is the specific inductive capacity

$$4\pi\sigma = \left[K(lX + mY + nZ)\right]_1^2,$$

where the expression on the right-hand side of this equation denotes the excess of the value of $K(lX+mY+nZ)$ in the outer medium over its value in the inner. But if ϕ_e is the electrostatic potential, then

$$4\pi\sigma = -\left[K\left(l\frac{d\phi_e}{dx} + m\frac{d\phi_e}{dy} + n\frac{d\phi_e}{dz}\right)\right]_1^2.$$

From these conditions we see from equations (1) that

$$\nabla^2(\phi_m+Fu+Gv+Hw)$$
$$=\frac{d}{dx}(cv-bw)+\frac{d}{dy}(aw-cu)+\frac{d}{dz}(bu-av),$$

and
$$\left[K\left(l\frac{d}{dx}+m\frac{d}{dy}+n\frac{d}{dz}\right)(\phi_m+Fu+Gv+Hw)\right]_1^2$$
$$=\left[K\{l(cv-bw)+m(aw-cu)+n(bu-av)\}\right]_1^2.$$

From these equations ϕ_m is uniquely determined, for we see that $\phi_m+Fu+Gv+Hw$ is the potential due to a distribution of electricity whose volume density is

$$-\frac{1}{4\pi}\left\{\frac{d}{dx}(cv-bw)+\frac{d}{dy}(aw-cu)+\frac{d}{dz}(bu-av)\right\},$$

together with a distribution whose surface density is

$$-\frac{1}{4\pi}\left[K\{l(cv-bw)+m(aw-cu)+n(bu-av)\}\right]_1^2.$$

Having thus determined ϕ_m and deducing ϕ by the process exemplified in the preceding examples we can determine ϕ_e.

438.] The question as to whether or not the equations (1) are true for moving insulators as well as for moving conductors, u, v, w being the components of the velocity of the insulator, is a very important one. The truth of these equations for conductors has been firmly established by experiment, but we have, so far as I am aware, no experimental verification of them for insulators. The following considerations suggest, I think, that some further evidence is required before we can feel assured of the validity of the application of these equations to insulators. We may regard a steady magnetic field as one in which Faraday tubes are moving about according to definite laws, the positive tubes moving in one direction, the negative ones in the opposite, the tubes being arranged so that as many positive as negative tubes pass through any area. When a conductor is moved about in such a magnetic field it disturbs the motion of the tubes, so that at some parts of the field the positive tubes no longer balance the negative and an electromotive intensity is produced in such regions. To assume the truth of equations (1), whatever the nature of the moving body may be, is, from this point of view, to assume that the effect on these tubes is the same whether the moving body be a conductor or an insulator of large or small specific inductive capacity. Now it is quite conceivable that though a conductor, or a dielectric with a considerable inductive capacity, might when in motion produce a considerable disturbance of the Faraday tubes in the ether in and around it, yet little or no effect might be produced by the motion of a substance of small specific inductive capacity such as a gas, and thus it might be expected that the electromotive intensity due to the motion of a conductor in a magnetic field would be much greater than that due to the motion of a gas moving with the same speed.

439.] As one of the most obvious methods of determining whether or not equations (1) are true for dielectrics is to investigate the effect of rotating an insulating sphere in a magnetic field: we give the solution of the case similar to the one discussed in Art. 434, with the exception that the metallic rotating sphere of that article is replaced by an insulating one, specific inductive capacity K_0, of the same radius. Using the notation of that article, we easily find that in this case

$$P\left\{\frac{3K_2}{\mathbf{b}^4}+\frac{2\mathbf{b}(3K_1+2K_0)K_1-6K_1(K_1-K_0)\mathbf{a}^5/\mathbf{b}^4}{2(K_1-K_0)\mathbf{a}^5+(3K_1+2K_0)\mathbf{b}^5}\right\}$$
$$=cK_1A\left\{\frac{1}{\mathbf{b}^2}-\frac{5\mathbf{a}^2\mathbf{b}K_1}{2(K_1-K_0)\mathbf{a}^5+(3K_1+2K_0)\mathbf{b}^5}\right\}.$$

When $\mathbf{b}-\mathbf{a}$ is small, this becomes

$$P=-\tfrac{1}{3}\frac{2K_0}{3K_2+2K_0}c\omega\mathbf{a}^5.$$

So that in this case the components of electromotive intensities in the region at rest are equal to the differential coefficients with respect to x, y, z of the function

$$\tfrac{1}{3}\frac{2K_0}{3K_2+2K_0}\frac{c\omega\mathbf{a}^5}{r^3}Q_2,$$

and thus, by Art. 435, bear to the intensities produced by the rotating conductor the ratio of $2K_0$ to $3K_2+2K_0$.

Thus, if equations (1) are true for insulators, a rotating sphere made of an insulating material ought to produce an electric field comparable with that due to a rotating metallic sphere of the same size.

The greatest difficulty in experimenting with the insulating sphere would be that it would probably get electrified by friction, but unless this completely overpowered the effect due to the rotation we ought to be able to distinguish between the two effects, since the rotational one is reversed when the direction of rotation is reversed as well as when the magnetic field is reversed.

In deducing equations (2) of Art. 434, we assumed that equations (1) held in the medium between the fixed and moving surfaces, the general equations will therefore only be true on this assumption. In the special case, however, when the layer of this medium is indefinitely thin, the results will be the same whether this medium is an insulator or conductor, so that the results in this special case would not throw any light on whether equations (1) do or do not hold for a moving dielectric.

Propagation of Light through a Moving Dielectric.

440.] We might expect that some light would be thrown on the electromotive intensity developed in a dielectric moving in a magnetic field by the consideration of the effect which the motion of the dielectric would have on the velocity of light passing through it. We shall therefore investigate the laws of propagation of light through a dielectric moving uniformly with the velocity components u, v, w.

In this case, since we have only to deal with insulators, all the currents in the field are polarization currents due to alterations in the intensity of the polarization. When the dielectric is moving we are confronted with a question which we have not had to consider previously, and that is whether the equivalent current is to be taken as equal to the time rate of variation of the polarization at a point fixed in space or at a point fixed in the dielectric and moving with it; i.e. if f is the dielectric polarization parallel to x, is the current parallel to x

$$\frac{df}{dt},$$

or

$$\frac{df}{dt} + u\frac{df}{dx} + v\frac{df}{dy} + w\frac{df}{dz}?$$

In the first case we should have, if α, β, γ are the components of the magnetic force,

$$4\pi\frac{df}{dt} = \frac{d\gamma}{dy} - \frac{d\beta}{dz}; \tag{3}$$

in the second,

$$4\pi\left(\frac{df}{dt} + u\frac{df}{dx} + v\frac{df}{dy} + w\frac{df}{dz}\right) = \frac{d\gamma}{dy} - \frac{d\beta}{dz}. \tag{4}$$

This point seems one which can only be settled by experiment. It seems desirable, however, to look at the question from as many points of view as possible; the equation connecting the current with the magnetic force is the expression of the fact that the line integral of the magnetic force round any closed curve is equal to 4π times the rate of increase of the number of Faraday tubes passing through the curve. We saw in Chapter I. that this was equivalent to saying that a Faraday tube when in motion gave rise to a magnetic force at right angles to itself, and to the direction in which it is moving and proportional to its velocity at right angles to itself.

When the medium is moving, the question then arises whether this velocity to which the magnetic force is proportional is the velocity of the tube relative (1) to a fixed point in the region under consideration, or (2) relative to the moving dielectric, or (3) relative to the ether in this region. If the first supposition is true we have equation (3), if the second equation (4), if the third an equation similar to (4) with the components of the velocity of the ether written for u, v, w. I am not aware of any experiments which would enable us to decide absolutely which, if any, of the assumptions (1), (2), (3) is correct; *a priori* (3) appears the most probable.

If X, Y, Z are the components of the electromotive intensity; a, b, c those of magnetic induction; f, g, h those of electric polarization, and F, G, H those of the vector potential, then we have

$$\left.\begin{aligned} X &= \frac{4\pi}{K} f = cv - bw - \frac{dF}{dt} - \frac{d\psi}{dx}, \\ Y &= \frac{4\pi}{K} g = aw - cu - \frac{dG}{dt} - \frac{d\psi}{dy}, \\ Z &= \frac{4\pi}{K} h = bu - av - \frac{dH}{dt} - \frac{d\psi}{dz}. \end{aligned}\right\} \qquad (5)$$

Then, since the dielectric is moving uniformly, we have

$$\begin{aligned} \frac{4\pi}{K}\left(\frac{df}{dy} - \frac{dg}{dx}\right) &= u\frac{dc}{dx} + v\frac{dc}{dy} + w\frac{dc}{dz} + \frac{dc}{dt} \\ &= \left(\frac{d}{dt} + u\frac{d}{dx} + v\frac{d}{dy} + w\frac{d}{dz}\right) c. \end{aligned} \qquad (6)$$

Now if equation (3) is true

$$\frac{df}{dt} = \frac{1}{4\pi\mu}\left(\frac{dc}{dy} - \frac{db}{dz}\right),$$

with similar equations for dg/dt, dh/dt; hence from (6) we have

$$\frac{1}{K\mu}\nabla^2 c = \frac{d}{dt}\left(\frac{d}{dt} + u\frac{d}{dx} + v\frac{d}{dy} + w\frac{d}{dz}\right)c. \tag{7}$$

If, on the other hand, equation (4) is true, we get

$$\frac{1}{K\mu}\nabla^2 c = \left(\frac{d}{dt} + u\frac{d}{dx} + v\frac{d}{dy} + w\frac{d}{dz}\right)^2 c, \tag{8}$$

with similar equations for a and b.

Let us apply these equations to a wave of plane polarized light travelling along the axis of x, the dielectric moving with velocity u in that direction. In this case equation (7) becomes

$$\frac{1}{K\mu}\frac{d^2c}{dx^2} = \frac{d^2c}{dt^2} + u\frac{d^2c}{dx\,dt}. \tag{9}$$

Let $c = \cos(pt - mx)$; then if V is the velocity of light through the dielectric when at rest, equation (9) gives

$$V^2m^2 = p^2 - upm,$$

or

$$\frac{p^2}{m^2} - \frac{up}{m} = V^2.$$

Since u is small compared with V, we have approximately

$$\frac{p}{m} = \tfrac{1}{2}u + V.$$

Thus the velocity of light through the moving dielectric is increased by half the velocity of the dielectric.

If we take equation (8), then

$$V^2\frac{d^2c}{dx^2} = \left(\frac{d}{dt} + u\frac{d}{dx}\right)^2 c,$$

or putting as before,

$$c = \cos(pt - mx),$$

$$V^2m^2 = (p - mu)^2,$$

hence

$$\frac{p}{m} = V + u;$$

so that in this case the velocity of the light is increased by that of the dielectric.

If we suppose that the condition (3) is the true one, viz., that

$$4\pi\mu\left(\frac{df}{dt} + u_0\frac{df}{dx} + v_0\frac{df}{dy} + w_0\frac{df}{dz}\right) = \frac{dc}{dy} - \frac{db}{dz},$$

where u_0, v_0, w_0 are the components of the velocity of the ether, then, when equations (1) are supposed to hold, the relation between p and m for the plane polarized wave is easily found to be

$$V^2m^2 = (p-mu)(p-mu_0),$$

or if u and u_0 are small compared with V,

$$\frac{p}{m} = V + \tfrac{1}{2}(u+u_0),$$

so that in this case the velocity of the light is increased by the mean of the velocities of the dielectric and the ether.

Fizeau's result that the increase in the velocity of light passing through a current of air is a very small fraction of the velocity of the air, shows that all of the preceding suppositions are incorrect.

Thus, if we retain the Electromagnetic Theory of Light, we must admit that equations (1) do not represent the electromotive intensities in a dielectric in motion if u, v, w are the velocities of the *dielectric itself*.

If we suppose that in these equations u, v, w ought to refer to the velocity of the *ether and not of the dielectric*, then the preceding work shows that if supposition (1) is true, the velocity of light passing through moving ether is increased by one half the velocity of the ether, while if supposition (3) is true it is increased by the velocity of the ether.

As we could not suppose that the motion of the dielectric makes the ether move faster than itself, the discovery of a case in which the velocity of light was increased by more than half the velocity of the dielectric would be sufficient to disprove supposition (1).

Currents induced in a Rotating conducting Sphere.

441.] When the external magnetic field is not symmetrical about the axis of rotation electric currents will be produced in the sphere. These have been discussed by Himstedt (*Wied. Ann.* 11, p. 812, 1880), and Larmor (*Phil. Mag.* [5], 17, p. 1, 1884). We can find these currents by the methods given in Chapters IV and V for dealing with spherical conductors.

From equations (1) we have, since

$$\frac{da}{dx} + \frac{db}{dy} + \frac{dc}{dz} = 0,$$

$$\frac{dX}{dy}-\frac{dY}{dx}=u\frac{dc}{dx}+v\frac{dc}{dy}+w\frac{dc}{dz}+c\left(\frac{du}{dx}+\frac{dv}{dy}+\frac{dw}{dz}\right)$$
$$-\left(a\frac{dw}{dx}+b\frac{dw}{dy}+c\frac{dw}{dz}\right), \qquad (10)$$

with similar equations for

$$\frac{dZ}{dx}-\frac{dX}{dz}, \quad \frac{dY}{dz}-\frac{dZ}{dy}.$$

If the sphere is rotating with angular velocity ω about the axis of z,

$$u=-\omega y, \quad v=\omega x, \quad w=0;$$

so that equation (10) becomes

$$\frac{dX}{dy}-\frac{dY}{dx}=\omega\left(x\frac{dc}{dy}-y\frac{dc}{dx}\right). \qquad (11)$$

If σ is the specific resistance of the sphere, μ its magnetic permeability, $\mathbf{p}$, $\mathbf{q}$, $\mathbf{r}$ the components of the current,

$$\left.\begin{aligned} X=\sigma\mathbf{p}&=\frac{\sigma}{4\pi\mu}\left(\frac{dc}{dy}-\frac{db}{dz}\right),\\ Y=\sigma\mathbf{q}&=\frac{\sigma}{4\pi\mu}\left(\frac{da}{dz}-\frac{dc}{dx}\right),\\ Z=\sigma\mathbf{r}&=\frac{\sigma}{4\pi\mu}\left(\frac{db}{dx}-\frac{da}{dy}\right).\end{aligned}\right\} \qquad (12)$$

If we substitute these values for X and Y, equation (11) becomes

$$\frac{\sigma}{4\pi\mu}\nabla^2 c=\omega\left(x\frac{dc}{dy}-y\frac{dc}{dx}\right),$$

similarly

$$\left.\begin{aligned} \frac{\sigma}{4\pi\mu}\nabla^2 b&=\omega\left(x\frac{db}{dy}-y\frac{db}{dx}\right)-\omega a,\\ \frac{\sigma}{4\pi\mu}\nabla^2 a&=\omega\left(x\frac{da}{dy}-y\frac{da}{dx}\right)+\omega b.\end{aligned}\right\} \qquad (13)$$

From these equations we find by the aid of (12)

$$\frac{\sigma}{4\pi\mu}\nabla^2\mathbf{p}=\omega\left(x\frac{d\mathbf{p}}{dy}-y\frac{d\mathbf{p}}{dx}\right)+\omega\mathbf{q},$$

$$\frac{\sigma}{4\pi\mu}\nabla^2\mathbf{q}=\omega\left(x\frac{d\mathbf{q}}{dy}-y\frac{d\mathbf{q}}{dx}\right)-\omega\mathbf{p},$$

$$\frac{\sigma}{4\pi\mu}\nabla^2\mathbf{r}=\omega\left(x\frac{d\mathbf{r}}{dy}-y\frac{d\mathbf{r}}{dx}\right).$$

Hence

$$\frac{\sigma}{4\pi\mu}\nabla^2(x\mathbf{p}+y\mathbf{q}+z\mathbf{r}) = \omega\left(x\frac{d}{dy} - y\frac{d}{dx}\right)(x\mathbf{p}+y\mathbf{q}+z\mathbf{r}). \quad (14)$$

Let $$x\mathbf{p}+y\mathbf{q}+z\mathbf{r} = F(r)\,Y_n^s\epsilon^{\iota s\phi},$$

where r, θ, ϕ are the polar coordinates of a point, θ being measured from the axis of z. $Y_n^s\epsilon^{\iota s\phi}$ is a surface harmonic of degree n. Substituting this value in (14), we find

$$\frac{d^2F}{dr^2} + \frac{2}{r}\frac{dF}{dr} - \left(\frac{n(n+1)}{r^2} + \frac{4\pi\iota\mu s\omega}{\sigma}\right)F = 0.$$

The solution of this is, Art. 308,

$$F(r) = S_n(kr),$$

where $$k^2 = -4\pi\mu\iota s\omega/\sigma.$$

Thus $$x\mathbf{p}+y\mathbf{q}+z\mathbf{r} = AS_n(kr)\,Y_n^s\epsilon^{\iota s\phi},$$

where A is a constant.

Now $x\mathbf{p}+y\mathbf{q}+z\mathbf{r}$ is proportional to the current along the radius, and this vanishes at the surface of the sphere where $r = \mathbf{a}$; hence we have $AS_n(k\mathbf{a}) = 0$, but since the roots of $S_n(x) = 0$ are real, and k is partly imaginary, $S_n(k\mathbf{a})$ cannot vanish, thus A must vanish. In other words, the radial currents must vanish throughout the sphere; the currents thus flow along the surfaces of spheres concentric with the rotating one.

Since $$x\mathbf{p}+y\mathbf{q}+z\mathbf{r} = 0,$$

we may by Art. 370 put

$$\left.\begin{aligned}\mathbf{p} &= f_n(kr)\left(y\frac{d}{dz} - z\frac{d}{dy}\right)\omega_n,\\ \mathbf{q} &= f_n(kr)\left(z\frac{d}{dx} - x\frac{d}{dz}\right)\omega_n,\\ \mathbf{r} &= f_n(kr)\left(x\frac{d}{dy} - y\frac{d}{dx}\right)\omega_n;\end{aligned}\right\} \quad (15)$$

where $$f_n(kr) = \frac{S_n(kr)}{(kr)^n},$$

$$k^2 = -4\pi\mu\iota S\omega/\sigma,$$

and ω_n is a solid spherical harmonic of degree n.

By Art. 372, α, β, γ, the components of magnetic force, will be given by

$$\alpha = \frac{4\pi}{(2n+1)k^2}\left\{(n+1)f_{n-1}(kr)\frac{d\omega_n}{dx} - nk^2r^{2n+3}f_{n+1}(kr)\frac{d}{dx}\left(\frac{\omega_n}{r^{2n+1}}\right)\right\}, \quad (16)$$

with similar expressions for β and γ.

Now the magnetic force may be regarded as made up of two parts, one due to the currents induced in the sphere, the other to the external magnetic field; the latter part will be derived from a potential. Let Ω_n be the value of this potential in the sphere; we may regard Ω_n as a solid spherical harmonic of degree n, since the most general expression for the potential is the sum of terms of this type. If a_1, β_1, γ_1 are the components of the magnetic force due to the currents, a_0, β_0, γ_0 those due to the magnetic field, then

$$a = a_1 + a_0 = a_1 - \frac{d}{dx}\Omega_n.$$

Hence in the sphere

$$a_1 = \frac{d\Omega_n}{dx} + \frac{4\pi}{(2n+1)k^2}\left\{(n+1)f_{n-1}(kr)\frac{d\omega_n}{dx} - nk^2r^{2n+3}f_{n+1}(kr)\frac{d}{dx}\left(\frac{\omega_n}{r^{2n+1}}\right)\right\}, \quad (17)$$

with similar expressions for β_1 and γ_1.

Outside the sphere the magnetic force due to the currents will (neglecting the displacement currents in the dielectric) be derivable from a potential which satisfies Laplace's equation; hence outside the sphere we may put, if ω_n' represents a solid harmonic,

$$a_1 = \mathbf{a}^{2n+1}\frac{d}{dx}\frac{\omega_n'}{r^{2n+1}},$$

with similar expressions for β_1 and γ_1, where $\mathbf{a}$ is the radius of the sphere. The magnetic force tangential to the sphere due to these currents is continuous, as is also the normal magnetic induction; hence, μ being the magnetic permeability of the sphere, we have

$$\Omega_n + \frac{4\pi}{(2n+1)k^2}\{(n+1)f_{n-1}(k\mathbf{a})\,\omega_n - nk^2\mathbf{a}^2 f_{n+1}(k\mathbf{a})\,\omega_n\} = -\omega_n',$$

$$\mu(n\,\Omega_n) + \frac{\mu n(n+1)\,4\pi}{(2n+1)k^2}\{f_{n-1}(k\mathbf{a})\,\omega_n + k^2\mathbf{a}^2 f_{n+1}(k\mathbf{a})\,\omega_n\} = (n+1)\,\omega_n'.$$

Solving these equations, we find at the surface of the sphere

$$4\pi\omega_n = -\frac{(2n+1)(\mu n+n+1)k^2\Omega_n}{(n+1)\{(\mu n+n+1)f_{n-1}(k\mathbf{a}) + n(\mu-1)k^2\mathbf{a}^2 f_{n+1}(k\mathbf{a})\}}, \quad (18)$$

$$\omega_n' = -\frac{n(2n+1)\mu k^2\mathbf{a}^2 f_{n+1}(k\mathbf{a})\,\Omega_n}{(n+1)\{(\mu n+n+1)f_{n-1}(k\mathbf{a}) + n(\mu-1)k^2\mathbf{a}^2 f_{n+1}(k\mathbf{a})\}}. \quad (19)$$

If we substitute these values of ω_n, ω_n' in equations (15) and (17), we get the currents induced in the sphere and the magnetic force produced by those currents.

442.] We shall consider in detail the case when $n = 1$, i.e. when the sphere is rotating in a uniform magnetic field. Let the magnetic potential of the external field be equal to the real part of

$$Cr\cos\theta + Br\sin\theta\epsilon^{\iota\phi},$$

where C is the force parallel to z and B that parallel to x.

Then in the sphere

$$\Omega_1 = \frac{3}{\mu+2}(Cr\cos\theta + Br\sin\theta\epsilon^{\iota\phi}).$$

We shall first consider the case when kr is very small, so that approximately by Art. 309

$$f_0(kr) = 1 - \tfrac{1}{6}k^2r^2, \quad f_1(kr) = -\tfrac{1}{3}, \quad f_2(kr) = \tfrac{1}{15}.$$

Substituting these values in (18) and (19) and retaining only the lowest powers of k, we find

$$4\pi\omega_1 = -\tfrac{9}{2}\frac{k^2}{\mu+2}\left(1 + \frac{\mu+4}{10(\mu+2)}k^2\mathrm{a}^2\right)Br\sin\theta\epsilon^{\iota\phi},$$

$$\omega_1' = -\frac{3k^2\mathrm{a}^2}{10(\mu+2)^2}Br\sin\theta\epsilon^{\iota\phi}.$$

The term $Cr\cos\theta$ in Ω does not give rise to any terms in ω_n, ω_n' since s and therefore k vanishes for this term. Substituting these values we get by equations (15)

$$\left.\begin{aligned} \mathrm{p} &= -\tfrac{3}{2}\frac{\mu\omega}{(\mu+2)\sigma}zB, \\ \mathrm{q} &= 0, \\ \mathrm{r} &= \tfrac{3}{2}\frac{\mu\omega}{(\mu+2)\sigma}xB. \end{aligned}\right\} \qquad (20)$$

Thus the currents flow in parallel circles, having for their common axis the line through the centre of the sphere which is at right angles both to the axis of rotation and to the direction of magnetic force in the external field. The intensity of the current at any point is proportional to the distance of the point from this axis.

The components of the magnetic induction in the sphere are given by the equations

$$\left.\begin{aligned} a &= -\frac{3\mu B}{\mu+2}\left(1+\frac{2\pi\mu\omega}{5\sigma}xy\right), \\ b &= \frac{3\pi\mu^2\omega B}{(\mu+2)\sigma}\left(\tfrac{4}{5}r^2-\tfrac{2}{5}y^2-\tfrac{2}{5}\frac{\mu+4}{\mu+2}\mathrm{a}^2\right), \\ c &= -\frac{3\mu}{\mu+2}\left(C+\frac{2\pi\mu\omega}{5\sigma}Byz\right). \end{aligned}\right\} \qquad (21)$$

Thus the magnetic force due to currents consists of a radial force proportional to yr, together with a force parallel to y proportional to $2r^2-(\mu+4)\,\mathrm{a}^2/(\mu+2)$.

Outside the sphere the total magnetic potential is

$$(Cz+Bx)\left(1-\frac{(\mu-1)}{\mu+2}\frac{\mathrm{a}^3}{r^3}\right)-\frac{6\pi B}{5\,(\mu+2)^2}\frac{\mu^2\omega\mathrm{a}^5}{\sigma}\frac{y}{r^3}.$$

Thus the magnetic effect of the currents at a point outside the sphere is the same as that of a small magnet at the centre, with its axis at right angles to the axis of rotation and the external magnetic field, and whose moment is

$$\frac{6\pi B}{5\,(\mu+2)^2}\frac{\mu^2\omega\mathrm{a}^5}{\sigma}.$$

443.] Let us now consider the case when $k\mathrm{a}$ is large, since, when $s=1$

$$k^2 = -\frac{4\pi\mu\omega\iota}{\sigma},$$

we have

$$k = \sqrt{2}K\epsilon^{-\frac{\iota\pi}{4}},$$

where

$$K^2 = \frac{2\pi\mu\omega}{\sigma},$$

thus the real part of $\iota k\mathrm{a}$ is positive and large; hence we have approximately

$$f_0(k\mathrm{a}) = \frac{\epsilon^{\iota k\mathrm{a}}}{2\iota k\mathrm{a}},$$

$$f_1(k\mathrm{a}) = \frac{\epsilon^{\iota k\mathrm{a}}}{2k^2\mathrm{a}^2},$$

$$f_2(k\mathrm{a}) = -\frac{\epsilon^{\iota k\mathrm{a}}}{2\iota k^3\mathrm{a}^3}.$$

Hence we find

$$4\pi\omega_1 = -3\iota k^3\mathrm{a}\epsilon^{-\iota k\mathrm{a}}Br\sin\theta\,\epsilon^{\iota\phi},$$

$$\omega_1' = \tfrac{3}{2}\frac{\mu}{\mu+2}Br\sin\theta\,\epsilon^{\iota\phi},$$

so that by (15)

$$
\left.\begin{aligned}
\mathrm{p} &= -\frac{3\sqrt{2}K}{8\pi}\frac{z\mathrm{a}}{r^2}B\epsilon^{-K(\mathrm{a}-r)}\cos\left\{K(\mathrm{a}-r)+\frac{\pi}{4}\right\},\\
\mathrm{q} &= -\frac{3\sqrt{2}K}{8\pi}\frac{z\mathrm{a}}{r^2}B\epsilon^{-K(\mathrm{a}-r)}\sin\left\{K(\mathrm{a}-r)+\frac{\pi}{4}\right\},\\
\mathrm{r} &= \frac{3\sqrt{2}K}{8\pi}\frac{\mathrm{a}}{r^2}B\epsilon^{-K(\mathrm{a}-r)}\left[x\cos\left\{K(\mathrm{a}-r)+\frac{\pi}{4}\right\}+y\sin\left\{K(\mathrm{a}-r)+\frac{\pi}{4}\right\}\right].
\end{aligned}\right\}
$$

The total components of the magnetic induction inside the sphere are given by

$$
\left.\begin{aligned}
a &= -\mu B\frac{\mathrm{a}}{r}\epsilon^{-K(\mathrm{a}-r)}\cos K(\mathrm{a}-r) - \tfrac{1}{2}\mu B\epsilon^{-K(\mathrm{a}-r)}\,\mathrm{a}r^2\cos K(\mathrm{a}-r)\frac{d}{dx}\frac{x}{r^3}\\
&\qquad - \tfrac{1}{2}\mu B\epsilon^{-K(\mathrm{a}-r)}\,\mathrm{a}r^2\sin K(\mathrm{a}-r)\frac{d}{dx}\frac{y}{r^3},\\
b &= -\mu B\frac{\mathrm{a}}{r}\epsilon^{-K(\mathrm{a}-r)}\sin K(\mathrm{a}-r) - \tfrac{1}{2}\mu B\epsilon^{-K(\mathrm{a}-r)}\,\mathrm{a}r^2\cos K(\mathrm{a}-r)\frac{d}{dy}\frac{x}{r^3}\\
&\qquad - \tfrac{1}{2}\mu B\epsilon^{-K(\mathrm{a}-r)}\,\mathrm{a}r^2\sin K(\mathrm{a}-r)\frac{d}{dy}\frac{y}{r^3},\\
c &= -\frac{3\mu C}{\mu+2} - \tfrac{1}{2}\mu B\epsilon^{-K(\mathrm{a}-r)}\,\mathrm{a}r^2\cos K(\mathrm{a}-r)\frac{d}{dz}\frac{x}{r^3}\\
&\qquad - \tfrac{1}{2}\mu B\epsilon^{-K(\mathrm{a}-r)}\,\mathrm{a}r^2\sin K(\mathrm{a}-r)\frac{d}{dz}\frac{y}{r^3},
\end{aligned}\right\}
$$

while the magnetic potential outside due to the currents in the sphere is

$$
\frac{3}{2}\frac{\mu B}{\mu+2}\mathrm{a}^3\frac{x}{r^3}. \tag{24}
$$

If we compare these results with those we obtained when $k\mathrm{a}$ was small, we see that they differ in the same way as the distribution of rapidly varying currents in a conductor differs from that of steady or slowly varying ones. When $k\mathrm{a}$ is small the currents spread through the whole of the sphere, while when $k\mathrm{a}$ is large they are, as equations (22) show, confined to a thin shell. The currents flow along the surfaces of spheres concentric with the rotating one, and the intensity of the currents diminishes in Geometrical Progression as the distance from the surface of the sphere increases in Arithmetical Progression.

The magnetic field due to these currents annuls in the interior of the sphere, as equation (23) shows, that part of the external magnetic field which is not symmetrical about the axis of rotation. Thus the rotating sphere screens its interior from all but symmetrical distributions of magnetic force if $\{4\pi\mu\omega/\sigma\}^{\frac{1}{2}}\mathbf{a}$ is large.

A very interesting case of the rotating sphere is that of the earth; in this case

$$\mathbf{a} = 6\cdot37 \times 10^8, \quad \omega = 2\pi/(24 \times 60 \times 60),$$

so that approximately

$$\{4\pi\mu\omega/\sigma\}^{\frac{1}{2}}\mathbf{a} = 2 \times 10^7 \sigma^{-\frac{1}{2}}.$$

Thus if σ is comparable with 10^8, which is of the order of the specific resistance of electrolytes, $k\mathbf{a}$ will be about 2000, and this will be large enough to keep the earth a few miles below its surface practically free from the effects of an external unsymmetrical magnetic field.

Again, we have seen, Art. 84, that rarefied gases have considerable conductivity for discharges travelling along closed curves inside them. For gases in the normal state this conductivity only manifests itself under large electromotive intensities, but when the gas is in the state similar to that produced by the passage of a previous discharge, it has considerable conductivity even for small electromotive intensities. We see from the preceding results that if there were a belt of gas in this condition in the upper regions of the earth's atmosphere, and if the part of the solar system traversed by the earth were a magnetic field, this gas would screen off from the earth all magnetic effects which were not symmetrical about the axis of rotation. Thus the magnetic field at the earth's surface would, on this hypothesis, resemble that which actually exists in being roughly symmetrical about the earth's axis. The thickness of a shell required to reduce the magnetic field to $1/\epsilon$ of its value at the outer surface of the shell is $\{4\pi\omega/\sigma\}^{-\frac{1}{2}}$, or if $\sigma = 10^8$, about two miles. The result mentioned in Art. 470 of Maxwell's *Electricity and Magnetism*, that by far the greater part of the mean value of the magnetic elements arises from some cause inside the earth, shows, however, that we cannot assign the earth's permanent magnetic field to this cause.

444.] The total magnetic potential outside the sphere is, when $k\mathbf{a}$ is large, by equation (24),

$$Cz\left(1-\frac{\mu-1}{\mu+2}\frac{\mathrm{a}^3}{r^3}\right)+B\left(x-\frac{\mu-1}{\mu+2}\frac{\mathrm{a}^3}{r^3}x+\tfrac{3}{2}\frac{\mu}{\mu+2}\frac{\mathrm{a}^3}{r^3}x\right)$$
$$=Cz\left(1-\frac{\mu-1}{\mu+2}\frac{\mathrm{a}^3}{r^3}\right)+Bx\left(1+\tfrac{1}{2}\frac{\mathrm{a}^3}{r^3}\right).$$

Thus the effect of the rotating sphere on the part of the external magnetic field which is unsymmetrical about the axis of rotation, i.e. upon the term $Br\sin\theta\,\epsilon^{\iota\phi}$, is exactly the same as if this sphere were replaced by a sphere of diamagnetic substance for which $\mu = 0$; in other words, the rotating sphere behaves like a diamagnetic body. Thus we could make a model which would exhibit the properties of a feebly diamagnetic body in a steady field, by having a large number of rotating conductors arranged so that the distance between their centres was large compared with their linear dimensions.

Couples and Forces on the Rotating Sphere.

445.] We shall now proceed to investigate the couples and forces on the sphere caused by the action of the magnetic field on the currents induced in the sphere.

If X, Y, Z are the components of the mechanical force per unit volume, then (Maxwell's *Electricity and Magnetism*, vol. ii. Art. 603, equations C)

$$X = c\mathbf{q}-b\mathbf{r},$$
$$Y = a\mathbf{r}-c\mathbf{p},$$
$$Z = b\mathbf{p}-a\mathbf{q}.$$

The couple on the sphere round the axis of z is

$$\iiint (Yx-Xy)\,dx\,dy\,dz,$$

the integration extending throughout the sphere.

Substituting the preceding values for Y and X, we see that this may be written

$$\iiint (\mathbf{r}\,(ax+by+cz)-c\,(\mathbf{p}x+\mathbf{q}y+\mathbf{r}z))\,dx\,dy\,dz.$$

But since the radial current vanishes,

$$\mathbf{p}x+\mathbf{q}y+\mathbf{r}z = 0;$$

thus the couple round z reduces to

$$\iiint \mathbf{r}\,Rr\,dx\,dy\,dz,$$

where R is the magnetic induction along the radius.

Similarly the couple round x is equal to

$$\iiint \mathbf{p} R r \, dx\, dy\, dz,$$

while that round y is

$$\iiint \mathbf{q} R r \, dx\, dy\, dz.$$

From equation (16) we see that

$$Rr = \frac{4\pi\mu}{(2n+1)k^2} n(n+1)\{f_{n-1}(kr) + k^2 r^2 f_{n+1}(kr)\}\, \omega_n.$$

Now by (4), Art. 370,

$$f_{n-1}(kr) + k^2 r^2 f_{n+1}(kr) = -(2n+1) f_n(kr),$$

so that

$$Rr = -\frac{4\pi\mu}{k^2} n(n+1) f_n(kr)\, \omega_n, \tag{25}$$

or by (15)

$$Rr = -\frac{\sigma}{\omega s^2} n(n+1)\, \mathbf{r}.$$

Thus the couple around z is

$$-\frac{\sigma}{\omega s^2} n.(n+1) \iiint \mathbf{r}^2 \, dx\, dy\, dz.$$

When ω is small we find, by substituting the value of $\mathbf{r}$ given in equation (20), that when the sphere is rotating in a uniform magnetic field the couple tending to stop it is

$$\frac{6\mu^2}{5(\mu+2)^2} B^2 \frac{\omega}{\sigma} \pi \mathbf{a}^5.$$

446.] We see by equation (25) that the normal component of the magnetic force is proportional to $f_n(kr)$, while by (16) the other components contain terms proportional to $f_{n-1}(kr)$, but when $k\mathbf{a}$ is very large we have approximately

$$f_{n-1}(k\mathbf{a}) = \tfrac{1}{2}\, \iota^{n-2} \frac{\epsilon^{\iota k \mathbf{a}}}{(k\mathbf{a})^n},$$

$$f_n(k\mathbf{a}) = \tfrac{1}{2}\, \iota^{n-1} \frac{\epsilon^{\iota k \mathbf{a}}}{(k\mathbf{a})^{n+1}}.$$

Thus when $k\mathbf{a}$ is very large $f_n(k\mathbf{a})$, and near the surface of the sphere $f_n(k\mathbf{r})$, is very small compared with $f_{n-1}(k\mathbf{a})$, so that by (25) the magnetic force along the normal to the sphere vanishes in comparison with the tangential force, in other words the magnetic force is tangential to the surface.

This result can be shown to be true, whatever the shape of the body, provided it is rotating with very great velocity. If we consider the part of the magnetic field which is not symmetrical about the axis of rotation we have the following results:—

Since the magnetic potential outside the rotating bodies is determined by the conditions (1) that it should have at an infinite distance from these bodies the same value as for the undisturbed external field, and (2) that the magnetic force at right angles to these bodies should vanish over their surface, we see that the magnetic force at any point will be the same as the velocity of an incompressible fluid moving irrotationally and surrounding these bodies supposed at rest, the velocity potential at an infinite distance from these being equal to the magnetic potential in the undisturbed magnetic field.

447.] If we substitute the value of R, given by equation (25), in the expression for the couple round z, we find that if we neglect powers of $1/k\mathbf{a}$ the couple vanishes. Thus the couple vanishes when $\omega = 0$ and when $\omega = \infty$, there must therefore be some intermediate value of ω for which the couple is a maximum.

Let us now consider the forces on the sphere. The force parallel to x is equal to

$$\iiint (c\mathbf{q} - b\mathbf{r})\, dx\, dy\, dz$$

$$= \frac{1}{4\pi}\iiint \left\{ c\left(\frac{da}{dz} - \frac{d\gamma}{dx}\right) - b\left(\frac{d\beta}{dx} - \frac{da}{dy}\right)\right\} dx\, dy\, dz$$

$$= \frac{1}{4\pi}\iint \{ a(la + mb + nc) - \tfrac{1}{2} l(a\alpha + b\beta + c\gamma)\}\, dS,$$

where dS is an element of the surface, and l, m, n the direction cosines of the outward drawn normal. The forces parallel to y and z are given by similar expressions. We see that the force is equivalent to a tension parallel to the magnetic force inside the sphere and equal to

$$\frac{1}{4\pi}(\alpha^2 + \beta^2 + \gamma^2)^{\frac{1}{2}} R$$

per unit of surface, R being the magnetic induction along the outward normal; and to a normal pressure equal to

$$\frac{1}{8\pi}(a\alpha + b\beta + c\gamma).$$

When the sphere is rotating so rapidly that $k\mathbf{a}$ is very large

R vanishes, and the force on the rotating sphere is that due to a pressure

$$\frac{\mu}{8\pi}(a^2+\beta^2+\gamma^2);$$

this pressure will tend to make the sphere move from the strong to the weak places of the field. We see, therefore, that not only does the rotating sphere disturb the magnetic field in the same way as a diamagnetic body, but that it tends to move as such a body would move, i. e. from the strong to the weak parts of the field.

448.] If instead of a rotating sphere in a steady magnetic field we have a fixed sphere in a variable field, varying as $\epsilon^{\iota pt}$, the preceding results will apply if instead of putting $k^2 = -4\pi\mu\omega\iota s/\sigma$ we put $k^2 = -4\pi\mu\iota p/\sigma$, and neglect the polarization currents in the dielectric. We can prove this at once by seeing that the equations for a, b, c in the two cases become identical if we make this change.

The results we have already obtained in this chapter, when applied to the case of alternating currents, show that in a variable field when ka is large the currents and magnetic force will be confined to a thin layer near the surface, and that a conductor will act like a diamagnetic body both in the way it disturbs the field and the way it tends to move under the influence of that field. The movement of currents from the strong to the weak parts of the field has been demonstrated in some very striking experiments made by Professor Elihu Thomson, *Electrical World*, 1887, p. 258 (see also Professor J. A. Fleming on 'Electromagnetic Repulsion,' *Electrician*, 1891, pp. 567 and 601, and Mr. G. T. Walker, *Phil. Trans.* A. p. 279, 1892). The correspondence of the magnetic force to the velocity of an incompressible fluid, flowing round the conductors, is more complete in this case than in that of the rotating sphere, inasmuch as we have not to except any part of the magnetic potential, whereas in the case of the rotating sphere we have to except that part of the magnetic potential which is symmetrical about the axis of rotation.

APPENDIX.

In Art. 201 of the text there is a description of Perrot's experiments on the electrolysis of steam. As these experiments throw a great deal of light on the way in which electrical discharges pass through gases I have, while this work has been passing through the press, made a series of experiments on the same subject.

The apparatus I used was the same in principle as Perrot's. I made some changes, however, in order to avoid some inconveniences to which it seemed to me Perrot's form was liable. One source of doubt in Perrot's experiments arose from the proximity of the tubes surrounding the electrodes to the surface of the water, and their liability to get damp in consequence. These tubes were narrow, and if they got damp the sparks instead of passing directly through the steam might conceivably have passed from one platinum electrode to the film of moisture on the adjacent tube, then through the steam to the film of moisture on the other tube and thence to the other electrode. If anything of this kind happened it might be urged that since the discharge passed through water in its passage from one terminal to the other, some of the gases collected in the tubes *gg* (Fig. 84) might have been due to the decomposition of the water and not to that of the steam.

To overcome this objection I (1) removed the terminals to a very much greater distance from the surface of the water and placed them in a region surrounded by a ring-burner by means of which the steam was heated to a temperature of 140° C to 150° C. (2) I got rid of the narrow tubes surrounding the electrodes altogether by making the tubes through which the steam escaped partly of metal and using the metallic part of these tubes as the electrodes.

Instead of following Perrot's plan of removing the mixed gases from the collecting tubes *ee* (Fig. 84) and then exploding them in a separate vessel, I collected the gases on their escape from the discharge tubes in graduated eudiometers provided with platinum terminals, by means of which the mixed gases were exploded *in situ* at short intervals during the course of the experiments.

Description of Apparatus.

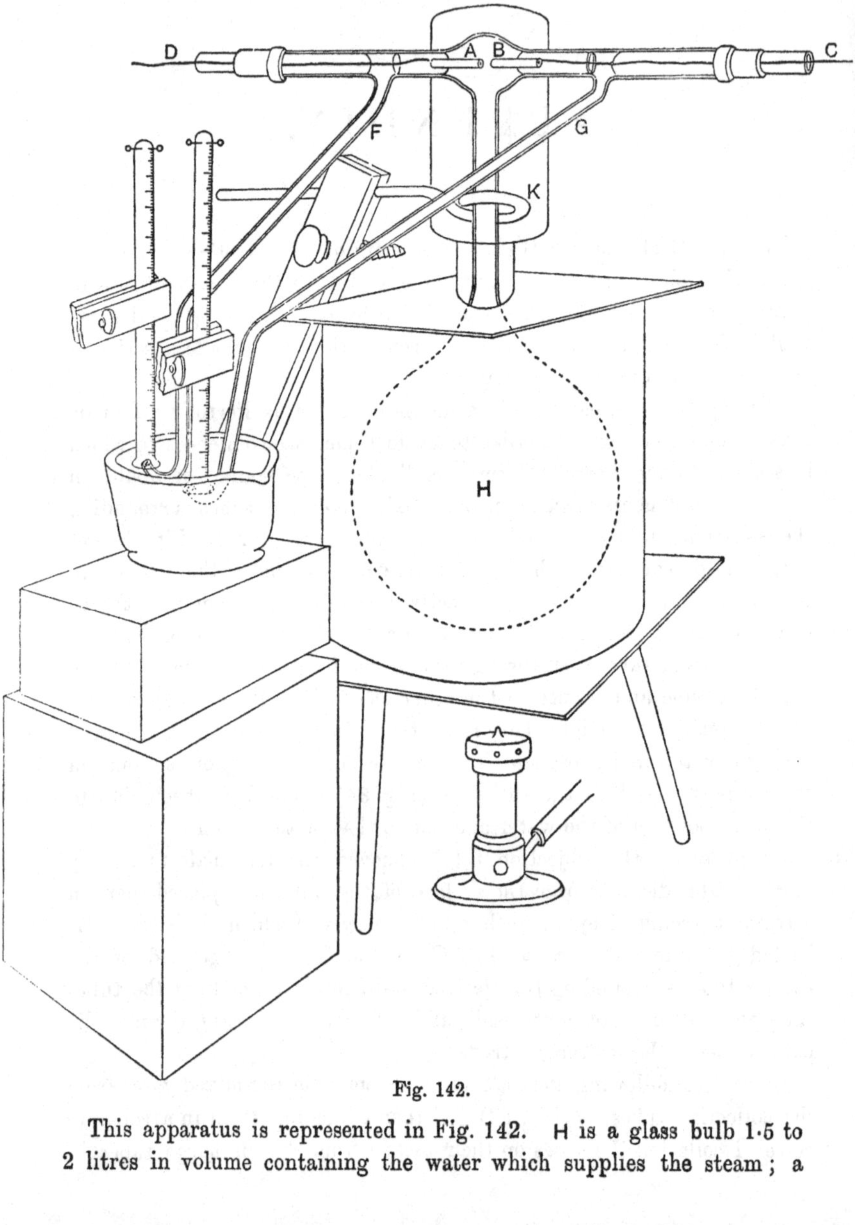

Fig. 142.

This apparatus is represented in Fig. 142. H is a glass bulb 1·5 to 2 litres in volume containing the water which supplies the steam; a glass tube about ·75 cm. in diameter and 35 cm. in length is joined on to this bulb, and the top of this tube is fused on to the discharge tube CD; this tube is blown out into a bulb in the region where the sparks pass, so that when long sparks are used they may not fly to the walls of the tube. This part of the tube is encircled by the ring-burner K by means of which the steam can be superheated.

The electrodes between which the sparks pass are shown in detail in Fig. 143; A, B are metal tubes, these must be made of a metal which does not oxidise. In the following experiment A, B are either brass tubes

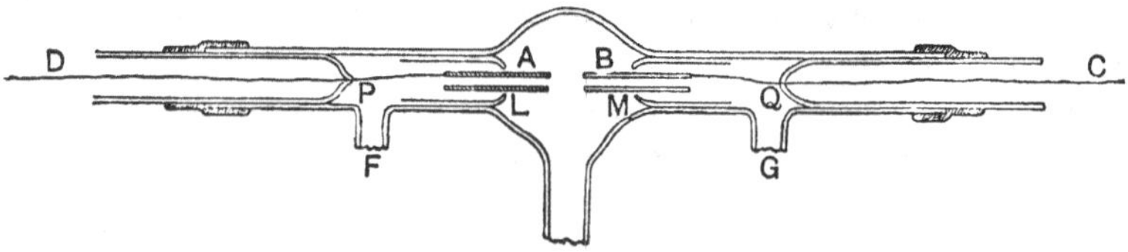

Fig. 143.

thickly plated with gold, or tubes made by winding thick platinum wire into a coil. These tubes are placed in pieces of glass tubing to hold them in position. These tubes stop short of the places F, G where the delivery tubes join the discharge tube. The discharge tube is closed at the ends by the glass tubes P and Q, and wires connected to the electrodes A and B are fused through these tubes.

The delivery tubes which terminate in fine openings were fused on to the discharge tube at F and G.

To get rid of the air which is in the apparatus or which is absorbed by the water, the apparatus is filled so full of water at the beginning of the experiment that when the water is heated it expands sufficiently to fill the discharge tube and overflow through the delivery tubes. The water is boiled vigorously for 6 or 7 hours with the ends of the delivery tubes open to the atmosphere. The eudiometer tubes filled with mercury are then placed over the ends of the delivery tubes, so that if any air is mixed with the steam it will be collected in these tubes. The sparking is not commenced until after the steam has run into the delivery tubes for about an hour without carrying with it a quantity of air large enough to be detected.

The sparks are produced by a large induction coil giving sparks about 5 cm. long when the current from five large storage cells is sent through the primary. When a condenser of about 6 or 7 micro-farads capacity is added to that supplied with the instrument a current was produced which, when the distance between the electrodes A and B in the discharge tube is not more than about 4 mm., will liberate about 4 c.c. of hydrogen per hour in a water voltameter placed in series with the discharge tube.

Method of making the Experiments.

When it had been ascertained that all the air had been expelled from the vessel and from the water, and that the rates of flow of the gases through the delivery tubes were approximately equal, the eudiometer tubes filled with mercury were placed over the ends of the delivery tubes, a water voltameter was placed in series with the steam tube, and the coil set in action.

The steam which went up the eudiometer tubes condensed into hot water which soon displaced the mercury; the mixture of oxygen and hydrogen produced by the spark went up the eudiometer tubes and was collected over this hot water and exploded at short intervals of time by the sparks from a Wimshurst machine. The gases did not disappear entirely when the sparks passed; a small fraction of the volume remained over after each explosion, and the volume which remained was greater in one tube than the other. The residual gas which had the greatest volume was found on analysis to be hydrogen, the other was oxygen. When a sufficient quantity of the residual gases had been collected they were analysed. The result of the analysis was that when the sparks were not too long the residual gas in one tube was pure hydrogen, that in the other pure oxygen; if any other gases were present their volume was too small to be detected by my analyses. When the sparks were very long there was always some other gas (nitrogen?) present, sometimes in considerable quantities.

Results of the Experiments.

The results obtained by the preceding method varied greatly in their character with the length of the spark, I shall therefore consider them under the heads—'short sparks,' 'medium sparks,' and 'long sparks.'

The lengths at which a spark changes from 'short' to 'medium' and then again to 'long' depend on the intensity of the current passing through the steam, and therefore upon the size of the induction coil and the battery power used to drive it. The limits of 'short,' 'medium,' and 'long' sparks given below must therefore be understood to have reference to the particular coil and current used in these experiments. With a larger coil and current these limits would expand, with a smaller one they would contract.

Short Sparks.

These sparks were from 1·5 mm. to 4 mm. long. The appearance of the spark showed all the characteristics of an arc discharge, it was a thickish column with ill-defined edges and was blown out by a wind to a broad flame-like appearance. For these arcs the following laws were found to hold:—

1. That within the limits of error of the experiments the volumes of the excesses of hydrogen in one tube and of oxygen in the other which remain after the explosion of the mixed gases are respectively equal to the volumes of the hydrogen and oxygen liberated in the water voltameter placed in series with the steam tube.

2. The excess of hydrogen appears in the tube which is in connection with the *positive* electrode, the excess of oxygen in the tube which is in connection with the *negative* electrode.

It thus appears that with these short sparks or arcs the hydrogen appears at the *positive* electrode instead of as in ordinary electrolysis at the *negative*.

The following table contains the results of some measurements of the relation between the excesses of hydrogen and oxygen in the eudiometer tubes attached to the steam tube and the quantity of hydrogen liberated in a water voltameter placed in series with the discharge tube. The ordinary vibrating break supplied with induction coils was used unless the contrary is specified:—

Spark length in millimetres.	Metal used for electrodes.	Excess of *H* in tube next + electrode.	Excess of *O* in tube next − electrode.	*H* liberated in water voltameter.	Duration of experiment in minutes.
1·5	Gold	3·25 c.c.	1·5 c.c.	3·2 c.c.	40
1·5	Platinum	2·8	1·6	3	30
1·5	Gold	1·7	·8	1·8	20
2	Gold	2	1·08	1·95	30
2	Gold	3·25	1·75	3·2	60
2	Platinum	1·8	Tube broke	2	Not noted
2	Platinum	3	1·5	3	60
2	Gold	2·5	1·5	3	60
3	Gold	1·8	Not noted	1·8	Not noted
3[1]	Gold	·7	·4	.8	90
3[2]	Gold	1·6	Not noted	1·75	Not noted
4	Gold	·9	.37	·7	20
4	Gold	2·75	1·25	2·7	60
4[2]	Gold	1·0	Not noted	1·25	Not noted
4	Gold	2·5	1·25	2·3	45

[1] In this experiment a slow mercury break, making about four breaks a second, was used.

[2] In these experiments Leyden jars were attached to the electrodes.

The results tabulated above show that the excesses of hydrogen and oxygen from the steam are approximately equal to the quantities of hydrogen and oxygen liberated in the water voltameter.

Medium Sparks.

When the spark length is greater than 4 mm. the first of the preceding results ceases to hold. The second, that the hydrogen comes off at the positive electrode, remains true until the sparks are about 11 mm. long, but the hydrogen from the steam, instead of being equal to that from the voltameter, is, when the increase in the spark length is not too large, considerably greater.

The following are a few instances of this:—

Spark length.	Hydrogen from steam in c.c.	Hydrogen from voltameter in c.c.
5 mm.	1·8	1·2
5 mm.	3·75	3
5 mm.	4·4	2·1
6 mm.	4	1·6
7 mm.	4·25	3
7 mm.	3·75	2
8 mm.	3·75	2·6

The increase in the ratio of the hydrogen from the steam to that from the voltameter does not continue when the spark length is still further increased. When the spark length exceeds 8 mm. this ratio begins to fall off very rapidly as the spark length increases, and we soon reach a critical spark length at which it seems almost a matter of chance whether the hydrogen from the steam appears at the positive or the negative electrode.

Long Sparks.

When the spark length is increased beyond the critical value, the excess of hydrogen instead of appearing at the positive electrode as with shorter sparks changes over to the negative, the excess of oxygen at the same time going over from the negative to the positive electrode. Thus the gases, when the spark length is greater than its critical value, appear at the same terminals in the steam tube as when liberated from an ordinary electrolyte, instead of at the opposite ones as they do when the sparks are shorter.

The critical length depends very largely upon the current sent through the steam; the smaller the current the shorter this length. It also depends upon a number of small differences, some of which are not easily specified, and it will sometimes change suddenly without any apparent reason. I have found, however, that this capriciousness disappears if

Leyden jars are attached to the terminals of the steam tube or if an air-break is placed in series with that tube.

It will be seen that the results when the spark length is greater than the critical length agree with those obtained by Perrot (Art. 201) and Ludeking (Art. 210), as both these observers found that the hydrogen appeared at the negative, the oxygen at the positive electrode. Ludeking worked with long sparks, so that his results are quite in accordance with mine. In Perrot's experiments the spark length was 6 mm. I have never been able to reduce the critical length quite so low as this, though I diminished the current to the magnitude of that used by Perrot; I have, however, got it as low as 8 mm., and it is probable that the critical length may not be governed entirely by the current.

I was not able to detect any decided change in the appearance of the spark as the spark length passed through the critical value. My observations on the connection between the appearance of the discharge and the electrode at which the hydrogen appears may be expressed by the statement that when the discharge is plainly an arc the hydrogen appears at the positive electrode, and that when the hydrogen appears at the negative electrode the discharge shows all the characteristics of a spark. It however looks much more like a spark than an arc long before the spark length reaches the critical value.

With regard to the ratio of the quantities of hydrogen liberated from the steam tube and from the water voltameter, I found that when the spark length was a few millimetres greater than the critical length the amount of hydrogen from the steam was the same as that from the voltameter. The following table contains a few measurements on this point:—

Spark length.	Hydrogen from steam in c.c.	Hydrogen from voltameter in c.c.
10 mm.	·7	·8
12 mm.[1]	·75	·9
14 mm.	·8	1·1

When the sparks were longer than 14 mm. the amount of hydrogen from the steam was no longer equal to that from the voltameter. The results became irregular, and there was a further reversal of the electrode at which the hydrogen appeared when the spark length exceeded 22 mm. In this case, however, the current was so small that it took several hours to liberate 1 c.c. of hydrogen in the voltameter. With these very long sparks the proportion between the hydrogen from the steam and that from the voltameter was too irregular to allow of any conclusions being drawn.

[1] In this experiment there was an air break 9 mm. long in series with the steam tube.

We see from the preceding results that in the electrolysis of steam, as in that of water, there is a very close connection between the amounts of hydrogen and oxygen liberated at the electrodes and the quantity of electricity which has passed through the steam, and that this relation for certain lengths of arc is the same for steam as for water. There is, however, this remarkable difference between the electrolysis of steam and that of water, that whereas in water the hydrogen always comes off at the negative, the oxygen at the positive electrode, in steam the hydrogen and oxygen come off sometimes at one terminal, sometimes at the other, according to the nature of the spark.

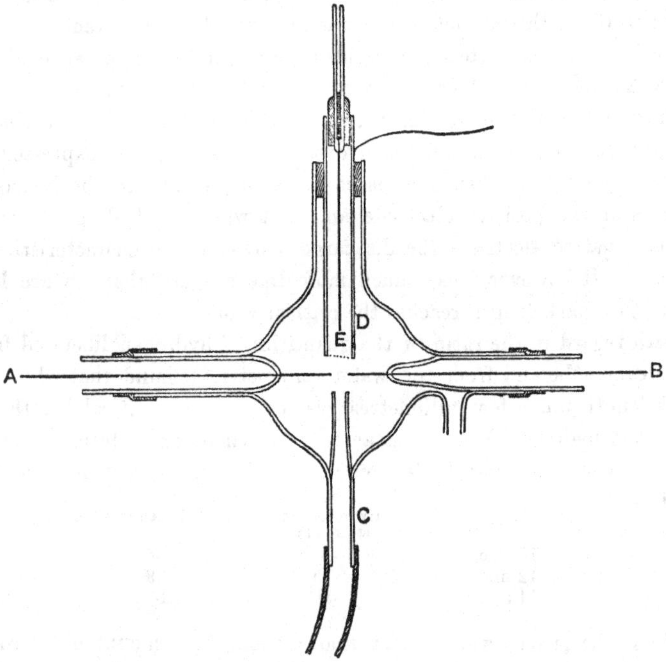

Fig. 144.

The results obtained when the discharge passed as an arc, i.e. that the oxygen appears at the negative electrode, the hydrogen at the positive, is what would happen if the oxygen in the arc had a positive charge, the hydrogen a negative one. With the view of seeing if I could obtain any other evidence of this peculiarity I tried the following experiments, the arrangement of which is represented in Fig. 144.

An arc discharge between the platinum terminals A, B was produced by a large transformer, which transformed up in the ratio of 400 to 1; a current of about 40 Ampères making 80 alternations per second was sent through the primary. A current of the gas under examination entered the discharge tube through a glass tube C and blew the gas in the neighbourhood of the arc against the platinum electrode E, which was connected to one quadrant of an electrometer, the other quadrant of which was connected to earth. To screen E from external electrical influences it was enclosed in a platinum tube D, which was closed in by fine platinum wire gauze, which though it screened E from external electrostatic action, yet allowed the gases in the neighbourhood of the arc to pass through it. This tube was connected to earth. The electrode E after passing out of this tube was attached to one end of gutta-percha covered wire wound round with tin-foil connected to earth.

The experiments were of the following kind. The quadrants of the electrometer were charged up by a battery, the connection with the battery was then broken and the rate of leak observed. When the arc was not passing the insulation was practically perfect. As soon, however, as the arc was started, and for as long as it continued, the insulation of the gas in many cases completely gave way. There are, however, many remarkable exceptions to this which we proceed to consider.

Oxygen.

We shall begin by considering the case when a well-developed arc passed through the oxygen.

If the electrode E was charged negatively, it lost its charge very rapidly; it did not however remain uncharged, but acquired a positive charge, this charge increasing until E acquired a potential V; V depended greatly upon the size of the arc and the proximity to it of the electrode E, in many of my experiments it was as large as 10 or 12 volts.

When E was charged positively to a high potential the electricity leaked from it until the potential fell to V; after reaching this potential the leak stopped and the gas seemed to insulate as well as when no discharge passed through it. If the potential to which E was initially raised was less than V (a particular case being when it was without charge to begin with) the positive charge increased until the potential of E was equal to V, after which it remained constant. Thus we see (1) that an electrode immersed in the oxygen of the arc can insulate a small positive charge perfectly, while it very rapidly loses a negative one; (2) that an uncharged electrode immersed in this gas acquires a positive charge.

When the distance between the electrodes A, B was increased until the discharge passed as a spark then the electrode E leaked slowly, whether charged positively or negatively. The rate of leak in this case was however exceedingly small compared to that which existed when the discharge passed as an arc.

Hydrogen.

When similar experiments were tried in hydrogen the results were quite different. When the *arc* discharge passed through the hydrogen the electrode E always leaked when it was positively electrified, and it did not merely lose its charge but acquired a negative one, its potential falling to $-U$, where U is a quantity which depended upon the size of the arc and on its proximity to the electrode E. In my experiments 5 to 6 volts was a common value of U.

When the electrode E was initially uncharged it acquired a negative charge, the potential falling to $-U$; when it was initially charged negatively, it leaked if its initial negative potential was greater than U until its potential fell to this value, when no further leak occurred. When the initial negative potential of E was less than U the negative charge increased until the potential had fallen to $-U$.

It is more difficult to get a good arc in hydrogen than in oxygen, so that the experiments with the former gas are a little more troublesome than those with the latter. When short arcs are used the electrode E must be placed close to the arc.

The following experiment was made to see if the charging up of the electrode was due to an electrification developed by the contact of the gas in the arc with the electrode, or whether this gas behaved as if it had a charge of electricity independent of its contact with the metal of the electrode. If the electrification were due to the contact of the gas with the electrode it would disappear if the electrode were covered with a non-conducting layer; if however the gas in the arc behaved as if it were charged with electricity, then even though the electrode were covered with a non-conducting layer the electrostatic induction due to the charge in the gas ought to produce a deflection of the electrometer in the same direction as if the electrode were uncovered. To test this point the electrode E was coated with glass, with mica, with ebonite, and with sulphur; in all these cases the needle of the electrometer was deflected as long as the arc was passing, and the deflection corresponded to a positive charge on the gas when the arc passed through oxygen and to a negative one when it passed through hydrogen; this deflection disappeared almost entirely as soon as the arc stopped.

In another experiment tried with the same object the arc was surrounded by a large glass tube coated inside and out with a thin layer of sulphur so as to prevent conduction over the surface. A ring of tin-foil was placed outside the tube so as to surround the place where the arc passed, and this ring was connected with one of the quadrants of an electrometer. As a further precaution against the creeping of the electricity over the surface of the tube two thin rings of tin-foil connected to the earth were placed round the ends of the tube. When the arc passed through oxygen the quadrants of the electrometer connected with the ring of tin-foil were positively electrified by induction, when the arc passed through hydrogen they were negatively charged.

These experiments show that the oxygen in the arc behaves as if it had a *positive* charge of electricity, while the hydrogen in the arc behaves as if it had a *negative* charge.

In all the above experiments the electrodes were so large that they were not heated sufficiently by the discharge to become luminous.

Elster and Geitel found (Art. 43) that a metal plate placed near a red-hot platinum wire became positively electrified if the wire and the plate were surrounded by oxygen, and negatively electrified if they were surrounded by hydrogen. If we suppose that the effect of the hot wire is to put the gas around it in a condition resembling the gas in the arc, Elster and Geitel's results would be explained by the preceding experiments, for these have shown that when this gas is oxygen it is positively electrified, while when it is hydrogen it is negatively electrified.

These experiments suggest the following explanation of the results of the investigation on the electrolysis of steam. We have seen (Art. 212) that when an electric discharge passes through a gas the properties of the gas in the neighbourhood of the line of discharge are modified, and (Art. 84) that this modified gas possesses very considerable conductivity. When the discharge stops, this modified gas goes back to its original condition. If now the discharges through the gas follow one another so rapidly that the modified gas produced by one discharge has not time to revert to its original condition before the next discharge passes, the successive discharges will pass through the modified gas. If, on the other hand, the gas has time to return to its original condition before the next discharge passes, each discharge will have to make its way through the unmodified gas.

We regard the arc discharge as corresponding to the first of the preceding cases when the discharge passes through the modified gas, the spark discharge as corresponding to the second case when the discharge passes through the gas in its unmodified condition.

From this point of view the explanation of the results observed in the electrolysis of steam are very simple. The modified gas produced by the passage of the discharge through the steam consists of a mixture of hydrogen and oxygen, these gases being in the same condition as when the arc discharge passes through hydrogen and oxygen respectively, when, as we have seen, the hydrogen behaves as if it had a negative charge, the oxygen as if it had a positive one. Thus in the case of the *arc* in steam the oxygen, since it behaves as if it had a positive charge, will move in the direction of the current and appear at the *negative* electrode; the hydrogen will move in the opposite direction and appear at the *positive* electrode.

The equality which we found to exist between the quantities of hydrogen and oxygen from the electrolysis of the steam and those liberated from the electrolysis of water by the same current, shows that the charges on the atoms of the modified oxygen and hydrogen are the same in amount but opposite in sign to the charges we ascribe to them in ordinary electrolytes.

In the case of the long sparks when the discharge goes through the steam itself, since the molecule of steam consists of two positively charged hydrogen atoms and one negatively charged oxygen atom, when this splits up in the electric field the hydrogen atoms will go towards the negative, the oxygen atom towards the positive electrode, as in ordinary electrolysis. The experiments described on page 565 show that with these long sparks the hydrogen appears at the negative, the oxygen at the positive electrode.

INDEX.

The numbers refer to the pages.

THE END.

Printed in the United States
By Bookmasters